The Sound Studio

I dedicate this Seventh Edition to my son Guy
rather belatedly, since he trod his own distinctive path towards the virtual studio and can now tell me a few things about what goes on there. I guess he doesn't really need this book – but it's the thought that counts.

The Sound Studio

Audio techniques for radio, television, film and recording

Seventh edition

Alec Nisbett

AMSTERDAM BOSTON HEIDELBERG LONDON NEW YORK OXFORD
PARIS SAN DIEGO SAN FRANCISCO SINGAPORE SYDNEY TOKYO

Focal Press
An imprint of Elsevier
Linacre House, Jordan Hill, Oxford OX2 8DP
200 Wheeler Road, Burlington MA 01803

First published 1962
Second edition 1970
Third edition 1972
Fourth edition 1979
Fifth edition 1993
Sixth edition 1995
Reprinted 1995, 1999, 2000
Seventh edition 2003

British Library Cataloguing in Publication Data
Nisbett, Alec
The Sound studio: audio techniques for radio, television, film and recording. – 7th ed.
1. Sound – Recording and reproducing
I. Title
621.3′893

Library of Congress Cataloguing in Publication Data
A catalogue record for this book is available from the Library of Congress

ISBN 0 240 51911 6

For information on all Focal Press publications visit our website at:
www.focalpress.com

Typeset by Integra Software Services Pvt. Ltd, Pondicherry, India
www.integra-india.com
Printed and bound in Great Britain by Biddles Ltd. *www.biddles.co.uk*

Contents

Introduction

The audio revolution is over – digital is king! One of my principal aims for this Seventh Edition has been to make sure that current practice in digital audio is fully represented. This will include such topics as: the virtual studio, along with aids like Pro Tools and reattached MIDI keyboards; 5.1 surround sound; hard drive mixers and multichannel recorders, plus DVD and CD-RW and . . . you name it (although on second thoughts you may not have to: computers seem to have a whole naming industry to support them).

And yet . . . a curious thing happened. I found that in the first eight chapters (and in several of the others) there was surprisingly little to say about this new digital world we all live in. So – why was that?

The short answer is simply that sound is not digital. Your ears and voice are not digital. The way that your brain processes sound is rather odd – but not digital. Nor are the loudspeakers you listen to, or (even more to the point) the microphones you set up to capture sound. If you encounter something called a digital microphone, what that really means is that the analogue electrical signal generated within it is turned into digital information before being sent down a cable to the digital processor or recorder downstream . . . but it will still have to be converted back to analogue sound before you can hear it. What we really live in today is a hybrid analogue-and-digital world. Some of the analogue parts are unavoidable; the digital bits are optional.

Where digital processes actually can be used they are often better. Digital information can be passed over greater distances and through more pieces of equipment that do things to it (process it), with far smaller losses of information along the way. Digital recording is more faithful, and digital recording media can hold ever more information. From the outside that might seem like revolution enough, but even more profound are the increases in processing power and processing speed that mean more can be accomplished (and completed in real time) than was possible before. This change is not just quantitative, it is qualitative: in this new digital world you can do things that were not possible, or not practicable, before.

This does not mean that all processes or processing have to be digital all the time. Many people who use digital processes most of the time like to return occasionally to older analogue devices or manual techniques in order to experience their 'hands-on' qualities more directly. Or, for a particular application, they replace smoothly operating new technology by clunky old gear, to bring back some perceived special sound quality that came with it. I am not a great fan of this aspect of 'retro', as it may be called; after all, in this new digital world you can always simulate any older kind of 'clunky' alongside an endless range of the new.

There is another reason why, in this book, I still describe some apparently archaic techniques. It is because many operations are designed to do the same thing whether they are achieved by analogue or digital means. Often, the terminology is unchanged and can be understood better in its original context, even though a computer may now do it for you.

In this book I will also make some short cuts – by, for example, putting in very little about particular techniques that have been painstakingly developed and exhaustively analysed, even to become mainstream for some of their users, when there are competitive techniques that in my view do the same job as well or better, and seem to have wider application. An example is 'dummy head' and related stereo microphone techniques. You may meet (or even work alongside) people who swear by these: if so, they will direct you to books that describe them and explain their justification in more detail.

All this remains consistent with the original purpose of this book: to explain the subject simply, in terms that are acceptable both to the creative people in production – including writers and performers – and to those who deal operationally with the balance and recording of sound or with its engineering aspects. By exploring the ground common to all these groups, the work of each may be interpreted to the other. It is also designed for those who may want to explore audio techniques that have been developed in fields that are remote from their own, whether this be radio, music recordings, television or film sound.

In these aims I am aided by the same digital advances that now pervade audio. This Seventh Edition is in part a computer construct, derived initially by optical character recognition from the Sixth Edition, then revised topic by topic and scrutinized and updated paragraph by paragraph in far greater detail than ever before. In consequence, this new version should be not only more finely attuned to current practice, but also (I hope) even more readable.

Several features of this book remain almost unchanged in this edition. One is its structure and the underlying logic, with each chapter in turn introducing ideas that are used later throughout the book. Another is that despite all the technical advances in equipment that are touched on, it remains a book about audio techniques, not audio engineering. No previous scientific, mathematical or engineering knowledge is assumed.

Since its first edition, my emphasis has always been on general principles rather than rules of thumb, though the shorter routes to high-quality results are not neglected. I still insist on the paramount importance of subjective judgement, of learning to use our ears, of exercising well-developed aural perception, allied to a strong critical faculty. Throughout, I describe studio and location techniques in terms of what each operation does to the sound and why, and *what to listen for*.

Of all the chapters in this book, that which has changed least is the last, on communication in sound. Happily, in the years since it was first written, neither scientific study of this human capacity nor the evolution of the media themselves has invalidated its observations. But some reviewers have suggested that I have put this last chapter in the wrong place: they strongly suggest that you read it first!

About the author

If you want to know what prejudices to expect, or why the advice or terminology here may be a little different from that in some other book you have seen, it may help you to know where the writer is coming from. So ... this is where I came from.

I began my working life as a studio manager in BBC Radio – for which they did not want 'engineers'. In fact, the entry qualifications seemed to be 'anything unusual, but different from the last successful applicant', of which about 50% were women and a few were gay. Oddly, being a male mathematical physicist did not seem to disqualify me. My earliest faltering footsteps (the 'spot effects' later called 'foley' by others) included horses' hoof effects generated with the aid of coconut shells. I went on to learn and use more of the basics – such as 'balance' – that you will read about here. I also helped to outline what would be needed for a 'Radiophonic Workshop' – now gone, but an early precursor to the virtual studios that generate much of today's music.

Later I moved to television, making both studio productions and 'outside broadcasts' (as we called them). I became a filmmaker and specialized in science documentaries such as *Horizon* (the British precursor to the American *Nova* series on PBS) of which I made 42, probably more than anyone else alive. Throughout all this, I maintained a particular concern for the audio input: voice, effects, and both conventional and synthesized music. I led co-productions with lots of other broadcasters in many other countries, collected a bunch of the usual awards and have long been a life member of the British Academy of Film and Television Arts.

Wee have also Sound-Houses, *wher wee practise and demonstrate all* Sounds, *and their* Generation. *Wee have* Harmonies *which you have not, of* Quarter-Sounds, *and lesser* Slides *of* Sounds. Diverse Instruments *of* Musick *likewise to you unknowne, some Sweeter then any you have; Together with* Bells *and* Rings *that are dainty and sweet. Wee represent* Small Sounds *as* Great *and* Deepe; *Likewise* Great Sounds, Extenuate *and* Sharpe; *Wee make diverse* Tremblings *and* Warblings *of* Sounds, *which in their* Originall *are* Entire. *Wee represent and imitate all* Articulate Sounds *and* Letters, *and the* Voices *and* Notes *of* Beasts *and* Birds. *Wee have certaine* Helps, *which sett to the* Eare *doe further the* Hearing *greatly. Wee have also diverse* Strange *and* Artificiall Eccho's, Reflecting *the* Voice *many times, and as it were Tossing it; And some that give back the* Voice Lowder *then it came, some* Shriller, *and some* Deeper; *Yea some rendring the* Voice, Differing *in the* Letters *or* Articulate Sound, *from that they receyve. Wee have also meanes to convey* Sounds *in* Trunks *and* Pipes, *in strange* Lines, *and* Distances.

From *The New Atlantis* by Francis Bacon
1624

Chapter 1

Audio techniques and equipment

If you work in a radio or recording studio, or with television or film sound; if you are in intimate daily contact with the problems of the medium; if you spend hours each day consciously and critically listening, you are bound to build up a vast body of aural experience. Your objective will be to make judgements upon the subtleties of sound that make the difference between an inadequate and a competent production, or between competent and actively exciting – and at the same time to develop your ability to manipulate audio equipment to mould that sound into the required shape.

The tools for picking up, treating and recording sound continue to improve, and it also becomes easier to acquire and apply the basic skills – but because of this there is ever more that can be done. Microphones are more specialized, creating new opportunities for their use. Studio control consoles can be intimidatingly complex, but this only reflects the vast range of options that they offer. Using lightweight portable recorders or tiny radio transmitters, sound can be obtained from anywhere we can reach and many places we cannot. By digital techniques, complex editing can be tackled. In fact, home computers can now simulate many of the functions of studios.

But however easy it becomes to buy and handle the tools, we still need a well-trained ear to judge the result. Each element in a production has an organic interdependence with the whole; every detail of technique contributes in one way or another to the final result; and in turn the desired end-product largely dictates the methods to be employed. This synthesis between means and end may be so complete that the untrained ear can rarely disentangle them – and even critics may disagree on how a particular effect was achieved. This is because faults of technique are far less obvious than those of content. A well-placed microphone in suitable acoustics, with the appropriate fades, pauses, mix and editing, can make all the difference to a radio production – but at the end an appreciative listener's search for a favourable comment will probably land on some remark about the subject. Similarly, where technique is faulty it is often, again, the content that is criticized. Anyone working with sound must develop a faculty for listening analytically to the operational aspects of a production, to judge technique independently of content, while making it serve content.

In radio, television or film, the investment in equipment and the professionals' time makes slow work unacceptable. Technical problems must be anticipated or solved quickly and unobtrusively, so attention can be focused instead on artistic fine-tuning, sharpening performance and enhancing material. Although each individual concentrates on a single aspect, this is geared to the production as a whole. The first objective of the entire team is to raise the programme to its best possible standard and to catch it at its peak.

This first chapter briefly introduces the range of techniques and basic skills required, and sets them in the context of the working environment: the studio chain or its location equivalent. Subsequent chapters describe the nature of the sound medium and then elements of the chain that begins with the original sound source and includes the vast array of operations that may be applied to the signal before it is converted back to sound and offered to the listener – who should be happily unaware of these intervening stages.

Studio operations

The operational side of sound studio work, or for sound in the visual media, includes:

- Choice of acoustics: selecting, creating or modifying the audio environment.
- Microphone balance: selecting suitable types of microphone and placing them to pick up a combination of direct and reflected sound from the various sources.
- Equalization: adjusting the frequency content.
- Feeding in material from other places.
- Mixing: combining the output from microphones, replay equipment, devices that add artificial reverberation and other effects, remote sources, etc., at artistically satisfying levels.
- Control: ensuring that the programme level (i.e. in relation to the noise and distortion levels of the equipment used) is not too high or too low, and uses the medium – recording or broadcast – efficiently.
- Creating sound effects (variously called foley, hand or spot effects) in the studio.
- Recording: using tape, disc and/or hard disk (including editing systems and back-up media) to lay down one, two or more tracks.

- Editing: tidying up small errors in recordings, reordering or compiling material, introducing internal fades and effects.
- Playing in records and recordings: this includes recorded effects, music discs or tapes, interviews and reports, pre-recorded and edited sequences, etc.
- Feeding output (including play-out): ensuring that the resultant mixed sound is made available to its audience or to be inscribed upon a recording medium with no significant loss of quality.
- Documentation: keeping a history of the origins, modification and use of all material.
- Storage: ensuring the safety and accessibility of material in use and archiving or disposing of it after use.

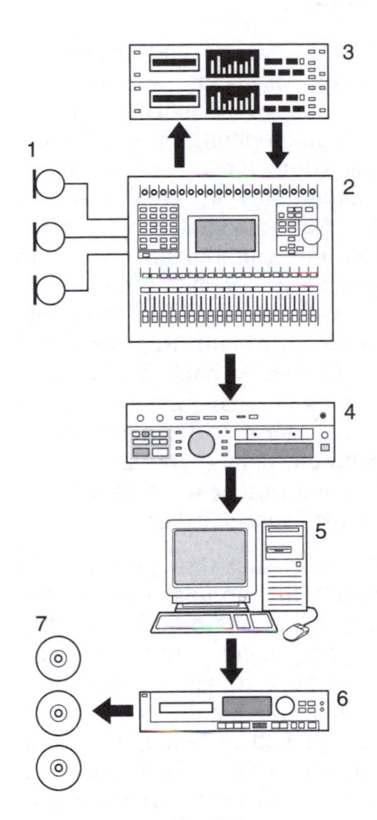

[1.1] **Digital studio chain**
1, Microphones. The electrical signal from a microphone is analogue, so it has to be fed through an analogue-to-digital (A/D) converter, after which it remains digital throughout the chain. 2, Digital console (mixer). An alternative is a personal computer (often, for media applications, a Mac) with audio software that includes a pictorial representation of faders, plus 'plug-ins'. 3, Modular digital multitrack eight-track recorders (MDMs): these can be linked together to take up to 128 tracks, although 32 are generally more than enough. Here the alternative is hard disk recording. 4, Digital audio tape (DAT) recorder. 5, Digital audio workstation (DAW – more software on a personal computer). 6, CD-R (or DVD-R) usually built in to the computer. 7, The end-product on CD.

In small radio stations in some countries, a further responsibility may be added: supervising the operation of a transmitter at or near the studio. This requires certain minimal technical qualifications.

The person responsible for balancing, mixing and control may be a sound or programme engineer, or balance engineer (in commercial recording studios); in BBC radio is called a studio manager; in television may be a sound supervisor; on a film stage is a sound recordist (US: production sound recordist); or when re-recording film or video, a re-recording mixer (US) or dubbing mixer (UK).

Although in many parts of the world the person doing some of these jobs really is an engineer by training, in broadcasting – and for some roles in other media – it is often equally satisfactory to employ men and women whose training has been in other fields, because although the operator is handling technical equipment and digital operations, the primary task is artistic (or the application of craftsmanship). Real engineering problems are usually dealt with by a maintenance engineer.

Whatever title may be given to the job, the sound supervisor has overall responsibility for all sound studio operations and advises the director on technical problems. Where sound staff work away from any studio, e.g. on outside broadcasts (remotes) or when filming on location, a higher standard of engineering knowledge may be required – as also in small radio stations where a maintenance engineer is not always available. There may be one or several assistants: in radio these may help lay out the studio and set microphones, create sound effects, and make, edit and play in recordings. Assistants in television and film may also act as boom operators, positioning microphones just out of the picture: this increasingly rare skill is still excellent training for the senior role.

The sound control room

In the layout of equipment required for satisfactory sound control, let us look first at the simplest case: the studio used for sound only, for radio or recording.

The nerve centre of almost any broadcast or recording is the control console – also known as desk, board or mixer. Here all the different sound sources are modified and mixed. In a live broadcast it is here that the final sound is put together; and it is the responsibility of the sound supervisor to ensure that no further adjustments are necessary before the signal leaves the transmitter (except, perhaps, for any occasional automatic operation of a limiter).

The console, in most studios the most obvious and impressive capital item, often presents a vast array of faders, switches and indicators. Essentially, it consists of a number of individual channels, each associated with a particular microphone or some other source. These are all pre-set (or, apart from the use of the fader, altered only rarely), so that their output may be combined into a small number of group controls that can be operated by a single pair of hands. The groups then pass through a main control to become the studio output. That 'vast array' plainly offers many opportunities to do different things with sound. Apart from the faders, the function of which is obvious, many are used for signal routing. Some allow modification of frequency content (equalization) or dynamics (these include compressors and limiters). Others are concerned with monitoring. More still are used for communications. Waiting to be inserted at chosen points are specialized devices offering artificial reverberation and further ways of modifying the signal. All these will be described in detail later.

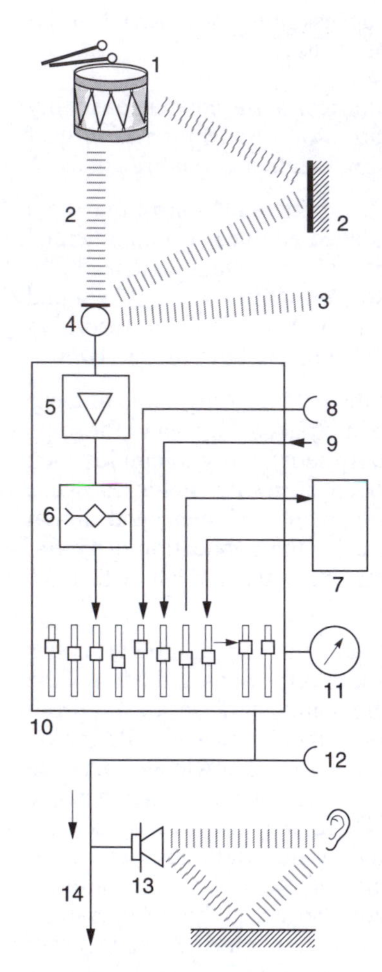

[1.2] **Components of the studio chain**
1, The sound source. 2, Direct and indirect sound paths. 3, Noise: unwanted sound. 4, Microphone. 5, Preamplifier. 6, Equalizer. 7, Reverberation, now usually digital. 8, Tape head or disc player. 9, External line source. 10, Mixer. 11, Meter. 12, Recorder. 13, Loudspeaker: sound reaches the ear by direct and reflected paths. 14, Studio output.

In many consoles, the audio signal passes first through the individual strips ('channels'), then through the groups, then to the main control and out to 'line' – but in some this process is partly or wholly mimicked. For example, a group fader may not actually carry the signal; instead, a visually identical voltage control amplifier (VCA) fader sends messages which modify the audio in the individual channels that have been designated as belonging to the group, before it is combined with all other channels in the main control. In some consoles, the pre-set equalization, dynamics and some routing controls are located away from the channel strips, and can all be entered separately into the memory of a computer: this greatly reduces the (visual) complexity and total area of the layout.

In the extreme case, the actual signal does not reach the console at all: the controls send their messages, via a computer, to racks in which the real signal is modified, controlled and routed. What then remains of an original console layout depends on the manufacturer's judgement of what the customer will regard as 'user friendly': it should be possible for the majority of users to switch between analogue and fully digital systems (and back again) without having to go through a major course of re-education.

Apart from the microphones and control console, the next most important equipment in a studio is a high-quality loudspeaker system – for a radio or recording studio is not merely a room in which sounds are made and picked up by microphone, it is also the place where shades of sound are judged and a picture is created by ear.

The main thing that distinguishes a broadcasting studio from any other place where microphone and recorder may be set up is that two acoustically separate rooms must be used: one where the sound may be created in suitable acoustics and picked up by microphone, and the other in which the mixed sound is heard. This second room is often called the control room – although in organizations where 'control room' means a line-switching centre it may be called a 'control cubicle' instead. Tape, disc and hard-drive recorders and reproducers and editing systems, and the staff to operate them, are also in the control area, as are the producer or director (in some cases these two roles are combined) and the production secretary (who checks running and overall timing, etc.).

In the simplest television studios, sound and picture are controlled side by side, but more often there is an acoustically separate sound room with much the same physical layout as it would have in a radio studio (except that it will have picture monitors, including one for that which is currently selected and one for preview). The television sound supervisor really does need to listen to programme sound in a separate cubicle because the director and his assistant must give a continuous stream of instructions on an open microphone, which means that their programme sound loudspeaker has to be set at a lower level than is necessary for the judgement of a good balance, and at times their voices obscure it. Their directions are relayed to the sound control room on an audio circuit, but at a reduced volume.

Studio layout for recording

It is no good finding a recording fault ten minutes after the performer has left the studio. If possible, a recording should be monitored *from the recording itself* as it is being made. On simple recordings the operational staff can do both jobs, operating the recorder and monitoring the output from a separate replay head. Another safeguard is backup: for example, a hard disk recording might be backed up both within the computer (recorded automatically in two places at once) and also on tape – but even this provides no guarantee that it was actually recorded anywhere, unless at least one is checked. Each track of a multitrack (tape or digital) recording cannot realistically be checked continuously, but a series of recordings can usually be retained for a while. When recording a live public event the options are more limited still: it is vital not only to have backup but also to listen continuously for any possible recording problem.

When a recording made on analogue tape is monitored while it is being recorded, there is a delay of perhaps one-fifth of a second that corresponds to the 3 inches of tape (8 cm) between recording and reproducing heads at 15 inches per second (in/s) (38 cm/s) or a delay of two-fifths of a second at 7.5 in/s (19 cm/s). This kind of delay makes it unsatisfactory to combine monitoring a recording with any other job when the production has any degree of complexity.

Units of measurement
Tape widths and speeds are commonly described by their Imperial measurements: 'quarter-inch tape', '7½ inches per second', and so on. This is a rare exception to the general rule in science and engineering that measures are presented in metric (or 'SI') units. In this book Imperial measures are usually given as well, for those who still prefer to use them. Measurements such as tape speeds are precise, so conversions must be given fairly accurately, but in other cases throughout this book, measures are intended only as ballpark figures, and conversions are rounded up or down as anything more precise would give a false impression.

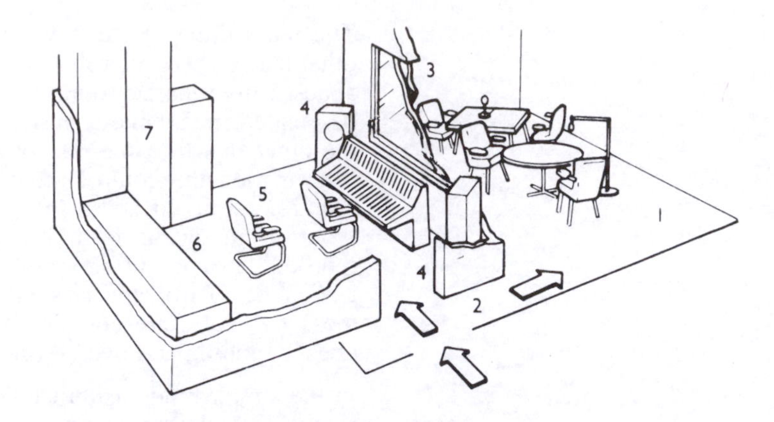

[1.3] **Radio studio layout**
1, Studio. 2, Lobby. 3, Double or triple glazed window for acoustic separation. 4, Control console and monitoring loudspeakers. 5, Producer's position (ideally, on same line as balancer). 6, Recording and replay equipment. 7, Equipment racks.

One-fifth of a second can completely destroy timing. The perfect fade on a piece of music, for example, may have to be accurate to less than one-twentieth of a second. These requirements suggest that, ideally, for complex recordings there might be a third acoustically separate room in which a recording engineer can check the quality of the recorded sound – although backup is generally cheaper than the cost of extra people.

Many productions are prepared (or rehearsed) and then recorded section by section in random order. Each segment of a few minutes is separately discussed, parts or all of it run through, then recorded (together with any necessary retakes), and the results are edited together later. In the simplest arrangement, segments are recorded in the control room, then checked and edited (using headphones) while the next piece is being prepared. Using digital (non-destructive) editing systems, material can even be rough-edited as the recording continues – to mark in and out points, and delete preliminary and intervening production instructions, gaps and errors – to create a rough assembly.

An increasingly rare alternative is more like a stage play or 'live' broadcast. This is based on the idea that the dramatic contour of a complex work is likely to have a better shape if it is fully rehearsed and then recorded continuously as a whole, leaving any faults to be corrected by retakes. Studio time may be divided into two parts, with the recordings limited to a relatively short session.

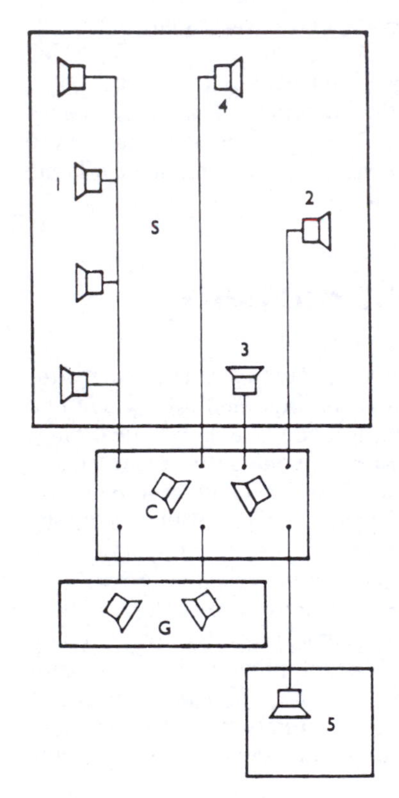

[1.4] **Loudspeakers in a television musical production**
C, Sound control room. G, Gallery. S, Studio. 1, Audience (public address) system. 2, Foldback of recorded sound effects. 3, Special feed of vocalist's sound to conductor. 4, Special feed of rhythm group to vocalist. 5, Remote studio.

Studio communications

In both radio (and other recording studios) and television there are sound circuits that are used for local communication only.

Headphones can be used by performers to hear a separate talkback circuit from the control room. Their use may range from full-scale relay of instructions (as in television productions), through feeding questions to an interviewer, to monitoring a feed of some event in order to give news of it. There may be both talkback (to the studio) and reverse talkback. The latter is an independent feed from the studio, often associated with a microphone-cut key controlled by the presenter in order to get off-air and ask for information or instructions (or another simply to get off-air and cough).

When a studio microphone is faded up it is usually arranged that the studio loudspeaker be cut off automatically (except for talkback purposes during rehearsal). Headphones or an earpiece may be used in the studio to hear talkback when the loudspeaker has been cut, which may be just before a cue.

In television the complexity of the communication system is greater, because there are so many more people involved, and also because they need to occupy a range of acoustically separate areas. The director speaks to the floor manager over a local radio circuit; the floor manager hears the director on a single earpiece; camera operators and others in the studio listen on headphones.

In the sound control room the sound supervisor can hear high-level and high-quality programme sound. In addition, a second, much smaller loudspeaker (so that it is distinctly different in quality) relays both the director's instructions and reverse talkback from sound assistants on the studio floor.

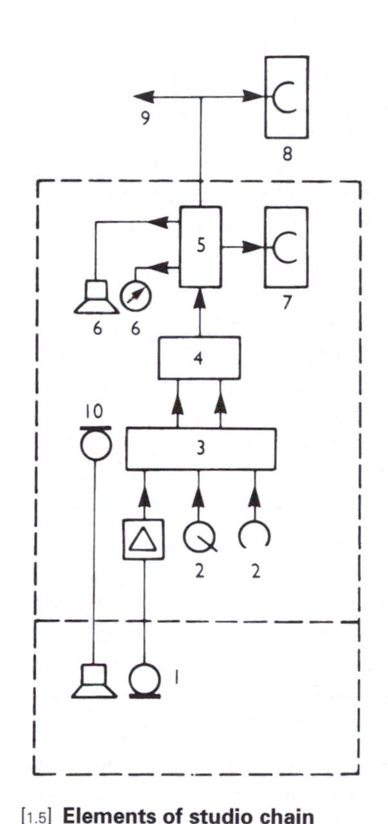

[1.5] **Elements of studio chain**
1, Microphones: low-level sources, fed through amplifier. 2, High-level sources (replay machines, etc.). 3, Microphone and line-input channels in control console. 4, Group channels in control console. 5, Main (or master) control. 6, Monitoring. 7, Studio recorder. 8, External recorder. 9, Studio output. 10, Talkback circuit.

Lighting and camera control also generally have a (third) acoustically separate area. They have their own talkback system to appropriate points on the studio floor, and camera operators in their turn have reverse talkback to them. In the control gallery a technical manager has, in effect, a small telephone exchange for communication with people at all points concerned with technical aspects of the production. This includes control lines to remote sources, and also to video recording and play-out and telecine operators (who, in their own areas, hear the director's circuit as well as programme sound or their own output).

Television outside broadcast or *remote* trucks sometimes have sound control, director and camera control all in the same tiny area of the vehicle: this is best for sports and other unrehearsed events where side-by-side working helps the operators to synchronize their actions. But for rehearsed events, and productions built on location sequence by sequence, good sound conditions are more important, and these demand an acoustically separate area. This may be either in the same mobile control vehicle or (for music, where it is better to have more space for monitoring) in a separate truck or, alternatively, 'derigged' and set up in some convenient room.

In both radio and television outside broadcasts the sound staff provide communication circuits between operational areas (e.g. commentary points) and the mobile control room. This often takes the form of local telephone circuits terminating at special multipurpose boxes that also provide feeds of programme sound and production talkback. On television outside broadcasts (remotes), talkback is primarily for the stage manager. In addition, field telephones may be required at remote points. Wireless links are also widely used.

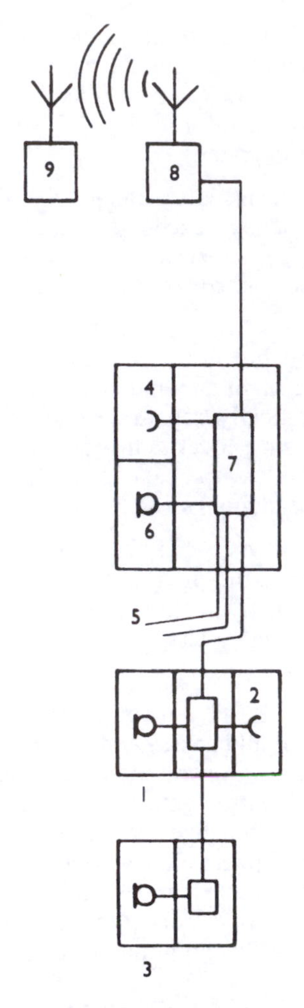

[1.6] **The broadcasting chain**
The studio output may go to a recorder or it may provide an insert into a programme compiled in another studio – or it may be fed direct to a programme service continuity suite. Between studio, continuity and transmitter there may be other stages: switching centres, boosting amplifiers and frequency correction networks on landlines, etc. 1, Studio suite. 2, Recording area or machine room. 3, Remote studio. 4, Machine room. 5, Other programme feeds. 6, Continuity announcer. 7, Continuity mixer. 8, Transmitter. 9, Receiver.

In broadcasting, the studio suite (studio and control room) may be only the first link in a complex chain. In a station with mixed output, the next link is a continuity suite where the entire output is assembled. The feed from individual studios is fed in live, and linked by station identification and continuity announcements. This announcer may have executive responsibility for the service as a whole, to intervene if any contribution under-runs, over-runs, breaks down, or fails to materialize. The announcer also cues recorded programmes, presents trailers (promotional material) and commercials where appropriate.

In small, specialized stations such as those of local radio, virtually the whole of the output emerges from a single room – though a separate studio may be available for news and discussion programmes or for recording commercials. In this 'electronic office' the presenter controls the microphone and plays in records, recorded commercials, etc. For this, there should be a range of CD and MiniDisc players, DAT, reel-to-reel tape decks and cassette players, together with digital audio editing: sufficient to allow commercials, promotional trailers, production effects and public service announcements to be played one after the other ('back-to-back'). Even with a hard-drive play-out system there may be backup from MiniDiscs, Zip files or other recordings reproduced from 'carts' or stacks. In addition to microphone and reproduction equipment and telephones (including broadband), the presenter may also have two-way communication with news or traffic cars or helicopters, which can be arranged to record automatically while on air with other material.

Alternatively, much of this may be automated, with output fed from digital playlists. Audio links are recorded in advance, including programmed overlaps to previously recorded or musical items. In this kind of scheduling the studio has a smaller role to play, as more of the assembly, editing, editorial changes and monitoring of work in progress are done in an office environment at any convenient time ahead of transmission – in particular, without having to wait for a last-minute run-through to check the result.

In large networks the basic service is put together in either way, but with greater emphasis on precision timing if the output of the network continuity appears as a source in the regional or local continuity that feeds the local radio transmitter. The links between the various centres may be dedicated landlines or radio links. Broadband (ISDN) telephone lines are increasingly used for some of these feeds.

In television the broadcasting chain is substantially the same as for radio, except that in the presentation area there may be both a small television studio and a separate small booth for out-of-vision announcements. Use of the sound booth means that rehearsals of more complex material in the main studio do not have to stop for programme breaks, whether routine or unscheduled.

Chapter 2

The sound medium

Some readers of this book will already have a clear understanding of what sound is and how it behaves. Others will know very little about the physics of sound and will not wish to know, their interest being solely in how to use the medium. These two groups can skip this section and move directly to the operational techniques, creative applications and practical problems and their solutions that are the main theme of this book. Non-technical readers should have little difficulty in picking up much of what is needed, perhaps with an occasional glance at the Glossary. Most of the terms used should be understandable from their context, or are defined as they first appear.

On the other hand, to help make the right choices on questions like where to place a microphone or what may be done to modify the signal from it, it will help to know a little about the sound medium and also how musical sounds are produced by instruments, including the human voice. In addition, a description of the curious manner in which our ears make sense of sound may explain some of the very strange ways in which we measure sound and treat it in our studio equipment.

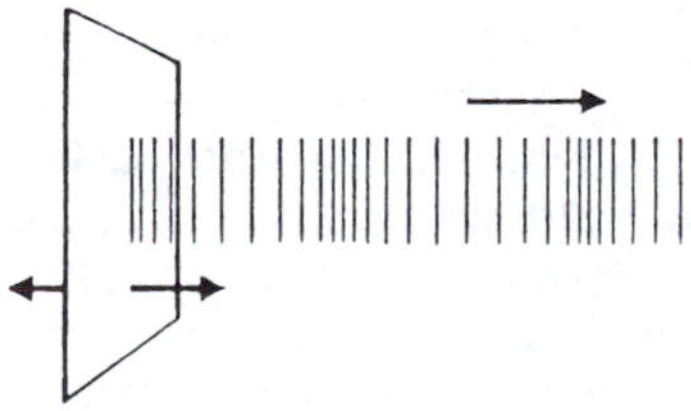

[2.1] **Sound wave**
A vibrating panel generates waves of pressure and rarefaction: sound. A tuning fork is often used for this demonstration, but in fact tuning forks do not make much noise. They slice through the air without moving it much. A vibrating panel is more efficient.

The nature of sound

Sound is produced by materials that vibrate. If a panel of wood vibrates, the air next to it is pushed to and fro. If the rate of vibration is somewhere between tens and tens-of-thousands of excursions per second, the air has a natural elasticity which we do not find at slower speeds. Wave your hand backwards and forwards once a second and the air does little except get out of its way; it does not bounce back. But if you could wave your hand back and forth a hundred times every second, the air would behave differently: it would compress as

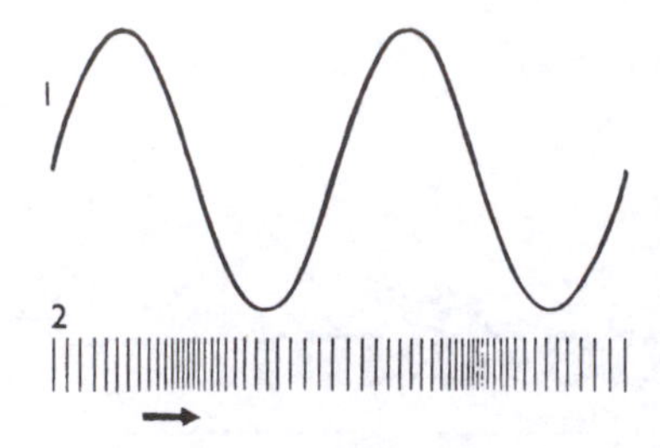

[2.2] **Waves**
1, Sound waves are usually represented diagrammatically as transverse waves, as this is convenient for visualizing them. It also shows the characteristic regularity of the simple harmonic motion of a pure tone. Distance from the sound source is shown on the horizontal axis, and displacement of individual air particles from their median position is shown on the vertical axis. 2, Because the particles are moving backwards and forwards along the line that the sound travels, the actual positions of layers of particles are arranged like this. It also shows the travel of pressure waves.

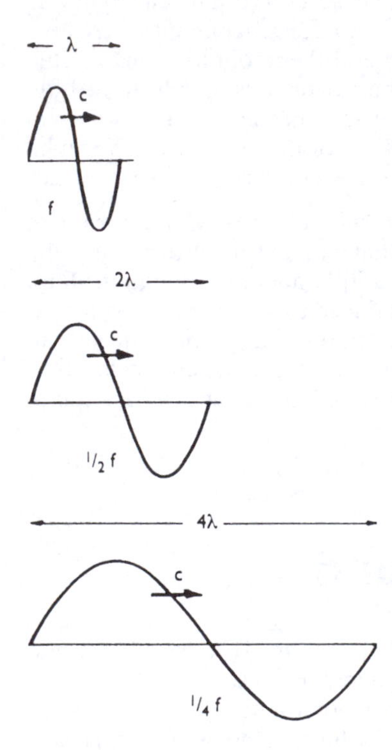

[2.3] **Frequency and wavelength**
All sound waves travel at the same speed, *c*, through a given medium, so frequency, *f*, is inversely proportional to wavelength, λ.

the surface of the hand moves forward and rarefy as it moves back: the natural elasticity of the air takes over. As the surface moves forward, each particle of air pushes against the next, so creating a pressure wave. As the surface moves back, the pressure wave is replaced by a rarefaction, which travels outward as the air next to it draws back. And that is followed by another pressure wave, and so on.

It is a property of elastic media that a pressure wave passes through them at a speed that is a characteristic of the particular medium. The speed of sound in air depends on its temperature as well as the nature of air. In cool air – at 15°C (59°F) – sound travels about 340 metres in a second (about 1120 feet per second). At 20°C (68°F) this increases to 344 m/s (1130 ft/s).

This speed is independent of the rate at which the surface generating the sound moves backward and forward. The example was 100 excursions per second, but it might equally have been 20 or 20 000. This rate at which the pressure waves are produced is called the *frequency*, and this is measured in cycles per second, or *hertz*: 1 Hz is 1 cycle per second.

Let us return to that impossibly energetic hand shaking back and forth at 100 Hz. It is clear that this is not a perfect sound source: some of the air slips round the sides as the hand moves in each direction. To stop this happening with a fluid material like air, the sound source would have to be much larger. Something the size of a piano sounding board would be more efficient, losing less round the edges. But if a hand-sized vibrator moves a great deal faster, the air simply does not have time to get out of the way. For very high frequencies, even tiny surfaces are efficient radiators of sound.

In real life, sounds are produced by sources of all shapes and sizes that may vibrate in complicated ways. In addition, the pressure waves bounce off hard surfaces. Many sources and all their reflections may combine to create a complex field of intersecting paths. How are we to describe, let alone reproduce, such a tangled skein of sound? Fortunately, we do not have to.

Consider a single air particle somewhere in the middle of this field. All the separate pressure waves passing by cause it to move in various directions, to execute a dance that faithfully describes every characteristic of all of the sounds passing through. All we need to know in order to describe the sound at that point is what such a single particle is doing. The size of this particle does not matter so long as it is small compared with the separation of the successive pressure waves that cause the movement. It does not have to be very small; in fact, we want to measure an average movement that is sufficiently gross to even out all the random tiny vibrations of the molecules of air – just as the diaphragm of the ear does. And thinking about the ear leads us to another simplification. A single ear does not bother with all the different directional movements of the air particle; it simply sums and measures variations in air pressure. This leads us to the idea that a continuously operating pressure-measuring device with a diaphragm something like the size of an eardrum should be a good instrument for describing sound in practical audio-engineering terms. What we have described is the first requirement of an ideal microphone.

Wavelength

Pressure waves of a sound travel at a fixed speed for a given medium and conditions, and if we know the frequency of a sound (the number of waves per second) we can calculate the distance between matching points on successive waves – the *wavelength*. (Wavelength is often symbolized by the Greek letter lambda, λ.)

Taking the speed of sound to be 340 m/s (1120 ft/s), a sound frequency of 340 Hz (or cycles per second) corresponds to a wavelength of 1 metre (just over 39 inches). Sometimes it is more convenient to think in terms of frequency and sometimes wavelength.

Because the speed of sound changes a little with air temperature, any relationship between wavelength and frequency that does not take this into account is only approximate. But for most practical purposes, such as calculating the thickness of padding needed to absorb a range of frequencies or estimating whether a microphone diaphragm is small enough to 'see' very high frequencies, an approximate relationship is all we need. Particular frequencies also correspond to notes on the musical scale:

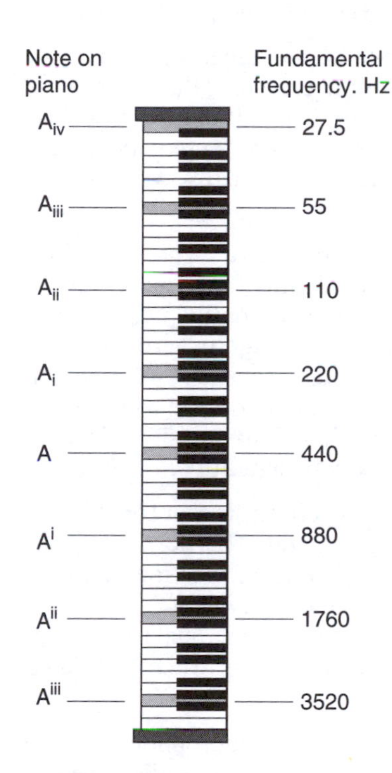

[2.4] **Piano keyboard**
Note and frequency of fundamental, octave by octave.

The correspondence of frequencies to musical notes

Note on piano	*Frequency (Hz)*	*Wavelength*	
–	14 080	2.5 cm	1 in
A^{iii}	3 520	9.5 cm	3.75 in
A^{ii}	1 760	19 cm	7.5 in
A^{i}	880	38 cm	1 ft 3 in
A	440	75 cm	2 ft 6 in
A_{i}	220	1.5 m	5 ft
A_{ii}	110	3 m	10 ft
A_{iii}	55	6 m	20 ft
A_{iv}	27.5	12 m	40 ft

Here the speed of sound is taken to be 335 m/s (1100 ft/s), as it may be in cool air. Note that the length conversions and shorter wavelengths are approximate only.

The first thing to notice from this is the great range in physical size these different frequencies represent. Two octaves above the top A of the piano, the wavelength is approximately 25 mm (1 in). As it happens, this is close to the upper limit of human hearing; and the corresponding size is found in the dimensions of high-quality microphones: they must be small enough to *sample* the peaks and troughs of this wave. At the lower end of the scale we have wavelengths of 12 m (40 ft). The soundboard of a piano can generate such a sound, though not so efficiently as some physically much larger air cylinders in a pipe organ. Below the 20 Hz lower limit of human hearing, the deepest note of an organ may be 16 Hz. Only its overtones are audible as sound; the fundamental is sensed as vibration.

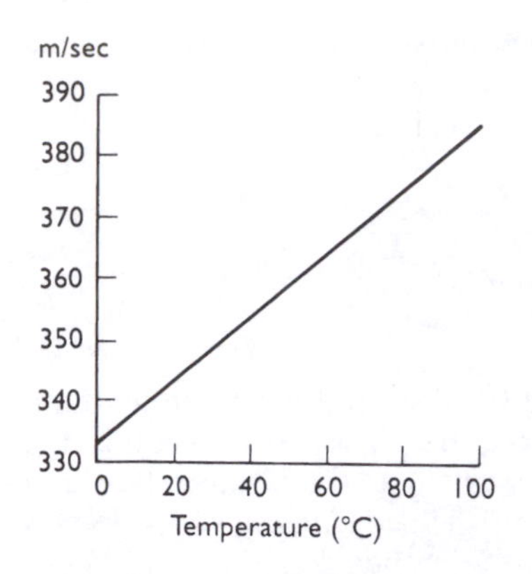

[2.5] **Velocity of sound in air**
The velocity of sound increases with temperature.

The second point to notice, and the reason for the great range of sizes, is that for each equal musical interval, a rise of one octave, the frequency doubles (and the wavelength is halved).

Much less obvious is the fact that as the pitch of a sound depends on the size of the object making it, and because that size remains nearly constant as temperature increases, then because the velocity of sound increases with temperature, the pitch goes up. Metal strings expand, so go a little flat, while wind instruments go sharp. For strings, the remedy is simple: they can easily be tuned, but the vibrating column of air in most wind instruments cannot be changed so easily. A flute, for example, sharpens by a semitone as the air temperature goes up by 15°C. (As the metal of the instrument itself warms up and so lengthens by a tiny amount, this means that it also flattens a little, but this is not by enough to make much difference.) To compensate for weather, hall and instrument temperature changes, the head joint of the flute can be eased in or out a little – in an orchestra, matching its pitch to that of the traditionally less adaptable oboe.

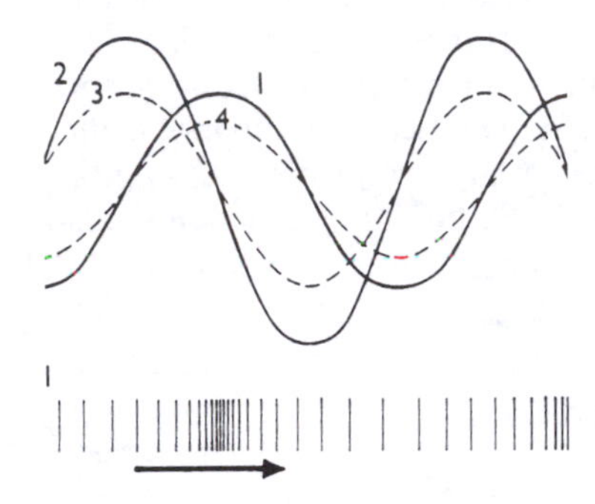

[2.6] **Waveform relationships**
1, Pressure wave. 2, Displacement wave. 3, Pressure gradient. 4, Particle velocity.

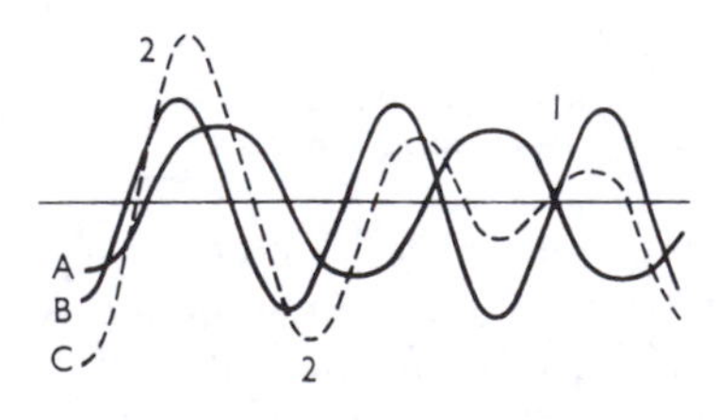

[2.7] **Adding sound pressures**
At any point the sound pressure is the sum of the pressures due to all waves passing through that point. If the simple waves A and B are summed this results in the pressure given by the curve C. 1, Here there is partial cancellation. 2, Here the peaks reinforce. The resultant curve is more complex than the originals.

Waves and phase

Still considering a pure, single-frequency tone, we can define a few other variables and see how they are related to the pressure wave.

First we have *particle velocity*, the rate of movement of individual air particles. As this is proportional to air pressure, the waveforms for pressure and velocity are similar in shape. But not only that: where one has a peak the other has a peak; where there is no excess of pressure, there is no velocity, and so on. The two waveforms are said to be *in phase* with each other.

The next variable is *pressure gradient*, the rate at which pressure changes with distance along the wave. Plainly, this must have a waveform derived from that of pressure, but where the pressure is at a peak (maximum or minimum) its rate of change is zero. Again it turns out that for a *pure tone* (which has the shape of what in mathematics is called a '*sine wave*') the shape of the wave is the same, but is displaced a quarter of a wavelength to one side. The two waveforms are said to be a quarter of a wavelength *out of phase*.

Another variable, again with a similar waveform, is that for *particle displacement*. This is a graph showing how far a particle of air has moved to one side or the other of its equilibrium position and how this changes in time. (This is, for sound, the equivalent of the wave we see moving over water.) Displacement is proportional to pressure gradient, and is therefore in phase with it. It is directly related to *amplitude*.

These terms are all encountered in the theory of microphones. Some (such as the moving-coil and ribbon types that will be described in Chapter 5) are devices that measure the velocity of the air particles next to them. Others (condenser or crystal types) measure the amplitude of the sound wave. A curious effect of this is that, while the information they transmit is virtually identical, the electrical waveforms that carry them are slightly different. More will, however, be heard of pressure and pressure-gradient modes of operation, as these characteristics of microphones lead to important differences in the way they can be used.

Because of these differences, the *phase* of the signal as it finally appears in the microphone output is not always the same as that of the sound pressure measured by the ear. Fortunately the ear is not usually interested in phase, so it does not generally matter whether a microphone measures the air pressure or pressure gradient. The only problem that might occur is when two signals very similar in form but different in phase are combined. In an extreme case, if two equal pure tones that are exactly half a wavelength out of phase are added, the output is zero. Normally, however, sound patterns are so complex that microphones of completely different types that are sufficiently far apart may be used in the same studio, and their outputs mixed, without much danger of differences of phase having any noticeable effect.

So far we have considered simple tones. If several are present together, a single particle of air can still be in only one place at a time: the displacements are added together. If many waves are superimposed, the particle performs the complex dance that is the result of adding all of the wave patterns. Not all of this adding together results in increased pressure. At a point where a pressure wave is superimposed on a rarefaction, the two – opposite in phase at that point – tend to cancel each other out.

Waves and obstacles

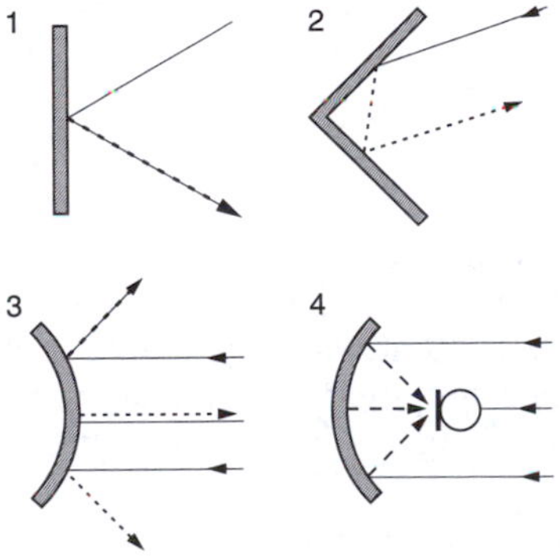

1 2 3 4

[2.8] **Reflected sound**
1, The angle of incidence is the same as the angle of reflected sound, but there is some loss as sound is absorbed. 2, At a 90° corner, sound is reflected in the direction it came from. 3, A convex surface disperses and so diffuses sound. 4, A concave surface concentrates sound; a parabolic reflector focuses it.

Close to most sources sound radiates outward in all directions that are open to it, in the simplest configuration as a *spherical wave*. Close to a large, flat sound source, sound radiates initially as a *plane wave*. In addition, a spherical wave gradually loses curvature as it gets further from the source, becoming more and more like a plane wave. Spherical and plane waves behave differently when they encounter microphones (see Chapter 5) or meet obstacles in a studio.

Sound waves *reflect* from hard surfaces in the same way that light reflects from shiny objects, and the familiar optical rules apply. Sound approaching a flat surface at an angle is reflected at the same angle, apart from some proportion that is absorbed. This is one of the basic features of studio acoustics (see Chapter 4). Double reflections, in a right-angled corner, send it back in the direction it came from.

If sound encounters a surface that bows outward, it will *disperse*. This is one way of creating sound *diffusion*; others are to place uneven surfaces or a variety of objects in its path (diffusion is another important feature of studio acoustics and also of reverberation). If sound encounters a concave surface it is focused and

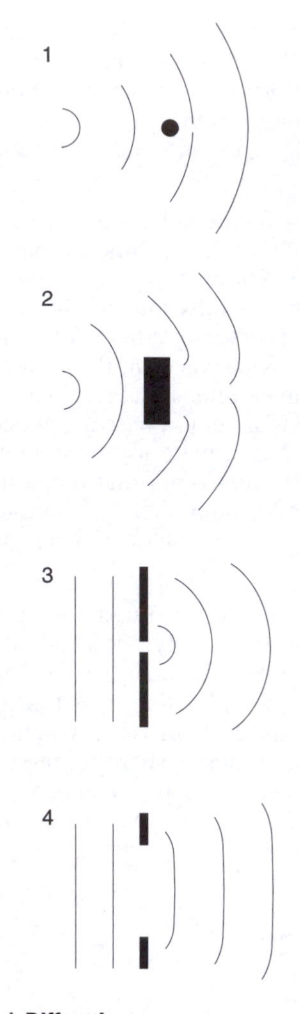

[2.9] **Diffraction**
1, An obstacle that is smaller than a wavelength leaves the wave undisturbed. 2, An obstacle that is broader than a wavelength casts a shadow. 3, A small (or narrow) gap acts as a new source. 4, A broad gap allows a wave to pass undisturbed, with shadows on either flank.

concentrated, making it louder without the need for added power. A plane wave going into a parabolic reflector is focused to a point (so a microphone at the focus can pick up extremely faint, distant sounds).

Diffraction is another effect that is like those that can be produced with light. If a surface has a hole in it that is small compared to the wavelength of sound approaching it, on the other side the hole radiates sound like a new point-source. This radiation is so faint that it is of no practical importance in studio acoustics – but this is also what happens inside some small microphone capsules.

Of greater practical importance is diffraction around an object in the path of a wave. If the wavelength is long compared with the width of the object, the wave is virtually undisturbed: it is as though the wave does not 'see' the object (but to reach the back of the object it has to travel further than if it could go right through it). If the obstruction is wider than the wave is long, it casts a shadow.

Energy, intensity and resonance

There are several more concepts that we need before going on to consider the mechanics of music.

First, the *energy* of a sound source. This depends on the amplitude of vibration: the broader the swing, the more *power* (energy output per second) it can produce. The sound *intensity* at any point is then measured as the acoustic energy passing through unit area per second. But to convert the energy of a sound source into acoustic energy in the air, we have to ensure that the sound source is properly coupled to the air, so that its own vibrations are causing the air to vibrate with it. Objects that are small (or slender) compared with the wavelength in air associated with their frequency of vibration (e.g. tuning forks and violin strings) are able to slice through the air without giving up much of their energy to it: the air simply slips around the sides of the prong or the string.

If a tuning fork is struck and then suspended loosely it goes on vibrating quietly for a long time. If, however, its foot is placed on a panel of wood, the panel is forced into vibration in sympathy and can transmit vibration to the air. The amplitude of vibration of the tuning fork then goes down as its energy is lost (indirectly) to the air.

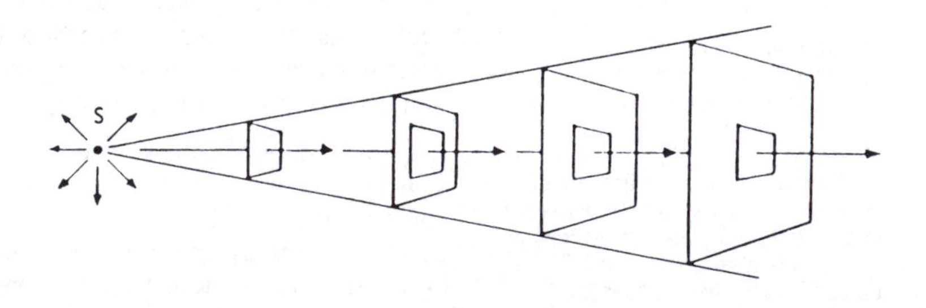

[2.10] **Sound intensity**
This is the energy passing through unit area per second. For a spherical wave (i.e. a wave from a point source) the intensity dies away very rapidly at first. The power of the sound source S is the total energy radiated in all directions.

[2.11] **Tuning fork**
1 and 2, Each fork vibrates at a specific natural frequency, but held in free air radiates little sound.
3, Placed against a wooden panel, the vibrations of the tuning fork are coupled to the air more efficiently and the fork is heard clearly.
4, Placed on a box having a cavity with a natural resonance of the same frequency as the tuning fork, the sound radiates powerfully.

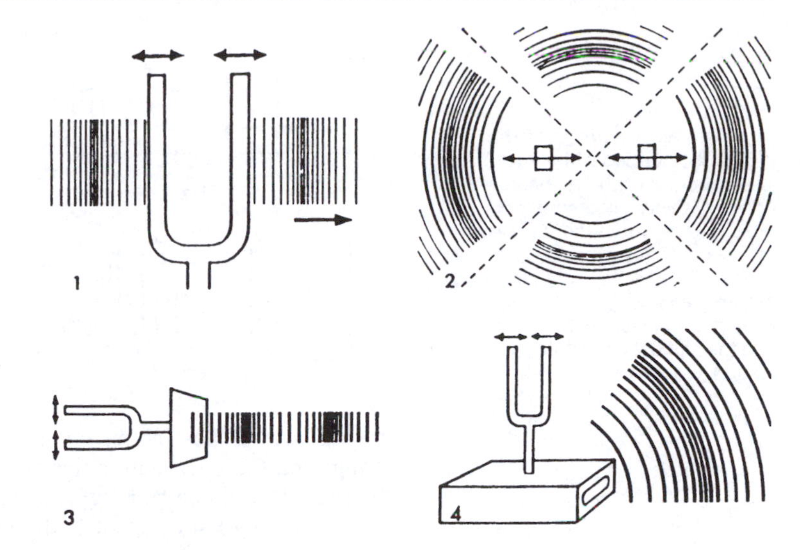

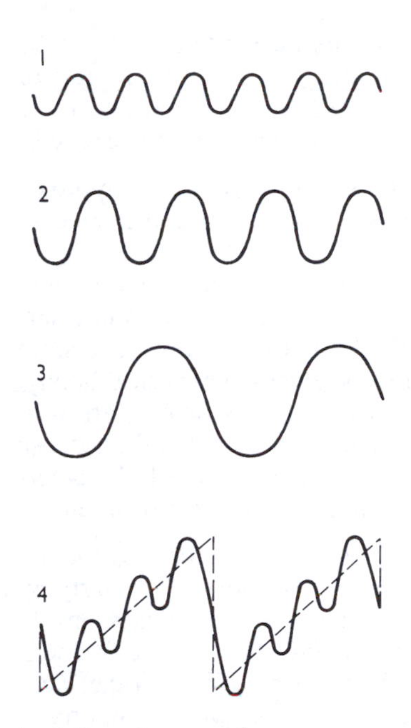

[2.12] **Complex waveforms**
Pure tones (1, 2 and 3) may be added together to produce a composite waveform (4). In this case the first two overtones (the second and third harmonics) are added to a fundamental. If an infinite series of progressively higher harmonics (with the appropriate amplitudes and phases) were added, the resultant waveform would be a 'saw-tooth'. When the partials are related harmonically, the ear hears a composite sound having the pitch of the fundamental and a particular sound quality due to the harmonics. But when the pitches are unrelated or the sources are spatially separated, the ear can generally distinguish them readily: the ear continuously analyses sound into its components.

If the panel's natural frequency of vibration is similar to that of the tuning fork, the energy can be transferred and radiated much faster. In fact, wooden panels fixed at the edges may not appear to have clear, specific frequencies at which they resonate: if you tap one it may make a rather dull sound, though perhaps with some definable note that seems to predominate. In the violin, the wooden panels are very irregular, which ensures that over a whole range of frequencies no single one is emphasized at the expense of others. The string is bowed but transmits little of its energy directly to the air. Instead, the energy is transferred through the bridge that supports the string to the wooden radiating belly of the instrument.

At this stage an interesting point arises: the radiating panels are not only smaller (in length) than the strings themselves, but also considerably smaller than some of the wavelengths (in air) that the strings are capable of producing. The panels might respond well enough to the lowest tones produced, but they are not big enough to radiate low frequencies efficiently. This leads us to consider several further characteristics of music and musical instruments.

Overtones, harmonics and formants

Overtones are the additional higher frequencies that are produced along with the *fundamental* when something like a violin string or the air in an organ pipe is made to vibrate.

On a string the overtones are all *harmonics*, exact multiples of the fundamental, the lowest tone produced. As in most musical instruments, it is this lowest tone that defines its *pitch*. If the string were bowed in the middle, the fundamental and odd harmonics would be

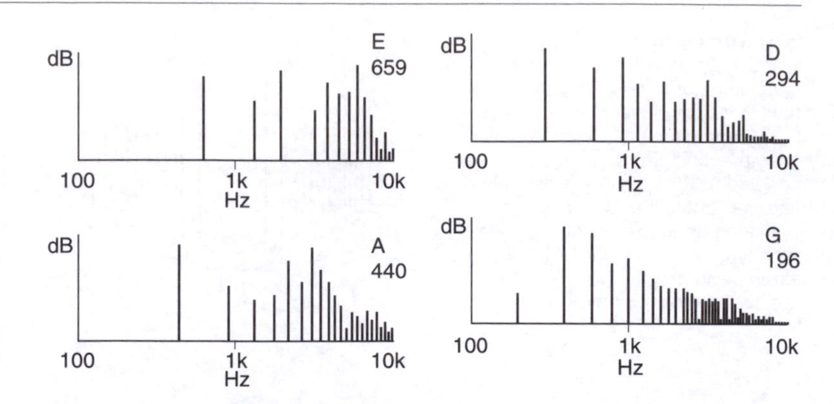

[2.13] **The open strings of the violin: intensities of harmonics** The structure of the resonator ensures a very wide spread, but is unable to reinforce the fundamental of the low G, for which it is physically too small. Note that there is a difference in quality between the lowest string (very rich in harmonics) and the highest (relatively thin in tone colour).

emphasized, as these all have maximum amplitude in the middle of the string; and the even harmonics (which have a node in the middle of the string) would be lacking. But the string is, in fact, bowed near one end so that a good range of both odd and even harmonics is excited.

There is one special point here: if the string is bowed at approximately one-seventh of its length it will not produce the seventh harmonic, as this would have a node, a point where the string is not displaced, at that point. So, for the most harmonious quality, that is a good place to bow at, because as it happens the seventh harmonic is the first that is not musically related to the rest. The sixth and eighth harmonics are both full members of the same musical family, while higher harmonics (whether family members or not) are produced in such rich profusion that their main role is to add a dense tonal texture high above the fundamental. These are all characteristics of what we call 'string quality', in any stringed instrument.

The character of the instrument itself – violin, viola, cello or bass – is defined by the qualities of its resonator, and most particularly by its size. The shape and size of the resonator superimposes on the string quality its own special *formant* characteristics. Some frequencies, or ranges of frequencies, are always emphasized; others are always discriminated against. Formants are very important: obviously, in music they are a virtue, but in audio equipment the same phenomenon might be called 'an uneven frequency response' and would be regarded as a vice – unless it were known, calculated and introduced deliberately with creative intent.

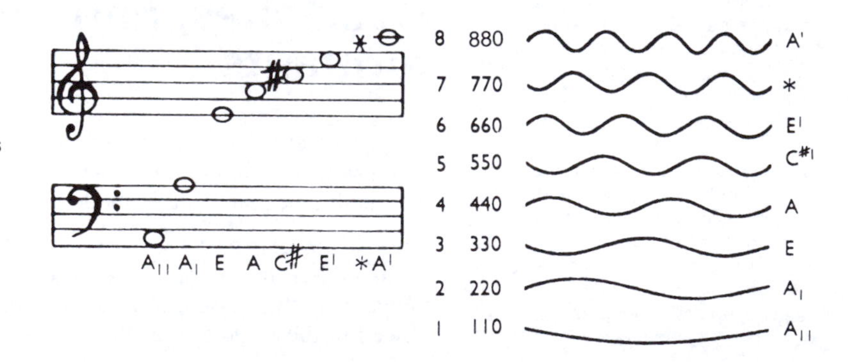

[2.14] **The first eight string harmonics**
The fundamental is the first harmonic: the first overtone is the second harmonic, and so on. They are all notes in the key of the fundamental, except for the seventh harmonic, which is not a recognized note on the musical scale (it lies between G and G#). If the violinist bows at the node of this harmonic (at one-seventh of the length of the vibrating string) the dissonance is not excited. The eleventh harmonic (another dissonant tone) is excited, but is in a region where the tones are beginning to cluster together to give a brilliant effect.

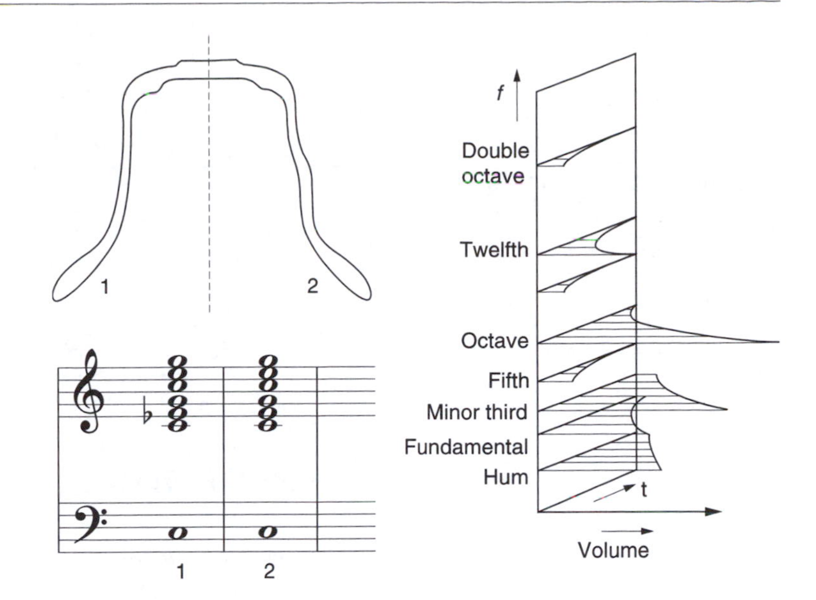

[2.15] **Bell overtones**
Left: Profiles and tones produced by 1, a minor-third bell and 2, a major-third bell. *Right*: The relative volumes and rates of decay of the tones present in the sound of a minor-third bell.

On some instruments, such as bells, getting the principal overtones to have any musical relationship to each other is a major manufacturing problem (in fact, the lowest note of a bell is not even distinguished by the name of fundamental). In drums, the fundamental is powerful and the overtones add richness without harmonic quality, while from triangles or cymbals there is such a profusion of high-pitched tones that the sound blends reasonably well with almost anything.

An important feature of any musical instrument is that its vibrating surfaces or columns of air do not radiate equally efficiently in all directions. Different frequencies, too, may radiate more or less powerfully in different directions, so that the mixture of tones changes with angle.

There are many other essential qualities of musical instruments. These may be associated with the method of exciting the resonance (bowing, blowing, plucking or banging); or with qualities of the note itself, such as the way it starts (the *attack*, which creates an irregular *transient*), is *sustained* and then *released*, changing in volume as the note progresses (to create its *envelope*).

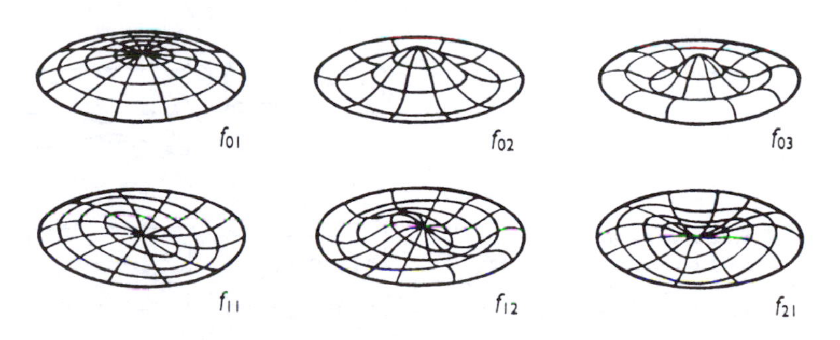

[2.16] **Vibration of a drumskin**
(a stretched circular membrane, clamped at the edges) The suffixes refer to the number of radial and circular nodes (there is always a circular node at the edge of the skin). The overtones are not harmonically related. If $f_{01} = 100$ Hz, the other modes of vibration shown here are: $f_{02} = 230$, $f_{03} = 360$, $f_{11} = 159$, $f_{12} = 292$, $f_{21} = 214$ Hz.

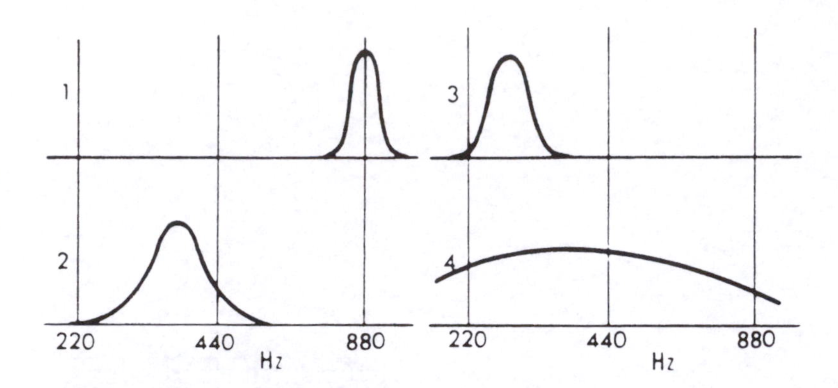

[2.17] **Wind instrumental formants**
1, Oboe. 2, Horn. 3, Trombone. 4, Trumpet. The formants, imposed by the structural dimensions of parts of the instruments, may be broad or narrow. They provide an essential and recognizable component of each instrument's musical character.

Air resonance

Air may have *dimensional resonances* very much like those of a string of a violin, except that whereas the violin string has *transverse waves*, those in air, being composed of compressions and rarefactions, are *longitudinal waves*. Radiated sound moves through air in the form of *progressive waves*, but dimensional resonances stand still: they form *stationary* or *standing waves*.

These stationary waves can again be represented diagrammatically as transverse waves. The waveform chosen is usually that for displacement amplitude. This makes intuitive sense: at *nodes* (e.g. at solid parallel walls if the resonance is formed in a room) there is no air-particle movement; while at *antinodes* (e.g. halfway between the walls) there is maximum movement of the air swinging regularly back and forth along a line in the room. Harmonics may also be present.

Standing waves are caused when any wave strikes a reflecting surface at a right angle and travels back along the same path. A double reflection in a right-angled corner produces the same effect, although reduced by the greater loss in two reflections.

The air in a narrower space (such as an organ pipe) can also be made to resonate. Where an end is closed, sound will reflect, as at any other solid surface. But if the pipe is open at the end, resonance can still occur, because in a tube that is narrow compared with the length of the wave

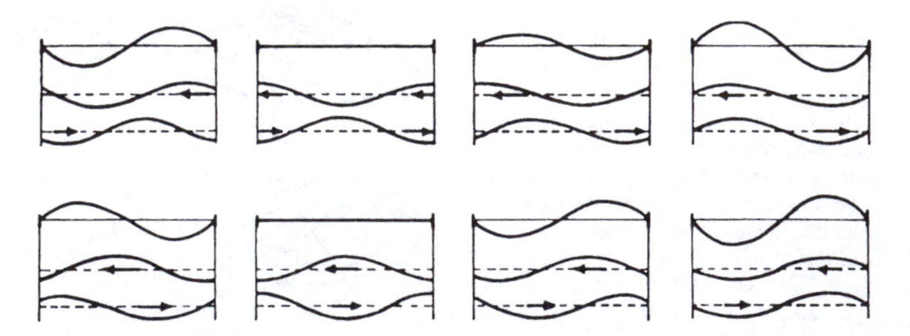

[2.18] **Stationary wave**
Stationary waves are formed from progressive waves moving in opposite directions.

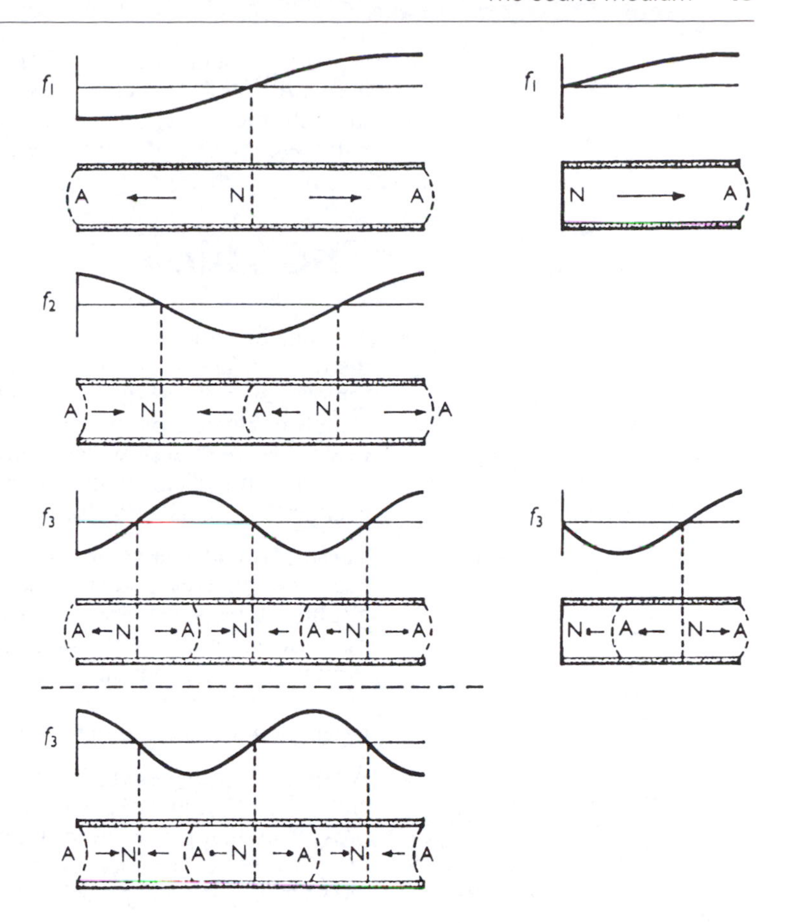

[2.19] **Vibration of air columns** *Left*: For an open pipe, the fundamental frequency, *f*, is twice the length of the pipe. There is a node N at the centre of the pipe and antinodes at the open ends. The second and third harmonics are simple multiples of the fundamental: $f_2 = 2f_1$, $f3 = 3f_1$, etc. *Left below*: The third harmonic, half a cycle later. Air particles at the antinodes are now moving in the opposite direction, but the air at the nodes remains stationary. *Right*: A pipe closed at one end (i.e. stopped). The wavelength of the fundamental is four times the length of the pipe. The first overtone is the third harmonic; only the odd harmonics are present in the sound from a stopped pipe: *f*, 3*f*, 5*f*, etc. In both types of pipe most of the sound is reflected back from an open end for all frequencies where the aperture is small compared with the wavelength.

the sound has difficulty in radiating to the outside air. The energy has to go somewhere: in practice, a pressure wave reflects back into the tube as a rarefaction, and vice versa. In this case the fundamental (twice the length of the pipe) and all of its harmonics may be formed. If one end is open and the other closed, only the fundamental (the wavelength of which is now four times the length of the tube) and its odd harmonics are formed. The tone quality is therefore very different from that of a fully enclosed space, or again, from that of a pipe that is open at both ends.

Wind instruments in the orchestra form their sounds in the same way: the length of the column may be varied continuously (as in a slide trombone), by discrete jumps (e.g. trumpet and horn), or by opening or closing holes along the length of the body (e.g. flute, clarinet or saxophone). Formants depend on the shape of the body and of the bell at the open end – though for many of the notes in instruments with finger holes the bell makes little difference, as most of the sound is radiated from the open holes.

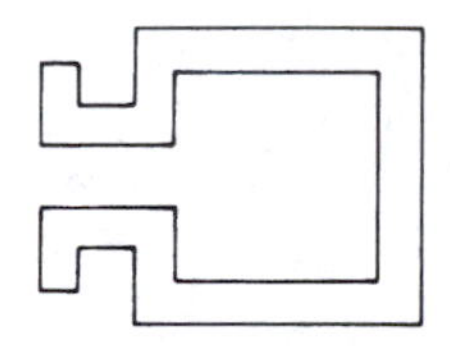

[2.20] **Helmholtz or cavity resonator** Air inside the cavity vibrates like a weight hanging on a spring.

Another important way in which air resonance may govern pitch is where a volume of air is almost entirely enclosed and is connected to the outside through a neck. Such a device is called a *Helmholtz* or *cavity resonator*. It produces a sound with a single, distinct frequency, an example being the note obtained by blowing across the mouth of a bottle. In a violin the cavity resonance falls within the useful range of the instrument and produces a 'wolf-tone' that the violinist has to treat with care, bowing it a great deal more gently than other notes.

The voice

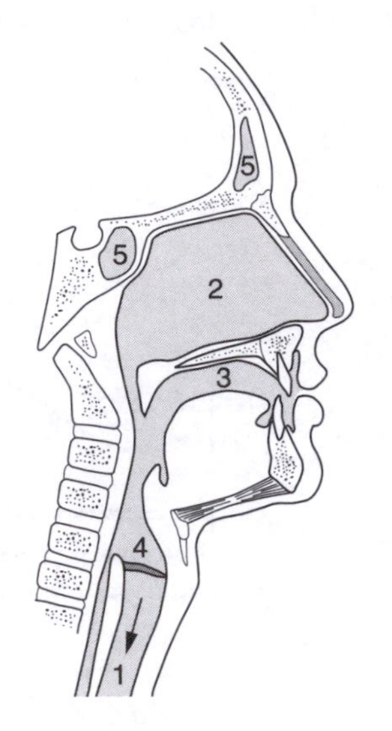

[2.21] **Vocal cavities**
1, Lungs. 2, Nose. 3, Mouth. This is the most readily flexible cavity, and is used to form vowel sounds.
4, Pharynx, above the vocal cords.
5, Sinuses. Together, these cavities produce the formants characteristic of the human voice, emphasizing certain frequency bands at the expense of others.

The human voice is a great deal more versatile than any musical instrument. This is because the cavities of the mouth, nose and throat impose such dramatically variable formant characteristics on the sounds modulated by the vocal cords. It is as though a violin had five resonators, several of which were continuously changing in size, and one (the equivalent of the mouth) so drastically that it completely changes the character of the sound from moment to moment.

These formant characteristics, based on cavity resonance, are responsible for vowel sounds and are the main vehicle for the intelligibility of speech. Indeed, sometimes they are used almost on their own: an example is a robot speech effect created by extracting formants from a human voice (or simulating them by computer) then using them to modulate some simple continuous sound, the nature of which is chosen to suggest the 'personality' of the machine.

A number of other devices are used in speech: these include sibilants and stops of various types which, together with formant resonances, provide all that is needed for high intelligibility. A whisper, in which the vocal cords are not used, is clear and understandable; in a stage whisper, intelligibility carries well despite lack of vocal power.

The vibrations produced by the vocal cords add volume, further character and the ability to produce song. For normal speech the fundamental may vary over a range of about 12 tones and is centred somewhere near 145 Hz for a man's voice and 230 Hz for a woman's. The result of this is that, as the formant regions differ little, the female voice has less harmonics in the regions of stronger resonance; so a woman may have a deep-toned voice, but its quality is likely to be thinner (or purer) than a man's. For song, the fundamental range of most voices is about two octaves, although, exceptionally, it can be much greater.

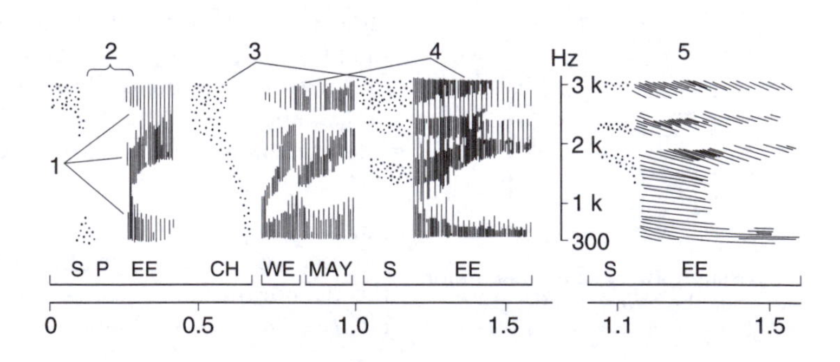

[2.22] **Human speech analysed to show formant ranges**
1, Resonance bands. 2, Pause before plosive. 3, Unvoiced speech. 4, Voiced speech. These formants arise from resonances in nose, mouth and throat cavities. They are unrelated to the fundamental and harmonics, which are shown in the second analysis of the word 'see'. 5, The vocal harmonics fall as the voice is dropped at the end of the sentence. At the same time, the resonance regions rise, as the vocal cavities are made smaller for the 'ee' sound.

The human ear

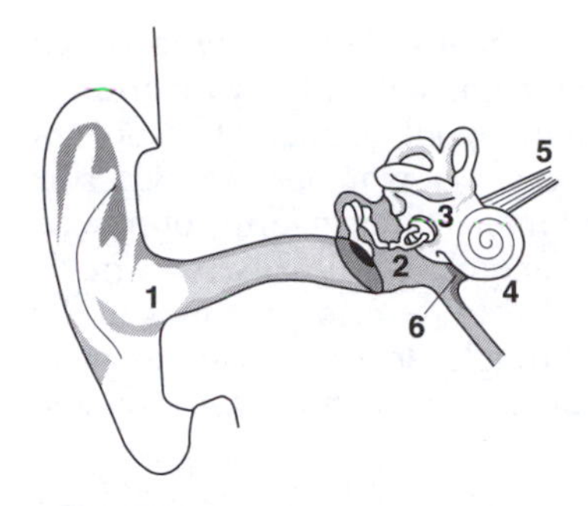

[2.23] **The ear**
1, Sound enters via the outer ear and the auditory canal. This channel has a resonance peak in the region of 3–6 kHz. At the end of the auditory canal is the eardrum, which vibrates in sympathy with the sound but is not well adapted to following large low-frequency excursions. 2, The middle ear contains air, the pressure in which is equalized through the Eustachian tube, a channel to the pharynx, behind the tonsils. With air on both sides, the eardrum vibrates to feed the sound mechanically across the middle ear along three small bones, ossicles, that form an impedance-matching lever mechanism: this converts the acoustic energy of the air to a form suitable for transmission through the fluid of the inner ear. Small muscles attached to the ossicles tighten up in response to loud noise: this aural reflex temporarily reduces the effect of any immediate repetition. 3, The sound is pumped via a membrane called the oval window into the tiny delicate channels of the inner ear, the cochlea. 4, The cochlea is shaped like a shell, with two channels along its length that get narrower until they join at the far end. Distributed along a vibrating membrane between the channels are hair-like cells that may respond to particular frequencies: when any 'hair' is bent a nerve impulse is fired. The further it lies along the canal, the lower the frequency recorded.
5, A bundle of 4000 nerve fibres carries the information to the brain, where it is decoded. 6, The pressure wave returns via another membrane, the round window, to dissipate back in the middle ear.

One of the functions of the outer ear is to act as a direction finder: we will return to this when we look at how stereo works. For the moment, just follow the sound in to the part that makes sense of it.

The part of the ear that senses sound is a tiny spiral structure called the cochlea. It is a structure that gets narrower to one end of the coil, like the shell of a snail. But unlike the shell, it is divided lengthways into two galleries, which join only at the narrowest inner end. The entire structure is filled with fluid to which vibrations may be transmitted through a thin membrane or diaphragm called the oval window. The acoustic pressures may then travel down one side of the dividing partition (the basilar membrane) and return up the other, to be lost at a further thin diaphragm, the round window.

All along one side of the basilar membrane are 'hairs' (actually elongated cells) that respond to movements in the surrounding fluid. Each of these hairs standing in the fluid acts as a resonator designed to respond to a single frequency (or rather, to a very sharply tuned narrow band). The hairs therefore sense sound not as an air particle sees it, as a single, continuous, very complex movement, but as a very large number of individual frequencies. The hairs are so arranged as to give roughly equal importance to equal musical intervals in the middle and upper middle ranges, but the separation is greater at very low frequencies: this is why it is more difficult to distinguish between very low pure tones. So (apart from indicating this lack of interest in low notes) a study of the mechanism of hearing confirms that we were right to be concerned with musical intervals that are calculated by their ratios rather than on a linear scale.

Growth of this sort is called *exponential*: it doubles at equal intervals, and a simple graph of its progress rapidly runs out of paper unless the initial stages are so compressed as to lose all detail. To tame such growth, mathematicians use a *logarithmic* scale, one in which each doubling is allotted equal space on the graph. The frequencies corresponding to the octaves on a piano progress in the ratios 1:2:4:8:16:32:64:128. The logarithms (to base 2) of these numbers are in a much simpler progression – 0, 1, 2, 3, 4, 5, 6, 7 – and these equal intervals are exactly how we perceive the musical scale.

Frequency is not the only thing that the ear measures in this way: changes of sound volume follow the same peculiar rule.

Sound volume and the ear

The hairs on the basilar membrane vibrate in different degrees, depending on the loudness of the original sound. Here again they measure not by equal increases in loudness but by ratios of loudness.

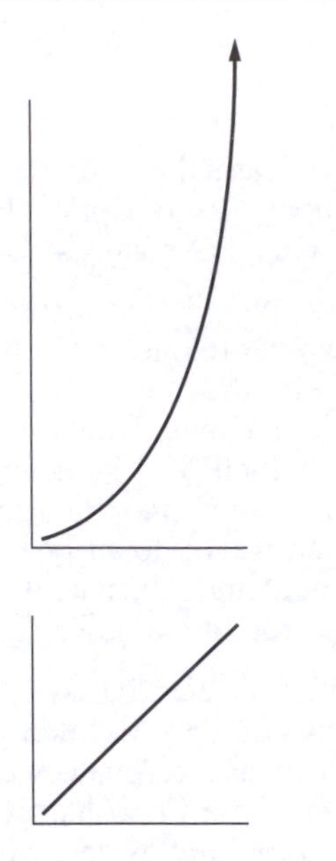

[2.24] **Exponential and logarithmic scales**
1, Exponential growth rapidly goes off the top of the scale (in this case a long way off the top of the page). 2, The corresponding logarithmic growth stays within reasonable bounds.

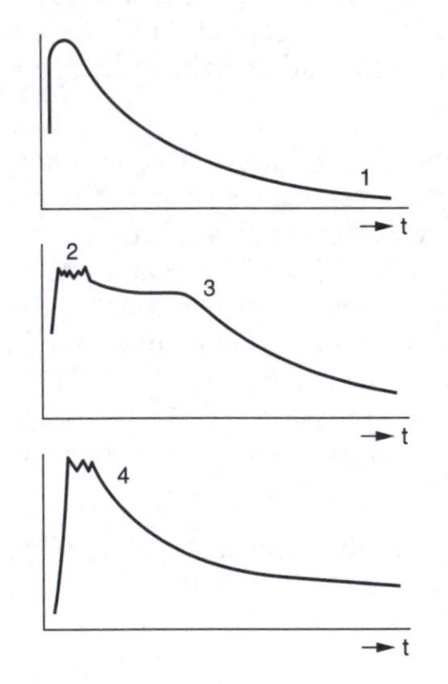

[2.25] **Sound decay**
1, Logarithmic decay, compared with examples of possible real sounds. 2, Attack, followed by sustained note then 3, release and rapid decay of reverberation. 4, Attack followed by rapid decay then sustained reverberation.

Now we have a choice (which is again because of the way in which ratios behave, and can be manipulated mathematically by logarithms – a point that those who really hated school mathematics do not really need to understand). The choice is whether we describe changes in loudness in terms of sound intensity or in terms of power. As it happens, for most purposes, the choice has already been made for us – because engineers have decided to describe changes in the volume of an audio signal passing through electrical equipment in terms of its power. (There is a slight difference in the way power and sound intensity are calculated.)

In our hearing the practical effect of the way this works is that *each doubling in power sounds roughly as loud again as the last*. It is therefore convenient once again to use a logarithmic scale, and the measure that is employed is the *decibel*, or dB. This is based on a unit called the bel (from the name of the pioneer audio engineer, Alexander Graham Bell), which corresponds to a decimal decrease or increase. As it turns out, the bel is inconveniently large, because the ear can detect much smaller changes. So an interval a tenth of the size – the *decibel* – has been adopted.

One decibel represents a ratio of about 1.26:1 (because 1.26 is approximately the tenth root of 10). This is just about as small a difference in power as the human ear can detect in the best possible circumstances.

As it happens, a power ratio of 3 dB is almost exactly 2:1. This is convenient to remember, because if we double up a sound source we double its power. So if we have one soprano singing with gusto and another joins in, equally loudly, the sound level will go up 3 dB (not all that much more than the minimum detectable by the human ear). But to raise the level by another 3 dB, two more sopranos are needed. Another four are needed for the next 3 dB – and so on. Before we get very far, we have to add sopranos at a rate of 64 or 128 a time to get any appreciable change of volume out of them. To increase intensity by increasing numbers soon becomes extremely expensive, so if volume is the main thing you want it is better to choose a more powerful source to begin with – such as a pipe organ, a trombone, or a bass drum!

Loudness, frequency and hearing

The ear does not measure the volume of all sounds by the same standards. Although for any particular frequency the changes in volume are heard more or less logarithmically, the ear is more sensitive to changes of volume in the middle and upper-middle frequencies than in the bass.

The range of hearing is about 20–20 000 Hz for a young person, but the upper limit of good hearing falls with age to 15 000 or 10 000 Hz

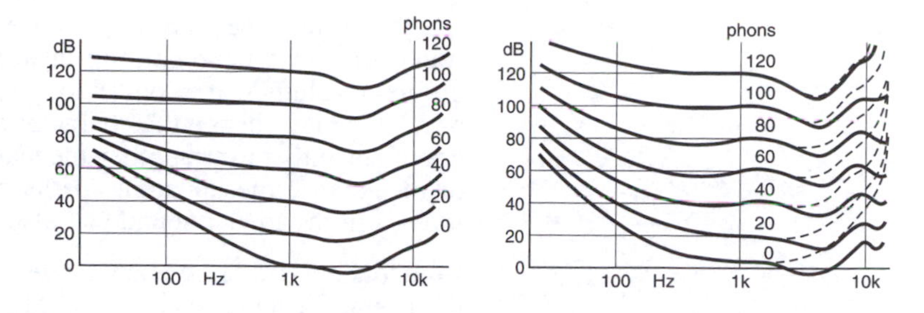

[2.26] **Equal loudness contours**
Left: The classical curves of Fletcher and Munson, published in 1933. Intensity (dB) equals loudness in phons at 1 kHz.
Right: The curves measured by Robinson and Dadson take age into account and show marked peaks and dips in young hearing. The unbroken lines are for age 20; the broken lines show typical loss of high-frequency hearing at age 60. The lowest line in each set represents the threshold of hearing.

(and for many people this fall-off is much steeper). Sensitivity is greatest at 1000 Hz and above, and best at 2000–3000 Hz: the auditory canal between the outer ear and the eardrum helps this by having a broad resonance in the region of 2000–6000 Hz.

Obviously *loudness* (a subjective quality) and the measurable volume of a sound are not the same thing, but for convenience they are regarded as being the same at 1000 Hz. Perceived loudness, originally described in units called *phons*, can then be calculated from actual sound volume by using a standard set of curves representing typical good human hearing. From these curves it is also clear that people with hearing loss get a double benefit if sound equipment is turned up louder: at progressively higher volumes, more of the high (and low) frequencies are heard, which profoundly affects the intelligibility of speech.

Figures for such things as levels of *noise* are also *weighted* to take hearing into account. This may refer to electrical as well as acoustic noise. For example, some types of microphone produce more noise in the low-frequency range: this is of less importance than if the noise were spread evenly throughout the entire audio range. Today, engineers use the unit 'dB(A)' rather than the phon.

The lower limit is called the *threshold of hearing*. It is convenient to regard the lower limit of good human hearing at 1000 Hz as zero on the decibel scale. There is no natural zero: absolute silence would be minus infinity decibels on any practical scale. (The zero chosen corresponds to an acoustic pressure of 2×10^{-5} pascals, or Pa for short. One pascal is a force of one newton per square metre.)

The upper practical limit is set by the level at which sound begins to be physically felt – a *threshold of feeling* at around 120 dB. In some experiments, 50% of the subjects reported feeling sound at 118 dB and that it became painful at 140 dB – a volume well above the level at which repeated or lengthy exposures can damage hearing.

When we hear two loud tones together, and they are separated by less than about 50 Hz, we also hear a 'beat', an oscillation in volume at a frequency equivalent to the difference between the two original tones. If one of the tones is variable, this effect can be used to synchronize pitch. If the difference between the two loud tones increases to more than 50 Hz, the beat changes to an additional difference tone. Increase the separation further, beyond a semitone, and this spurious combination tone gradually disappears.

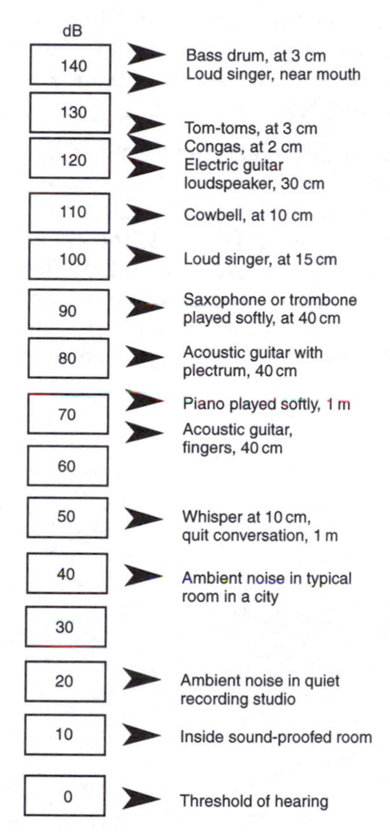

[2.27] **Sound pressure levels**
Examples of sounds over the wide range that may reach microphone or ear.

Another effect we may experience is called masking. When the sound from two neighbouring instruments contains similar tones at different volumes, the louder tone may completely mask the softer one. This may change the character of the quieter sound – requiring treatment, either to strengthen the masked component (see Chapter 8) or to increase the apparent spatial separation. Note that loud harmonics in the louder sound can also cause masking.

Both loudness and frequency range are matters of vital importance in the design of loudspeakers, though for those working in the sound medium this is mainly restricted to monitoring loudspeakers. But the frequency range that human ears can respond to has important applications for most other components of the chain, and especially to microphones.

Frequency range is particularly important to broadcasting transmission standards. An individual may be happy to spend money on equipment with a frequency response that is substantially level from 20 to 20 000 Hz, knowing this goes beyond the limits most listeners will require, but for a broadcasting organization this may be too high. For most people hearing is at best marginal at 15 000 Hz, and many television receivers produce a whistle only a little beyond this. So it is wasteful to go above 15 000 Hz – particularly as the amount of information to be transmitted does not observe the same logarithmic law that the ear does. The octave between 10 000 and 20 000 Hz is by far the most 'expensive' in this respect.

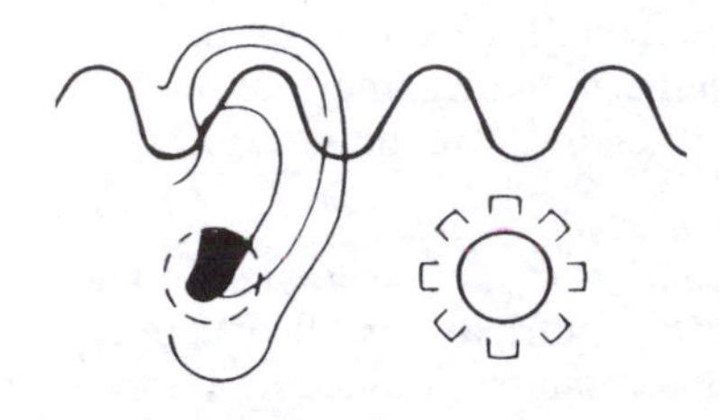

[2.28] **Eardrum and microphone diaphragm**
The shortest wavelength that young ears can hear defines the size of diaphragm that is needed for a high-quality pressure microphone.

Most people seem to expect that recording equipment (and many other items in an audio chain) should respond to the full range from 20 Hz to 20 kHz (although few loudspeakers or microphones can do this). Some people experience hums below 20 Hz, and find them deeply unpleasant; plainly, these frequencies should be avoided. The most controversial range is 20 Hz to 100 kHz: some people report that whether they can 'hear' that or not, its presence adds to their listening pleasure. But a practical range for many broadcasting transmitters is 30–15 000 Hz.

Chapter 3

Stereo

The object of stereo is to lay before (or around) the listener a lifelike array of sound – both direct sound and reflections from the walls of the original studio – to recreate some real or simulated layout in space.

In its simplest form this *sound stage* is created by two loudspeakers, each of which has a separate signal fed to it. Sound that is reproduced by a single loudspeaker appears to come directly from it. If, however, the sound is split evenly and comes equally from the two speakers it seems as though it is located halfway between them. Other mixtures being fed to the two loudspeakers simulate other directions on the sound stage. It is usual to refer to the left-hand loudspeaker as having the 'A' signal and that on the right as having the 'B' signal. (Some manufacturers called them 'X' and 'Y'.)

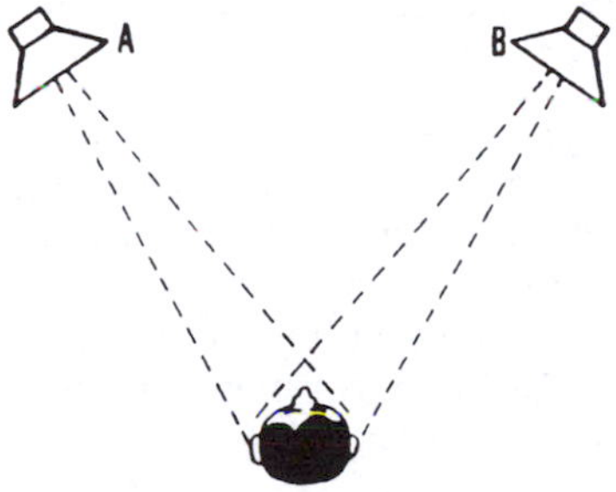

A serious disadvantage of mono is that reverberation is superimposed on the same point source of sound, reducing the clarity of music and intelligibility of speech. In stereo, the reverberation is spread over the full width of the sound stage – a vast improvement, although in two-channel stereo it still does not surround the listener.

A

B

[3.1] **Hearing stereo**
The position of apparent sources is perceived as the brain processes time, amplitude and phase differences between signals received by the left and right ears, and also (at high frequencies) by the effects of shadowing due to the head. For any particular frequency the phase at each ear is given by the sum of the sounds arriving from the two loudspeakers.

Two loudspeakers

A stereo pair of loudspeakers must work together in every sense. The minimum requirement is that if the same signal is fed to both, the cones of both loudspeakers move forward at the same time; in other words, the loudspeakers must be wired up to move in phase with each other. If they move in opposite senses an effect of sorts is obtained, but it is not that intended. For example, a sound that should be in the centre of the sound stage, the most important place, is not clearly located in the middle, but has an indeterminate position.

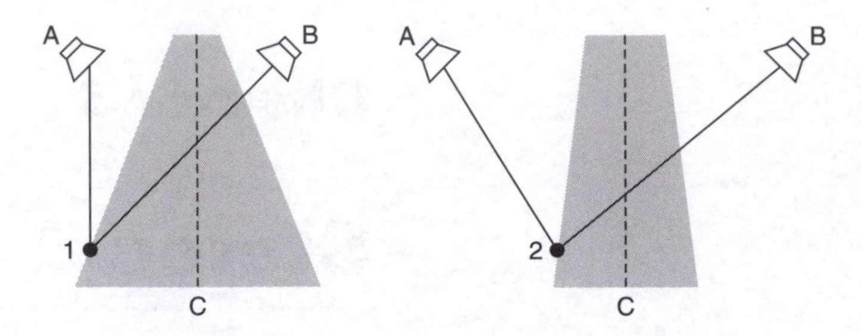

[3.2] **Stereo loudspeakers: spacing**
With the closer loudspeakers (*left*), the spread of the stereo image is reduced, but there is a broad area of the listening room over which reasonable stereo is heard. With broader spacing (*right*), the stereo image has more spread, but for the benefit of a relatively narrow region in which it can be heard at its best. For listeners off the centre line C, the growing difference in distance rapidly 'kills' the stereo, due to the Haas effect. Listening position 1 (with the narrower spacing of speakers) has the same loss of stereo effect as position 2 (with the broader spacing).

A second requirement is that the loudspeakers be matched in volume and frequency response. If the volume from one is set to be greater than that from the other, the centre is displaced and the sound stage distorted. If the difference reaches 20 dB, the centre is displaced to the extreme edge of the sound stage. For most listeners the exact matching of frequency responses is less important. Probably the most important effect of a mismatch is that it could draw attention to the loudspeakers themselves and diminish the illusion of reality.

In the studio, anyone who must listen critically to sound needs to work in fully standardized conditions. This generally means sitting on the centre line of a pair of matched high-quality loudspeakers placed about 2.5 m (8 ft) apart. It is also recommended that a home listener should have loudspeakers 2–3 m (between 6 and 10 ft) apart, depending on the size of the room. With less separation the listener should still hear a stereo effect, but the width of the sound stage is reduced. On the other hand, the speakers should not be too far apart. For normal home listening, if one person is on the centre line, others are likely to be 60 cm (2 ft) or more to the side, and the further the loudspeakers are apart, the greater the loss of stereophonic effect to the outer members of the group. If a listener is as little as 30 cm (1 ft) closer to one loudspeaker, phase effects that distort the sound stage are already marked (see later), and as the difference in distance grows, the nearer source progressively appears to capture the whole signal.

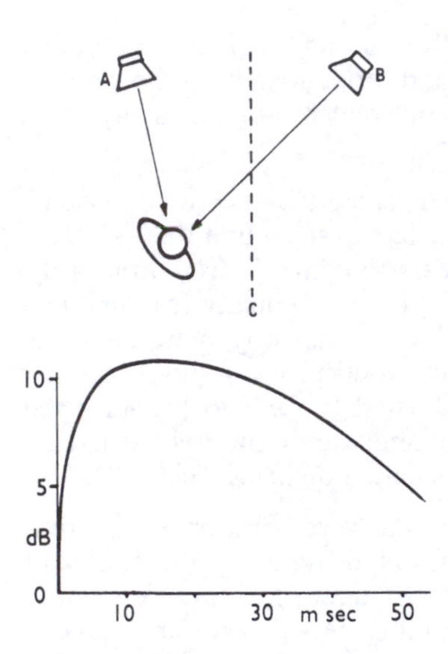

[3.3] **Haas effect**
For a listener away from the centre line C, the difference in distances means that the sound from the farther loudspeaker arrives at the listener's ears later than that from the nearer one. The delay is 1 ms for every 34 cm (13.4 in) difference in path length. For identical signals coming from different directions, a difference of only a few milliseconds makes the whole sound appear to come from the direction of the first component to arrive. Increasing the volume of the second signal can reverse the effect: the curve shows the increase in volume required to recentre the image.

This second distortion of the sound stage – indeed, its near collapse – is induced by the *Haas effect*. This is one of the ways in which the ear can tell the source of a sound: it can distinguish direct from reflected sound that is delayed by as little as a few thousandths of a second. This happens even in a small room: in 1 ms (one millisecond), sound travels just over 34 cm (13½ in). Sit much closer to one of the loudspeakers and the Haas effect will fool you.

The Haas effect is at its strongest for delays of 5–30 ms, which are produced by a difference in path lengths of about 1.7–10 m (roughly 6–33 ft). The way ears respond to this effect diminishes a little with increasing delay, but remains effective until eventually the reflection of a staccato sound can be distinguished as a separate echo.

What happens at the lower end of this scale of delays and their corresponding distances is even more important for our perception of stereo. As the difference in distance from the two loudspeakers grows from zero to about 1.5 m (5 ft), the Haas effect grows rapidly and continuously to approach its maximum, at which it would take

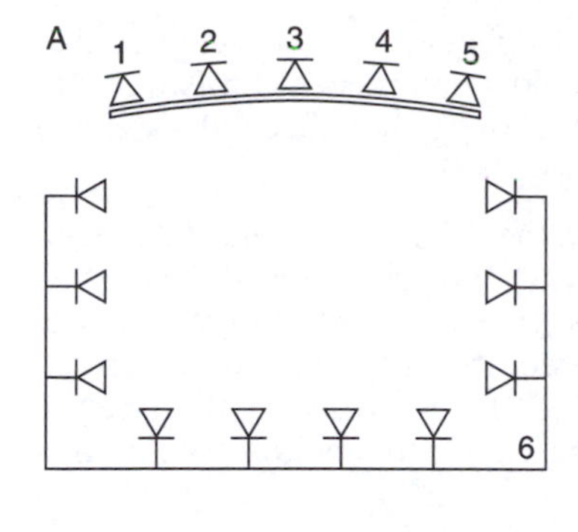

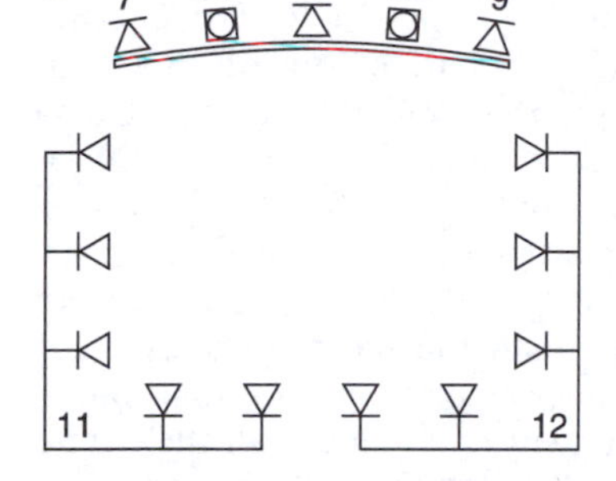

[3.4] **Early stereo surround for 70 mm film**
A, To cover its full width evenly, 1950s 70 mm film had six magnetic tracks. 1–5, Five loudspeakers, spread across the picture. 6, One track fed an array of loudspeakers around the auditorium. B, In a variant of this the six tracks fed: 7, Left. 8, Centre. 9, Right. 10, Low-frequency loudspeakers. 11, Left surround. 12, Right surround.

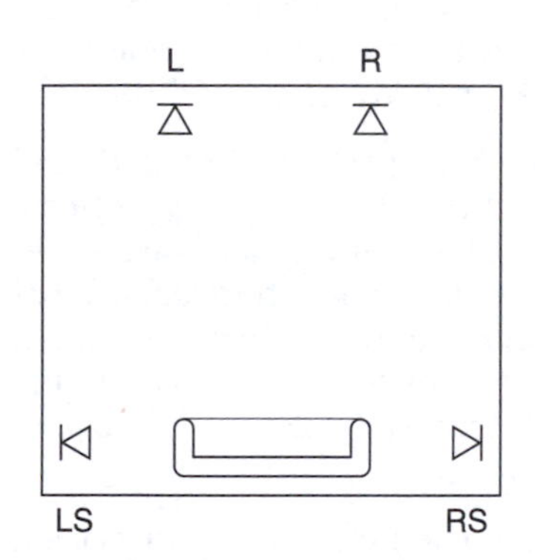

[3.5] **Quadraphonic loudspeakers in the home**
A commercially short-lived system using four loudspeakers: L, Left. R, Right. LS, Left surround. RS, Right surround.

an increase of 10 dB in volume from the farther loudspeaker for it to begin to recapture the source. Plainly, an optimum stereo effect is achieved only in a narrow wedge-shaped region along the centre line, and this is narrower still if the loudspeakers are widely separated.

Between the reduced spatial effect of narrow separation and the reduced optimum listening area created by a broad layout, it may seem surprising that stereo works at all. The fact is, most casual listeners happily accept a sound picture that is spatially severely distorted. This is their prerogative: the professional should be aware that listeners often prefer their own seating plan to a clinical ideal.

Reduced loudspeaker separation

Many people accept a reduced loudspeaker separation in their own homes, often from portable or compact audio equipment that has loudspeakers built in or attached on either side. Television or video stereo is also usually heard through loudspeakers within the box itself. This reduces the perceived sound stage to a fraction of that available to the music balancer. For television, it may match the size of the visual image, but limits any suggestion of off-screen sound sources.

Fortunately there are compensations. One is that the Haas effect is reduced, permitting off-centre-line viewing, as is necessary for family groups. There is still some separation of point sources and, equally important, reverberation is spread away from the original source, which increases intelligibility and is more pleasant to listen to.

Alternatives to two loudspeakers

The simplest alternative to two loudspeakers is to dispense with them entirely and to wear stereo *headphones* instead. People with impaired hearing who resort in desperation to headphones (and to restore peace with those around them) get an unexpected benefit as the stereo spreads to a full 180°, and sound effects are disentangled from speech. Headphones are also convenient where people of different tastes are crowded together, for example on aircraft.

With headphones the (broad) spread of the image is the same for everyone, but several characteristics of the two-speaker system are lost. One is that sound from *both* loudspeakers should reach *both* ears, and do so from somewhere to 'the front'; another is that the sound stage is no longer static, so that if the listener turns it lurches sideways. Headphones offer a different aural (and social) experience.

The sound balancer should not use headphones to monitor stereo that will be heard by loudspeaker. For the location recordist this may not be avoidable, but the sound will usually be checked later on loudspeakers.

A variation on two-speaker stereo is the 2.1 or 3.1 sound (also sold to upgrade sound on home computers). The first figure designates the number of frontal loudspeakers: in 3.1 a third is in the centre. From conventional two-channel stereo, its output has to be derived from the left and right components; if taken from 5.1 the track is already available. The centre track helps to anchor sound to a screen image.

5.1 *surround sound* is a full 'jump' up from regular stereo. In addition to left, centre and right at the front (LCR) it has extra channels, left and right surround (LS and RS), towards the rear. For the home this is supplied on CD or DVD-Video (both with a data compression system that slightly reduces sound quality), or DVD-Audio or SACD, Super Audio CD (both with full quality on a suitable player). 5.1 requires a special sound mix (see Chapter 6), comparable to the six-channel stereo originally developed for 70 mm film and just a little more complex than the techniques used for quadraphonics ('quad'), the old four-speaker system that failed commercially. Surround then re-emerged in cinema with the centre track to lock sound to picture. To allow for theatres that are very wide at the back a further feed (or two) is added by matrixing extra signals to those for the two rear-side speakers to fill in the rear from an extra speaker or two. This is 6.1 or 7.1 surround.

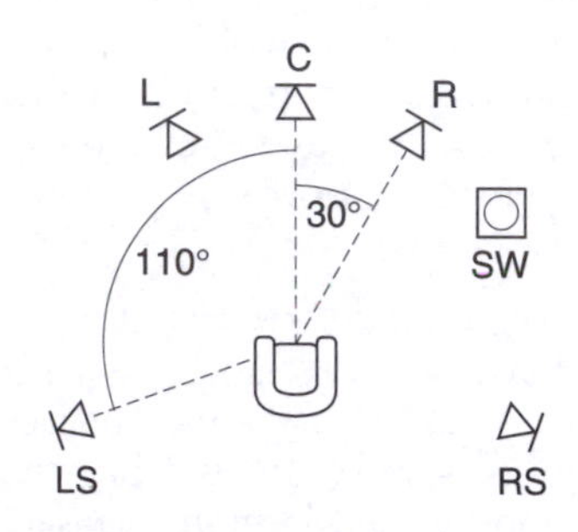

[3.6] **Loudspeaker array for 5.1 surround**
L and R are equivalent to AB two-channel stereo from the front, and LS and RS (S for surround) are placed at the rear sides of the listening room: these four loudspeakers are similar in function to those of 'quad'. C is added in the centre to coincide with the screen for film, television or video, or (if there is no picture) to give a firmer location for sources in the middle of the sound stage. Taking C to be at 0° from the middle of the room, recommended angles are ±30° for L and R, and ±110° (or perhaps a little more) for LS and RS. The position of the deep bass subwoofer SW is not critical, as our perception of direction is poor at very low frequencies, and in any case it will resonate throughout the enclosed space. In theatres it will usually be at the front. 2.1 has just L, R and SW. 3.1 has L, C, R and SW. 7.1 has speakers on each side, like 5.1, and two more at the back.

The '0.1' (for 'one-tenth of the full frequency range') simply provides deep (and potentially loud) bass – typically 120 Hz and below – from a single, more powerful 'subwoofer' loudspeaker. This can therefore be of a separate design from the other five, which in turn benefit by their isolation from its more extreme excursions and excessive harmonics. The ears make little use of frequencies below 120 Hz to localize sound sources, so a single speaker is sufficient for the sub-bass from any direction. It will usually be off-centre (because the centre line is at a node of many low frequencies), and often at the front, but otherwise position does not matter much, provided nobody gets too close.

For cinema, including home cinema, the main uses of surround sound are obvious: the subwoofer carries low-frequency effects (LFE). Where 5.1 is used for music or for any other sound-only application, the subwoofer receives extreme bass at frequencies that are too low for the other five to handle. For music there are two main ways in which the 'surround' can be created – just as for the quadraphony ('quad') that preceded it, and which has left a substantial library of recordings.

The more conventional use for the five main channels is to place a standard direct-sound stage between the three at the front, and spread indirect sound (reflected sound, reverberation) to all five. This offers the listener a sense of space that cannot be achieved by the acoustics of a small room. In fact, most listeners probably prefer that their room has this clear orientation, with a 'front' and a 'back'.

The more adventurous alternative is to surround listeners with direct sources, so they are placed in the middle of the action. To bring individual sources forward, these can also be 'placed' on the axis between pairs of diagonally opposed loudspeakers. The effect can be dramatic and exciting. Popular music can certainly make use of this. Little existing classical music requires it, although exceptions include the Berlioz *Requiem* (with its quartet of brass bands), off-stage effects in opera, and early antiphonal church music by composers such as Gabrieli and Monteverdi.

Finally, for a fully 3-D sound field a system called 'Ambisonics' has been developed. This requires specialized microphone placement: one design provides this within a single housing (the Soundfield microphone; see Chapter 5). To handle the signal, it also requires down-line techniques that are beyond the mainstream described in this book.

Hearing and stereo

How, in real life, do our ears and brain search out the position of a sound and tell us where to look for its source? First, the Haas effect allows us to distinguish between direct and reflected sound (except in the rare cases where the reflections are focused and their increased loudness more than compensates for the delay in their arrival). A second clue is offered by differences in the signal as it arrives at our two ears: a sound that is not directly in line with the head arrives at one ear before the other. At some frequencies the sounds are therefore partly out of phase, and this may give some indication of angle. Such an effect is strongest at about 500 Hz.

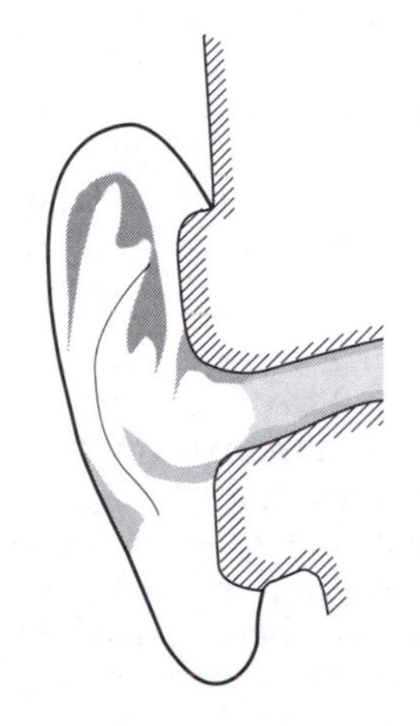

[3.7] **Outer ear (the pinna)**
This collects high-frequency sound from the front and obstructs that from the rear. In addition, asymmetrical ridges on the outer ear create faint reflections with delays of 80 µs or 100–330 µs in high-frequency sounds from different directions. Even with such tiny clues, a single ear can often judge the direction of a source, not just in a horizontal plane but also in three dimensions.

The physiology of the outer ear (the pinna) also helps to distinguish front and rear, partly because it screens sound from behind. In addition two ridges on the pinna create reflections with tiny delays that can be processed by the brain. One of the ridges helps to some extent to distinguish sound from above (providing some justification for Ambisonics). Note here that one ear on its own can pick up a significant amount of directional information: stereo is not restricted to people with two good ears.

But (using both ears again) the head itself begins to produce screening effects for frequencies at 700 Hz and above, resulting in differences in the amplitude of signals reaching the ears from one side or the other. Further aids to direction finding are provided by movements of the head, together with interpretation by the brain in the light of experience. Even with all these clues, with our eyes closed we might still be uncertain: we can distinguish, separate out and concentrate our attention on the contents of particular elements of a complex sound field far better than we can pinpoint their components precisely in space. The clues tell us where to look: we must then use our eyes to complete and confirm the localization – if, that is, there is any picture to see. If there is none, the fact that in real life we rely rather heavily on the visual cues is perversely rather helpful

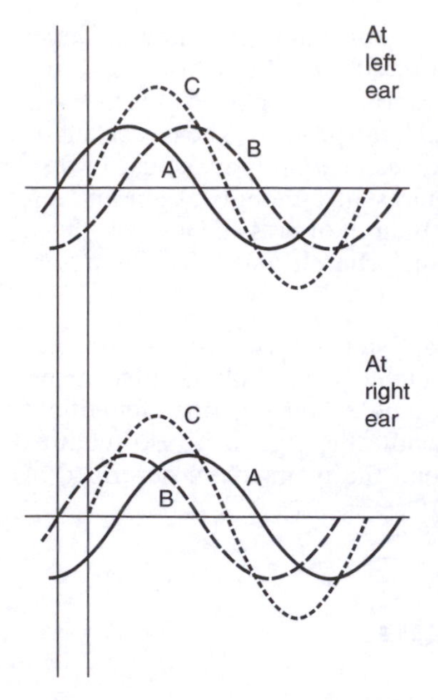

[3.8] **Two-loudspeaker stereo: how we perceive a central source**
The signals from the A and B loudspeakers are equal: the B signal takes longer to reach the left ear, and the A signal takes longer to reach the right. But the combined signal C is exactly in phase: the image is in the centre.

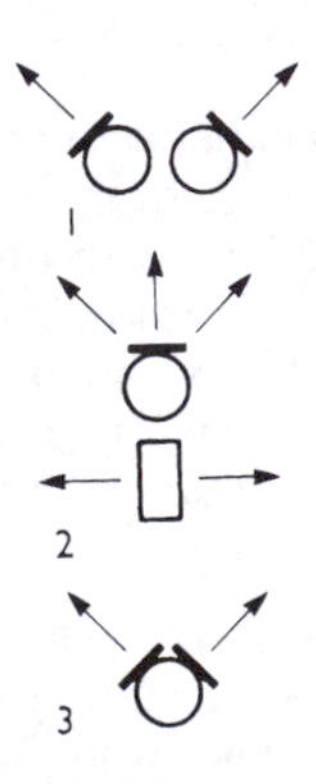

[3.9] **Stereophonic microphones**
One way of picking up stereophonic sound is the coincident pair: two directional microphones very close together (1) or in a common housing. The two components will not necessarily be at 90° to each other. 2, Main or 'middle' and side microphones used as a coincident pair. 3, Symbol used in this book for a coincident pair, which may be either of the above.

to us: it permits us to continue to enjoy a spatially distorted image even if we are a little off the stereo centre line.

Two-speaker stereo measures up to this specification quite well. To understand how it does this, assume first that the source is in the centre of the sound stage and that the listener is on the centre line. The A signal reaches the left ear slightly before the B signal arrives, and for middle and low frequencies the two combine to produce a composite signal that is intermediate in phase between them. For the right ear the B signal arrives first, but the combination is the same. The brain, comparing the two combined signals, finds them to be the same and places the source in the centre. If the amplitude of the sound from one speaker is now increased and the other reduced, the signals combine at the ears as before, but the resultant signals differ from each other in *phase* – in effect, it appears as though the same signal is arriving at the two ears at slightly different times. This was one of the requirements for directional information.

If the listener moves off the centre line, or if the volume controls of the two loudspeakers are not set exactly the same, the sound stage is distorted, but there is still directional information that matches the original to some reasonably acceptable degree.

Microphones for stereo

To create a stereo signal, the sound field has to be sampled by microphones in such a way as to produce two sets of directional information. There are several ways to do this. Given that the loudspeakers used to reproduce stereo are separated, it might seem logical to sample the field at points that are similarly spaced. However, the distance between the loudspeakers already produces phase cancellation, and separating the microphones can only add to this.

Far greater control is achieved by placing them as close together as possible, but pointing in different directions. They are then virtually equidistant from any sound source, and although (by their orientation) they may pick up different volumes, the two signals will be in phase for almost all the audio-frequency range. Only at extreme high frequencies will a slight separation produce any phase differences, and because this region is characterized by erratic, densely packed overtones, even drastic phase changes will be relatively unimportant.

This first arrangement is called the *coincident pair*, or *coincident microphone* technique. There is a range of variations to it, but typically the two microphones feeding the A and B information are mounted one above the other or closely side by side and are directional, so that the A microphone picks up progressively more of the sound on the left and the B microphone more on the right. They pick up equal sound from the centre, on a line halfway between their directional axes. The frequency characteristics of the two microphones must be the same.

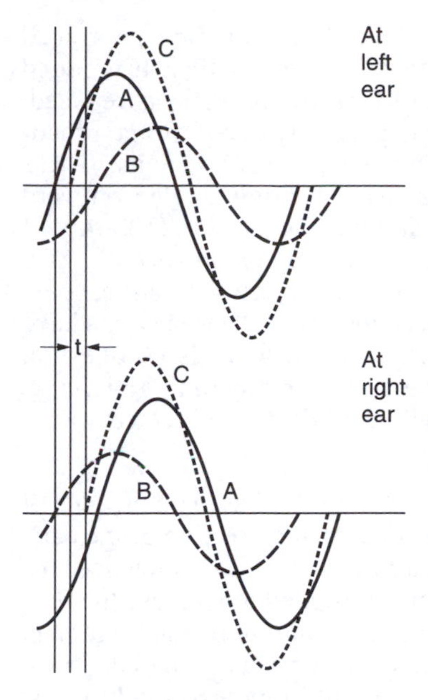

[3.10] **Two-loudspeaker stereo: how we perceive an offset source**
As the source is offset from the centre to the left, signal A is increased. The effect is to make the peak in the combined signal C at the left ear arrive a little earlier, and that at the right ear later. This creates a slight delay (*t*) between the two signals C. The resulting phase difference is perceived as a displacement of the image to the left of centre. In these diagrams it is assumed that the listener is on the centre line between A and B, and facing forward. Turning the head will cause the phase difference at the ears to change, but in such a way as to keep the image roughly at the same point in space. If the listener moves away from the centre line the image will be distorted, but the relative positions of image elements will remain undisturbed.

An alternative kind of coincident pair uses radically different microphones, one behind the other. The front microphone covers most of the sound field, including all required direct sound. It produces the *middle, main* or *'M' signal* – which, on its own, is also mono. The second microphone picks up sound from the *sides*, producing the *'S' signal*. This contains information required to generate stereo, but is not, by itself, a stereo signal. Sounds from the two sides are opposite in phase, so that if the M and S signals are added together they produce a signal that is biased to one side of the sound field – the left – and if S is subtracted from M it is biased the other way. Quite simply, $M + S = A$ and $M - S = B$. To convert back, add again: $A + B = 2M$ and $A - B = 2B$.

Whether by analogue or digital techniques, the conversion can be made easily; indeed, its very simplicity can lead to confusion. For example, if the two tracks of a stereo recording are added to make a single-audio-track copy, this will work well if the originals were A and B – but if they were M and S, half the sound will disappear.

There can be advantages to working with M and S signals, rather than A and B. In particular, analogue A and B signals taking similar but not identical paths may be changed differentially in volume or phase; if this happens to the M and S signals, the perceived effect is much less. Note also that because $A + B$ will normally be greater than $A - B$, M should be greater than S, offering a useful visual check (by meter) that the original signal is not grossly out of phase. For video or film sound from a source within the visual frame, M starts out at the centre of the sound stage, making it easy to shift by any small amount.

Coincident pairs are only one way to pick up stereo. They are often augmented by or may be replaced by spaced mono microphones. Such *spaced microphone* techniques are widely used, and can also give excellent results. Over the years, balancers who have been prepared to experiment with competitive techniques have, at best, improved the range of balances heard, but have also tended to produce a cycle of fashions in the kinds of balance that are currently in favour.

The only test that can be applied is practical, if subjective: does the balance sound right? Experiments with a well-spread subject such as an orchestra soon demonstrate that a single pair of widely spaced microphones does not sound right: there is plainly a 'hole in the middle'. Sources right at the front in the centre sound farther away than sources at the sides that are closer to the microphones. To combat this, it is usual to add a third forward microphone in the centre, the output of which can be split between the two sides. It also helps to pin down the centre in surround sound, where there is a central speaker.

One recording company's variation on this, the Decca 'tree', has employed three microphones in the middle: a splayed pair 2 m (80 in) apart, with another 1.5 m (60 in) in front of them, pointing forward.

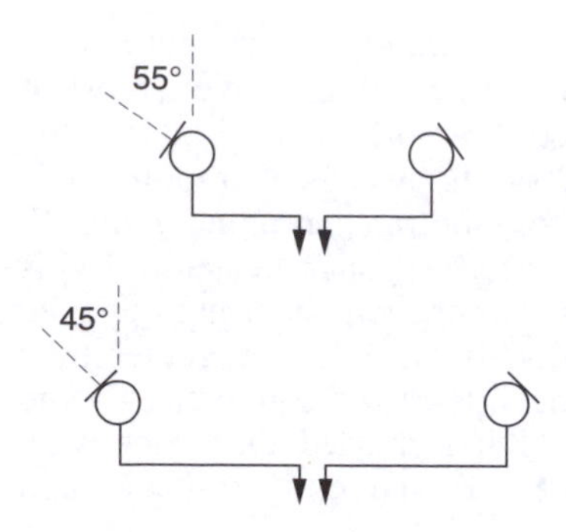

[3.11] **Near-coincident pairs**
Above: The ORTF array has microphones separated by 17 cm (6.7 in), about the width of a human head. *Below*: In the NOS array separation is 30 cm (12 in), about half the circumference of a human head.

An alternative way of avoiding any hole in the middle of a spaced pair is simply to move the microphones closer together. It is found that if the spacing is less than 1 m (40 in), the problem is reduced, and at two-thirds of that separation is generally cured. Some broadcasters have developed and used *near-coincident* layouts. These include the Dutch (NOS) array that has cardioids angled outward at 45° and spaced at 30 cm (12 in), and the French (ORTF) array of cardioids angled outward at 55° and spaced at 17 cm (6.7 in). Also on offer are other near-coincident arrays, for which separation, splay angle and microphone are all precisely specified. These layouts have been exhaustively analysed and justified, and there is no question that they produce stereophonic effects that many find satisfying. Anyone who works with a user will doubtless be directed to the appropriate literature.

Yet another technique sometimes encountered is based on the idea of a *dummy head* with pressure microphones in the position of the ears, and the '*sphere microphone*' is a variation on this. In practice, the 'head' is often replaced by a disc that is designed to increase the path difference for sound coming from the sides, and in this arrangement is called a '*baffled omni pair*'. There certainly is a spatial effect, because the A and B signals are slightly different, especially in the high frequencies at which the obstacle effect of a real head helps to define position. But the test is whether it works in practice and some would argue that (except when heard on headphones, as for airline in-flight entertainment) other techniques work better or are more flexible.

It may also be argued that either coincident stereo pairs or spaced mono microphones set closer in (or combinations of these) are all that most readers will need. These are the main techniques described here.

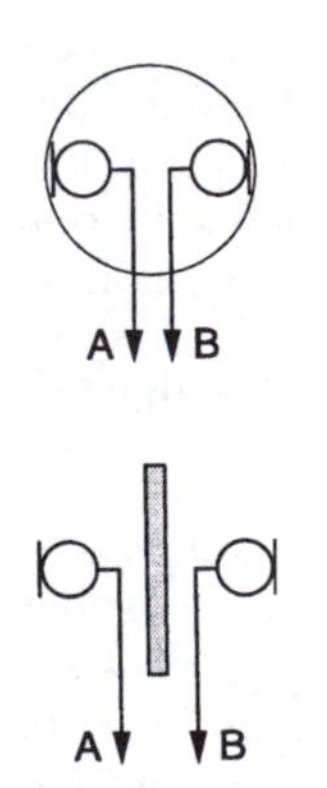

[3.12] **Baffled omni pair**
Above: Sphere microphone, a simplified 'dummy head'. This has two elements mounted in a spherical baffle. *Below*: A baffle replaces the sphere. In both arrangements the obstacle increases time delays for sound approaching from the side. Although not widely used, it does produce a stereo image.

For radio or television broadcasters whose audiences have a mixture of stereo or mono receivers, fully coincident stereo microphones have a further virtue: the stereo signal from them translates easily to mono. Spaced microphones are more problematical: as A and B signals of similar strength from any source are combined in a mono central channel, some frequencies (at regular intervals on a linear scale) are enhanced; those in between are reduced: this is *comb filter distortion*. In a complex sound, cancellations like these are so common that they can be difficult to spot, but it seems unwise to add to them.

Note also that microphones spaced in this way often transgress the so-called '3:1 rule', which proposes that they should be placed much further from each other than from the sources they cover. (The 3:1 ratio arises from calculations for omnidirectional microphones: it ensures that when the sources covered are equally loud, a signal that is picked up by the nearest neighbouring microphone will be at least 10 dB lower.) Balancers seem to observe this rule with more care when setting up spotting microphones for music (see Chapter 8).

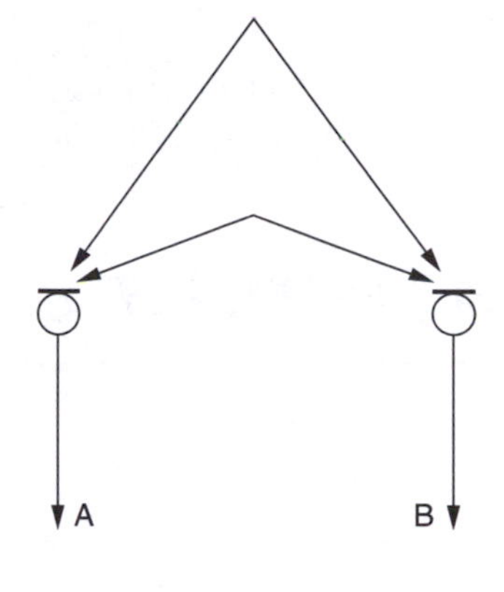

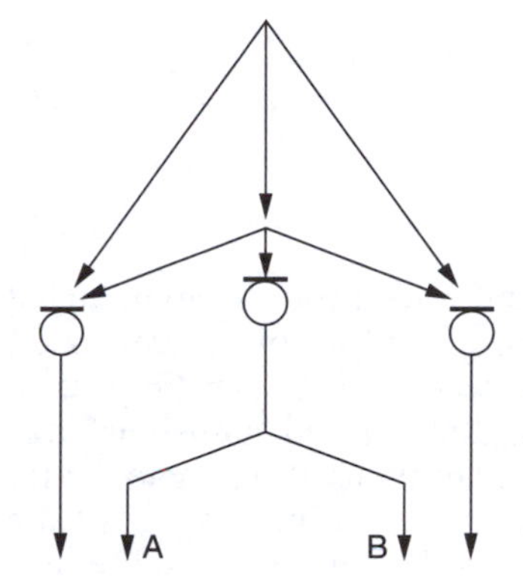

[3.13] **Spaced pair**
Above: Two microphones. For sources along the centre line their apparent distance from the front of the audio stage is greater than for sources in line with the microphones; and as the source moves forward, this 'hole-in-the-middle' effect is even more pronounced. *Below*: This is filled in by adding a third microphone in the centre.

Stereo compatibility

If a stereo signal produces acceptable mono sound when reproduced on mono equipment, it is said to be *compatible*. As we have seen, the M signal (or A + B) may be compatible, but this is not automatic: it would be easy to place an M and S microphone in such a way that it is not. To achieve compatibility, the M component should include all the required direct sound sources in acceptable proportions, usually together with enough indirect sound (reverberation) to produce a pleasing effect, but not so much as to obscure it.

Domestic listening conditions vary greatly: in particular, the stereo component may be debased or absent. To allow for this, compatibility is desirable. Only where the listening conditions can be controlled is compatibility dispensable, but this means also that the material cannot be transferred for other uses without further, special processing.

Historically, many analogue stereo systems have been arranged for automatic compatibility: for example, on stereo records. On a vinyl disc, A + B corresponds to lateral displacement of the stylus, so a mono signal is easy to pick up. A – B is recorded as vertical hill-and-dale signals: a mono stylus rides up and down on it but cannot convert that into an electrical signal. Similarly, on analogue magnetic tape the stereo A and B signals are side by side on separate tracks. A full-track head reproduces both together, or alternatively the combined signal can be obtained electrically by adding the two outputs.

Similar advantages are obtained from using A + B and A – B signals in radio. When stereo was introduced, the A + B signal occupied a place in a waveband previously used for the mono signal; the A – B signal was modulated on to a subcarrier above the main A + B band. Today, digital systems have replaced many of these intervening stages. Provided that the A and B or M and S signals are interlocked so they cannot be displaced, they can be converted back into an analogue system that preserves compatibility. For 5.1 surround sound to be compatible with stereo and mono it requires special features such as those devised by Dolby.

'Spatialization', using out-of-phase effects to increase the apparent width of stereo (to more than the speaker separation), is incompatible with mono because cancellation effects appear. It is not recommended here as a studio technique, but for domestic use adds spaciousness, at the cost of some confusion of the intended stereo image.

Chapter 4

Studios and acoustics

At the heart of most studios there is a dedicated acoustic environment within which sounds are created, then picked up by microphones. This space is designed or arranged to serve a particular purpose or function, with the layout, acoustics, treatments and furnishings that are required to achieve it. The studio may be used for radio, or for recordings (often of music), or for television. A film stage (not 'film studio', which is the whole organization) is dedicated to successive individual shots and has less electronic equipment, so although similar is somewhat simpler than the television studio. (Film and television re-recording theatres and MIDI-equipped 'virtual' studios, where acoustics play a lesser role, are described in later chapters.)

The music recording industry alone is so vast that it has generated its own great range of studio variants. Many of the more specialized books on professional audio are designed primarily to serve the practitioners of this one group; here, however, a broader range will be covered. But acoustics is such a broad field that its principles apply with only minor changes of application to almost any kind of studio.

The range and function of studios

Broadcasting studios vary according to the economic needs and the size of the country, region or community they serve, and with the way they are funded. The cost of television limits the number of well-equipped studios that a given population can support. A regional population of five to eight million is generally sufficient to sustain several centres, each with about three studios for major productions (including drama), for general purposes (including entertainment, with audience), and for speech only (presentation and news), plus an outside (remote) broadcast unit.

A smaller population makes it difficult to sustain even this number: there may be only two television studios or even one. There would also be less use for an outside broadcast unit, which has in any case been partly displaced by the use of individual video cameras (camcorders).

At the lower end of the scale, towns serving areas of a million or so may have single-studio stations (or more, in prosperous regions). But these studios will be small, big enough for only two or three setting areas of moderate size.

A national population the size of the UK or the USA is sufficient to sustain four or more national broadcast networks, augmented by a large, and seemingly ever-increasing, range of local and cable or satellite channels.

In radio (existing, in most countries, in the shadow of television), an even more complicated situation exists. In the USA there is a radio station to every 25 000–30 000 population. Of the vast total number, half subsist on an unrelieved output of 'middle-of-the-road', 'conservative' or familiar 'wall-to-wall' music in some narrowly defined style. Then there are the top 40, the country-and-western and the classical specialists, plus stations that offer just news, conversation (e.g. telephone chat) or religion. Educational stations have a broader range, broadcasting both speech and music. In addition, a few 'variety' stations still exist, segmented in the old style and containing an integrated range of drama, documentary, quiz, comedy, news and so on. There are also ethnic stations that serve the broader interests of a defined community. Most of these survive with low operational costs by using records or syndicated recordings for much of their output, and their studio needs are necessarily simple: perhaps no more than a room converted by shuttering the windows against the low-frequency components of aircraft noise that heavy drapes will not impede. Acoustic treatment can be added on the inside of the shutters. For some stations, a second, general-purpose studio may be partly justified by its additional use for board or staff meetings.

But looking again at the region with five to eight million people, we now have a population that can sustain a full range of studios for all of the specialized functions that the current purposes of radio permit. These are likely to be fewer in the USA and other countries that have followed its example in dedicating commercial stations to a narrow range of output, than in public service broadcasting, where a regional centre may have the following studios:

- speech only (several);
- pop music;
- light and orchestral music;
- light entertainment (with audience);
- general purpose, including drama and education.

Smaller regions (100 000 to one million population) are capable of sustaining vigorous, but more limited, radio stations. National populations are capable of sustaining networks in about the same number as for television (though these are much more fragmented in the USA than in the UK, where they all reach almost the entire population). The BBC has about 60 radio studios in London alone, though many are for overseas broadcasting. Some are highly specialized: for example, a few that are acoustically furnished for dramatic productions in stereo, a form that still exists in Britain.

Noise and vibration

To help in designing or choosing buildings for radio, television (or any other sound studio), the first requirement is acoustic isolation. Also, if noise cannot get in, neither will it get out and disturb the neighbours. Here are some general guidelines:

- Do not build near an airport, railway or over underground lines.
- Prefer massive 'old-fashioned' styles of construction to steel-framed or modern 'component' architecture. Thick brick or concrete walls are good; wood or plasterboard much less so (but if they are used, adhesives are better than nails).
- In noisy town centres, offices that are built on the outside of the main studio structure can be used as acoustic screening. These outer structures can be built in any convenient form, but the studio must be adequately insulated against noise from (or to) *them.*
- The best place to put a studio is on solid ground.
- If a radio or recording studio cannot be put on solid ground, the whole massive structure, of concrete floor, walls and roof, can be floated on rubber or other similar materials. The resonance of the whole floating system must be very low: about 10 Hz is reasonable. One inch of glass fibre or expanded polystyrene gives resonances at about 100 Hz, so care is needed. An alternative is to suspend the entire structure. If floors or false ceilings are floated separately, they should not be fixed to side walls.
- Airborne noise from ventilation is a major problem. Low-pressure ducts are broad enough to cause resonance problems if not designed with care: they also require acoustic attenuation within them to reduce the passage of sound from one place to another. Grilles can also cause air turbulence and therefore noise.
- High-speed ducts have been suggested as an answer to some of the problems of airborne noise, but would also require careful design, particularly at the points where the air enters and leaves the studio.
- Double doors with pressure seals are needed.
- Holes for wiring should be designed into the structure and be as small as possible. They should not be drilled arbitrarily at the whim of wiring engineers. Wiring between studio and listening room should pass round the end of the wall between them, perhaps via a corridor, lobby or machine room.
- Windows between studio and control should be triple or possibly double glazed, with several glass thicknesses, say 6 and 9 mm ($^1/_4$ and $^3/_8$ in), and the air space between pairs at least 100 mm (4 in). Contrary to what might be expected, it does not seem to

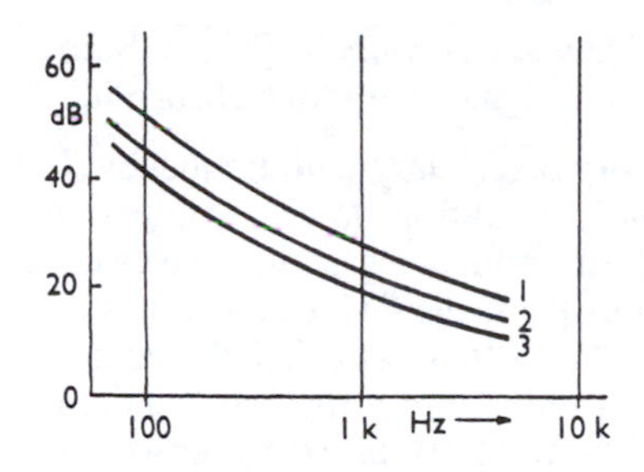

[4.1] **Permissible background noise in studios**
1, Television studios (except drama and presentation). 2, Television drama and presentation studios and all radio studios except those for drama. 3, Radio drama studios.

matter much whether they are parallel, although in practice one is often angled by 5° or more. Triple glazing may not be necessary if the loudest sounds are likely to be much the same on both sides of the glass. Prefer elastic (e.g. rubber) seals around the edges.

- Windows to areas outside the studio area *or to the control area if the loudspeaker is to be switched to replay during recordings* should be triple glazed if replay or noise levels are likely to be high. Measures to avoid condensation between the glass panes will be necessary only in the case of windows to the outer atmosphere, where there will be temperature differences.
- Any noisy machinery should be in a structurally separate area, and on anti-vibration mountings.

Massive structures are recommended, but to make them *extremely* massive may be disproportionally expensive. *Doubling* the thickness of a brick wall reduces transmission by only 5 dB.

For television studio roofs the lower limit for single-skin construction is about 25 cm (10 in) thick: this gives an average attenuation of 60 dB over the range of 100–3200 Hz. For noisy jet aircraft flying at 1000 ft, a minimum of 65 dB attenuation is desirable. Even low-flying helicopters are less of a problem than this (provided they do not land on the studio roof), but supersonic aircraft producing a sonic boom with overpressures of 9.6 kg/m^2 (2 lb/ft^2) would require 70 dB attenuation. Since a double skin is in any case desirable for more than 60 dB attenuation, it may be wisest to adopt this higher standard – of 70 rather than 65 dB attenuation. This results in a structure that is little heavier than a 60 dB single skin.

Subjective tests have been carried out using various types of aircraft noise, attenuated as if by different kinds of roof, then added to television recordings at quiet, tense points in the action (in the limiting case the action may be in period drama, where aircraft noise would be anachronistic). Obviously, permissible noise levels of any kind depend on the type of production. In television drama, background noise from all sources, including movement in the studio, should ideally not exceed 30 dB (relative to 2×10^{-5} Pa), at 500 Hz. For light entertainment the corresponding figure is 35 dB. However, in practice, figures up to 10 dB higher than these are often tolerated.

Reverberation

The basic studio structure is a bare empty box of concrete (or some similar hard and massive material). It will reflect most of the sound that strikes it, and its capacity to reflect will be almost independent of frequency, though very broad expanses of flat thin concrete may resonate at low frequencies and so absorb some of the low bass.

When typical furnishings are placed in a room they include carpets, curtains and soft chairs. These (together with people) act as *sound absorbers*. Other furnishings include tables, hard chairs, and other wooden and metal objects. These reflect much of the sound striking

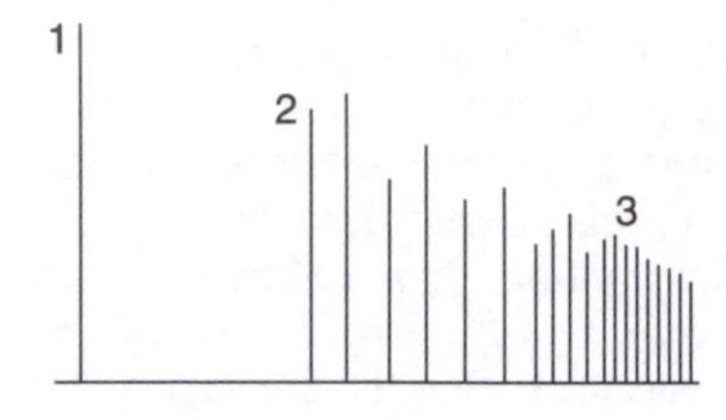

[4.2] **Early reflections**
After direct sound (1) is heard, there is a short delay, then a number of early reflections help to define the character of the acoustic environment (2). With more reflections, this develops into densely structured reverberation (3). If the gap before the first reflection is more than 55 ms, it will be heard separately as an echo, but if subsequent reflected sound has shorter delays the effect should be pleasant.

them, but break up the wave fronts. They act as *sound diffusers*. Some things do both: a bookcase diffuses sound and the books absorb it.

The absorption or diffusion qualities of objects vary with frequency. In particular, the dimensions of an object condition how it behaves. A small ornament diffuses only the highest frequencies; sound waves that are long in comparison to its size simply pass round it. The thickness of a soft absorber affects its ability to absorb long wavelengths, and so does its position. At the hard reflecting surface of a rigid wall there is no significant air movement, so 2.5 cm (1 in) of sound-absorbing material reaches out into regions of substantial air movement (and damps them down) only for the highest frequencies; to have any effect on lower frequencies, the absorber must be well away from the wall.

Sounds in an enclosed space are reflected many times, and some (great or small) part of the sound is absorbed at each reflection. The rate of decay of reverberation defines a characteristic for each studio: its *reverberation time*. This is the time it takes for a sound to die away to a millionth part of its original intensity, i.e. through 60 dB. Reverberation varies with frequency, and a studio's performance may be shown on a graph for all audio frequencies. Alternatively it may be given for, say, the biggest peak between 500 and 2000 Hz, or at a particular frequency within that range.

Reverberation time depends in part on the distance that sound must travel between reflections, so large rooms generally have longer reverberation times than small ones. This is not only expected but also, fortunately, preferred by listeners.

In between the first (direct) sound and the densely bunched much-reflected sound that we think of as reverberation, there is a period when the earliest reflections arrive from the main walls and other large surfaces on the studio. The time to the first reflection, and the way other early reflections are spread in time, also help to define studio quality. (They are also important when creating or choosing artificial reverberation.)

After a few reflections, reverberation is spread fairly evenly and therefore is at much the same volume throughout an enclosed space, in contrast to direct sound, which dies away as the inverse square of the distance from the source. In large reverberant halls, at some *critical distance* the direct sound is overwhelmed by reverberation. Beyond the critical distance, the intelligibility of speech is reduced, then lost, and melody is submerged in mush. Theatres need to be designed so that no one has to sit near or beyond this critical distance, and microphones set to balance direct and reverberant sound will always be well within it.

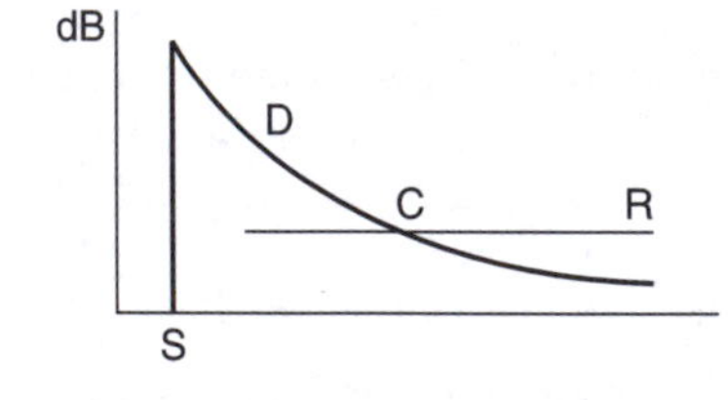

[4.3] **Critical distance**
S, Sound source. D, Direct sound dominant. C, Critical distance. R, Reverberation dominant.

Coloration

In a large room an *echo* may be detected. If there is little reverberation in the time between an original sound and a repetition of it, and if this time gap exceeds about an eighteenth of a second, which is equivalent to a sound path of some 18 m (60 ft), it will be heard as an echo.

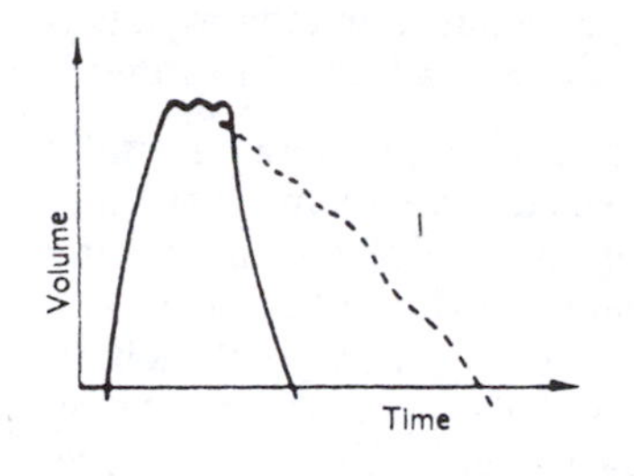

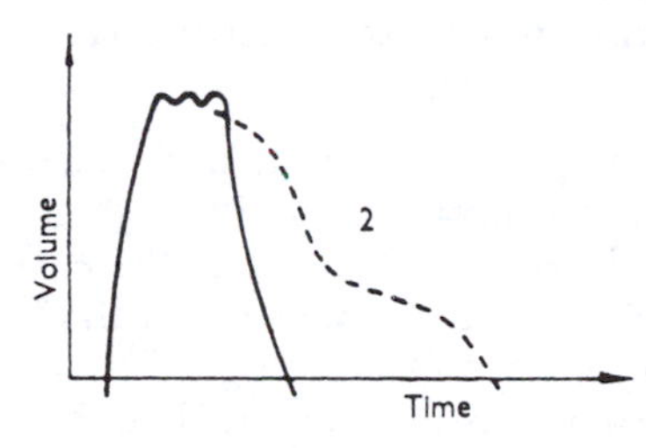

[4.4] **How sounds die away**
1, In a good music studio the sound dies away fairly evenly. 2, In a poor music studio (or with bad microphone placing) the reverberation may decay quickly at first, and then more slowly.

In a small room, *coloration* may be heard. This is the selective emphasis of certain frequencies or bands of frequencies in the reverberation. Short-path reflections (including those from the ceiling) with less than 20 ms delay interfere with direct sound to cause cancellations at regular intervals throughout the audio-frequency range, resulting in a harsh *comb filter* effect.

In addition, there may be *ringing* between hard parallel wall surfaces which allow many reflections back and forth along the same path: at each reflection absorption always occurs at the same frequencies, leaving others still clearly audible long after the rest have decayed.

Eigentones, the natural frequencies of air resonance corresponding to the studio dimensions, are always present, and in small rooms are spaced widely enough for individual eigentones to be audible as ringing. If the main dimensions are direct multiples of, or in simple ratios to, each other, these may be further reinforced.

The rate of absorption in different parts of a hall may not all be the same, giving rise to anomalous decay characteristics. For example, sound may die away quickly in the partially enclosed area under a balcony, but reverberation may continue to be fed to it from the main hall. More unpleasant (though less likely) might be the reverse of this, where reverberation 'trapped' in a smaller volume is fed back to a less reverberant larger area. Resonances may interchange their sound energies: this happens in the sounding board of a piano and is one of the characteristics that we hear as 'piano quality'. In a room it may mean that a frequency at which the sound has apparently disappeared may recur before decaying finally.

Not all of these qualities are bad if present in moderation. They may give a room or hall a characteristic but not particularly unpleasant quality: something, at any rate, that can be lived with. In excess they are vices that may make a room virtually unusable, or usable only with careful microphone placement.

The answer to many of these problems – or at least some improvement – lies, in theory, in better diffusion. The more the wave fronts are broken up, the more the decay of sound becomes smooth, both in time and in frequency spectrum.

In practice, however, if the studio geometry has no obvious faults and if acoustic treatment has been carried out and if panels of the various types that may be necessary for adequate sound absorption have been distributed about the various walls and on the ceiling, little more generally needs to be done about diffusion.

Studios for speech

Studios used only for speech are often about the same size as a living room in an ordinary but fairly spacious house. Where lack of space leads to smaller rooms being used, the result may be unsatisfactory, with awkward, unnatural-sounding resonances that even heavy acoustic treatment will not kill. And, of course, acoustic treatment that is effective at low frequencies reduces the working space still further.

Coloration, which presents the most awkward problem in studio design, is at its worst in small studios because the main resonances of length and width are clearly audible on a voice containing any of the frequencies that will trigger them. An irregular shape and random distribution of acoustic treatment on the walls both help to cut down coloration. But in small rectangular studios treatment may be ineffective, and this means working closer to the microphone. With directional microphones, bass correction must be increased and the working distance restricted more than usual.

Coloration is not necessarily bad (a natural-sounding 'indoor' voice always has some) but it has to be held to acceptable limits – which are lower for monophonic recordings. Even on a single voice there is great difference between mono and stereo, because with mono all of the reverberation and studio coloration is collected together and reproduced from the same apparent source as the speech. Note that even when the final mix is to be in stereo, pre-recorded voices are often taken in mono, so the acoustics of (for example) a commentary booth should allow for this.

One solution might appear to lie in creating an entirely 'dead' studio, leaving only the acoustics of the listener's room. But listening tests indicate that most people prefer a moderately 'live' acoustic, whether real or simulated. For normal living-room dimensions, 0.35–0.4 s seems about right for speech when using directional microphones; a smaller room would need a lower time, perhaps 0.25 s.

Given that reverberation time is defined as the time it takes for a sound to die away through 60 dB, you can use a handclap to give a rough guide to both duration and quality. In a room that is to be used as a speech studio the sound should die away quickly, but not so quickly that the clap sounds muffled or dead. And there must certainly be no 'ring' fluttering along behind it.

Listening rooms and sound-control cubicles should have similar acoustics, with a maximum of 0.4 s at frequencies up to 250 Hz, falling to 0.3 s at 8000 Hz. Treatment might be distributed to give good diffusion at the sides and greater absorption at the rear. Larger theatres (and particularly cinemas) need to be dead enough to ensure that sound at the back is not dominated by reverberation: in such conditions, film dialogue may be clearer than when heard in a more lively home acoustic.

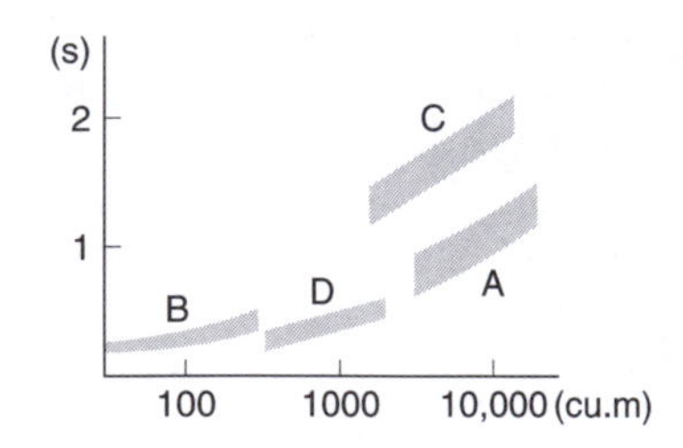

[4.5] **Radio studios**
Typical reverberation times for:
A, Programmes with an audience.
B, Talks and discussions. C, Classical music. D, Drama, popular music.

General-purpose sound studios

In some countries there is still a need for studios where a variety of acoustics is used for dramatic purposes. In others, such facilities can hardly be justified except, perhaps, to add interest to commercials. They may also be used for special projects to be issued as

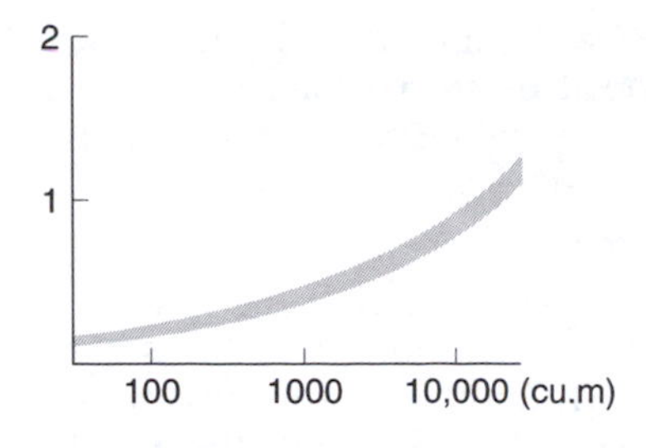

[4.6] **Television studios**
Typical reverberation times for general-purpose studios.

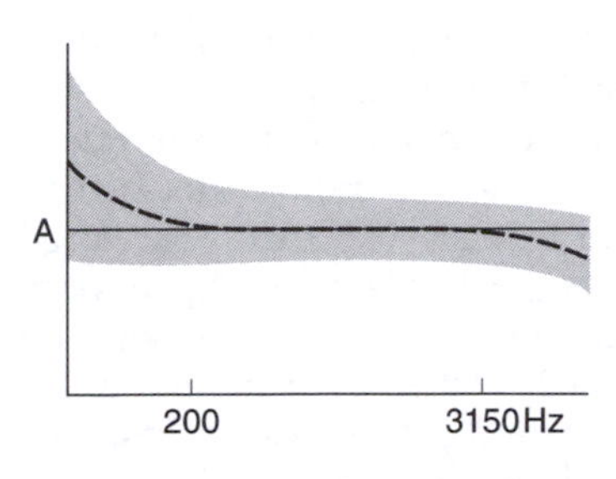

[4.7] **Range of reverberation times**
A, Median for any particular type. Below 200 Hz the reverberation in talks studios and their control rooms may rise; above 3150 Hz, for most studios it should fall by 10–15%. For a large music studio the tolerance is about 10% on either side of the average; for other studios and control rooms it may be a little wider.

recordings. This studio type is therefore still important, but less than when radio was the principal broadcasting medium.

In a dedicated drama studio there are two main working areas, usually separated by no more than a set of thick curtains. One part, the 'dead' end, has a speech acoustic that is rather more dead (by about 0.1 s) than would be normal for its size. The other end is 'live', so that when used for speech the acoustic is obviously different from the dead end. If the curtains are open, a speech microphone in the dead end can sometimes pick up a 'live' resonance – especially if a directional microphone is open towards the live end. This parasitic reverberation generally sounds unpleasant unless there is a good reason for it, so the curtains are usually drawn far enough to screen off the live end. But useful effects can sometimes be found near the middle of the studio.

The live end can also be used for small music groups working with speech taken from the dead part of the studio.

There may also be a very dead area, although if this is small it may sound like the inside of a padded coffin rather than the open air it is often intended to represent. For good results the type of treatment used in anechoic chambers for sound testing would be best for the walls and ceiling: these have wedges of foam extending out about 1 m (3 ft or more) from the walls. On the floor a good carpet (with underlay) is sufficient. Note that some performers find these conditions unpleasantly claustrophobic.

A studio for stereo speech (or any space within a studio that is used for this) needs to be more dead than other general-purpose studios. The reason for this is that where coincident microphones are used, voices must be placed physically farther back than for mono; otherwise, any movement will be exaggerated.

Music studios

For many audio professionals, the production of music, and very little else, is what studios are for. But 'music studio' means some very different things to different people: it can be live, making full use of rich acoustics; it may be dead, with the acoustics synthesized; it can even, in fact, not be there at all, present only as facilities in a 'virtual studio' (see Chapter 16). Many music studios have combinations of these characteristics. To some extent the choice depends on the type of music, but fashion also seems to play a part. At one time, almost all popular music recordings were made in very dead acoustics; since then, there has been a shift back to include some live acoustic components, and studio design now often reflects this.

A primary characteristic of any music studio with live acoustics is its reverberation time. Preference tests show that, for any given size of studio, music requires longer reverberation than speech. Such preferences are not absolute, but are valid for most conventional music and instruments. As we know them in the West today, these have developed in a very specialized way: in particular, our orchestral and chamber music has developed from that played in rooms within large houses in seventeenth- and eighteenth-century Europe. The original acoustics defined the music – and that music now, in turn, determines the acoustics needed for it.

In contrast, at one extreme, much Eastern music is designed for performance in the open air, as also is much military or brass-band music. Also, some modern music is written for drier acoustics than those that sound best with earlier classical works. Nevertheless, the acoustics we have been accustomed to have in their turn led to the design of instruments for which music of great beauty and power has been composed; so the studios we look for have certain specific qualities which can be judged by these subjective criteria.

Listening tests are not always the best indication of true value: they measure what we like best at the time of the test, and not what we could learn to like better if we chose to change our habits. But it does seem clear that the ideal (i.e. preferred) reverberation time varies with the size of studio: small studio – short reverberation; large studio – long reverberation. For the very smallest studios (e.g. a room set aside as a music room in a private house) the ideal is between 0.75 and 1 second. This is what might be expected if the room has no carpet and few heavy furnishings. Some authorities suggest that there should be more reverberation in the bass.

An interesting sidelight on this is that, on old mono recordings, it sounded better if the reverberation was about a tenth of a second less than that for live music. When reverberation came from the same apparent source as the direct sound, less seemed to be required.

Apart from reverberation, the principal quality of a good music studio is good diffusion, so that the sound waves are well broken up and spread. There should be no echoes due to domes, barrel vaults or other concave architectural features that can concentrate sound.

Dead pop music studios

There was a phase in popular music production when studio acoustics were little used. The internal balance of the music was deliberately surrendered in exchange for electronic versatility, and here nothing has changed. A celesta, piano or flute might be given a melody that has to sound as loud as brass, or other effects may be sought but can be achieved only if the sound of a louder instrument does not spill to the microphones set at high amplification for quiet instruments. Where separation is prized above natural acoustics, microphones are placed as close to instruments as may be practicable, and their directional properties are exploited to avoid picking up direct sound from distant loud sources, and to discriminate against reverberation from the open studio. In this musical style, reflected sound is just a nuisance.

A deliberately dead music studio was made a great deal more dead than most speech studios. Such an acoustic environment requires strong attenuation even at a single reflection from any wall, so that much less of the sound of (for example) the brass will bounce off the walls to reach the flute or celesta microphone from directions that cannot be discriminated against.

An aspect of acoustics that is important to pop music, not for its presence but again by its absence, is attenuation in air. At frequencies below about 2000 Hz, absorption by the air is negligible. But for 8000 Hz, air at 50% relative humidity has an absorption of 0.028 per 30 cm (per foot). This means that for high frequencies there is a loss of about 3.5 dB at a distance of 30 m (100 ft). For damper air the loss is a little lower, while for very dry air it rises sharply and begins to affect lower harmonics (although still mainly over 2 kHz) appreciably. For classical music this can make a big difference between performances using a distant balance in the same concert hall. But it does not affect close-balanced pop music, except to make instrumental quality *different* from that which may be heard acoustically: more of the higher harmonics of each instrument reach the microphone.

Television and film studios

Studios for television and film (and particularly the larger ones) are usually made very dead for their size. The primary reason is that if they are not to appear in vision, microphones must be farther away from sources than in other types of balance. But it also helps to reduce inevitable noise from distant parts of the studio. Noise arises in a variety of ways. Studio lamps produce heat, and the ventilation for that cannot be completely quiet. Scenery may have to be erected or struck during the action, or props set in. Performers and staff must walk in and out of the studio, and from set to set. Cameras move about close to the action, dragging cables. The floor of a television studio must allow the smooth movement of heavy equipment, so is acoustically reflective.

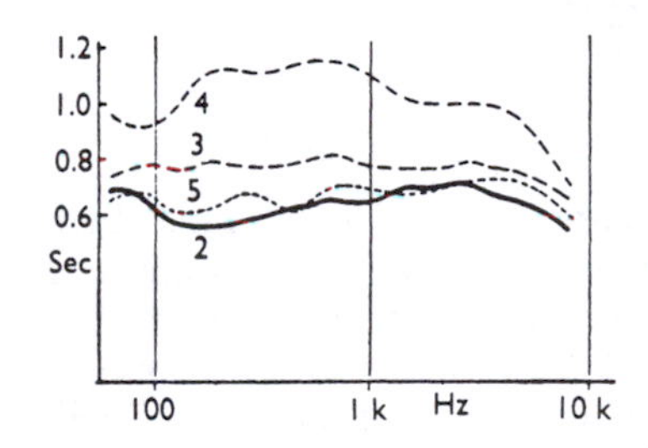

[4.8] **Four television studios: frequency response**
Variation in reverberation time with frequency. BBC Television Centre Studios 2 and 5 are both 3320 m^3 (116 000 ft^3); studios 3 and 4 are 10 230 m^3 (357 000 ft^3). Studio 4 was designed primarily for musical productions.

Sound that radiates upward may be reflected several times before it can reach a microphone, but sound that travels horizontally might be reflected only once. For this reason the most effective place for the absorption of sound is in the lower parts of the walls; more treatment is concentrated here than is placed higher up.

The set itself may reduce the overall reverberation a little: braced flats absorb low frequencies, while soft furnishings and thick drapes reduce the highs. In addition, the set may have unpredicted effects on the local acoustics: hard surfaces may emphasize high frequencies. A cloth cyclorama makes little difference, but if it is constructed from wood or plaster it reflects sound strongly (and curved parts focus it).

Television studios have acoustics approaching those used in 'dead' music studios, so they are suitable for music techniques of this type.

However, for the occasions when an orchestra has to be brought into the studio, the liberal application of unheard artificial reverberation is no help to musicians who find the surroundings acoustically unpleasant. In particular, string players need reflected sound to hear that they are producing good tone quality at adequate power. In its absence they may bow harder, changing the internal balance of the orchestra and making its tone harsh and strident. The timing of ensemble playing also suffers. An orchestral shell (an acoustically reflective surface behind and above) helps all the players, but reinforces the brass players (who do not need it) more than anyone else: a full shell turns a balance inside out and should not be used. It is best to design the set with reflecting surfaces to benefit the strings: this does nothing for the overall reverberation time, but helps musicians to produce the right tone.

Film studios ('stages') have in the past tended to be even deader than television studios. In addition to the other problems, film cameras are, by the nature of their intermittent action, noisy and difficult to silence effectively, while some HD video cameras (with nearby 'video villages') can be even worse.

Television sound control rooms and film dubbing theatres have the same characteristics as their counterparts in radio: 0.4 s at 250 Hz falling to 0.3 s at higher frequencies. The acoustics of the main production and lighting control areas should be relatively dead, in order that the director and his assistant and the technical director can be clearly audible on open microphones in areas that also have programme sound. In practice, however, absorption is limited by the necessarily large areas of hardware and glass.

Other technical areas, such as those for telecine and recording equipment, may need acoustic treatment: some machines can be noisy. If these are grouped in partly open bays, acoustic partitioning is desirable: aim for 20 dB separation between one bay and the next.

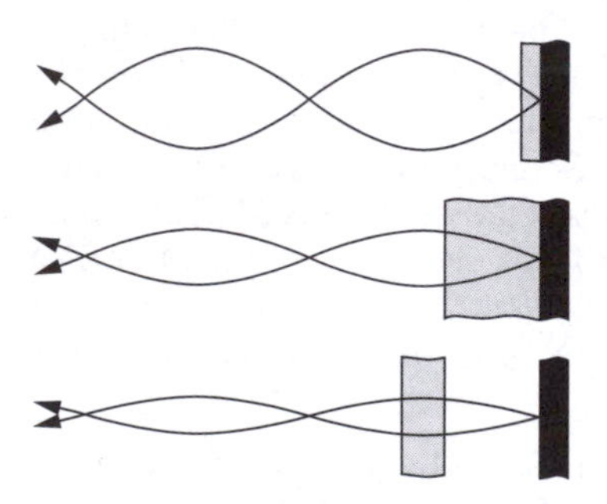

[4.9] **Sound absorbers**
Padding that is close to a reflecting surface is efficient only for short wavelengths. Screens with a thin layer of padding are poor absorbers at any but the highest frequencies. For greater efficiency in absorbing a particular wavelength, the absorber should be placed at a quarter wavelength from reflecting surfaces. In studios where space permits, boxes of different depths may be sunk into the wall, with quarter-wavelength absorber inside them, flush with the walls.

Acoustic treatment

There are several possible types of acoustic treatment:

Soft absorbers. These are porous materials applied to walls: their effectiveness depends on loss of sound energy as air vibrates in the interstices of the foam, rock wool or whatever is used. The method works well at high and middle frequencies, but for efficiency at low frequencies absorbers need to be about 4 ft (1.2 m) deep. So it is reasonable to lay absorbers at thicknesses that are efficient down to about 500 Hz, then become less so. Many types are commercially available. Control of high frequencies (for which absorption may be excessive) can be moderated by using a perforated hard surface (e.g. hardboard): with 0.5% perforation much of the 'top' is reflected; with 25% perforation much of it is absorbed. By changing the proportions of these two surfaces, the acoustics may be customized.

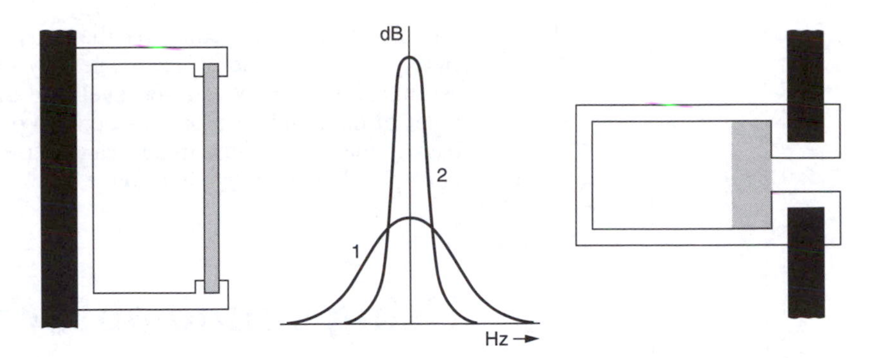

[4.10] **Absorbers**
The membrane absorber (*left*) is a box covered by a panel that is fixed at the edges and weighted with some material such as roofing felt. It absorbs a broad range of frequencies (1). The Helmholtz or damped cavity absorber (*right*) is in effect a bottle of air resonating at a particular frequency, with damping material inside it. It absorbs at the frequency of resonance (2).

Helmholtz resonators. These are not widely used, but the principle is available. Cavities open to the air at a narrow neck resonate at particular frequencies. If the interior is damped, they absorb at that frequency. Such resonators can be used for attacking dimensional resonances in studios or halls. A potential disadvantage is that it requires treatment to be fitted either at particular places that may already be occupied or some of the only places that are not.

Membrane absorbers. These have sometimes been used to cope with excessive low frequencies – although if low frequencies are a problem, the absorbers can be too. The idea is simple enough: a layer of some material or a panel of hardboard is held in place over an air space. The material moves with the sound wave like a piston, and this movement is damped so that low-frequency sound energy is lost.

A *combination absorber* might consist of units of a standard size: say, 60 cm × 60 cm × 18 cm deep (2 ft × 2 ft × 7 in). These are easy to construct and install; building contractors can be left to it with little supervision by acoustical engineers. A level response for a studio can be obtained by making (literally) superficial changes to this single design.

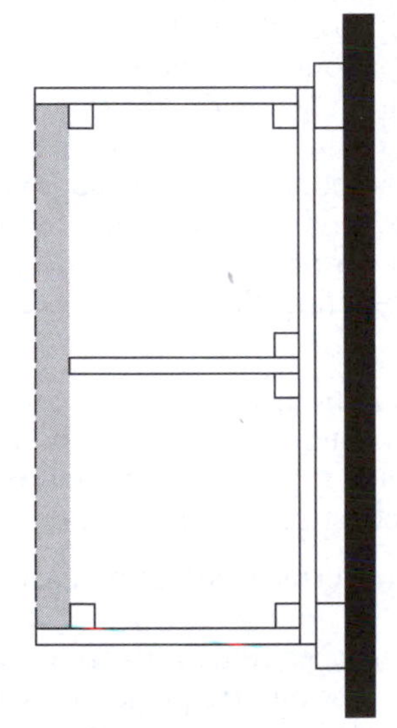

[4.11] **All-purpose absorber**
A plywood box mounted on battens fitted to the wall and covered with rock wool 3 cm thick (about 1¼ in) and perforated hardboard. The response is tuned by varying a single factor, the amount of perforation in the hardboard surface layer. Combinations of boxes with 0.5% and 20% perforations can be used as required.

The box is made of plywood, with a plywood back. Hardboard dividers are used to partition the interior into four empty compartments 15 cm (6 in) deep. Over this airspace, and immediately behind the face of the box, is laid 2.5 cm (1 in) of heavy-density rock wool. The face itself should be made of perforated hardboard or similar. If the open area perforation is very small there is a resonance peak of absorption at about 90 Hz. If the perforation area is large (20% or more), wideband absorption is achieved, but with a fall-off at 100 Hz and below. By using similar boxes with two kinds of facing, it is possible to achieve almost level reverberation curves.

Decorative *diffusers* can also be specially constructed, perhaps for the sides of a control room. A tray with sides and ends is made by using wood 15 or 20 cm deep by 25 mm thick (6–8 in × 1 in). The length of the tray could be rather more than half the height of the room. The width of the tray is divided at regular intervals by more strips of the same wood, to make a series of deep slots. The slots are given a range of different depths by separating the dividers with narrower strips of wood. In fact, the best way to make the tray might be to work from

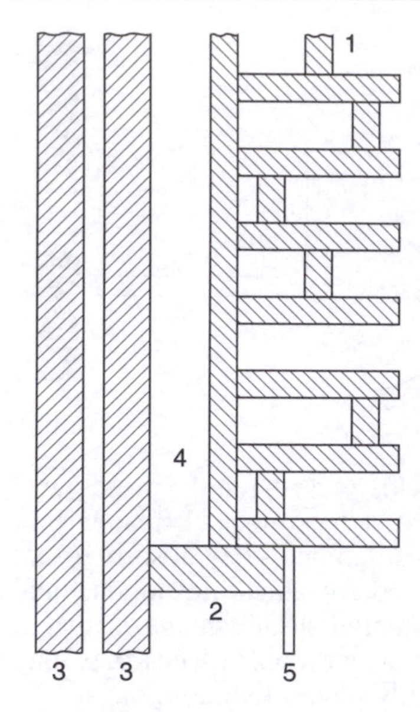

[4.12] **Diffuser**
1, Part of wooden diffuser (in plan view). 2, Support strut running from floor to ceiling. 3, Studio or control room wall. 4, Space for acoustic damping behind diffuser. 5, Front of absorber: this fills the spaces between neighbouring diffuser panels, and above and below them.

one side to the other, then fix the ends – i.e. top and bottom – and finally panel the back. The whole tray can then be supported about 30 cm (1 ft) or so above floor level, and away from the wall (so that a layer of damping absorber fits behind) on floor to ceiling battens set back at the sides. These in turn can be used to mount surface layers of absorbers below, above and to the sides of each diffuser.

Using sound absorbers

If the acoustics of a sound studio are fully designed, a variety of absorbers is likely to be used. To help evaluate them, their *absorption coefficients* are measured. That most commonly used is averaged over all angles of incidence, and over the range from 100 to 3150 Hz. Where there is total reflection, the absorption coefficient is zero; for total absorption it is unity. In practice the figure should therefore lie between zero and one. For example, free-hanging heavy drapes might be 0.8 for 500 Hz and above. But note that for relatively small areas absorbers may have a coefficient that is greater than one – an apparent impossibility that can be produced by diffraction effects.

Larger objects have sometimes been described in terms of the number of '*open window units*' (their equivalent in the number of square feet of total absorption) that they contribute. On this scale, individual, separated people check in at about four and a half units (at lower-middle to high frequencies), and a padded sofa may have 10 times the effect of the person sitting on it. For seated audiences, this would give too high a number, as sound does not approach each individual body from all angles, so it is necessary and in any case more convenient to calculate their absorption as though they are part of a flat surface: for occupied seating this gives a coefficient of 0.85.

Tables of these values are available: in addition, they give values for the studio structure as well and for the volume of air within it. The calculations are complex, but here are some rule-of-thumb ideas that might help when trying to correct the acoustics of a given room:

- apply some of each type of absorber to surfaces affecting each of the three dimensions;
- try to ensure that no untreated wall surfaces remain to face each other (note that a surface that is treated at one frequency may be reflective and therefore effectively untreated at another);
- put the high-frequency absorbers at head level in speech studios.

The cheapest and best treatment for the floors of speech studios is a good carpet and underlay (but the roof will need compensation in the bass). Wooden blocks are good on the floor of a (live) music studio.

Take equal care with furnishings. Plainly, it is unwise to get the acoustics right for an empty studio and then to bring in a lot of items (and particularly, any large soft absorbers) that change it. Furniture, decoration and even people must be included in the calculations from the start. But apart from this, it is not worth worrying too much about problems until you actually hear them. For example, if there is enough treatment in a room to get the reverberation low and level, and if the main rules for the layout of absorbers have been observed, diffusion of the sound will probably be all right anyway. So you can arrange the furniture for how it looks best, makes the occupants comfortable, and gives the most convenient layout of working areas.

The use of screens

Very often the acoustics that are built into a studio are not exactly those that are wanted for a particular use, so ways of varying studio quality have been devised. Sometimes the acoustic treatment of a studio is set in hinged or sliding panels, so an absorber can be replaced by a reflecting surface. But a more flexible approach relies on layouts of studio screens – in America, called flats or gobos ('go betweens').

If padded screens are to be easily movable (on small wheels), they must be light in weight (which in practice means they cannot be very thickly padded) and not too broad, often about 0.9 m (3 ft) wide. For stability they stand on cross-members at each end or on a tripod-like mount in the middle. Individual free-standing screens are effective only for sound of short wavelengths: low-frequency sound flows round and past them as though they were not there. But grouping screens edge to edge improves on this. At best, thin screens may reduce the sound flow between neighbouring areas by some 10–15 dB, while heavier screens with more absorber, say 10 cm (4 in) thick, will do better.

A common type of screen consists of a panel of wood with a layer of padding on one side. Placed with the dead side to the microphone, it will damp down the highs. With the bright (reflective) side forward, the highs are emphasized, while ambient sound in the studio is somewhat reduced by the absorbent back. In music studios with dead acoustics, large free-standing sheets of Perspex or wood are used in addition to the thickly padded screens that separate instruments and their associated microphones from each other.

Whatever the type, the effect of all screens is the same in several important respects:

- With screens to get in the way, there are shorter path lengths between reflections of the sound in the studio, so there are more reflections and more losses in a given time. This leads, in general, to a lower reverberation time for the studio as a whole.
- Sound from distant sources is reduced overall, but sound that does get round the screens is usually lacking in top. This also applies to any appreciable reverberation from the open studio.

- Coloration can be introduced by careless layout. If there are two parallel screens on opposite sides of the microphone, a standing-wave pattern will be set up. On the other hand, such distortion effects may be used constructively, though this is more often done by placing a microphone fairly close to a single screen.
- For the purposes of multimicrophone music balances, the separation of a large number of individual sources that are physically close together can be made more effective. Important enough in mono, this is vital in stereo.

In any studio it helps to have a few screens around. They may be needed to produce subtle differences of voice quality from a single acoustic; to separate a light-voiced singer from heavy accompaniment; or in classical music for backing and reinforcing horns.

[4.13] **'Flying saucer' diffusers** Royal Albert Hall, London: suspended convex reflectors of various sizes break up sound waves that might otherwise be focused by the interior of the dome.

Altering the acoustics of concert halls

Traditionally, the design of concert halls has been a fallible art rather than an exact science. In some existing halls of a suitable size it can be difficult to pick up good sound, and even where careful experiment has found a reasonable compromise for microphone placement, many of the audience (whose audible enjoyment may contribute to the sense of occasion) may still experience poor sound. Fortunately it is now possible to do something about many such halls (or studios). Among those that have had their acoustics physically altered are several of the main concert halls of London.

Ever since it was built in Victorian times, the Royal Albert Hall inflicted echoes on large areas of seating and the unfortunate occupants. This was partly due to its size: being very large, for many places the difference in path length between direct sound and a first reflection could be 30 m (100 ft), resulting in a delay of nearly 0.1 s, almost twice as long as required for a distinct echo. This would have mattered less if it had blended with other, similar reflected sound, but it did not. The hall was designed as a near-cylinder topped by an elegant dome. These concave surfaces focused the reflections, and some coincided with the seating, so the echo was amplified.

Apart from the gallery, where the sound was clean though distant, there were (as regular concert-goers claimed) only about two places in the hall that were satisfactory. Staccato playing from percussion or piano made it even worse. Also, the natural reverberation time of the hall is long. As a result, the range of music that could be played to good effect was limited virtually to a single type: romantic music with long flowing chords. Even this often lacked impact: it was simply not loud enough, because the hall is too big.

Attempts to improve the acoustics without changing the appearance of the hall failed. The problem was eventually tackled by suspending flights of 'flying saucers' 2–3.75 m (6–12 ft) in diameter: 109 of these were hung just above gallery level. Taken all together, they defined a new ceiling, slightly convex towards the audience and filling about 50% of the roof area.

These polyester-resin-impregnated glass-fibre diffusers (some with sound-absorbent materials on the upper or lower surfaces) did largely solve the problem of echo from above (if not from the oval-sectioned surrounding wall) and at the same time reduced the reverberation at 500 Hz to more reasonable proportions for a wider range of music. Even so, the design has been superseded by another, in which diffusers are also integrated into the ribs of the dome.

In London's post-war Royal Festival Hall the error was in the opposite direction. Musicians had been asked in advance what acoustics they would like to have, 'tone' or clarity. The answer came, 'tone', but unfortunately what they actually got was a remarkable degree of clarity. The sound was far more dead, particularly in the bass, than had been expected, largely due to the construction of too thin a roof shell. Minor measures, such as taking up carpets from walkways and making surfaces more reflective, had insufficient effect. While a cough sounded like a close rifle shot, bass reverberation remained obstinately at about 1.4 s. How, short of taking all the seats out and sending the audience home, could the reverberation time be lengthened?

The answer here was *assisted resonance*. Cavity resonators were used, but not in their normal form: these contained microphones, tuned to respond to the narrow bands of frequencies present and re-radiate sound through individual loudspeakers. A hundred channels could each have a separate frequency; allocating each a band of 3 Hz, the whole range of up to about 300 Hz could be boosted. In this way, resonance was restored at surfaces where, previously, sound energy had begun to disappear. A reverberation time of 2 s was now possible, but it could be varied and actually tuned at different frequencies.

Assisted resonance has since become available as a commercial package, prolonging a selected range of about 90 frequencies between 50 and 1250 Hz by half as long again. At the Festival Hall this dealt effectively with the balanced overall duration of reverberation, but loud brass or timpani could still overwhelm other instruments. Changes to the acoustics have continued.

In both of these venues, however, broadcasters had been able to get good or fairly good sound even before the alterations by their choice of microphones and careful placing, so the balance problems (other than for reverberation *time*) have not been so severely affected.

Recording companies working with London's many first-rate orchestras found their own solutions. Audiophiles around the world may not know it, but the acoustics they have often heard on record

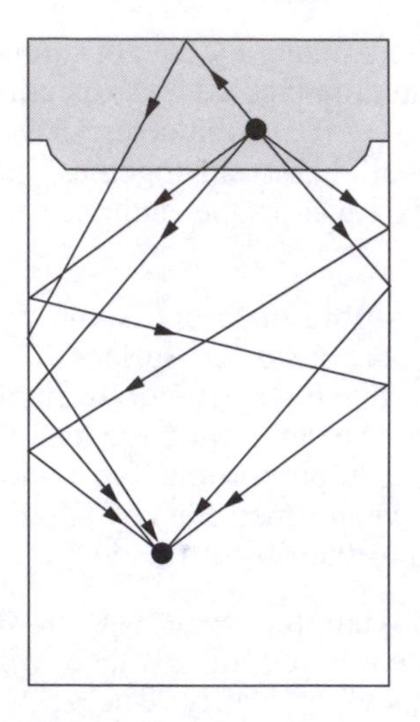

[4.14] **First reflections in a 'shoebox' concert hall**
Long parallel side walls create a rich array of early reflections between direct sound and reverberation.

are not those of well-known concert halls, but obscure suburban town halls in places like Walthamstow, Wembley, Watford and Hammersmith – narrow, high-ceilinged rectangular boxes that, by chance rather than musically dedicated design, have much in common with the Concertgebouw in Amsterdam, the Musikvereinsaal in Vienna and Symphony Hall in Boston, all favourites of musicians as well as audiences. All these shoebox-shaped halls have two distinctive advantages over their mid-twentieth-century fan-shaped successors. One is simply that the older halls were smaller. Less obvious was that the solid, narrow, parallel walls deliver a swathe of strong early reflections, which are now understood to add essential character to musical acoustics.

Forty years after the Royal Festival Hall, another British hall, the Symphony Hall in Birmingham, reverted to the size and parallel sides of the shoeboxes (though with rounded surfaces to add diffusion from smaller features within the space). But it was also designed for far greater physical control of its acoustics. It has concrete-walled reverberation chambers behind the orchestra and extending high along the sides (adding half as much again to the volume of the hall), with massive adjustable doors that can be moved at the touch of a button. There is a 32-tonne counterbalanced 'spaceship' canopy over the orchestra. This can be lowered and the chamber doors partly closed for smaller forces or a drier sound. To convert it into a conference hall or for commercial presentations (and also for loudspeaker-reinforced popular music), thick absorbent 'banners' and a rear curtain emerge, hanging well clear of the reflective walls for maximum effect.

Built only 30 m (100 ft) from a covered main-line railway, the hall is mounted on rubber cushions (as also is the rail track), and shielded from heavy traffic on the opposite side by double skins of concrete. Ventilation noise is cut by using many large, low-speed ducts. On stage, a heavy-duty wagon of orchestral risers wheels in, with built-in wooden sounding boards just for instruments such as cellos that need them.

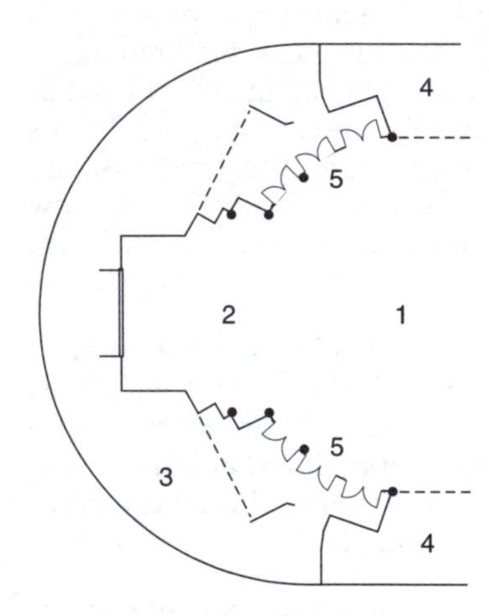

[4.15] **Controlled concert hall reverberation**
Symphony Hall, Birmingham:
1, Auditorium with parallel side walls. 2, Orchestra with space for organ behind. 3, Reverberation chamber. 4, Extensions to reverberation chamber at high level. 5, Massive doors opening into chamber.

Acoustic modelling of studios

The design of new studios – or redesign of defective ones – can be further refined by modelling their acoustics in advance of full-scale construction. Modelling reverberation is one of the fields of audio that has been totally transformed by computers (see Chapter 12), and studio design has inevitably followed the same path. It was not among the earliest advances in computer technology, and the reason for this is obvious as soon as we consider the computational problem for a reasonably realistic set of sound reflections.

From each source within the studio, sound travels out in a virtually infinite number of paths ('rays'), so the first digital compromise is to settle on a practicable distribution of rays. Then the path of each ray must be traced through each of many reflections in what may be a very irregularly shaped volume, with the absorption coefficient known at every point on its inner surface. After a large number of reflections the tiniest error in this calculation will be gross. Add into this the approximations required when modelling architectural and other details, together with the physical compromises required to allow for diffusion, diffraction and other features of enclosed spaces that have been described earlier, and the problem begins to appear intractable.

In fact, the first steps towards successful acoustic modelling were taken by building and testing physical scale models. In this earlier analogue world, there was no need for compromise on the number of rays or number of reflections, but it is obvious that here, too, there were bound to be enormous compromises. This was done as follows.

In a model made to a linear scale of one-eighth, sound path lengths are reduced in the same proportion. With absorbers scaled accordingly, reverberation is also one-eighth of that for a full-size studio. Surfaces are one-sixty-fourth life size and the volume about one-five-hundredth. Variations in design could therefore be explored at a cost that was modest compared with that of correcting problems in the real hall. A special acoustically dead selection of musical recordings was replayed into the model at eight times normal speed, and re-recorded with the acoustics of the model added. When this was replayed at normal speed, the results of various layouts and treatments could be compared.

The design of a loudspeaker to fill the model with sound at the required 400 Hz to 100 kHz was inevitably unorthodox. It had three elements: an 11 cm (4¼ in) thermoplastic cone (for 400–3000 Hz), a 20 mm (8 in) domed plastic diaphragm (for 3–21 kHz), and a cluster of 45 small electrostatic units mounted on a convex metal hemisphere giving wide-angle high-frequency radiation (at 21–100 kHz). Microphones to pick up the same raised frequency band ideally required a tiny diaphragm: the 6 mm (¼ in) diaphragm used was rather large for the highest frequencies. It was omnidirectional, so for a stereo pick-up a spaced pair was used. The (tape) recorder had to cover 50 Hz to 12.5 kHz at normal speeds and the 400 Hz to 100 kHz band when speeded up.

Structural components and miniature furniture were first acoustically tested in a small reverberation chamber. Materials with textures similar to the full-scale object were used where possible; e.g. velvet for carpets. Perforated absorbers were represented by scaled-down absorbers of similar construction. An orchestra is widely spread and mops up much of the sound it creates, so early experiments were unsuccessful until the musicians, too, were modelled – in expanded polystyrene with 3 mm (⅛ in) of felt stuck to their backs. The biggest problem was the air, which had to absorb eight times more sound in transit achieved by drying it to 4% relative humidity (compared with around 45% in normal air).

The first application of this was to redesign an orchestral studio, a former ice-rink that had served the BBC Symphony Orchestra for many years, but which had nagging deficiencies. Many experimental modifications were tried out on a model of it: absorbers were changed at one end, added under the roof, then covered with a low-frequency reflective material; the choir rostra were changed, then switched to a different position, and subsequently replaced by another design, while various shapes of wooden canopy were suspended over the orchestra. The results were compared with each other and (as was possible in this application) with the sound produced by playing the same test material through the real studio. The experiment saved more than its cost by just one result: the big reflective canopy would have been a waste of money. When built, the chosen design came within 10% of its predicted performance, an unusually good result in this problem-infested field.

Scale models that are smaller still (1:50) have also been used, though plainly with tighter limits on what can be tested. This ratio is too small for listening tests, but a spark simulates a gunshot and the reverberation can be measured using microphones with a 3 mm ($^1/_8$ in) diaphragm.

Digital alternatives to these physical models have already reached an astonishing degree of sophistication. Plainly, even an expensive software program costs much less than a complicated model and is far more versatile in its application.

The EASE software program is one of several designed to predict the acoustic performance of large auditoria such as theatres, arenas, stadiums and churches. Among its many modules are the 'tools' to build a room, with a range of surfaces, audience areas, and loudspeaker and listener positions. Wall materials can be defined, together with their absorption and diffusion characteristics. Loudspeaker data (at least, as supplied by their manufacturers) is included. Several modules are concerned with the design and performance of speaker systems and clusters giving, for example, a 3-D view of their coverage, a 2-D analysis of their effect on the room and a tool for rendering those effects in 3-D. A probe can be used to investigate the acoustical properties at any given point, and ray tracing can check for problem reflections (the more complex program traces up to 19 reflections). An optional EASE 'ears' auralization tool allows users to listen to the calculated response at any point in the room on their own PC.

Plainly, this is well adapted to the layout of rooms such as cinema auditoria (theatres), but can also offer many design and analysis tools to acousticians working on all kinds of halls. With any kind of acoustic modelling there are substantial savings of time and money in the preliminary design, and a better chance of getting good results later. It should be obvious that the best approach is always to model acoustics first and build the structure around it, but unfortunately this is not always standard architectural practice.

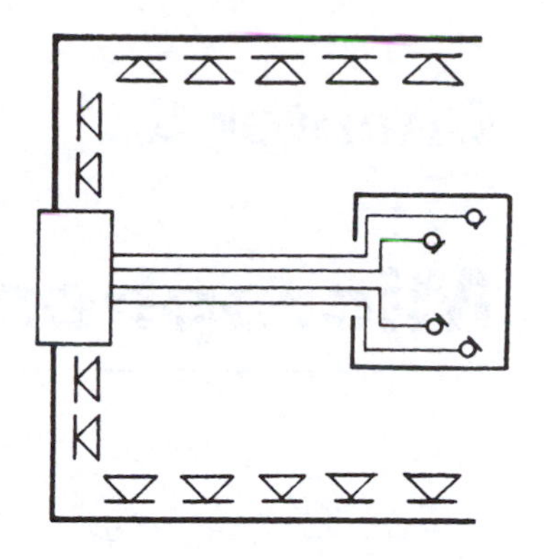

[4.16] **Acoustical control system (ACS)**
Output from microphones above the performers is fed to a complex digital system that mixes, delays and feeds the recombined signals to loudspeakers (more than shown here) around the hall. Feedback between loudspeakers and microphones is kept as low as possible. The aim is to simulate the sound field of a good music studio, but with variable reverberation time.

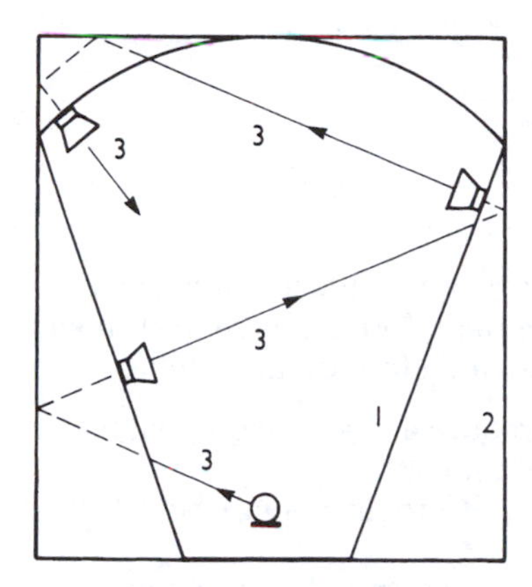

[4.17] **'Acoustic holography' principles**
In this example, the original walls (1) carry speech rapidly to the back of the fan-shaped auditorium, so that people at the rear can hear individual speech syllables clearly. At the back wall, sound must be absorbed so that it does not reduce clarity at the front. 2, Ideal music auditorium simulated by the acoustical control system. 3, Simulated sound path using three of the loudspeakers lining the walls at the sides and rear. In practice, there will be many loudspeakers in a line along these walls.

Acoustic holography

One further way to create an all-purpose auditorium is to use *acoustic holography*. Within an existing acoustically rather dry hall, this aims to construct a range of richer sound fields that are also rewarding to perform and listen in. One such arrangement is called ACS (Acoustical Control System). The audience sits between rows of loudspeakers that run along the side and rear walls. The sound wave from what is in effect a horizontal line source radiates up and down but does not fan out horizontally to the sides. In consequence, speakers at the side that are close to the stage radiate only weakly towards it. Only a little of their sound is picked up by microphones in the stage area that are in any case set up to discriminate against sound from the direction of the hall.

ACS employs a matrix of electronic delays and filters to interconnect all microphones and loudspeakers. This reduces acoustic feedback even further to such a safe, low level that it plays virtually no part in the final sound, and coloration that may accompany it is also avoided. In the absence of feedback between auditorium and stage, the speaker system is designed to recreate an ideal wave front in the audience area. The computer processing that this requires became practicable in the mid-1980s.

The reverberation time can be increased six-fold to make a small, relatively dead theatre sound like a cathedral. This allows an auditorium designed for speech to be used for many different kinds of music, and its acoustics can be instantly reset by the control computer between pieces, or even between announcements and music.

Chapter 5

Microphones

A microphone converts sound energy into its electrical analogue – and should do this without changing the information content in any important way. A microphone should therefore do three things:

- for normal sound levels it must produce an electrical signal that is well above its own electrical noise level;
- for normal sound levels the signal it produces must be substantially undistorted;
- for a particular sound source it should respond almost equally to all significant audio frequencies present.

In current high-quality microphones the first two objectives are easily met, although to achieve a signal that can be transmitted cleanly even along the lead of the microphone may need a transformer or amplifier so close to the head that it becomes part of the microphone. The third requirement is more modest, and deliberately so: any claim that sound above the 20 kHz range of human hearing somehow '*feels*' better smacks of mysticism and, in fact, the practical limit to broadcast systems may be as low as 15 kHz. But for a source of restricted range, even less is required of the microphone: a full audio-frequency range response may pick up unwanted high-frequency noise or catch spill from neighbouring sources that have a more extended output.

Microphones differ from one another in the way that air movement is converted into electrical energy. The most important types in professional use are condenser microphones (also called capacitor or electrostatic microphones, and including electrets), moving-coil (often called 'dynamic') microphones and, returning to favour after a period in the wilderness, ribbons. Other principles have also been employed in the past, including piezoelectric (crystal), carbon, inductor, moving-iron, ionic and magnetostriction, but these are not described here: most have practical disadvantages that have limited their development.

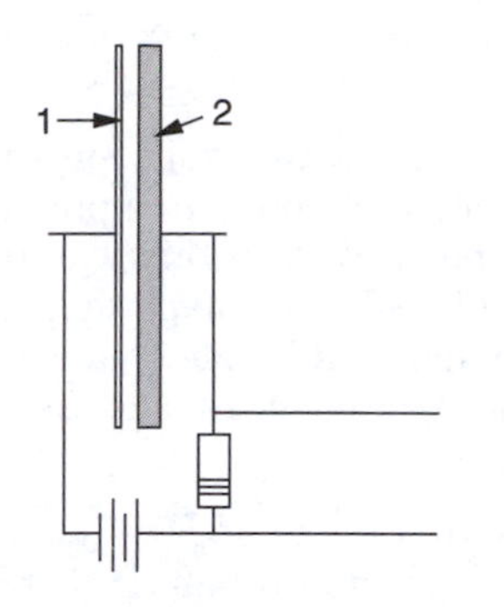

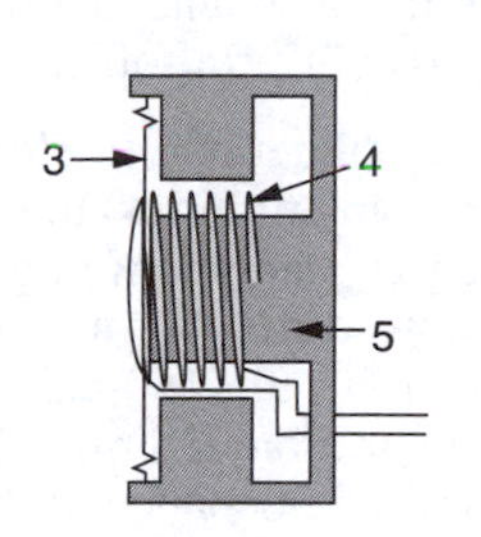

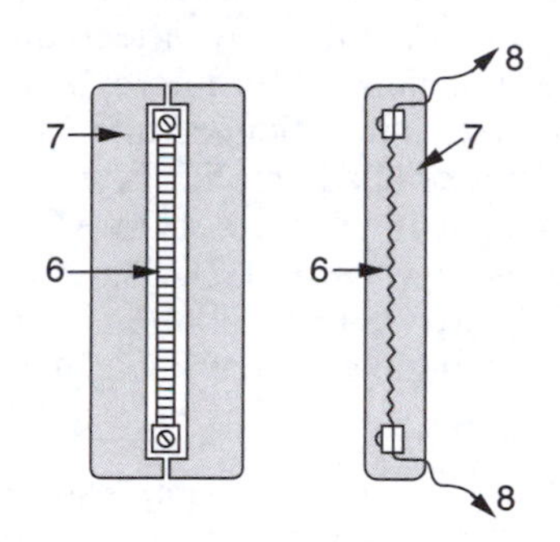

[5.1] **Types of microphone**
Above: Condenser (electrostatic). 1, Foil diaphragm. 2, Back-plate. An amplifier is needed so close to the head as to be considered part of the microphone. *Centre*: Moving coil. 3, Thin diaphragm, typically of Mylar. 4, Voice coil fixed to diaphragm. 5, Permanent magnet. *Below*: Ribbon microphone. 6, Light, conductive foil ribbon, corrugated for high flexibility. 7, Permanent magnet, with pole pieces on either side of ribbon. 8, Transformer (one range has a head amplifier in place of this). Ribbons, less popular for some years because of their limited high-frequency response and archaic appearance, have since returned to greater favour. For both moving coil and ribbon the magnet still adds weight, and must also be kept away from magnetic recordings.

Nearly all microphones have a *diaphragm*, a surface that is physically shifted in response to sound waves (although another possible principle employs a pressure-sensitive semiconductor in an integrated circuit). But there are two fundamentally distinct ways in which air pressure is converted into movement of the diaphragm. These are distinguished by differences in their directional characteristics.

Although the output of a microphone may be designed to change with the angle from a given axis, its frequency response, ideally, should not. Unfortunately, this ideal is almost impossible to achieve, so it is usual to define a 'useful angle' within which this condition is more or less met, or to devise ways in which the deficiency is turned to advantage.

Remember, this 'electrical analogue' really is analogue. So it is still subject to the kinds of distortion and noise that afflict analogue signals. To convert it to 'digital' requires a separate, later process.

Microphone properties

The range of properties affecting choice of microphone includes:

Sensitivity. This is a measurement of the strength of the signal that is produced. In practice, most professional microphones of a particular type have output voltages within a range of perhaps 10 dB. Taking moving coils as 'average', condenser types can be higher because they need a head amplifier anyway. Ribbons tend to be lower, unless they, too, have a head amplifier.

Characteristically, 60 or more dB (further) amplification is required to reach a level suitable for input to a mixer. The upper limit to sensitivity is dictated by overload level, beyond which unacceptable harmonic distortion is generated (or by potential damage). Moving coils are relatively free of this, unlike condensers, which also produce more electronic noise (hiss). The limits for both of these may be described in terms of sound pressure levels (SPL). For example, a microphone with a stated upper limit of 130 dB is claiming low distortion up to levels that approach the threshold of pain, though this may still be less than that close to the mouth of a loud singer. Moving-coil and condenser types can both have a wide range, as can some modern ribbons.

Robustness. Moving-coil microphones in particular stand up well to the relatively rough treatment a microphone may accidentally suffer when used away from the studio. Condensers and ribbons require more care.

Sensitivity to handling, e.g. by pop singers. Moving-coil microphones perform well, and condensers with mechanical insulation between the casing and microphone capsule and electronics are also good. A reduced low-frequency response helps. Older ribbon microphones are unsuitable, as their diaphragms may have a resonant frequency of about 40 Hz that is easily excited by any movement or even a puff of air. Some modern ribbons are engineered to reduce

[5.2] **Microphone sensitivities**
These are usually given in dB relative to 1 volt per pascal. This corresponds to a (linear) scale in millivolts per pascal, as follows:

dB rel. to 1 V/Pa	*mV/Pa*
−20	100
−25	56
−30	31.6
−35	17.8
−40	10.0
−45	5.6
−50	3.16
−55	1.78
−60	1.00
−65	0.56

Other units that may be encountered are dB relative to 1 volt per dyne/cm^2: this gives figures that are 20 dB lower; 1 mV/Pa is the same as 1 mV/10 μbar or 0.1 mV/μbar.

handling problems. Cables and connectors should also be checked for handling noise.

Sensitivity to wind and blasting. Gusting produces erratic pressure gradients, so microphones that are designed to respond to a pressure gradient are at a disadvantage. Most ribbons work in this way (see below), so older designs that have a big, floppy diaphragm are doubly at risk; modern ribbons are smaller, stiffer and may have built-in screens. The shape of the microphone head and use of windshields also need to be considered.

Sensitivity to rumble. Rumble is vibration at 3–25 Hz that can modulate higher frequencies, making them sound rough. If there is any danger of rumble, avoid microphones with an extended low-frequency response – or use a shock mount or possibly a roll-off filter.

Impedance. An impedance much below 50 Ω is vulnerable to electrical mains hum, so it needs a transformer (or head amplifier) to lift it. The high (20–50 kΩ) impedance of a condenser diaphragm might pick up electrostatic noise from motors or fluorescents, but a head amplifier prevents that. A moving coil is already in the 150–250 Ω range that allows a cable run of 300 m (1000 ft) or more. Nominally a microphone should be rated appropriately for the impedance of the system into which its signal is fed, otherwise the interface between components reflects some of the signal and this may affect the frequency response. To avoid this, the impedance of a microphone should be less than a third of that of the system it is plugged into, and a ratio of 1:10 is safer. After a transformer or head amplifier (where required), many microphones have an output impedance of 200 Ω or less (although some moving coils are as high as 600 Ω). The preamplifiers built into consoles generally have a 1500–3000 Ω input impedance; high-quality preamplifiers often have source impedance matching.

Sensitivity to temperature and humidity variations. Location recording may be subject to extreme climatic conditions, and heat and humidity can also cause trouble high among the lights in a television or film studio. If water vapour condenses on to a lightweight diaphragm, this increases its inertia and temporarily wrecks its response. A condenser microphone that is transported or stored in the cold is vulnerable to this, and also to any moisture in its insulation or head amplifier.

[5.3] **Microphone performance**
Figures sometimes quote maximum sound pressure levels (usually for 1% harmonic distortion). A conversion to corresponding linear pressure levels in pascals is:

SPL (dB)	*Pressure (Pa)*
140	200
134	100
128	50
94	1
0	0.000 02

Zero SPL approximates to a young listener's threshold of hearing.

Transient response. Speech and music contain important transients that are 10–100 ms in duration. High-quality condenser microphones, with diaphragms less than 0.04 mm (0.0015 in) thick are excellent, with a response of less than 1 ms, and ribbons are also good. Moving coils, limited by the mass of diaphragm and voice coil, are less clear.

Flexibility of frequency response. Some microphones have a variable polar response, and the switches for this may be located on the microphone head, or on a separate control unit, or remotely on the control console. Another switch may control bass response.

Electronic noise. Microphones generally produce little noise, but as later stages have improved (especially by the use of digital recording media), even a tiny effect becomes more noticeable. In condenser microphones the head microphone is again a potential problem, for example during very quiet passages in classical music. *Equivalent noise rating* is the electronic counterpart of studio noise. A 20 dB limit means that the electrical microphone noise is no higher than the 20 dB (sound pressure level) ambient noise of a quiet recording studio.

Power requirements. Moving-coil and (most) ribbon microphones have no power supply, but condensers need one, both for the charge on the diaphragm (except for permanently charged electrets) and for a small head amplifier. Some early condensers had a plug-in power supply unit. Today, power is usually fed back along the cable from the console as a 'phantom' supply. (But note that any so-called 'digital' microphone also needs a supply to power its internal analogue-to-digital converter.)

Shape, size and weight. Appearance matters in television, and also size and weight in microphones that are hand-held or attached to clothing.

Cost. Electrets (permanently polarized electrostatic microphones) can be cheap, and good moving-coil microphones moderately so; some high-quality condensers and ribbons are more expensive.

In a given situation any one of these criteria may be dominant, perhaps even dictating choice of microphone. Manufacturers may emphasize particular characteristics at the expense of others, and offer them for some particular application. Popular music attracts a wide range of different types, some to cope with extreme physical conditions, others on account of frequency characteristics that are appropriate to individual instruments or for their 'perceived' tonal qualities.

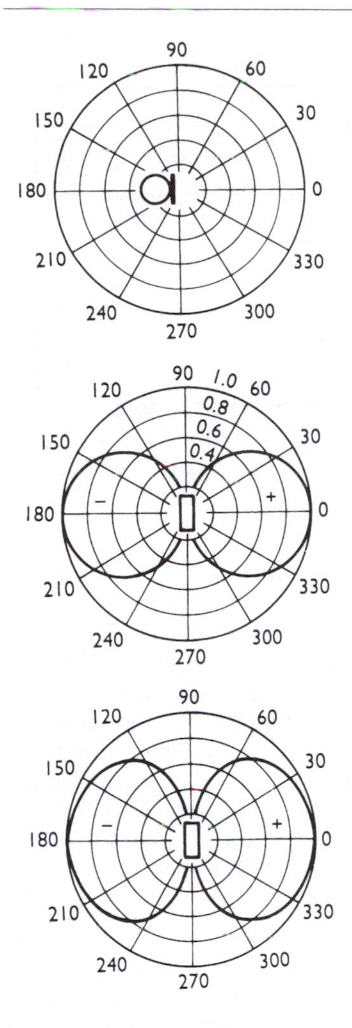

[5.4] **Polar diagrams**
Field patterns of *omnidirectional* and *bidirectional* microphones. The scale from centre outwards measures sensitivity as a proportion of the maximum response, which is taken as unity. *Above*: A perfect omnidirectional response would be the same in all planes through the microphone (this diagram is a section through a sphere). *Centre*: A perfect bidirectional response is the same in all planes through the 0–180° axis. At 60° from this axis the output is reduced to half, and at 90° to zero. The response at the back of the microphone is opposite in phase. This diagram has a linear scale, reducing to zero at the centre, and the figure 'eight' is perfect. *Below*: The figure-of-eight, shown on a more conventional decibel scale, is fatter, but the decibel scale does not have a zero at the centre.

Directional response

Microphones fall into several main groups, according to their directional characteristics. Their *field patterns* are best visualized on a *polar diagram*, a type of graph in which the output in different directions is represented by distance from the centre of the diaphragm.

Omnidirectional microphones. These, ideally, respond equally to sounds coming from all directions. Basically, they are devices for measuring the *pressure* of the air, and converting it into an electrical signal. The simplest designs of moving-coil and condenser microphone work in this way. The diaphragm is open to the air on one side only. This type of operation is also commonly called 'dynamic'.

Bidirectional microphones. These measure the difference in pressure (*pressure gradient*) between the two sides of a diaphragm, and so at two successive points along the path of a sound wave approaching from the front or back. If the microphone is placed sideways to the path of the sound, the pressure is always the same on both sides of the diaphragm, so no electrical signal is generated. The microphone is therefore *dead* to sound from the side and *live* to the front and rear.

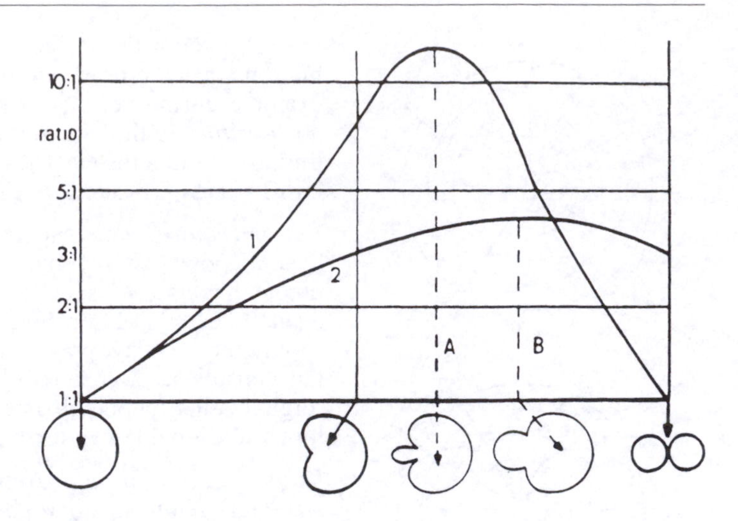

[5.5] **Directivity**
The range from pure pressure operation (omnidirectional) to pure pressure gradient (bidirectional) operation. Curve 1 shows the ratio of sound accepted from the front and rear. The most unidirectional response is provided by the supercardioid (A), for which the microphone is, essentially, dead on one side and live on the other. Curve 2 shows discrimination against ambient sound. With the hypercardioid response (B), reverberation is at a minimum and the forward angle of acceptance is narrower. In the range between omnidirectional and cardioid, direct sound from the front and sides is favoured, so a microphone of this type could be placed close to a broad source. (Note that the close balance will already discriminate against reverberation, and that this microphone will reduce long path reverberation still further.) Another measure used is directivity index (DI). For the omnidirectional response of 360° the DI is zero; for cardioid, with a solid angle of 131° the DI is 1.7, with 4.8 dB rejection of reverberation; supercardiod, 115°, DI 1.9, 5.7 dB rejection; and hypercardioid, 105°, DI 2.0 with a 6 dB reduction of indirect sound.

Moving from the front to the side of the microphone, the response becomes progressively smaller then rises again, and a graph of the output looks like a *figure-of-eight* or *figure-eight* (variations on a term sometimes used to describe this kind of microphone). The *live angle* is generally regarded as being about 100° on each face. For sound reaching the microphone from the rear, the electrical output is similar to that for sound at the front, but is exactly opposite in phase. The simplest designs of ribbon microphone employ this principle. They respond to the difference in pressure on the two faces of a strip of aluminium foil. Condenser figure-of-eight microphones have two diaphragms, one on either side of a charged backplate.

Cardioids (as the name implies) have a heart-shaped response. This is obtained if the electrical output of a pressure-operated microphone is added to that of a pressure gradient microphone with a response of similar strength along its forward axis. The output is doubled at the front, falls to that of the omnidirectional microphone on its own at 90° to either side and then to zero at the rear, where the two are antiphase and cancel out. Various ways of combining the two principles are possible: early cardioid microphones actually contained a ribbon and a moving coil within a single case. There was substantial pick-up of sound on the front and some round to the side, but progressively less beyond that.

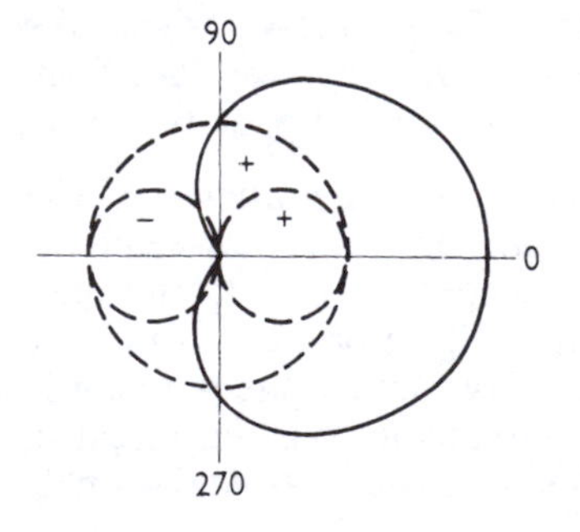

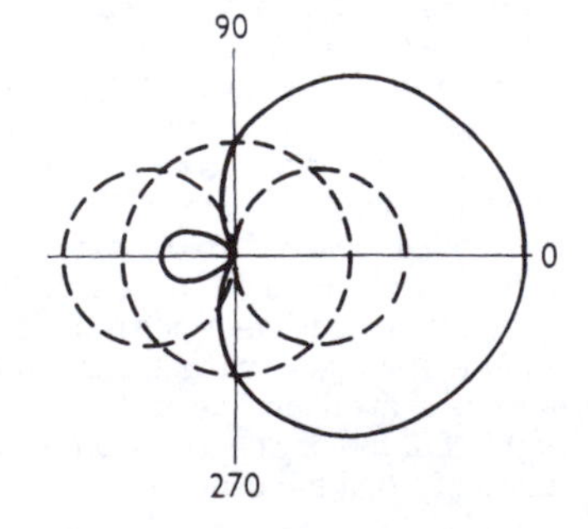

[5.6] **Cardioid response**
Left: The sum of omnidirectional and figure-of-eight pick-up patterns when the maximum sensitivity of the two is equal. The front lobe of the 'eight' is in phase with the omnidirectional response and so adds to it; the back lobe is out of phase and is subtracted. *Right*: The supercardioid is a more directional version of the cardioid microphone. It is one of a continuous range of patterns that can be obtained by combining omni- and bidirectional response in various proportions. As the bidirectional component increases further, the rear lobe grows, to make a cottage-loaf-shaped or hypercardioid response.

Supercardioids and hypercardioids. If the pressure and pressure-gradient modes of operation are mixed in varying proportions, a range of polar diagrams is produced, passing from omnidirectional through cardioid, then supercardioid to hypercardioid, and finally bidirectional. The directional qualities within this range may be described in two ways. One is by the degree of *unidirectional response*, indicating the ratio of sound accepted at front and back of the microphone (i.e. before and behind a plane at right angles to its directional axis). This is 1:1 for omnidirectional and bidirectional responses (the two extremes) and maximum, reaching a peak ratio of 13:1, at an intermediate stage. Another attribute of a microphone's directivity is its *discrimination* against indirect sound: in practice, this too is expressed as a ratio, here describing the proportion of the total solid angle over which the microphone is effectively sensitive to sound. Again it is 1:1 for omnidirectional microphones; it rises to 3:1 (reverberation is 4.8 dB down) for both cardioid and bidirectional pick-up, and is higher (a little over 4:1 with 6 dB rejection of reverberation) for a response halfway between cardioid and bidirectional.

The combination of the pressure and pressure-gradient principles in different proportions is useful for cutting the amount of reverberation received and for rejecting unwanted noises. Some microphones can be switched to different polar diagrams; others are designed for one particular pattern of response. When choosing between different field patterns, note that the breadth of pick-up narrows as the pressure-gradient component gets stronger. A *hypercardioid* (also called *cottage loaf*) pattern is most evident in the range where the discrimination against ambient sound is at its greatest; the term 'unidirectional' may be applied to the range between cardioid and hypercardioid.

Highly directional microphones. These are substantially dead to sound from the side or rear. They are characterized by their physical size. A large parabolic dish can be used to concentrate sound at a microphone placed near its focus, but more commonly a long 'interference' tube extends forward from the microphone capsule. For both, at high frequencies the response is concentrated in a narrow forward lobe. This degenerates into a broader response at wavelengths that are greater than the major dimension of the microphone. Even more strongly directional is an 'adapted array' microphone in which a digital signal processor compares the output of five capsules (one with an interference tube).

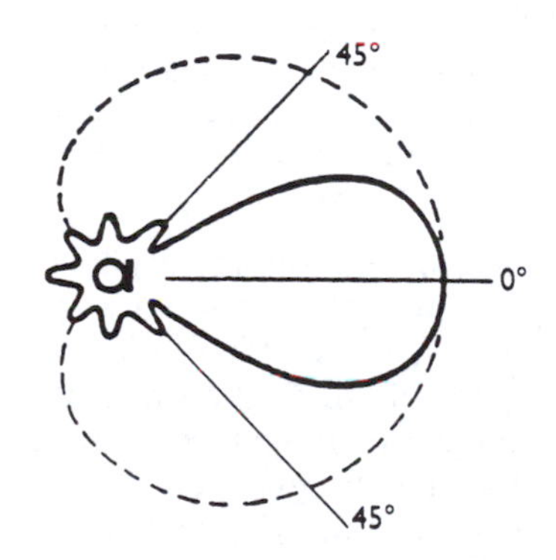

[5.7] **Highly directional microphone** The unbroken line shows a polar response for medium and high audio frequencies. At low frequencies (broken line) the response degenerates to that of the microphone capsule (in this example, cardioid).

The frequency response of practical microphones

Practical, high-quality professional microphones do not have a response that is perfectly flat. In older microphones peaks and troughs of up to 34 dB were common. This was not just bad workmanship: at one time it was difficult to design a microphone for a

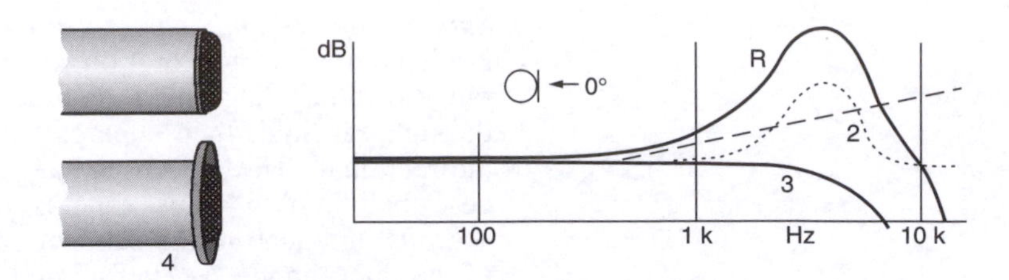

[5.8] **High-frequency effects on a large diaphragm**
The response R at 0° (i.e. on-axis) is the sum of several components including: 1, Obstacle effect (which sets up a standing wave in front of the diaphragm). 2, Resonance in cavity in front of the diaphragm. 3, Inertia of the diaphragm as its mass becomes appreciable compared with high-frequency driving forces. 4, This effect can also be created on a microphone with a smaller diaphragm by fitting a ring-baffle around the tip, and can be increased further by switching in high-frequency shelving.

level response at all audio frequencies, and even today some deviations may be helpful. In addition, an individual sample may vary by 2 dB or so from the average for its type. Fortunately, few people notice variations of this order: in a reverberant balance they are easily overshadowed by variations due to the acoustics. Sometimes a high-frequency peak is permitted on-axis, to allow for a more level response at a known angle to it.

Some microphones have a response that falls away in the bass enough to make them unsuitable for particular orchestral instruments (and certainly for a full orchestra). Much more common is loss of top, from about 12 kHz on some microphones. Others have a high-frequency response that goes well beyond this, but with these the angle at which the microphone is used may be critical: for some, to obtain the most even response the axis of the diaphragm should point directly at the sound source. A common defect – a serious one for high-quality work – is an erratic response in the upper-middle range. Except in individual cases where a continuous, measured trace of the frequency response is provided with each microphone, manufacturers tend to publish data and diagrams that somewhat idealize the performance of their products. This defect may not be apparent from publicity material.

The size and mass of a microphone (such as a moving coil) affects its frequency response. In general, the larger the diaphragm and the larger the case, the more difficult it is to engineer a smooth, extended top response (but the easier it is to obtain a strong signal). However, even though smaller-tipped moving coils are available, there has been a resurgence of interest in the older 'peaky' high-frequency models.

For extended, smooth high frequencies, condenser microphones have particular advantages. The diaphragm has an extremely low mass, so it responds readily to minute changes of air pressure. Further, because it must have one stage of amplification near the microphone capsule (in order to change the impedance before the signal is sent along the cable), the diaphragm itself can be small. The precision engineering and head amplifier add to the cost, but on account of its quality (especially in its frequency response) and sensitivity, it is widely used in studios. The weakest link may be the head amplifier: at high volumes, while the capsule is still within its working limits the amplifier may overload. For this reason many condensers have switchable attenuation (a 'pad') on the microphone. Always remember to switch it off again after use.

[5.9] **Ribbon microphone: live arc**
This older ribbon microphone has a ribbon that is 64 mm long by 6.4 mm wide ($2\frac{1}{2}$ in × $\frac{1}{4}$ in). Arcs of approximately 100° at front and back provide two 'live' working areas. The frequency response at 60° to the axis is halved, but begins to depart from the common curve at about 2000 Hz, where high-frequency effects begin. For a ribbon that is in a vertical line, those for variation in the vertical plane (v) are more pronounced than those for the horizontal plane (h).

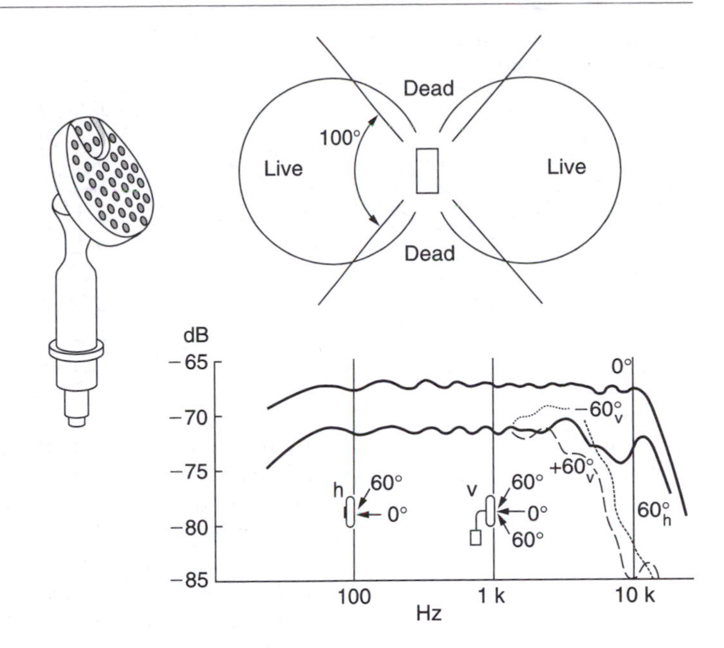

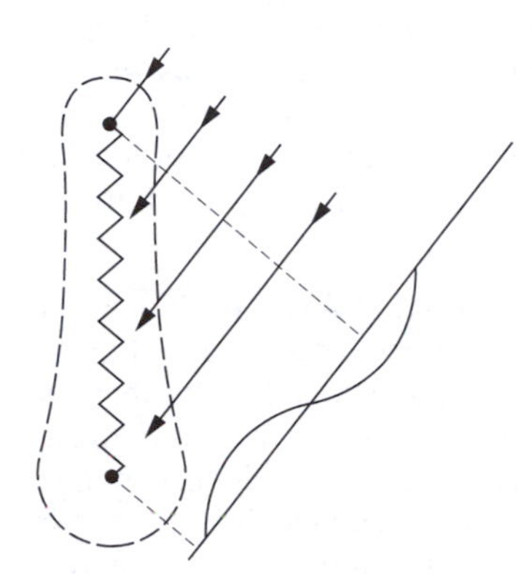

[5.10] **Ribbon microphone: vertical response**
There is partial cancellation of very short wavelengths (i.e. high frequency) sounds along the length of the ribbon.

Proximity effect

The pressure-gradient mode of operation, by its very nature, produces a distortion of the frequency response called *proximity effect* or *bass tip-up*: when placed close to a source the microphone exaggerates its bass. The distance at which this begins to set in depends on the difference in path length as the signal reaches first the front then the rear of the diaphragm. Bass tip-up occurs when the sound source is close enough for the path difference to be a significant proportion of that distance, and is due to the fact that, as it radiates outward, sound dies away. For spherical waves it diminishes in proportion to the square of the distance. It is most noticeable on voices (including song).

All directional microphones that employ a pressure gradient are affected to some degree. Cardioid and hypercardioid microphones begin to show this effect, and a pure bidirectional response shows it

[5.11] **Smaller ribbon design**
The 25 mm (1 in) ribbon (1) and pole pieces (2) are enveloped by a spiral of soft iron (3), which acts as a magnetic return path. Its smaller size increases the high-frequency response, reduces bass tip-up and lowers the resonant frequency of the foil (to 10–25 Hz, damped further by the very narrow spacing between foil and pole pieces). The almost spherical shape of the magnet system allows simpler protection from wind effects (4).
A substantial transformer (5) is still required. The high-frequency response (6) is extended, but still rolls off a little (at about 12 kHz).

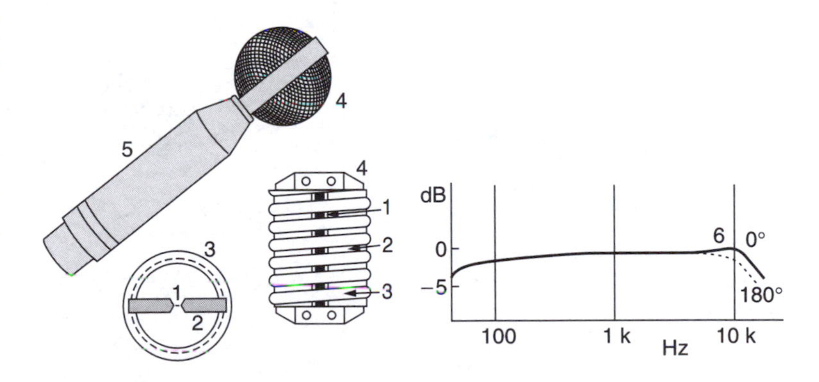

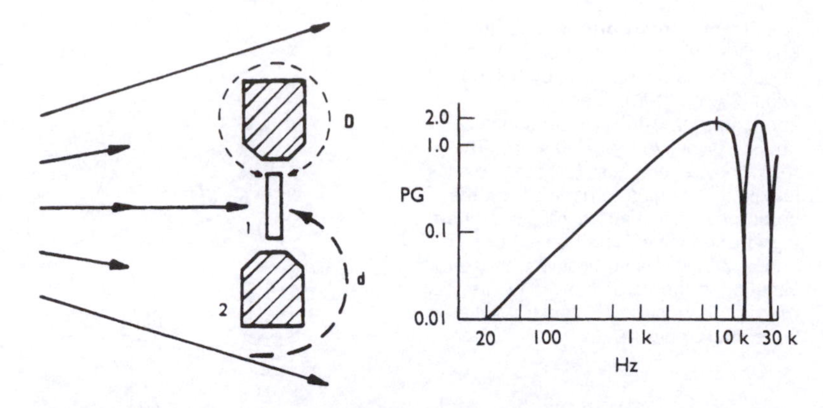

[5.12] **Ribbon microphone: dimensional effects**
Left: The sound wave travels farther to reach the back of the ribbon (1) than it does to the front. The effective path difference *d* is roughly equivalent to the shortest distance *D* round the pole piece (2). *Right*: Variation of pressure gradient (PG) with frequency. For a large, old design of ribbon the effective path difference, *D*, from front to rear of ribbon is 2.5 cm (1 in). At this wavelength the pole pieces create a shadow, and the principle switches over to pressure operation.

more. Older (big) ribbon microphones offer the most extreme form of bass tip-up, as they also require a substantial path difference to achieve adequate sensitivity. In some cases the effect can be heard as far out as 50 cm (20 in), becoming increasingly marked as the distance is reduced. In modern, smaller ribbons it is heard at 25 cm (10 in), and is also heard (though less marked) on cardioids and hypercardioids at that distance. In all types, with careful use of distance, it can be used constructively.

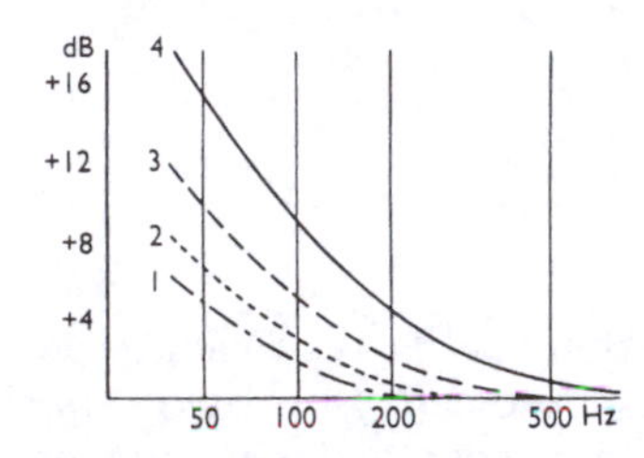

[5.13] **Proximity effect: variation with distance**
This increases as a directional microphone approaches a sound source. For a large bidirectional ribbon the curves 1–4 may correspond to distances of 60 to 15 cm (24 to 6 in); on more compact designs they are correspondingly shorter.

Bass roll-off can be introduced as equalization (EQ) later in the chain, but some directional microphones also have built-in switchable roll-off, which can be set to compensate for the increase that occurs at some particular working distance. However, any movement back and forth from this chosen position produces changes in the ratio between bass and middle frequencies that cannot be compensated by simple control of volume, and to make matters worse, such movements also introduce changes in the ratio of direct to indirect sound. If this becomes too intrusive it may be necessary to revert to an omnidirectional microphone.

At very low frequencies, another effect sets in: the output drops again because on a very long wave the difference in pressure is small.

When two bidirectional microphones are placed close together a check should be made that the two are in phase. If they are not, cancellation of direct sound occurs for sources equidistant from the two (although normal reverberation is heard) and severe distortion may be apparent for other positions of the source. The cure is simply

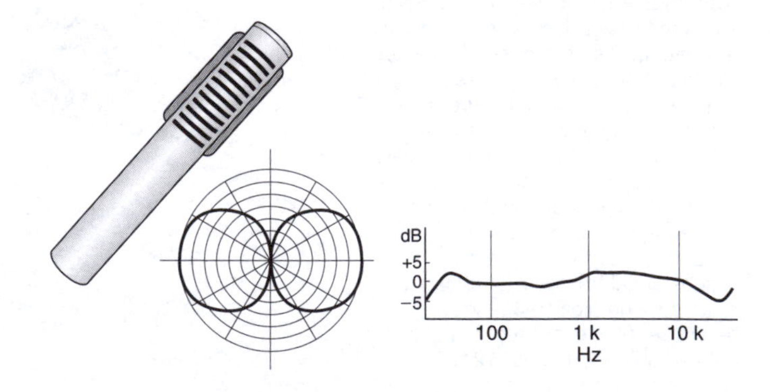

[5.14] **Compact ribbon microphone**
Much smaller than their forebears, later-generation ribbons require precision engineering. In some, an iron casing is the magnetic flux return path. The mellow bass and HF presence of this design make it suitable for pop music instruments. One version has a head amplifier in place of the more usual transformer.

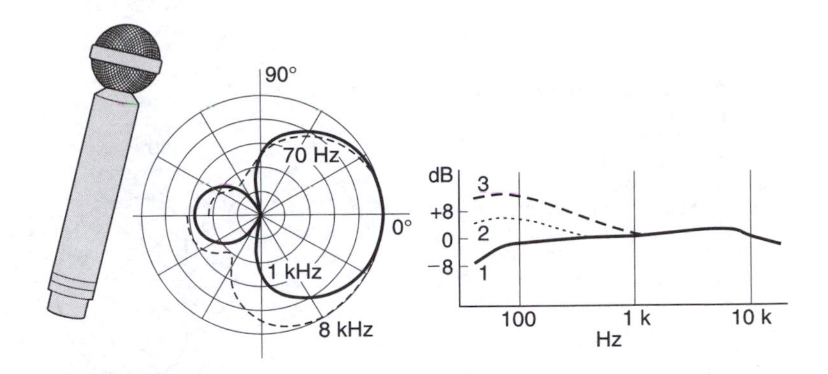

[5.15] **Double ribbon hypercardioid microphone**
Diameter of head: 38 mm (1½ in). Its end-fire response (with the band across the head) can be used as the M component of an MS pair. With gentle HF roll-off, and also at LF when set at 1 m or more, it is suitable for piano balance and several other instruments. 1, Response at 1 m. 2, Bass tip-up at 10 cm. 3, Increased bass at 2 cm.

to turn one of them round (or to reverse its leads). When bidirectional microphones are used to pick up two separate but fairly close sources, if they can be placed so that each is dead-side-on to the other source, this allows better control in mixing.

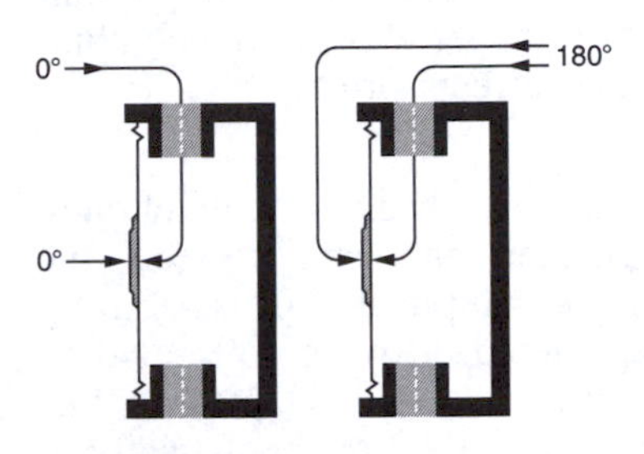

[5.16] **Condenser microphone with directional response**
For sound from the front, the path through the delay port is longer, so there is some pressure gradient effect. For sound from the back, the paths to the front and rear of the diaphragm are the same, so there will be some cancellation.

Cardioid and hypercardioid microphones

With true cardioid microphones the useful angle is a broad cone including at least 120° on the live side – but which may be taken as extending to a total included angle of about 180°, although so far round to the side the overall response is noticeably lower and in some designs the high-frequency response may be lower still. Continuing round to the back of the microphone, the output should in theory simply diminish gradually to nothing, but in practice, though at low level, it may be uneven, with an erratic frequency response.

The cardioid's coverage is much broader than the figure-of-eight, so it can be used where space is limited. When working sufficiently close there is the usual tendency to bass tip-up: in fact, it is half as much as with a similar figure-of-eight microphone. This reflects the combination of polar responses that makes a cardioid.

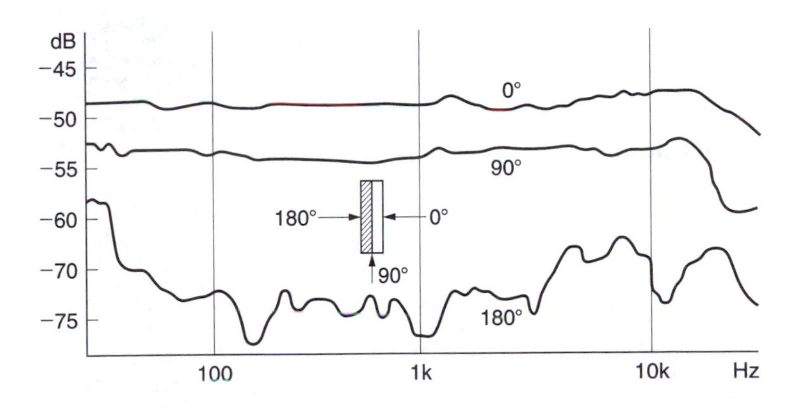

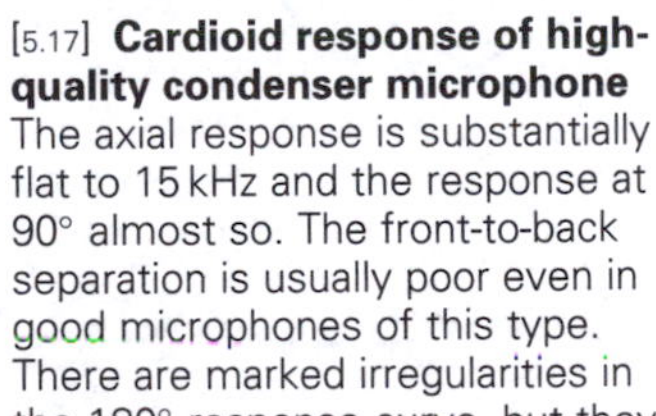

[5.17] **Cardioid response of high-quality condenser microphone**
The axial response is substantially flat to 15 kHz and the response at 90° almost so. The front-to-back separation is usually poor even in good microphones of this type. There are marked irregularities in the 180° response curve, but they are at a low level and therefore generally unimportant.

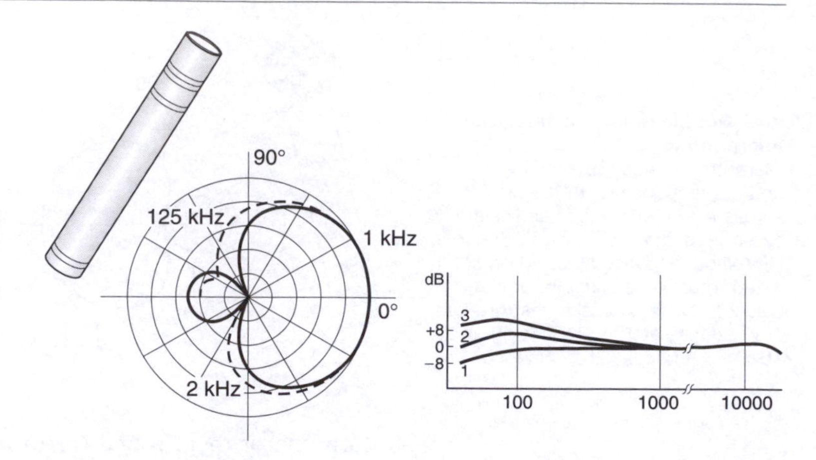

[5.18] **Hypercardioid speech studio microphone**
A moving-coil 'workhorse' of modest cost. With a diameter of 24 mm (1 in), the low mass diaphragm and coil allows a good transient response. In a radio studio several may be used, with different-coloured windshields for easy identification, one for each guest speaker. A good working distance for speech is 30 cm. 1, Frequency response at 1 m. 2, Bass tip-up at 10 cm. 3, Tip-up at 2 cm. Also used for snare drum, toms, hi-hat and percussion.

The most obvious way of producing a cardioid, by connecting omni- and bidirectional units in parallel, is no longer used, as the combination of different types with disparate effects produces an erratic response.

A better method is to have both front and back of a diaphragm open to the sound pressure, but to persuade the signal to reach the back with a change of phase. For this a complicated phase-shifting network of cavities and ports is engineered. Moving-coil, condenser and even ribbon microphones have all been adapted to this principle. Systems of this kind can be described mathematically in terms of acoustic resistances, capacitances and inductances. Using these, the pressures on both sides of the diaphragm can be calculated for a range of frequencies arriving from different directions, so that working back from the desired result, for example a cardioid response reasonably independent of frequency, it has been possible to design appropriate acoustic networks. However, it is difficult to engineer a really even response in such a complex design.

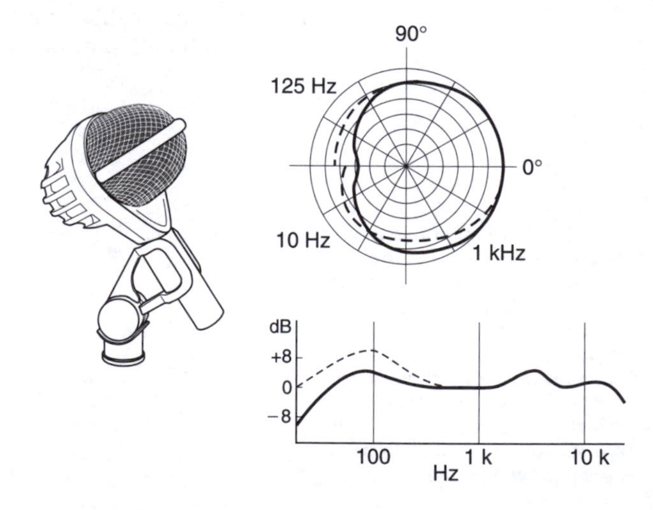

[5.19] **Rugged kick-drum microphone**
A cardioid moving coil designed to withstand high sound pressure levels with low distortion. For a close balance on bass drum and bass guitar amps the frequency response rises at 100 Hz and there is a presence peak (suitable for both) at 2–3 kHz. It can also be used for deep, powerful brass such as tuba and trombones.

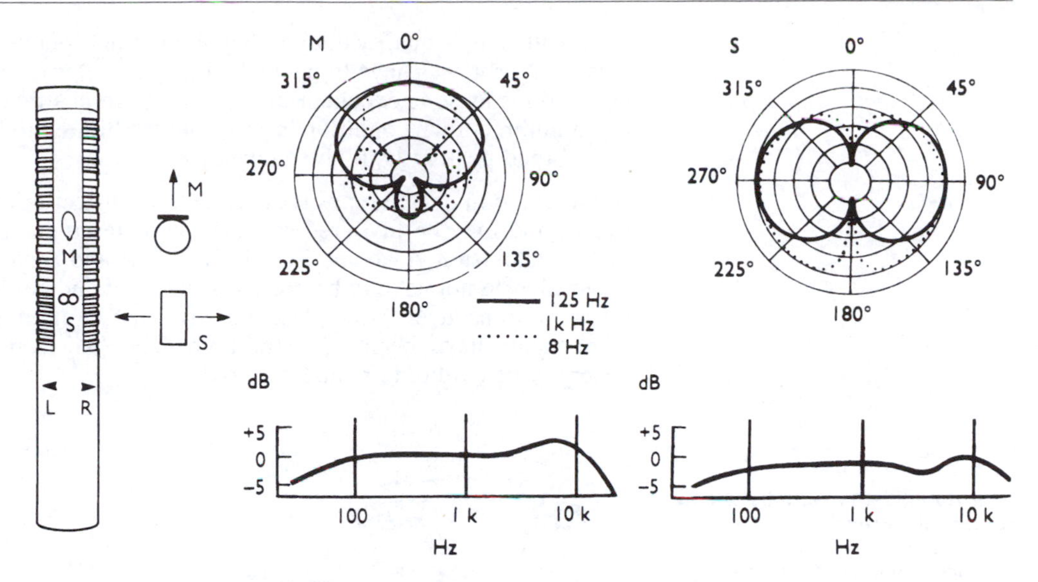

[5.20] **MS microphone** (*above*)
This microphone combines capsules employing two different principles in a single housing. The forward M or middle microphone is end-fire; behind this is a bidirectional S or side microphone. The M response has a 'presence' peak that is suitable for speech or actuality; that for S is flatter, but matches the effect of the peak. Stereo is obtained by combining the two signals.

Switchable microphones

In a switchable condenser microphone one diaphragm is constantly polarized and the other can have its polarization changed. If the two diaphragms are polarized in the same sense the microphone operates as a pressure capsule, as would two single-diaphragm cardioid condenser microphones placed back-to-back and wired to add outputs. But if the polarizing current on one diaphragm is decreased to zero (and the centre plate has suitable perforations) the polar response becomes cardioid.

Taking this a stage further, if the polarizing current is increased in the opposite sense, so the voltage on the centre plate is intermediate between those of the two diaphragms, the mode of operation becomes fully pressure gradient and the output bidirectional. This is very similar to the symmetrical layout in which a central diaphragm is balanced between two oppositely polarized perforated plates.

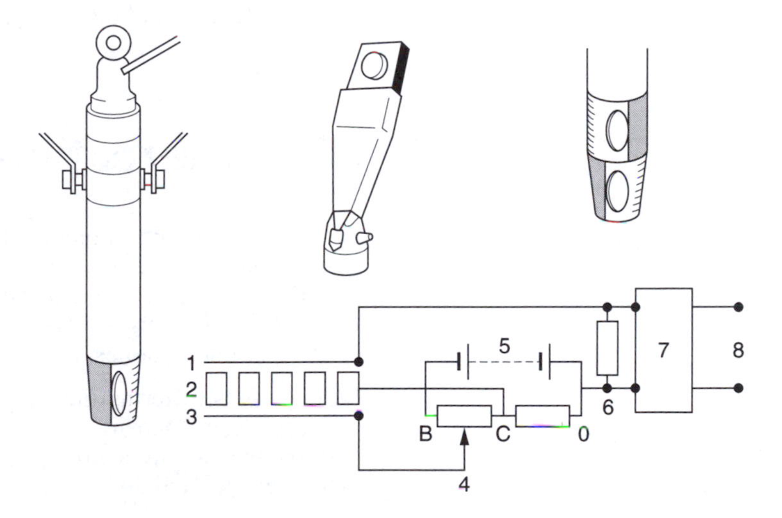

[5.21] **Switchable condenser microphone**
Left: The early version of a high-quality microphone with variable polar response and, *right centre*, a later, smaller version that replaced it. Its side-fire operation is more obvious. The response is varied by changing the polarizing voltage to one diaphragm. 1, Front diaphragm. 2, Rigid centre plate (perforated). 3, Rear diaphragm. 4, Multi-position switch and potentiometer. 5, Polarizing voltage. 6, High resistance. 7, Head amplifier. 8, Microphone output terminals. When the polarization is switched to position 0, the voltage on the front and back diaphragm is the same and above that of the centre plate: the capsule operates as an omnidirectional (pressure) microphone. At B, the capsule measures pressure gradient and is bidirectional. At C, the polar response is cardioid. *Far right*: Two such capsules in a single casing, making a coincident pair.

Even with a high-quality double-diaphragm microphone of this type, some compromises have to be made. In a particular microphone, the cardioid response is excellent, the intermediate positions useful and the omnidirectional condition is good, but the bidirectional condition has a broad peak in its high-frequency response.

A combination of two types is acceptable for direct middle and side stereo, and may be favoured for television stereo sound, including that for television video and film. Unlike an A and B pair, the M and S capsules do not have to be precisely matched and could even have different transducer principles, although more often electrostatic components have been used for both. The 'M' component will generally be cardioid or supercardioid.

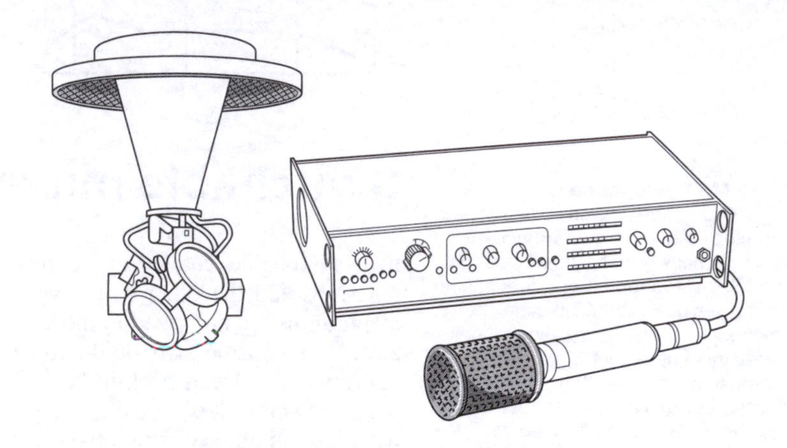

[5.22] **Soundfield microphone and its capsule array**
The variable polar response is switched remotely, as are the azimuth (horizontal response) and the angle up or down. A control for 'dominance' narrows the response, giving a zoom effect, within the usual limits (i.e. there is no highly directional response available). The signals from the four capsules can be recorded separately, and then fed back through the control box so that the directional response can be modified later.

A more exotic combination microphone has four identical condenser capsules, set close together in a tetrahedral arrangement. Named the Soundfield, its great flexibility in use allows a wide range of applications. For example, its response can be altered during a live event (or later, on a four-channel recording from it) to favour a source from some unexpected direction, which may be outside the plane set for a standard stereo microphone pair.

Highly directional microphones

For a more highly directional response, a microphone capsule (which may have any of the characteristics described above) is placed within an enhanced or focused sound field. The most common way of achieving this is by the *interference* method employed in *gun* or *shotgun* microphones, also called *line microphones*.

Early designs (offering the simplest way of understanding the principle) had many narrow tubes of different lengths forming a bundle, with the diaphragm enclosed in a chamber at the base. When these tubes were pointed directly at a subject, the sound pressures passing

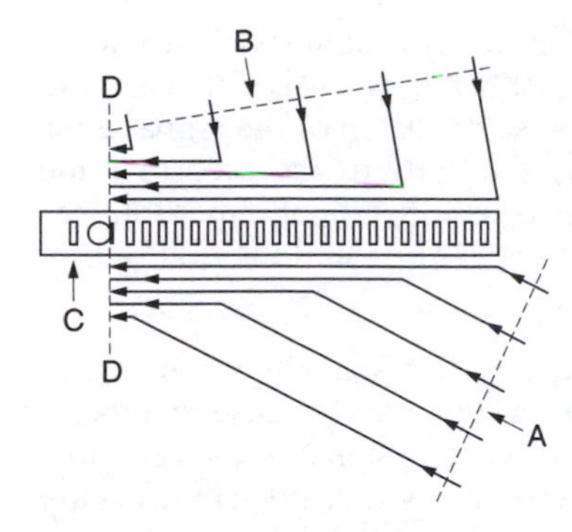

[5.23] **Gun (US: shotgun) microphone: interference effect**
Wave A, approaching from a direction close to the axis, reaches the microphone front diaphragm D by a range of paths that differ only slightly in length. Cancellation occurs only at very high frequencies. Wave B reaches the diaphragm by a much wider range of path lengths. Cancellation is severe at middle and high frequencies. Undelayed sound reaches the back of the diaphragm via port C. The polar diagram of the capsule without the acoustic delay tube in front is cardioid; with the tube it becomes highly directional.

along the different tubes all travelled the same distance, so the tubes had no significant effect on the sound. But sound that approached from an oblique angle divided to follow paths of different lengths, so when the pressures recombined in the cavity before the diaphragm, differences in phase caused cancellation over a wide range of frequencies. The further round to the side, the more extreme this cancellation became, until for sound from the rear the path lengths varied by up to twice the length of the 'barrel' of the 'gun'. But there was, and still is, a limit to effective cancellation and therefore to the directional response itself. The length of any gun microphone gives an immediate visual indication of the longest wavelength (and so the lowest frequency) at which its directional properties are effective.

In today's gun microphones the many tubes have been replaced by a single narrow barrel, to which the sound has access at many points along the length. The phase cancellation effect is the same.

Another early, extreme example of a 'rifle' microphone had a tube 2 m (6 ft 8 in) long. As an acoustic engineering aid it has been used to pinpoint the sources of echoes in concert halls. Similar microphones have in the past been used for picking up voices at a distance: for example, reporters in a large press conference. But their operation could be aurally distracting, as it was very obvious when they were faded up or down, or panned to find a person who was already speaking. The speech it picked up had a harsh, unpleasant quality – but was, at least, clearly audible.

Today, commonly-used gun microphones have interference tubes about 46 cm (18 in) or 25 cm (10 in) long, each facing a unidirectional condenser capsule. The off-axis cancellation is still limited to shorter wavelengths, with the switchover depending on length of tube. At wavelengths longer than the tube, the directional response reverts to the first-order effect of the capsule itself, with little loss of quality. Gun microphones are widely used by film and television recordists for location work. They are sensitive to low-frequency noise from the sides, but if bass-cut can be switched in, traffic rumble is reduced. They are supplied with a pistol grip that is attached below the microphone, so they can be pointed by hand.

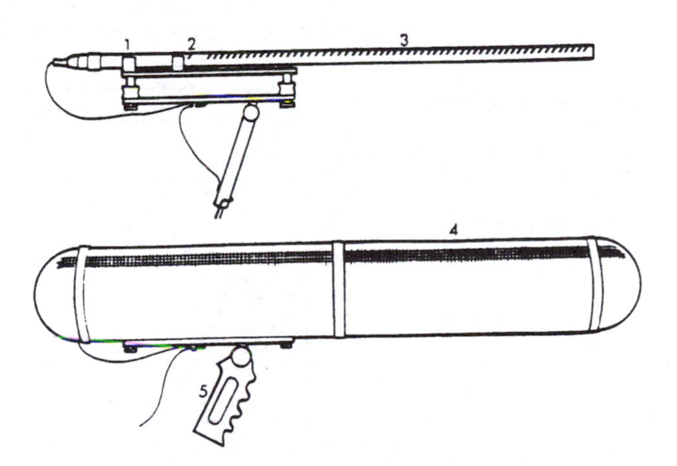

[5.24] **Gun microphone**
1, Housing for head amplifier. 2, Condenser capsule (cardioid operation). 3, Acoustic interference tube (with entry ports along upper surface). 4, Basket windshield for open-air use (a soft cover may be slipped over this for additional protection in high wind). 5, Hand grip.

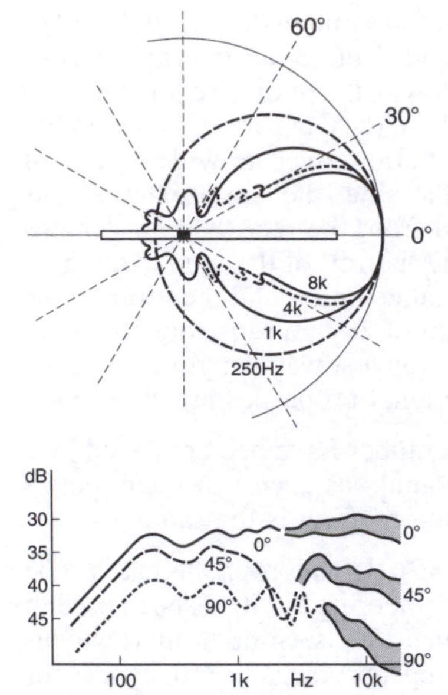

[5.25] **Response of gun microphone** This has a 45 cm (18 in) interference tube. For high frequencies the cone of acceptance encloses 30°; movement outside this will pass through regions of marked and changing coloration. The axial response is reasonably level, with some roll-off. One version has a broad peak in the range above 2 kHz to allow for air losses on distant sources; this can be switched to a flatter response for closer sources or where sibilance is noticeable.

A mesh windshield, often wrapped in an additional soft fleece, is used for outdoor work, but its bulk makes it obtrusive if it gets in vision. A strong, gusting wind can cause break-up of the signal even when the windshield is protected by the body of the recordist and other windbreaks. An alternative is to use 'personal' microphones (described later) but wind can afflict those too – and the gun microphone offers better quality outdoors.

When used indoors or in any acoustically live space the gun microphone suffers partial loss of its directional qualities because reverberation is random in phase, so is not subject to systematic cancellation. Indoors, therefore, the response may be no better, nor significantly worse, than that of the capsule itself. Unless the microphone is brought close to the subject (which defeats the object of using the 'gun') off-axis sound is likely to be unpleasantly reverberant. In television or film studios with dead acoustics it can be slung on a boom and set farther back than a first-order directional microphone, so it is useful when a wide, clear picture is required. Its discrimination in favour of high-frequency sound may improve intelligibility.

The short (25 cm, 10 in) gun can also be used indoors as a hand microphone. Here it may reintroduce an older problem of directional microphones, proximity effect, although in this design it is controlled in part by the distance between mouth and microphone that is created by the length of the tube, and switching in the appropriate bass roll-off can compensate the remainder. Some examples are sold in a kit that permits the use of the capsule either by itself as a condenser cardioid or with an interference tube. One of these is based on an inexpensive electret with a tiny head amplifier powered by a long-life battery.

An earlier method of achieving a highly directional response was the *parabolic reflector*. The principle whereby this achieves its special properties is obvious; equally clearly this, too, degenerates at low frequencies, for which the sound waves are too long to 'see' the dish effectively (they diffract round the outside instead of being reflected within it). For a reasonably manoeuvrable reflector of 1 m (39 in) diameter the directional response is less effective below 1000 Hz. Also, the microphone is subject to close unwanted noises at low frequencies unless a bass-cut is used.

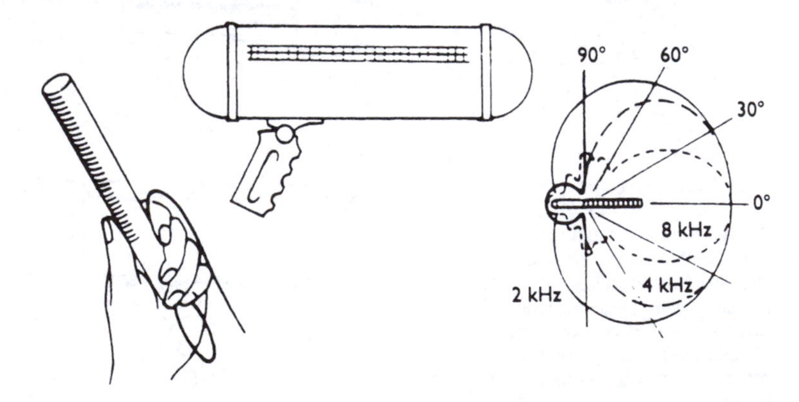

[5.26] **Short interference microphone** Overall length about 25 cm (10 in). At 2000 Hz and below, the response is cardioid. The length of the tube controls the minimum distance from capsule to mouth, so that there is little proximity effect (bass tip-up) even when the tip is held close to the mouth. Cancellation in the short interference tube makes the microphone highly directional only at high frequencies. It has excellent separation from surrounding noise when hand-held by solo singers or reporters, or when used as a gun microphone for film sound recording.

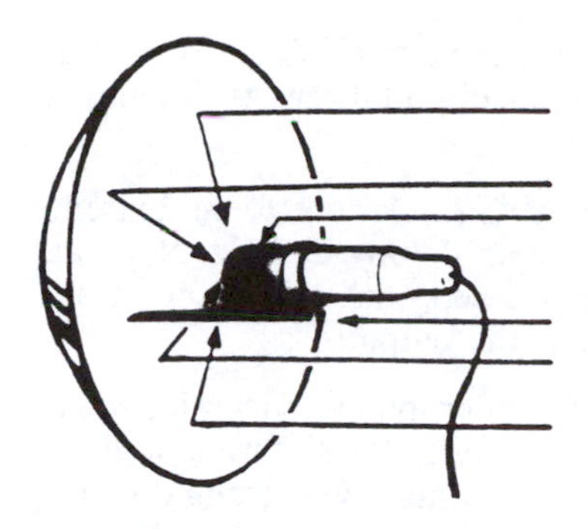

[5.27] **Microphone in parabolic reflector**
The bowl may be of glass fibre or metal (coated on the outside with a material to stop it 'ringing'). Alternatively, if it is made of transparent plastics the operator can see through it to follow action, e.g. at sports meetings. In another design the reflector is segmented and can be folded like a fan for travel and storage.

For picking up faint sounds, the parabolic reflector has one great advantage: its directional properties rely on acoustic amplification of the sound before it reaches the microphone, achieving perhaps a 20 dB gain. This offers a far better signal-to-noise ratio than could be obtained from a microphone that has a similar capsule but employs phase cancellation. The reflector is therefore ideal for recording high-pitched birdsong and should be considered for any specialized purpose where its bulk and poor low-frequency pick-up are acceptable: one sports use has been to pick up contact noise (the clash of helmets) in American football. It is of less use for the football (soccer) played elsewhere, as it does little for the lower pitched thump of boot on ball.

In a reflector, at high frequencies the sound can be concentrated to near-pinpoint accuracy, or defocused by moving the microphone along the axis to give a broader pick-up lobe: the latter is often preferred. An advantage over gun microphones is that reflectors work just as well indoors as out. Tested in highly reverberant conditions (an underground garage), a parabolic reflector picked up intelligible speech at twice the distance that could be achieved by the better of two gun microphones.

To allow for the contribution from the rim, the microphone capsule should have a cardioid response, pointing into the bowl. But listen for phase cancellation at wavelengths related to the dish diameter, for which the reflected signal is weak and out of phase with direct sound.

It is possible to avoid the limitations on frequency response imposed by the dimensions of the reflector or tube by resorting to more complex technology: a combination that has an array of four cardioid electrets plus a fifth operating as a gun microphone projecting from the centre of the array and a digital signal processor (DSP) to compare and combine their output. Inevitably the result will be heavier and more expensive than standard highly directional microphones, but there is a big jump in performance. The pick-up is far more sharply focused, with a 20° lobe that with the help of a DSP is available over a 200–2000 Hz working range, which makes it far more versatile than other highly directional microphones. Curiously, the angle of response opens out to approach 60° at 4 kHz and above, so at these frequencies it operates more like a normal gun microphone.

An array of microphones (like an array of loudspeakers) has a directional response: a row of capsules will cancel signals coming from the side of the array, or in the case of a plane array, from directions that are close to this plane. In this 'adaptive array' microphone the array forms a diamond pattern that is little more than 50 mm (2 in) across, so its unenhanced directionality would be limited to very high frequencies – however, that is not how it works; instead, it is used as a feed to the nearby DSP to radically modify the signal from the standard short gun microphone set at the centre of the diamond pattern. It can also be switched so that only two (the top and bottom) of the surrounding capsules are used, and with this arrangement the response opens out to a horizontal lobe of about 20° × 60°. Obviously, by rotating the array through 90° the lobe

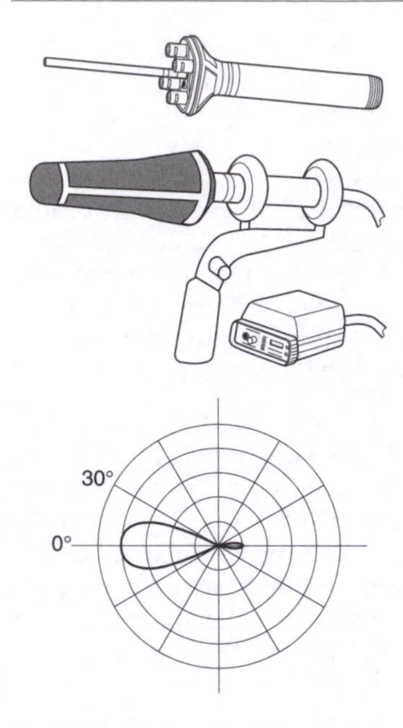

[5.28] **Adaptive array microphone** At about 35 cm (14 in), this is halfway between the short and long gun microphones in length, and 7 cm (2¾ in) across around the capsule array. Weight: 475 g (16¾ oz) plus almost as much again for pistol grip shockmount and windshield (alternatively, it can be attached to a pole). The digital control pack and batteries (500 g, 17½ oz) will hang on a belt. The mid-range polar response is much narrower than that of a gun microphone.

turns with it, so it is then broader in the vertical plane. If all four surrounding capsules are taken out of the circuit it reverts to standard gun microphone operation.

The DSP also has two built-in switchable filters, one to cut extreme low frequencies, the other to pass a range of 300–5500 Hz. Note that although the signal is digitally processed, the operation of the capsules is (as always) analogue, as is the output.

A huge advantage of the adaptive array microphone is its rejection of off-axis ambient noise (cutting it by about 70 dB) throughout the mid-frequency range. In a sports stadium, this gives strong discrimination against crowd noise, so it can pick up the thump of boot on football cleanly, or alternatively the voice of the manager on the sidelines. In a rodeo it separates out all of the action noises. It also allows a shore-based microphone following a rowboat race to hear oars cutting into the water, but virtually eliminates the far louder chase boat engines.

For speech, although most users would prefer a standard microphone for any normal balance, it can be hand-held for interviews. But the adaptive array comes into its own when used for full-frequency-range pick-up at a distance. It can match crowd noise to the actual section of crowd seen on a zoom lens, or (rather closer) will separate wanted dialogue from surrounding crowd chatter and noise – offering a real edge in crucial news operations that must be covered from a distance.

Noise-cancelling microphones

There are several ways of reducing noise: one is to put the microphone user, a sports reporter perhaps, in a 'soundproof' box. But such enclosures may not keep out the low-frequency components of crowd noise, and in any case can introduce unpleasant resonances if they are small. Another technique is to use an omnidirectional microphone close to the mouth, but in very noisy surroundings this may still not provide a good balance between voice and background, and if the noise is still too loud, there is no good way of correcting the mixture.

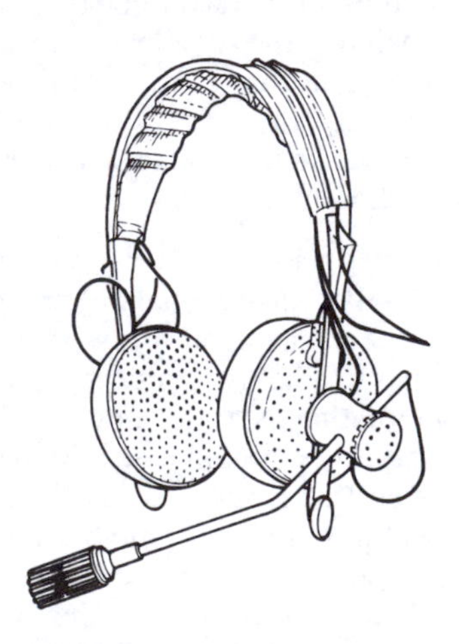

[5.29] **Headphone/microphone combination (out of vision)** Discrimination against noise is by distance.

An electret cardioid with its response corrected for use close to the mouth is much better. Some hand-held microphones have a switched response, so that they can be used either at a standard distance (60 cm, 2 ft) or very close (e.g. at 6 mm, ¼ in). Plainly, the user will need to control the distance in close use with some care, or the frequency response will be wrong. A head-worn high-quality electret microphone of this type can be fitted to the individual user, so that distance (35 mm, 1⅜ in) is controlled accurately.

In such cases, noise rejection depends on two factors. One is the extremely close position, which makes the voice much louder than ambient sound; the other is the microphone's natural bass tip-up

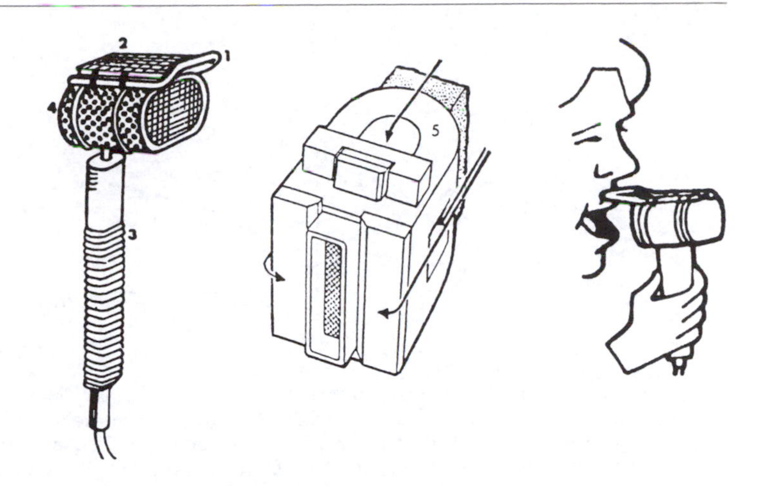

[5.30] **Lip-ribbon microphone**
1, The mouth guard is placed against the upper lip. 2, A stainless steel mesh acts as a windshield below the nose. 3, The handgrip contains a transformer. 4, Case containing magnet assembly. 5, Yoke of magnet is towards the mouth, and the ribbon away from it. The sound paths are marked.

(proximity effect) which, when compensated, reduces middle-to-low frequencies still further, as there is no bass tip-up on ambient sound arriving from a distance. Because bidirectional microphones have the most extreme bass tip-up, they must offer the greatest possible reduction: with its pressure-gradient operation, a ribbon is subject to extreme bass tip-up on sound from any very close source.

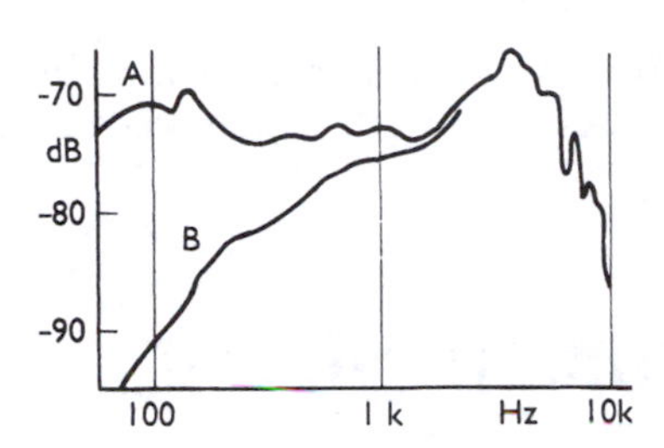

[5.31] **Response of lip-ribbon microphone**
A, Response to spherical wave (user's voice) at 54 mm (2$^1/_8$ in).
B, Response to plane wave (ambient sound), on-axis. This effect is achieved by equalization, without which B would be flat and A would be high at low frequencies, due to extreme bass tip-up at such a close working distance.

A design used successfully for many years by sports commentators, so that they can speak in a normal voice with the roar of the crowd on all sides, is a *lip ribbon*. Obviously, the closer to the mouth the ribbon is, the better – so long as the explosive sounds in speech and streams of air from the mouth and nose can be controlled. At a distance of 54 mm (2$^1/_8$ in) – fixed by placing a guard close to the upper lip – these problems can be reduced by windshields (which, as they are so close to the mouth or nose, need to be made of stainless steel or plastic) and by cupping the ribbon inside the magnet.

Some versions have a variety of equalization settings for different degrees of bass, but in general the reduction of noise is about 10 dB at 300 Hz, increasing to 20 dB at 100 Hz. High frequencies up to about 7 kHz are all that are needed for speech in these conditions, so above that the microphone's sensitivity falls away. And, unlike a cardioid, the figure-of-eight response discriminates against noise from the side.

The lip-ribbon design is not visually attractive because it obscures the mouth, but has proved effective in almost any conditions of noise yet encountered in radio, as well as for out-of-vision announcements

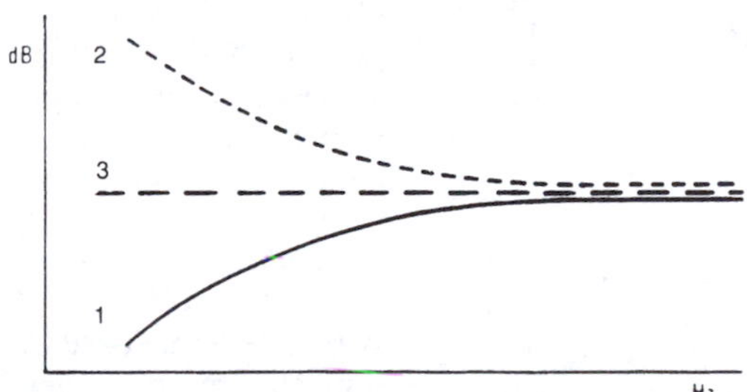

[5.32] **Compensation for proximity effect**
1, Some microphones have a reduced bass response that automatically compensates for the tip-up (2) due to close working. 3, With the microphone at a certain distance from the source, its response to direct sound is level.

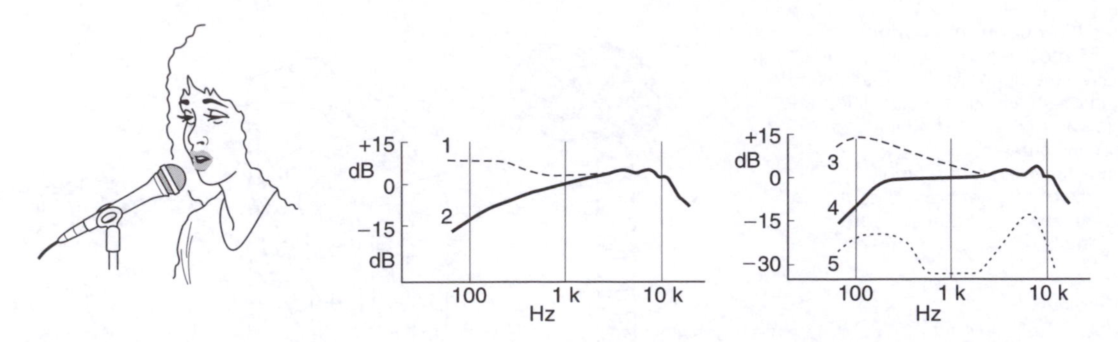

[5.33] **Noise-rejecting cardioid**
For use in vision, an electret microphone with built-in pop-shield that eliminates most noise, mostly by extreme close working, but also by further reduction of ambient low-frequency noise as bass tip-up is compensated. 1, Frequency response, on-axis, with mouth at 6 mm ($^1/_4$ in) from microphone. 2, Frequency response at 0° for sources 60 cm (2 ft) or more away. 3, Without bass compensation, on-axis at 60 cm (2 ft). 4, Working very close with compensation switched off, there is a big lift in the bass. 5, Response at rear of microphone at 60 cm (2 ft) or more.

over loud applause in television. For in-vision applications a more elegant cardioid, either in a hand-held or head-worn design, may be preferred.

Microphones for use in vision

Everything within a television or film picture, as much as everything in the sound, adds to the total effect. A microphone in the picture or a series of pictures for a long time is an important visual element. Accordingly, many modern microphones have been designed for good appearance as well as high-quality performance. They may be silver-grey or black, but always with a matt finish to diffuse the reflection of studio lights. If the capsule is not intrinsically small it should be slender, and performers will more readily understand how they should work to it if any directional properties it has are in line with the long axis (*end-fire*), so that in use it points towards them. A microphone looks better in the edge of frame at the bottom rather than intruding at the top. Hanging into the picture, it is untidy and attracts attention.

Table microphones. A pencil shape perhaps 1.5–2 cm (about $^3/_4$ in) in diameter has room for a directionally sensitive capsule pointing along the axis; the body of the pencil houses a condenser microphone's head amplifier. If the mounting is to be relatively heavy and stable and able to protect the microphone from vibration, that cannot also be made small – but it can be given clean lines or can be concealed from the camera in a sunken well in the table.

[5.34] **Flexible goose-neck connectors**
These are stiff enough to allow precise positioning of the microphone.

Microphones on floor stands. For someone standing up, a microphone may be on a full-height stand, which again should be strong, with a massive base. Many are telescopic. Whatever the design, in a long shot there is a vertical line very obviously in the picture, and the

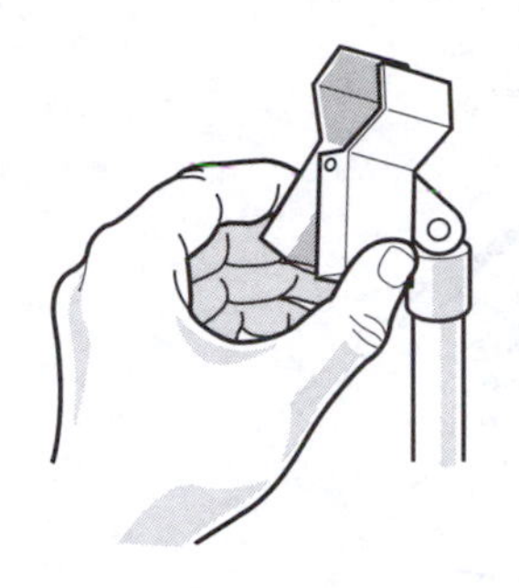

[5.35] **Spring-loaded clamp**
This requires two hands to release but is more secure than the easy-release clip.

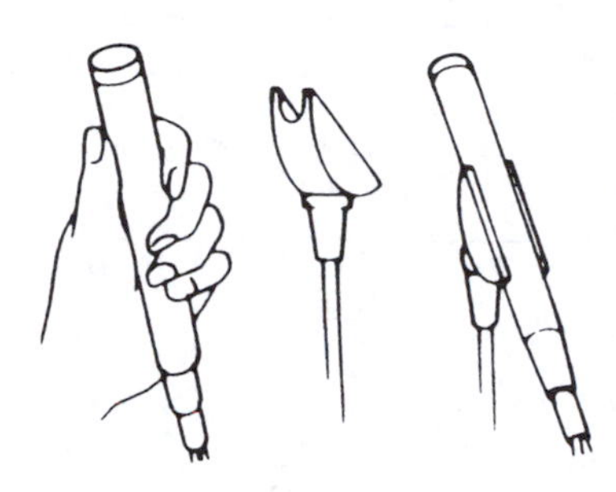

[5.36] **Easy-release clip on floor stand**
This padded clip allows the performer to take the microphone from the stand with one hand while in vision.

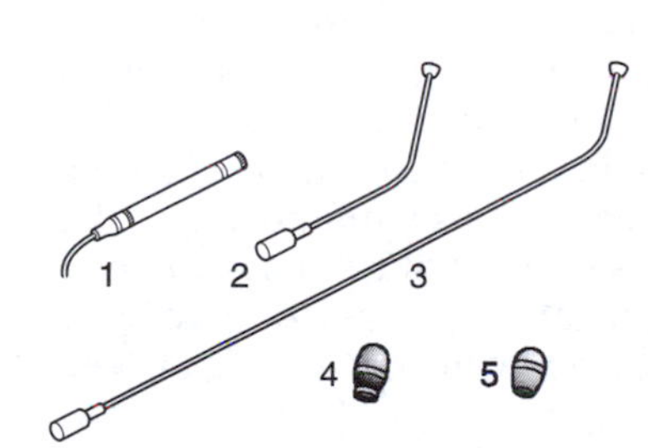

[5.37] **Condenser microphone kit**
This is supplied with alternative capsules for cardioid and omnidirectional operation. 1, Head amplifier and capsule. 2, 3, Extension pieces (these are fitted between head amplifier and microphone capsule). 4, 5, Windshields. *Right, above*: Response curves for omnidirectional head fitted directly to head amplifier (i.e. without the extension pieces). *Below*: Response obtained when cardioid head is used in combination with extension piece.

cleaner this is the better. Any clip that holds the microphone should be neat but easy to release by a performer in vision. Some microphones have only a condenser capsule at face level, plugged on a thin rigid extension tube that connects it to the head amplifier: the capacitance of the wiring is part of the design. Some microphone kits come with a choice of capsules for omnidirectional or cardioid response and different lengths of column, including a flexible 'goose-neck' section, so that the capsule can be angled precisely without the need for a separate arm.

Hand microphones. These generally have a 'stick' shape so they can be held easily. If its distance from the user's mouth cannot be predicted or controlled, variable proximity effects (bass tip-up) can be avoided by choosing one with a response that becomes omnidirectional at low frequencies. An additional advantage of this is that by its pressure operation at these frequencies the microphone is less sensitive to wind and breath noise. A hand microphone that becomes directional at middle or high frequencies should have an end-fire response (as it is natural to point the microphone toward the mouth) but the high-frequency pick-up lobe must be broad enough to allow a good range of angles. If the axis is angled more towards the mouth, there may be an increase in the high-frequency response. A skilled user may employ this effect creatively.

Robustness and lack of handling noise are essential, but sensitivity is less so: the microphone is often held close to the lips, which is visually intrusive but improves separation from other sound. To manage sound levels up to 135 dB it may need switched attenuation (a 'pad').

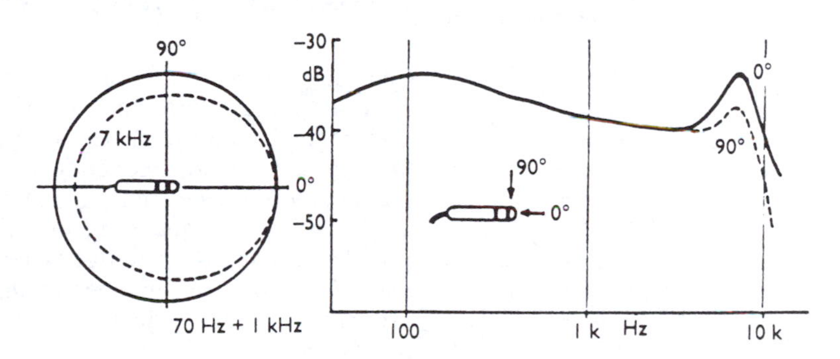

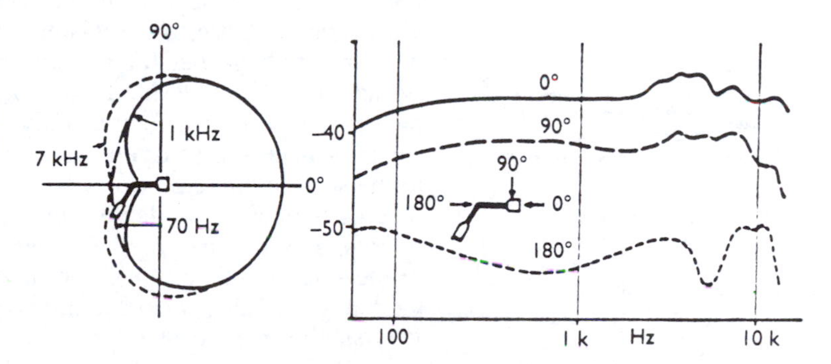

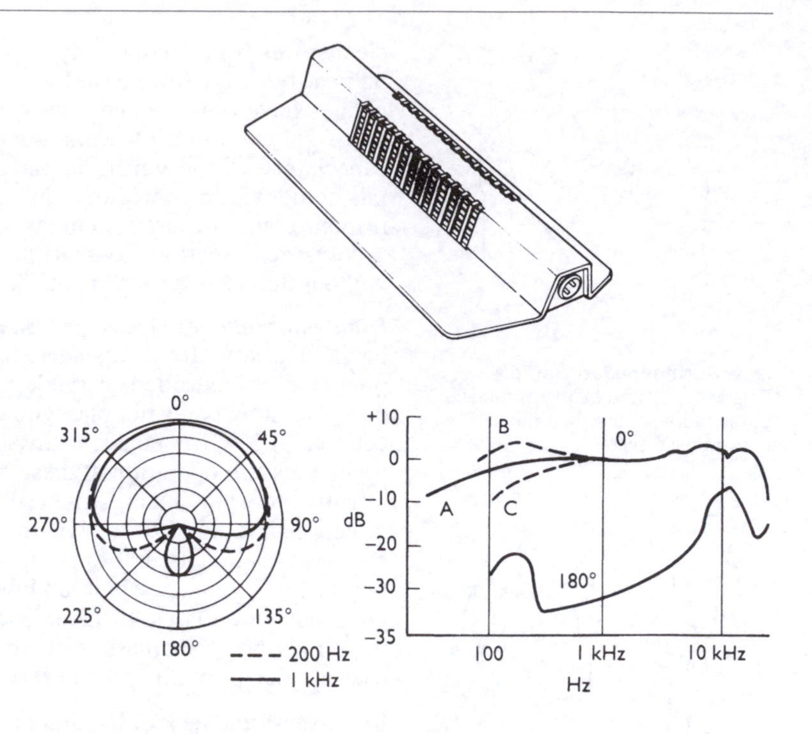

[5.38] **Pressure zone microphone with flanges**
The flanges may be fixed to a stage surface by tape of a matching colour, or by double-sided adhesive tape or adhesive putty. In this case the response is near-cardioid, and the response at low frequencies depends on the distance of the source. Spaced at intervals of 3–4 m (10–13 ft) along the apron of a stage, these have been used successfully for opera. The polar response (A – for a source at an angle of 30° above the stage) is unidirectional. There are controls to boost bass (B) and to cut it (C).

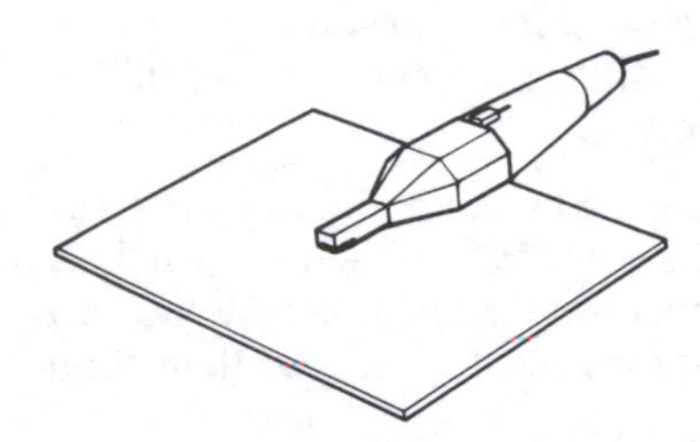

[5.39] **Pressure zone microphone on plate**
This is easily fixed to a surface by tape (on top) or double-sided adhesive (below). The tiny capsule receives sound through a gap that is close enough to the surface to avoid high-frequency roll-off within the audio range, then broadens out to allow for a conventional connector. The hemispherical polar response does not change with distance, a benefit when the distance of the source is unpredictable.

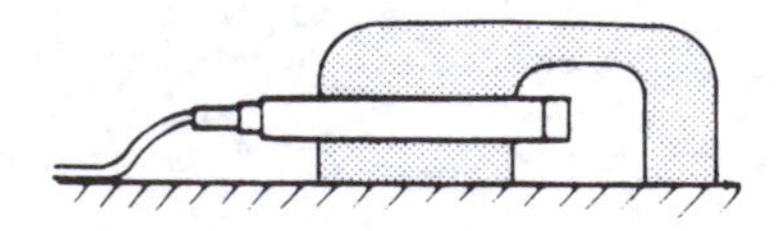

[5.40] **Microphone mouse**
Polyurethane foam is moulded with a tubular slot for the insertion of the microphone, and there is a cavity for the capsule, which must be very close to the floor.

Boundary microphones

Generally, microphones are kept away from reflective surfaces because of the interference fields these produce. However, if the diaphragm is moved closer to the surface this effect moves up through the sound frequency spectrum, until at a distance of 1.2 cm (½ in) it occurs only in the highest octave of human hearing, so that on speech and song the effect becomes unimportant. If it is moved even closer, the full range is restored. It becomes a *boundary microphone* or, in one commercial range, a *Pressure Zone Microphone* (PZM; which sounds better with the American 'zee' pronunciation).

For a sound wave approaching a rigid, solid surface at a right angle, a microphone close to it registers only pressure variations, not the pressure gradient, but for a sound travelling lengthways along the surface it can be measured by either method. A directional microphone is therefore effective only when it is parallel to it. This suggests an end-fire cardioid microphone with its diaphragm normal to the surface, directed towards action at a distance. Again, a commercial version is available, but an alternative is to put a standard microphone in a polyurethane pad with its diaphragm very close to the surface: this arrangement is called a *microphone mouse*. To avoid high-frequency cancellations, the microphone does have to be very small (or slender).

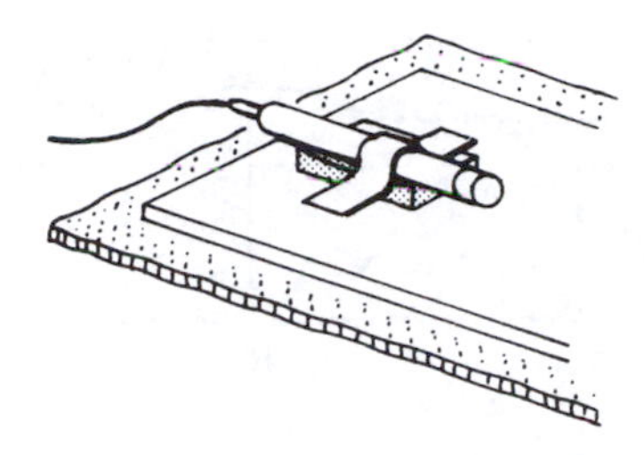

[5.41] **Reducing floor noise**
To sample the boundary layer close to a floor, the microphone may be insulated by a spongy pad and taped either to the floor or, if this radiates too much noise, to a board which is insulated from the floor by a piece of carpet or underlay, or both.

Commercial varieties can be engineered to be so flat that the greatest contribution to their thickness is the padding used to protect them from floor vibration. Some have pure pressure-operated capsules, so they are omnidirectional in principle, but in this housing have a hemispherical response. In all cases, the hard, reflective surface should, if possible, be clear for a metre or so (several feet) around.

Floor-mounted boundary microphones are in danger of picking up noise conducted along or through the floor. Near the edge of a stage, a simple way of minimizing directly conducted sound is to mount the microphone on a stand in an orchestra pit, with the capsule on a flexible swan-neck bending over the edge of the stage and angled almost to touch the surface. As before, the directional response will be horizontal. The final limitation will be noise radiated from the stage surface and travelling through the air to the microphone.

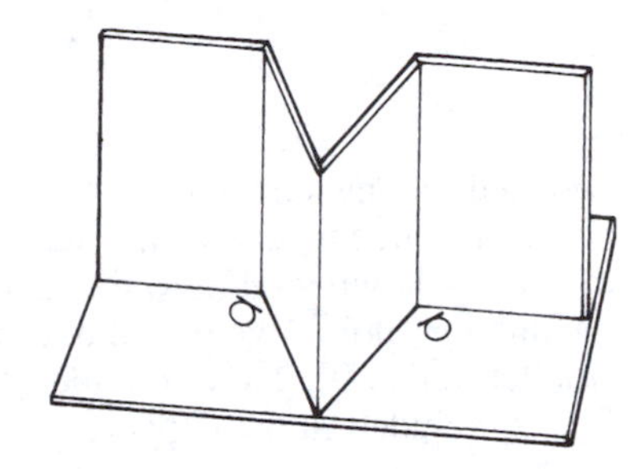

[5.42] **Floor-mounted boundary microphone stereo array**
One of many options for the application of the boundary principle in stereo, this home-made layout stands on a 60 cm × 60 cm (2 ft × 2 ft) base, with reflectors 30 cm (1 ft) high. The hinged 'V' forms two sides of a 20 cm (8 in) equilateral triangle.

Personal microphones

A *lavalier microphone* may be suspended round the neck in the manner of pendant jewellery – but rather less obtrusively. In this arrangement it may also properly be described as a *lanyard* or *neck microphone*. Alternatively, it may simply be called a *personal microphone*, a term with more general application, because the small capsules commonly used are more often simply attached to clothing. For uncompromised sound quality the microphone will be in vision; for neater appearance it can be concealed beneath a tie or some other light garment.

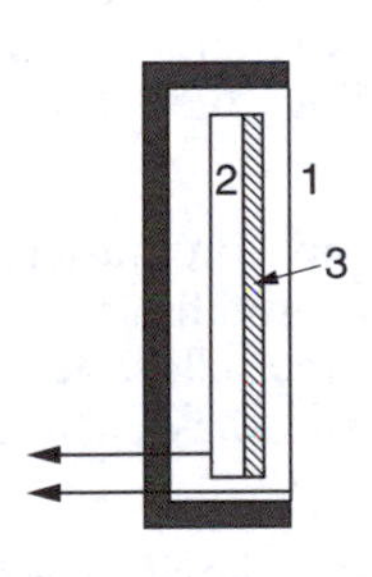

[5.43] **Electret: permanent electrostatic charge**
1, Metallic foil. 2, Conductive back-plate. 3, Coating of insulator with sealed permanent electrostatic charge. Here the coating is on the back-plate. It still works as an electret if the coating is on the foil instead, but this increases its mass, and therefore limits high-frequency response.

The microphone should be omnidirectional because, although it is situated fairly close to the mouth, it is not always at a predictable distance. Such microphones work reasonably well because the pattern of radiation from the mouth is also essentially omnidirectional, although there is some loss of high frequencies reaching the capsule, particularly if it is hidden in clothing. Some personal microphones have a raised high-frequency response; this can also be added or adjusted later in the chain by equalization. Other possible uses for a small unit of this type, e.g. held in the hand, or hidden in scenery to cover an otherwise inaccessible position in a set, require a level response.

Personal microphones must be very light in weight as well as being tiny, but also need to be robust, as they are easily bumped or

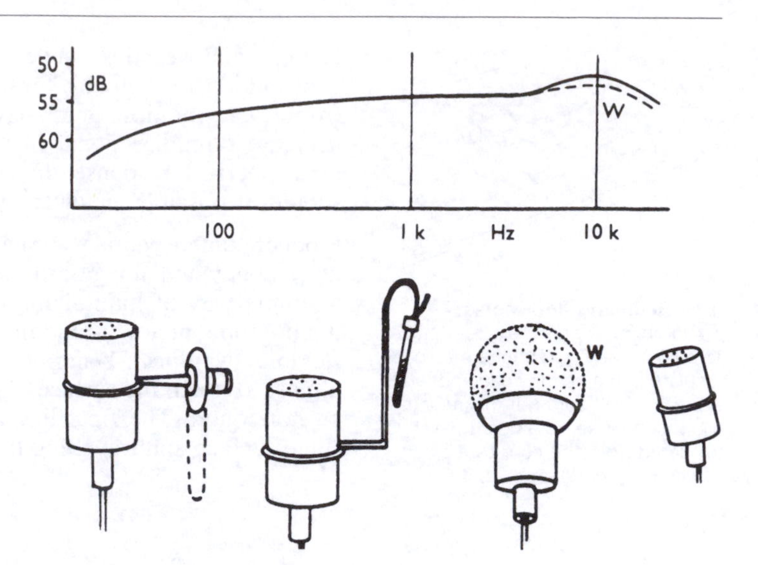

[5.44] **Compact personal microphone**
Electret capsule, connected by cable to head amplifier. The microphone can be used with one of several clip attachments or a windshield (W), which very slightly affects the high-frequency response.

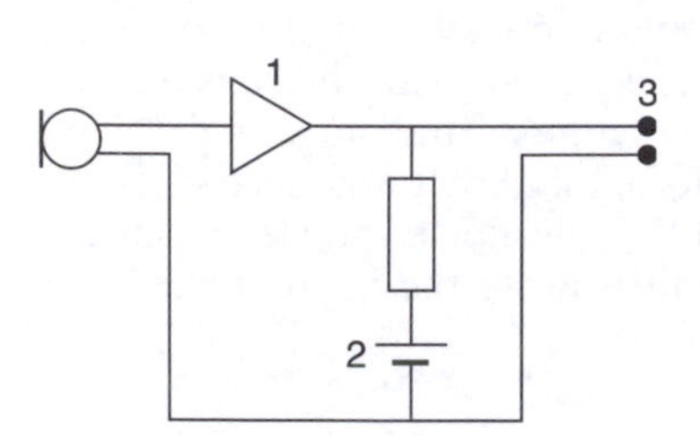

[5.45] **Electret circuitry**
Compared with earlier electrostatic (condenser) microphones this is very simple. 1, Integrated circuit amplifier. 2, Low-voltage d.c. power supply and resistor. Phantom power may also be used. 3, Output (cable).

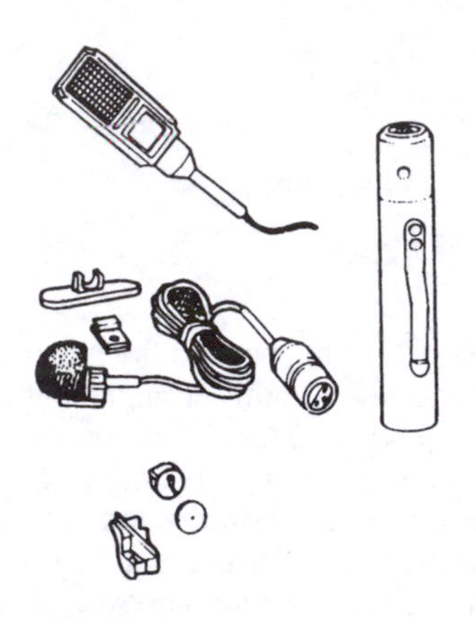

[5.46] **Versatile, tiny personal electret**
This measures 4.4 mm × 7 mm × 13 mm (0.175 in × 0.275 in × 0.5 in), and is supplied with a moulded connector and a range of holders – clip, tape down strip, tie-bar and tiepin – in black, grey, tan and white, or with a small windshield or pressure-zone adaptor. The response is level in the mid-range, with a smooth, extended presence peak of 3–4 dB in the 5–15 kHz range. Its relatively large power supply can be clipped to a waistband or put in a pocket.

knocked. Some condensers have a separate head amplifier, connected by a short cable: the wire is thin (to be less obtrusive) but inevitably is vulnerable to damage, and is also a possible source of conduction noise when rubbed by clothing. Other capsules are smaller still, with cable moulded to housing. In some, a tiny amplifier is included in the capsule.

Commonly, a tiny electret is used. In this less expensive variation on the condenser's electrostatic principle, the diaphragm has a charge sealed within it ('pre-polarized') during manufacture. This eliminates the power supply that would otherwise be required to charge the condenser, although a small d.c. battery or phantom power supply is still needed for an amplifier close to the diaphragm. Early models suffered some loss of high-frequency response as they aged, so this may need to be checked. Later variants have been designed to hold their polarizing charge better (often in polytetrafluorethylene), and the technology (particularly of sealing the surface) has improved.

For the electret, a variant on 'personal' is its use on musical instruments, to which it may be clipped, often on a small arm or gooseneck arranged to angle it in towards the source. Close to that, the reduced sensitivity and therefore higher noise level associated with a smaller diaphragm is acceptable.

In the studio and on location, a personal microphone may be used either with a cable (heavier beyond the head amplifier) or with a radio link.

Radio microphones

[5.47] **Singer's hand-held radio microphone**
The short stub of the UHF antenna is visually inconspicuous.

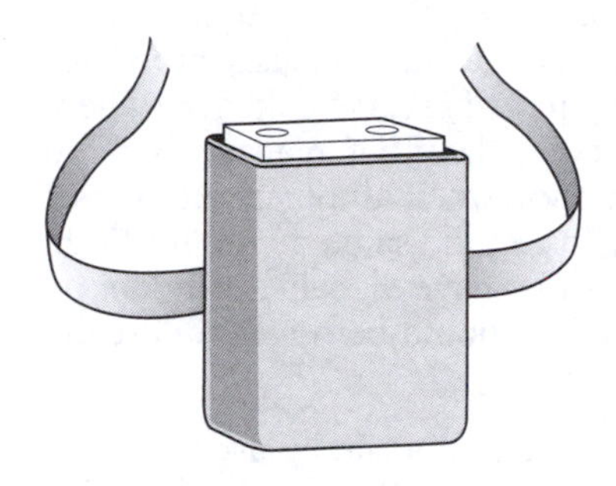

[5.48] **Communications pouch**
A radio microphone transmitter or talkback receiver can be carried in a pouch held by a tape in the small of the back.

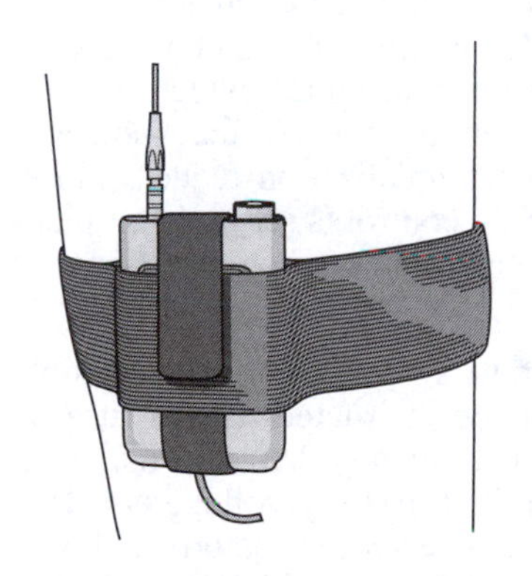

[5.49] **Garter transmitter pouch**
One of a variety of pouches that are commercially available.

In radio, television or film the user of a personal microphone may need to be completely free of the cable because:

- in a long shot the cable may be seen to drag – this is unacceptable in drama and distracting in other programmes;
- in a sequence involving moves, the cable may catch on the scenery or get tangled, or movement may be restricted to avoid such mishaps;
- the performer may have to move about so much in the intervals between rehearsals and takes that it is preferable to allow full freedom of movement rather than keep unplugging the microphone.

In these cases the cable may be replaced by a radio microphone (also called a wireless microphone). This has a battery-powered UHF/FM or sometimes a VHF/FM radio transmitter and aerial. Unless it is housed within the microphone casing, the transmitter pack should be small enough to slip into a pocket, be pinned inside clothing or hang at the waist beneath a jacket. Another option is to place it on a garter, under loose clothing. If simpler answers fail, put it in the small of the back, using a cloth pouch of the exact size, with tapes to hold and support it. (A similar pouch might be used for a television presenter's radio communication equipment.) A VHF aerial (a quarter wavelength long) may hang free from the waist or, if that would be in shot, might be drawn upward within outer clothing. It should be kept as straight as possible, away from the microphone lead, and clear of the body. A UHF aerial is shorter (only about 8 cm, 3 in) and can be housed in a stub on the end of a hand-held microphone. For this it is visually cleaner and easier to handle than a VHF aerial, but its range may be more confined. Miniature UHF transmitters are widely (and reliably) used for talk or game shows in the confined and controlled television studio environment.

Receiving antenna are usually placed where there are likely to be unobstructed electromagnetic paths from the performer. The equipment can 'see' through wood and fabric sets, but not through metal mesh, and are likely to be affected by other metallic objects such as camera dollies that move around during a production.

When setting up radio microphones, after everything else in the operating environment is in place but before they are handed out, check that the antennae are suitably placed. This takes two people: one walks wherever the performer might go, talking over the circuit to the other, who listens for dead spots. Alternatively, signal strength should be monitored (it is too low if a light on the receiver goes off). It may be possible to find a receiving position that works consistently well for any required transmitter position but, in any case, wherever possible, it is best to use 'diversity reception'. This has two receiving antennae spaced between a quarter and one wavelength apart, perhaps at either end of the receiver or at a higher point nearby; the receiver automatically checks and selects the strongest

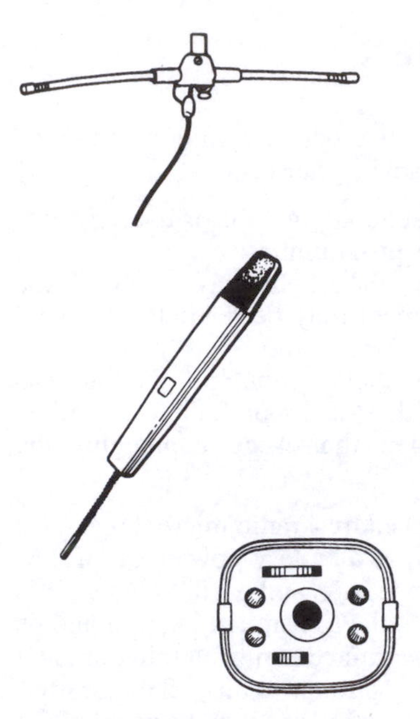

[5.50] **Radio microphone and dipole**
The microphone contains its own UHF transmitter and has a stub aerial. The base of the stick has switches including on/off, tone and bass roll-off (for cardioid close working), and LED condition and warning lights.

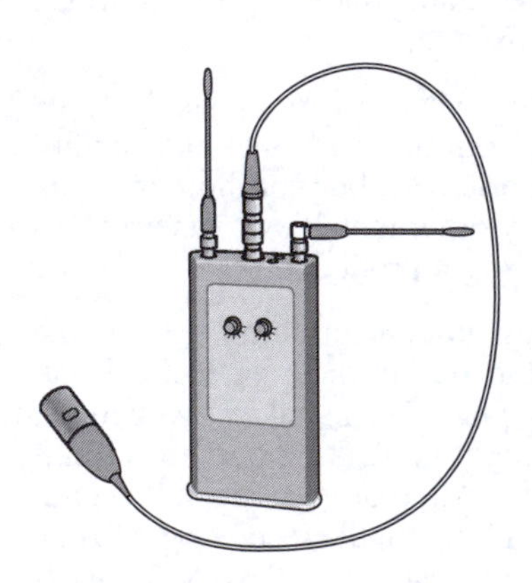

[5.51] **Small diversity receiver**
A compact pack that could be used with a mobile recorder or attached to a camera. This is far more convenient than a cable and current designs should be almost as reliable.

signal. This avoids multipath effects, distortion caused by a drop in level or even total loss of signal when the performer moves though a position where two radio paths of different lengths are set up, and interfere. Smaller diversity receivers can be attached to a camera or mobile recorder.

If several radio microphones are in use at the same time, this can cause intermodulation signals (audible 'birdies'). This can happen if transmitters get too close together, and is worse if the frequencies used are close. The cure is usually to change one of the transmission frequencies – but it might still be picked up on a third channel. Occasionally, other equipment in a studio interferes with the radio link: for example, a signal operating radio-controlled special effects. Listen also for electrical interference from machinery, vehicles, mobile phones, television monitors and (especially) disconnected video cables: if the cause cannot be eliminated, try moving the receiving antenna, or to discriminate against a distant source, use directional antennae.

In any case, it is safest to anticipate the unexpected with standby transmitter packs on different frequencies. Obvious frequencies to avoid are those used for production talkback continuously broadcast to the studio floor manager and for switched reverse talkback (where this is used). If it is of a suitable type, the floor manager's radio microphone is available in an emergency when all others are in use: it is therefore helpful to standardize, using the same model for both. As a final safety measure, if possible have a cable version on standby. For crucial live broadcasts this should be rigged and ready for use.

In a television studio, existing monitor circuits can be used to carry the signal from the dipoles to a radio receiver in the sound gallery. To smooth out unexpected signal variations, the receiver can sometimes be set for a modest amount of automatic gain (but not in cases where the performer is going to speak soft and loud for dramatic purposes). As the transmitter frequency and sensitivity controls have to be pre-set (where the performer cannot disturb them), the signal can easily be overloaded, causing distortion, or undermodulated, reducing the signal-to-noise ratio. These can be avoided by introducing compression, reducing an 80 dB range to 40 dB at the transmitter, which is compensated by expanding it back to 80 dB at the receiver: the whole acts as a *compander*. Matched pre- and post-equalization reduces noise. On the receiver a squelch-mute threshold control is set so that it picks up the signal when it is available, but mutes when the signal is switched off.

Transmitter power and microphone mute switches will be accessible to the user, and regular users should be instructed in their use. A good transmitter should be free from frequency drift and its signal should not 'bump' if it is tapped or jolted. Battery voltage can drop suddenly when the power source is exhausted, so batteries should be discarded well within their expected lifetime (which should itself be reasonably long). Most have automatic early warning of failure; if not, a log of hours of use can be used to schedule replacement.

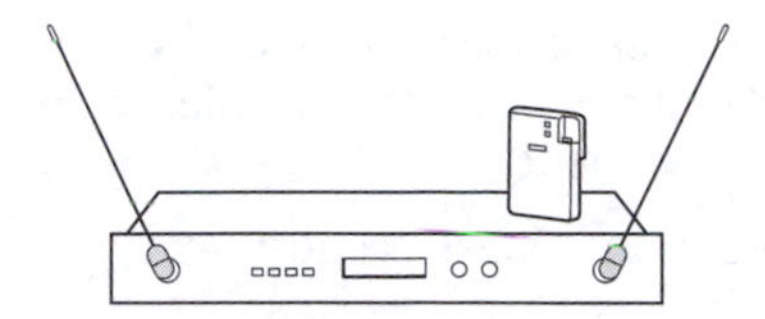

[5.52] **A radio microphone pair**
Small body pack transmitter and rack-mounted receiver with diversity receiver antennae.

Different wavebands are used in Europe and the USA, so equipment carried across the Atlantic must be reset (not always easy). The band used in the UK is allocated to television broadcasting in the USA.

Radio microphones can be used in filming (particularly for documentary or news reports), for roving reporters in television outside broadcasts, by studio programme hosts and for many other purposes. But even if the quality of the link is high it can still be compromised by the position of a personal microphone, and despite all the caveats listed above, clothing rustle is the main hazard. So alternatives should be considered: in the studio a boom, and on location a gun microphone.

Contact and underwater microphones

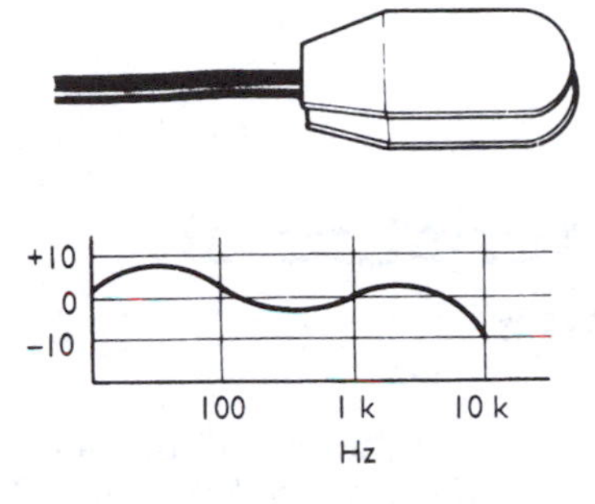

[5.53] **Contact microphone**
This can be attached to the soundboard on a guitar or other stringed instrument. The principle is electrostatic but is unusual in having a rubber-coated diaphragm, in order to give it relatively large reactive mass. The polar diagram is figure-of-eight.

Many solids and liquids are excellent conductors of sound – better than air. A device that picks up sound vibrations from a solid material such as the soundboard of an electric guitar is called a *contact microphone*. Because in this the whole microphone housing moves with the vibration of the solid surface, one way of generating a signal is for a suspended 'active mass' to remain still, by its own inertia. If the moving-coil principle is used, the mass of the magnet remains almost at rest, while the body of the microphone moves. For a massive source such as a girder or rail, the weight of the contact microphone does not matter much. For light sources, including most musical instruments, the mass of the moving capsule should be low, in order to avoid damping its motion. Where the condenser principle is used, the diaphragm requires added mass: in one design, this is achieved by coating it with rubber.

Where mass is an important part of the design, it is unreasonable to expect a flat, extended frequency response. A contact microphone that employs a piezoelectric transducer (e.g. by bending a strip of barium titanate to generate an electric signal) may emphasize the mid-frequency range: this may be accepted as part of the character of the sound or it may be equalized later. Another, an electret, has a broader response but still suffers a loss of high frequencies (above 4 kHz). This can be compensated by adding an acoustic electret to

[5.54] **Sound from solids: active mass**
1, Contact microphone. 2, Vibrating surface. 3, The mass of the suspended magnet holds it relatively still. The mass and suspension system can be arranged so that the resonant vibration of the magnet is at a very low frequency and heavily damped.
For a source in a fixed position, the microphone assembly can be mounted on a rigid arm (4).

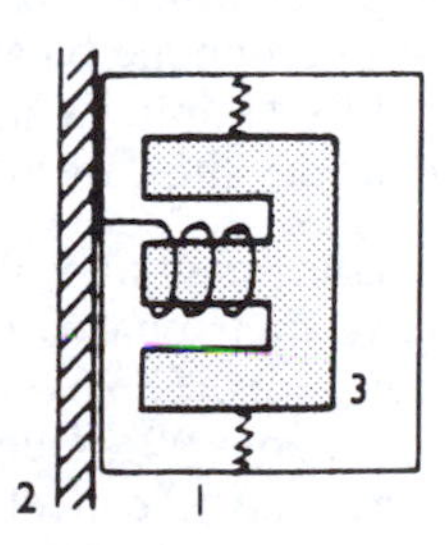

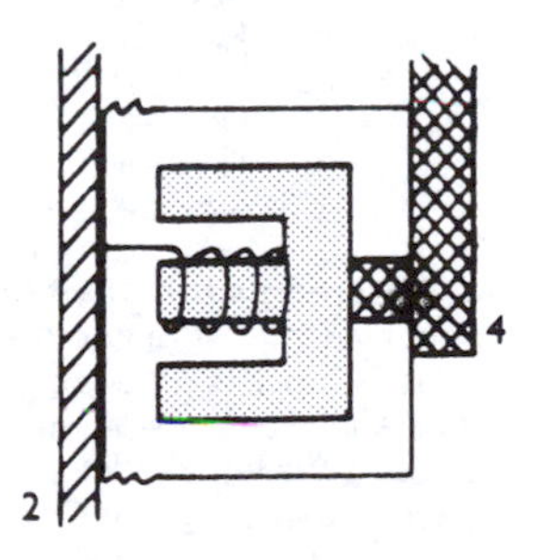

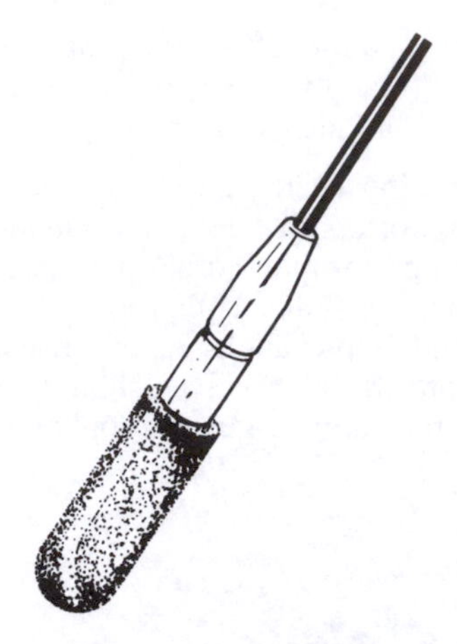

[5.55] **Miniature hydrophone**
The ideal underwater microphone, 9.5 mm (0.37 in) in diameter, with a rubber compound bonded directly onto a piezoelectric ceramic. This permits an excellent sound transmission from the water, with a frequency response that extends far beyond the range of human hearing. It can be used to capture the extremely high- or low-frequency signals used by some marine life (which we can hear by pitch and time changing). The response is flat over the normal audio-frequency range, in directions at right angles to the axis. There is a very slight reduction in some other directions.

pick up high-frequency sound directly from strings: in the case of the guitar, it can be fixed underneath the strings, in the air hole, from which it will receive an additional component.

Contact microphones can be attached with adhesive putty.

To pick up sound under water, one technique is to seal a conventional microphone in a rubber sheath (perhaps a condom). However, the sound must cross the barrier from water to air, an acoustic impedance that distorts frequency response and reduces sensitivity (totally apart from causing embarrassment, hilarity, or both).

It is better that water is separated from the transducer only by a material such as chloroprene rubber, which is of matched impedance. One miniature hydrophone has a frequency range of 0.1 Hz to 100 kHz or more, which may seem excessive in terms of human hearing, but not for that of whales or dolphins. By recording whale 'song' then replaying it at a different speed or putting the signal through a pitch-changer, the extended frequency range of underwater communication and sound effects can be matched to our human perception.

Directional characteristics of A and B stereo pairs

For stereo, an important technique for giving positional information is the coincident pair (see Chapter 3). If separate A (left) and B (right) microphones are used they should be matched by manufacture and model (at the very least), because even a minor difference in design may lead to big differences in the structure of the high-frequency response. An alternative to the matched pair is the double microphone, in which two capsules are placed in the same casing, usually with one diaphragm vertically above the other, so they have a common axis. In one modern paired ribbon design, the capsules are fixed so that the two figure-of-eight elements are at a perfect 90° to each other.

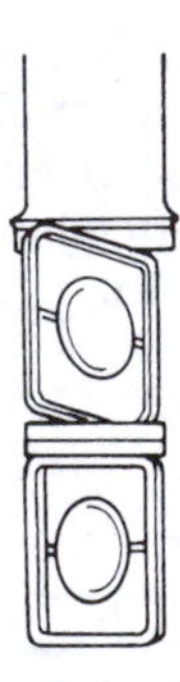

[5.56] **Stereo pair in one housing**
For convenience the two diaphragms are usually placed one above the other, and in this design the outer one can be swivelled to adjust the angle between them.

In some paired condensers the polar responses can be switched between figure-of-eight, hypercardioid, supercardioid and cardioid conditions (as well as omnidirectional, which has no direct use in this kind of stereo pick-up). The outer capsule can be rotated and the angle at which it is set is a matter of choice: it is part of the process of microphone balance. In practice, it may be as narrow as 60–70°, but for simplicity in describing how the signals combine and cancel, consider the case where they are set at 90°, a right angle.

A coincident pair of figure-of-eight elements set at 90° to each other have their phases the same (positive or negative) in two adjacent quadrants. The overlap of these allows a useful pick-up angle that is also about 90°. Outside this angle, the elements are out of phase, so they tend to cancel out. At the back there is a further angle of 90°

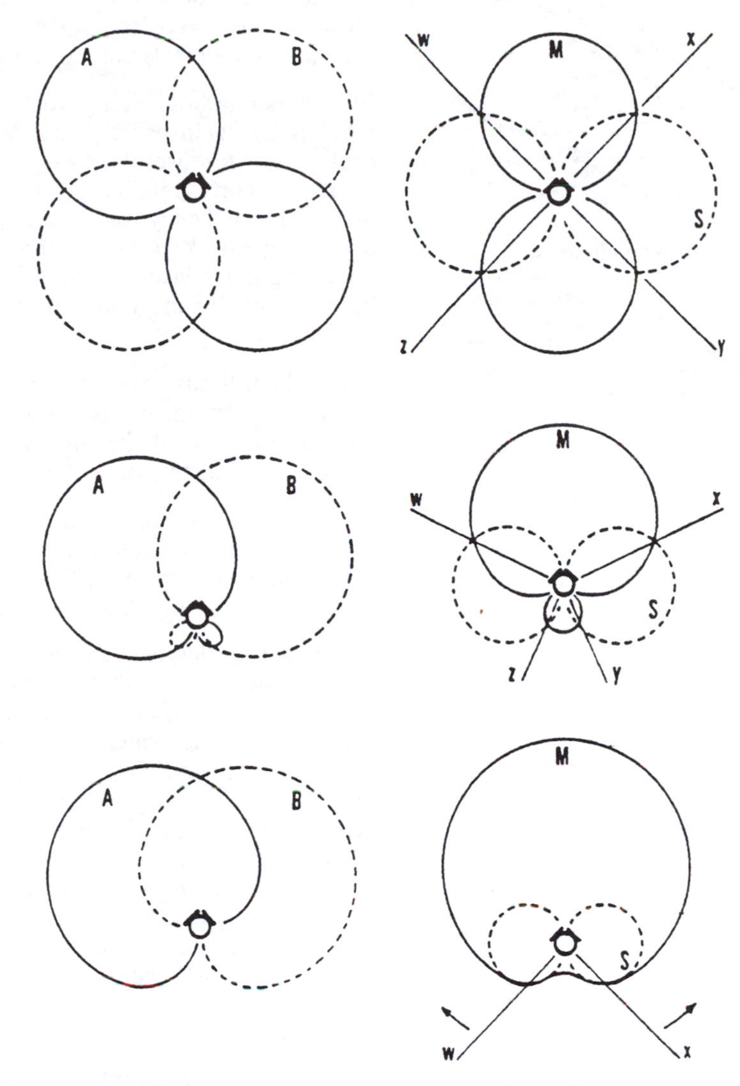

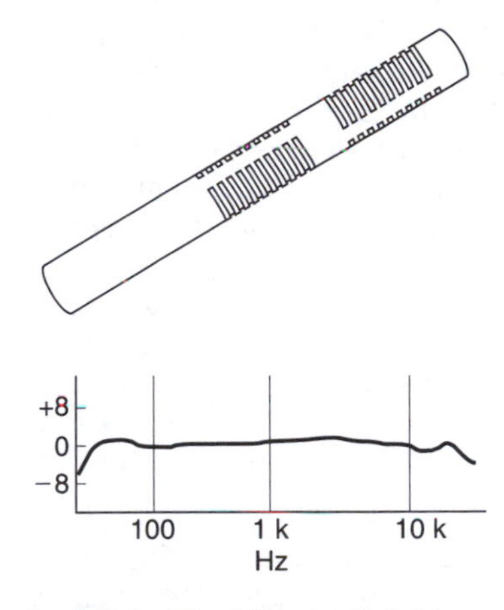

[5.57] **Coincident figure-of-eight stereo pair**
Two ribbon microphones in line, at a fixed angle of 90°. They have a smooth, matched response that is suitable for orchestra, piano or woodwind. An iron case completes the magnetic flux return circuit, helping to make this an unusually compact design.

[5.58] **AB polar diagrams for coincident pairs of microphones at 90°**
Above: Crossed figure-of-eight. *Centre*: Crossed supercardioid. *Below*: Crossed cardioid. The MS diagrams corresponding to these are shown opposite. Note that although the subject of crossed microphones is introduced here in terms of elements set at 90°, a broader angle is often better for practical microphone balances; 120° is commonly used.

[5.59] **MS diagrams for the same coincident pairs**
Above: For crossed figure-of-eight the useful forward angle (W–X) is only 90°. Sound from outside this (i.e. in the angles W–Z and X–Y) will be out of phase. The useful angle at the rear (Y–Z) will (for music) be angled too high for practical benefit. *Centre*: Crossed supercardioid microphones have a broader useful angle at the front (W–X). *Below*: The crossed cardioid has a 270° usable angle (between W and X, forward) but only about 180° is really useful, as to the rear the response is erratic.

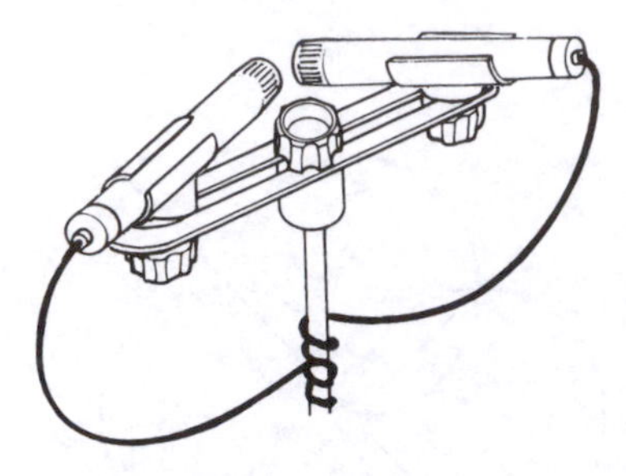

[5.60] **Stereo pair on bar**
This employs two directional monophonic microphones with their diaphragms close together. For nearby voices, it is usually better to have the capsules side by side rather than one above the other.

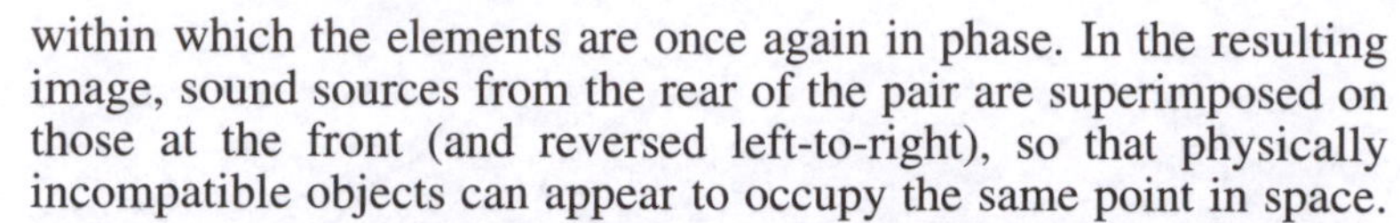

within which the elements are once again in phase. In the resulting image, sound sources from the rear of the pair are superimposed on those at the front (and reversed left-to-right), so that physically incompatible objects can appear to occupy the same point in space.

In a hypercardioid pair (crossed cottage-loaf response) the front lobes must again be in phase. With the diaphragms still at 90° to each other, the useful pick-up angle is greater, about 130° at the front. The corresponding 50° at the back is little use for direct pick-up, as the frequency response of the individual microphones degenerates at the back. However, if a source to be picked up on a separate microphone is placed within this angle, at least its position will not be distorted by the out-of-phase pick-up that would be found at the side.

Crossed cardioids have a maximum usable angle of up to 270°, but for direct pick-up the useful angle is only about 180°, as the frequency response of the microphone that is turned away from the sound source degenerates at angles that are farther round to the side.

As described in Chapter 3, there are two ways of looking at the polar response of a coincident pair. One uses the usual response curves for the A and B microphones. The other considers their derived M signal (obtained by adding A and B) and the S signal (their difference).

The M and S diagrams correspond to the polar responses of another pair of microphones that can be used to obtain them directly – a 'middle' (or main) one that has a range of different polar diagrams and a 'side' one which is figure-of-eight. Sometimes A + B and A – B stereo information is taken directly from main and side microphones, thereby cutting out several items of electrical equipment. A potential problem with this is that the combined frequency response is at its most erratic at the most important part of the sound stage, close to the centre – though the effect gets less as the S contribution gets smaller. Plainly, the quality of the S component must be high, and its polar response should not change with frequency. The 'M' capsule, which occupies the tip of the casing, works as an end-fire microphone, while the 'S' element, behind it, is side-fire.

AB pairs are widely used in radio and recording studios; MS microphones may be more practical in television and video recording, as the 'M' signal on its own will be in the same place as the picture.

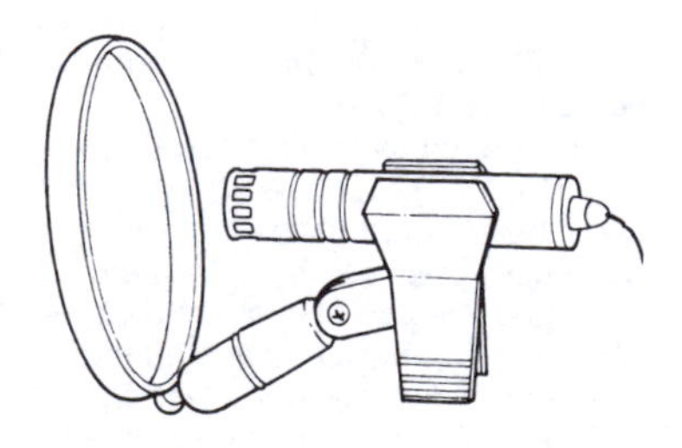

[5.61] **Pop-shield**
A framework covered with gauze protects a microphone from gusts of air caused by plosives, etc., in speech and song. An improvised version can be made from a nylon stocking and coat-hanger wire.

Windshields

The term *windshield* (US: windscreen) covers several distinct functions, for which different devices are used.

Some are used to reduce breath effects, particularly plosive 'p' and 'b' consonants, in close working. These are sometimes called *pop-shields* (US: pop-screens). Where obstruction to sightlines does not matter, a simple design has silk or gauze stretched over a wire frame

10 cm (4 in) in diameter: this is fixed between mouth and microphone.

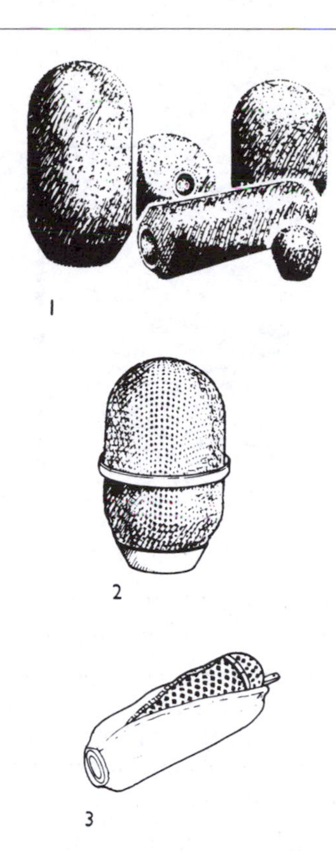

[5.62] **Windshields**
These come in a variety of shapes for different sizes of microphones and may be made of: 1, plastic foam; 2, foam-filled wire mesh; or 3, mesh with an outer protective cover and air space between the mesh and the microphone.

Some microphones have an integral windshield, typically two layers of acoustic mesh with a thin layer of a porous material between. So long as this protects the diaphragm and its cover from the direct airflow from mouth and nose the shape does not matter. The effect on frequency response should have been considered as part of the overall design. Some microphones, designed primarily for use without one, have separate compact metal-gauze shields made to slip on to them; alternatively, shields made of foamed material will stretch to fit over a variety of shapes. These may slightly affect high-frequency response, for which EQ may be used if desired. To restore the nominal frequency response, a shield should be taken off when the microphone is used at a greater distance.

For a talk studio, foamed shields may be supplied in several colours, so each speaker can be directed to sit at a clearly identified microphone, and will be reminded by its colour to work to that and no other.

The other main use for a shield is to cut down wind noise, which may be partly due to turbulence at sharp edges, or even sharply curved corners. Ideally, a smooth airflow round the microphone is required. If the direction of the wind were predictable, the best shape would be a teardrop, but in practice windshields are usually made either spherical or a mixture of spherical and cylindrical sections, typically 10 cm (4 in) in diameter. The framework is generally of metal or moulded plastic, covered by a fine and acoustically transparent mesh, often of wire and foamed plastic. This should reduce noise by more than 20 dB.

Mountings

In addition to the equipment already described, there are many other standard ways of mounting microphones, including a variety of booms. In concert halls, most microphones may in principle be slung by their own cable from the roof of the auditorium, but for additional safety many halls insist on separate slings, and in any case guying

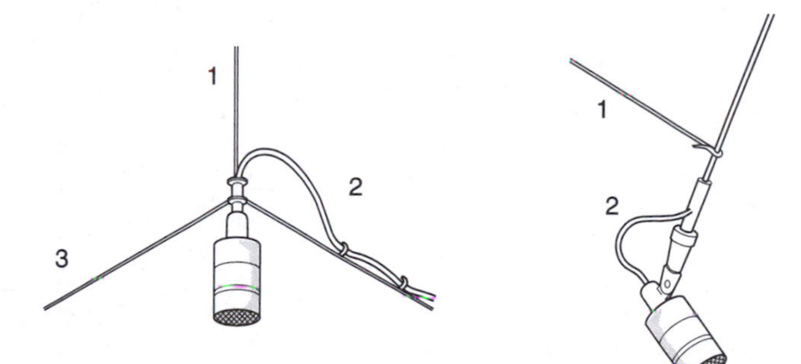

[5.63] **Control of microphone position**
1, Tie-cord. 2, Microphone cable. 3, Guy-wire for fine control of position.

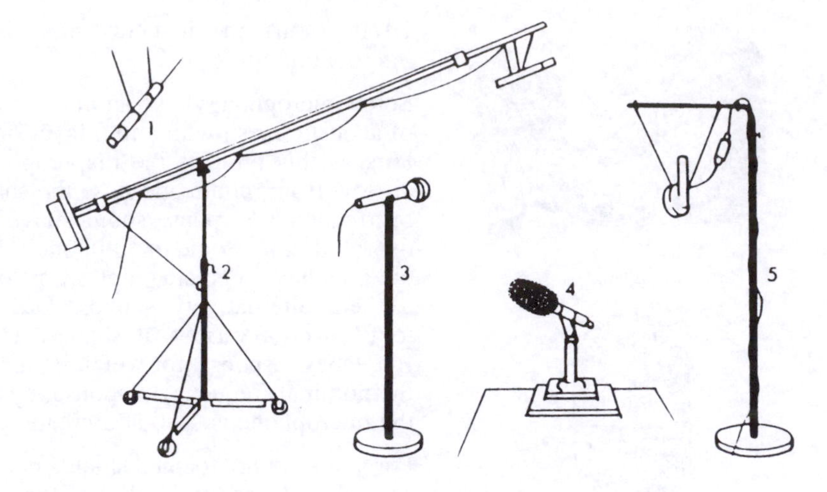

[5.64] **Stands and slings**
Methods of mounting microphones (used in radio studios). 1, Suspension by adjustable wires, etc. 2, Boom (adjustable angle and length). 3, Floor stand (with telescopic column). 4, Table stand (but 1, 2 or 5 are better for table work if there is any risk of table-tapping by inexperienced speakers). 5, Floor stand with bent arm. Most methods of mounting have shock absorbers or straps.

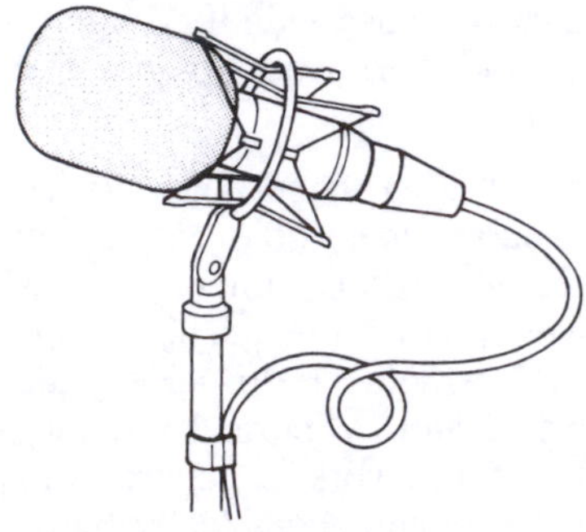

[5.65] **Shock-absorbing microphone cradle**
This is bulky and unsightly but efficiently guards against all but very low-frequency vibration. A loop of the cable should be clipped to the stand.

(by separate tie-cords) may be needed for an accurate final choice of position and angle. A radio-operated direction control may be added.

Floor stands and table mountings should have provision for shock insulation, to protect the microphone from conducted sound, vibration or bumps. Where appearance does not matter, this may be a cradle of elastic supports. In all cases the microphone must be held securely in its set position and direction.

A studio boom used in television and film has a telescopic arm that extends from about 3 to 6 m (10–20 ft). It can swing in almost any direction and tilt to 45°. The operator stands on the platform to the right of the boom, with the right hand holding the winding handle that racks the counterbalanced arm in or out, and with the left on a crank handle that rotates and tilts the microphone on the axis of its supporting cradle. The left arm also controls the angle of the boom.

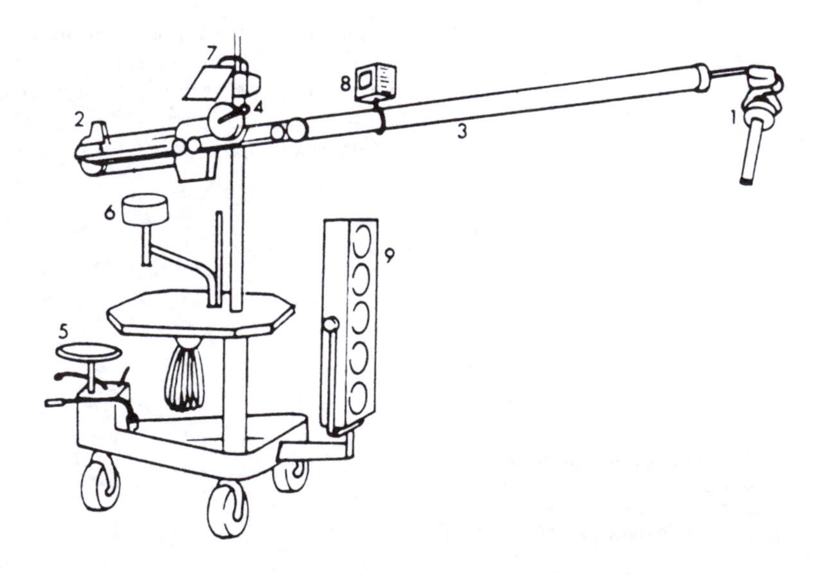

[5.66] **Studio boom**
1, The microphone cradle pivots about its vertical axis, controlled by the angle of the crank (2). 3, The arm pivots and tilts, and is extended by winding the pulley (4). The whole dolly may be tracked, using the steering controls (5), and the platform and arm fulcrum height may be raised and lowered. 6, The seat is used only during long pauses in action. The operator has a script rack and talkback microphone (7). If a floor monitor is difficult to see, a small personal monitor (8) might be attached to the boom arm. 9, The loudspeaker feeds foldback sound to the studio floor.

[5.67] **Articulated fishpole for film or video**
The angle between the two sections of this boom is fixed with a locking nut. The microphone hangs just out of shot without danger of a pole crossing the corner of the frame.

A seat is available but is used only in long pauses as it restricts mobility. Some television organizations provide the operator with a miniature monitor, fixed to the boom arm a little forward of the pivot. The platform and boom can be raised or lowered together: they are usually pre-set to a height such that the boom extends almost horizontally, swinging freely above pedestal cameras. The whole carriage can be steered, tracked or crabbed (moved sideways) by a second operator.

A hand-held boom, which may be telescopic or have extension pieces, is called a *fishing rod* or *fishpole*. It takes up much less space and can also be brought in from directions that are inaccessible to larger booms; for example, it may work close to the floor. In normal use, it can be supported on a shoulder or anchored beneath the shoulder to hold it steady. In exceptional circumstances it can be held high with both arms braced over the head, but directors should expect this only for short takes. A cranked boom rises at a more comfortable angle, and then a locked nut controls the angle at which the extension hangs down to keep the microphone clear of the film or television picture.

[5.68] **XLR 3 M**

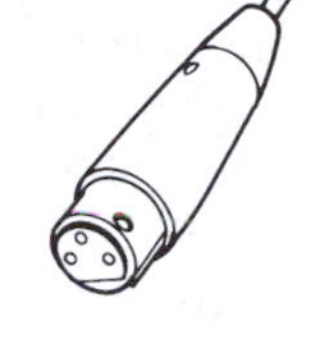

[5.69] **XLR 3 F**

The figure is the number of pins. 'M' is 'male', with pins pointing in the direction that the signal travels. The output from a microphone would normally have 'M' pins; the feed to the control console would go into an 'F' connector. A rugged, heavy-duty design that is used widely in studios.

Cables and connectors

The most common types of connectors are XLR, DIN, Lemo and simple jackplugs. The wiring that goes into them falls into two main categories: balanced and unbalanced lines.

To protect the low-level and therefore vulnerable signal from many microphones, a minimum of three conductors is generally employed. Two wires twisted together form a balanced pair that carries equal and opposite currents. Additional currents that are accidentally induced in them anywhere along the way should be in the same direction and can be made to cancel out. Because two microphones that are used close together must have the same polarity (to avoid phase cancellation), it is important to identify each wire in a balanced pair clearly: one is called 'hot' and one 'cold' (even though their signals are similar). It is also vital to ensure everything is wired up correctly; otherwise there will be an accidental polarity reversal. In addition to the balanced pair there is an earthed (US: grounded) sleeve that surrounds them and gives extra protection. This neutral wire is sometimes, confusingly, called 'common', a term that properly relates to unbalanced wiring.

[5.70] **XLR 3 F**
Wall mount with latch.

An unbalanced pair, used for high-level lines and much domestic equipment, has a 'hot' wire in the middle, and an earthed screen that is also used to complete the circuit is therefore called 'common'. If an unbalanced pair is wired into a three-pin plug the 'cold' and 'common' pins are wired together. There is no problem with polarity: if anything is miswired, it simply will not work. In professional

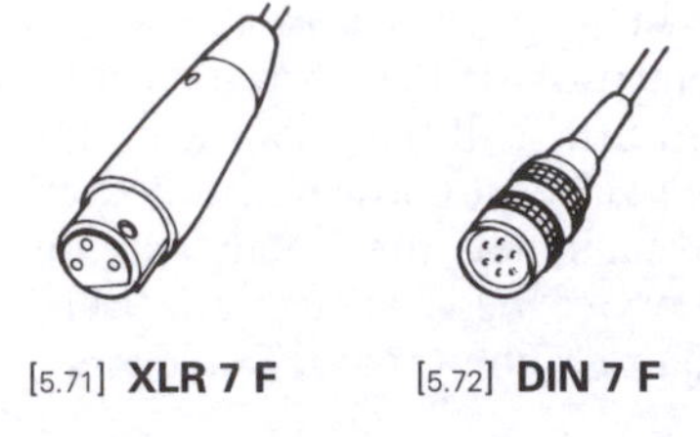

[5.71] **XLR 7 F** [5.72] **DIN 7 F**

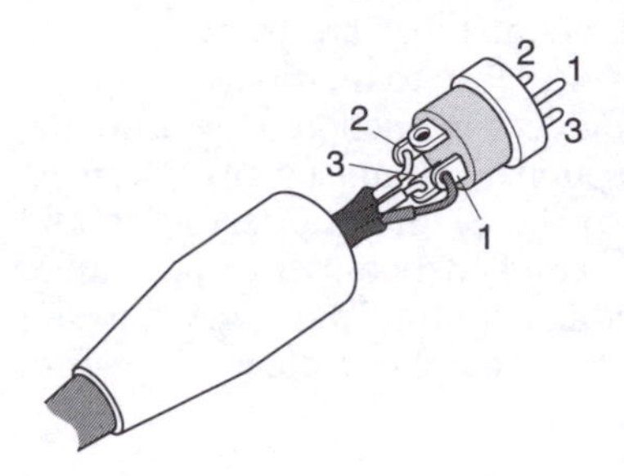

[5.73] **XLR 3 wiring**
In a balanced cable there are three wires: 1, Common. 2, Cold. 3, Hot. For an unbalanced cable (with hot wire and screen) the common and cold pins are connected together.

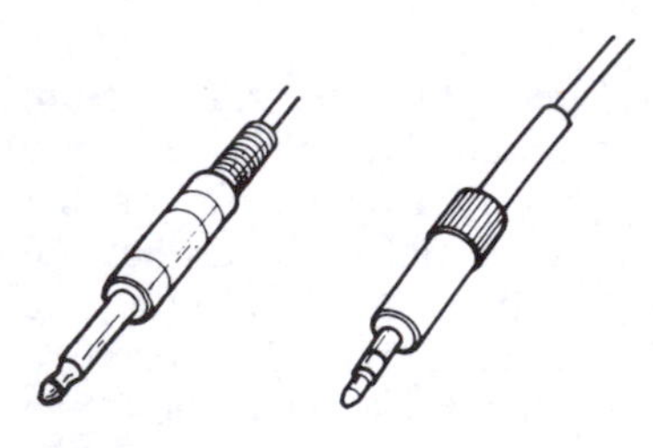

[5.74] **Mono jack** 6.35 mm ($\frac{1}{4}$ in). [5.75] **Stereo jack** 3.5 mm ($\frac{1}{8}$ in).

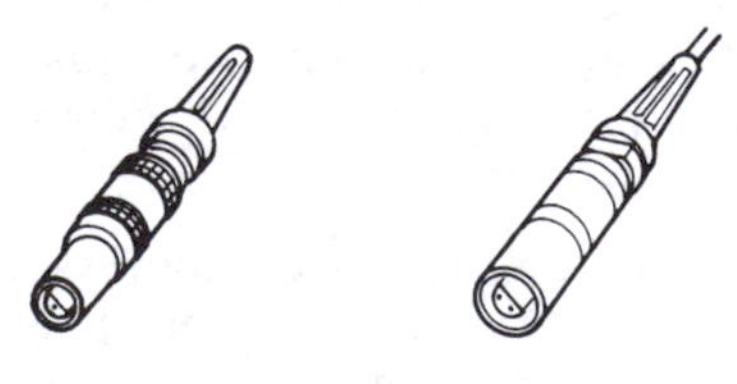

[5.76] **Lemo**
In some Lemo plugs and sockets the inner connectors have a half-moon overlap.

work, unbalanced lines are limited to short runs – rarely more than 3 m (10 ft). This is enough for the wiring between electric instruments and their 'amps'.

XLR connectors, widely used in studios, are virtually an industry standard. They are rugged and rather chunky in appearance, especially when compared with some of the delicate professional microphones equipped with them. The number of pins may be indicated by a suffix: XLR-3, the most common, has three. A cavity in the base of a microphone contains three projecting (i.e. male) pins: they point in the direction of signal output, with 'cold' off-centre and between the other two. Wiring is standardized, but unfamiliar equipment should still be checked for polarity. Most consoles have switches for polarity reversal in case of need, and separate adaptors are also available for this.

The cable has a female socket at the microphone end and another male plug to go into the wall socket (or preamplifier). All XLR connectors have a locating groove to ensure the correct orientation as the pins are pushed in. Some are secured by a latch that closes automatically, and released by pressing on the latch as it is pulled out. Other XLR arrangements are available, including right-angled plugs, splitters, and other numbers of pins. Non-standard pin directions are also sometimes encountered (more often in connectors to monitoring loudspeakers), so female-to-female cables may be required. The XLR system is robust and will withstand accidental bumps or rough use, but is also rather too massive for applications where low weight or small size are required.

Another common standard that is used for more compact and also much domestic equipment is DIN (the German Industry Standard). The pins are more delicate, and the plugs and sockets smaller. For connectors that must be secure, a screw collar may be provided. Amongst others, Lemo connectors (adopting another common standard) are tiny, with fine pins, but tough and reliable for use with (for example) compact radio microphone transmitters. Most jackplugs (which can be mono or stereo) have either 6.35 mm ($\frac{1}{4}$ in) or 3.5 mm (a little over $\frac{1}{8}$ in) central pins. Take care that jackplugs attached to microphones are not plugged into headphone or loudspeaker sockets by mistake, and consider replacing them with XLR connectors.

Three wires (and therefore three pins), one being earthed, are sufficient for most purposes, including both the signal and the power plus polarizing voltage for condenser microphones. The same pins can be used for all of these by raising both of the live signal wires to a nominal 48 V above ground: this system is called *phantom power*, phantom being a term previously used for grouped telephone circuits that shared a common return path, thereby saving wires. In practice, many modern condenser microphones will work with any phantom power from 9 to 52 V and some, alternatively, with batteries. The difference in voltage between the pair and the grounded third wire provides the polarization for condenser microphones that need it (electrets do not, but still need power for head amplifiers).

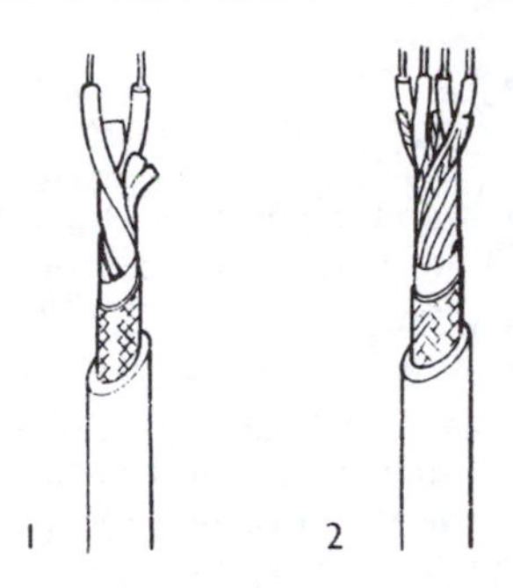

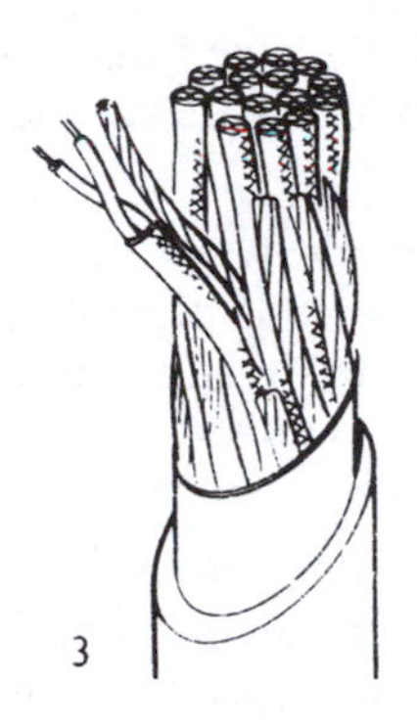

[5.77] **Cables**
1, Screened pair (good). 2, Star-quad cable (better). 3, Multichannel cable 'snake' with 12 or 16 pairs, each having a foil inner shield, within a single outer sheath.

More complex systems have more wires (and pins) – for example, for switching polar response or bass-cut, and also for radio microphones. Old designs of condenser microphones containing valves (vacuum tubes) – collectors' items or working replicas that are treasured by some balancers for their distinctive sound quality – require an additional 12 V supply to the heater.

Trailing or hard-wired audio cables will usually have substantial, but flexible, colour-coded conductors, with braided or lapped shielding to protect against electrical interference, and with a strong outer sheath. Somewhat more complex 'star-quad' cables have higher noise rejection: in a studio it will soon become apparent if these are needed. For longer distances, multichannel cables (with additional coding colours) will terminate in a junction box for the individual microphone cables that lead through a multi-pin connector to the control console. Fibre-optic cables are not affected by either electromagnetic or electrostatic fields and can be used for extremely long runs with virtually no loss, but when used in a studio require analogue-to-digital conversion at the output of each microphone.

A radio studio may be hard-wired to offer a choice of two sockets for each channel, one on each side of the studio. The sockets may be numbered 1 and 1A, 2 and 2A, and so on, to make the choices clear.

In television studios, routings are often more complex, as they may have to go round or through a set (which may include extensive backings) at suitable points, and also avoid camera tracking lines and areas of the studio floor that may appear in vision. So here a very large number of points is provided: some studios have over 50, some even as many as 100 or more. Mostly they are distributed around the walls in groups, but some are in the studio ceiling, so that cables can be 'flown' or microphones suspended from above. A few terminate in the gallery area. It is convenient to have several multi-way single-socket outputs at the studio walls so that a single cable can be run out to a terminating box on the floor, to which, say, 10 microphones can be routed. In order to minimize induction, sound wiring should be kept away from power cables (to lighting or other equipment) and should cross them at right angles. This is also necessary for wiring built in to the studio structure.

Since microphone cables often carry low-level analogue signals, they are particularly at risk from earth loops – also known as ground loops (one of several technical quality problems described in Chapter 9). If, by mistake, wiring is earthed at two or more points along its path, this creates a gigantic loop in which electricity can be generated by any power cable that by chance passes through it. Since this parasitic current passes along the screening wire, it immediately gets one layer closer to the signal and may be audible as mains hum. Unscreened wiring carries it directly. This is obvious when it happens, but may not strike until some item of equipment is switched on, possibly at just the wrong time. So take care when laying wiring: it is vital to prevent this 'enemy in the camp' from getting in.

Microphone checks

When a microphone has been plugged in – and where more than one is in use – a few simple checks need to be made. First, it must be positively identified at the mixer. Gently tap or scratch the casing while a second person listens, to be sure which channel carries the signal. If the console is already in operation, it may have pre-fade listen (PFL) facilities that can be used so that the main programme signal is not interrupted. The polar response is confirmed by speaking into the microphone from various sides, identifying them verbally in turn: a dead side will be obvious. For a stereo pair, these checks ensure that the 'A' channel is on the left, and that both are working. Checks can also be carried out by one person (such as a recordist on location) by using headphones.

In addition, check microphones for handling noise and decide whether a pop-shield is needed for close speech. Gently check cable connections and plugs by hand to see if these make any noise. Cable runs may need to be covered or taped down, not only to protect them but also for public safety. Gather up any excess into loops.

A new or unfamiliar microphone may require more exhaustive tests before use. The manufacturer's literature gives some indication of what to expect, but is not always completely clear. It is best to start by making a direct comparison with some similar microphone if one is available; failing that, another by the same manufacturer (for a common standard of documentation). This 'control test' should show up any obvious deficiency in sensitivity or performance. Try a variety of sound sources: speech, a musical instrument and jingling keys (for high frequencies). Check the effect of any switches – to change polar response, roll off bass or reduce output.

A large organization using many microphones may test them more comprehensively, using specialized equipment in an anechoic chamber – establishing its own standards, rather than relying on claims in various manufacturers' literature, which may not be directly comparable.

Lining up a stereo pair

If coincident microphones do not have the same output level, the positional information is distorted. A more detailed line-up procedure that can be carried out in a few minutes by two people, one at the microphones and one to listen at the console, is outlined below:

1. Check that the monitoring loudspeakers are lined up reasonably well, e.g. by speaking on a mono microphone (or one element of the pair) that is fed equally to A and B loudspeakers. The sound should come from the centre.

2. If microphones are switchable, for a coincident pair select identical polar diagrams. These may be the ones that will subsequently be used operationally, or if this is not known, figure-of-eight.
3. Visually set the capsules at about 90° to each other. Check and identify the left and right capsules in turn. Where microphones are mounted one above the other, there is no agreed standard on which capsule should feed which channel.
4. Visually set the pair at 0° to each other, making sure that microphones of symmetrical appearance are not back-to-back (there is always some identifying point, e.g. a stud or logo on the front, or mesh at the back of the housing that is coloured black).
5. Listen to A – B only, or if there is no control to select this, reverse the phase of one microphone and listen to A plus reversed B.
6. While one person speaks into the front of the microphones, the other adjusts for minimum output. If possible, this should be on a pre-set control for one microphone channel or the other; most conveniently, it may be a separate microphone balance or 'channel balance' control.
7. Restore the loudspeakers to normal stereo output. As the person in the studio talks and walks in a circle round the microphone, the reproduced voice should remain approximately midway between the speakers, whatever the position of the speaker. Except for minor variations when the speech is from near the common dead axis or axes of the microphones, if there are any major shifts in image position as the person speaks from different places there is something wrong with the polar response of one of the microphones; the test must be started again when the microphone has been corrected. The earlier part of the line-up procedure is then invalid. As the person walks round, loudness should vary in a way that coincides with the polar characteristic chosen.
8. Restore the angle between the capsules to 90° (or whatever angle is to be used operationally) and again identify left, centre and right. This completes the test.

Chapter 6

Microphone balance

Earlier chapters have described the physical nature of sound and how it behaves in an enclosed space, such as a studio. We have also seen how a microphone converts the energy of sound into its electrical analogue – which it should do faithfully (or with changes that are known, understood and acceptable) and at a level suited to onward transmission. Now we must put all of these elements together. This chapter describes the principles that govern the relationship between microphones and their acoustic environment: sound sources, ambient sound (reverberation) and unwanted sound (noise). The application of these principles to speech and music are described in Chapters 7 and 8, and to effects in Chapter 15.

Its surroundings may suggest the way in which a microphone is used. In a very noisy location, for example close to an aircraft that is about to take off or among a shouting crowd, the obvious place for a speech microphone is right by the mouth of the speaker. In these and many other cases the first criterion must be intelligibility. But if clarity is possible in more than one limiting position of the microphone (or with more than one type) there are choices to be made. In this book the exercise of these choices is called *microphone balance*. 'Balance' is a broader term than *microphone placement*, although placement is certainly a part of balance.

The first object of balance is *to pick up the required sound, but at the same time to discriminate against unwanted noise*. Beyond just finding a place to put a microphone, in professional practice this also means choosing the best type of microphone, with particular regard to its directional and frequency characteristics. Another choice to be made is whether one microphone will cover the sound source adequately, or if several would do it better. In addition, it is often possible to choose the position of the source or degree of spread of several sources, and to select or modify their acoustic environment.

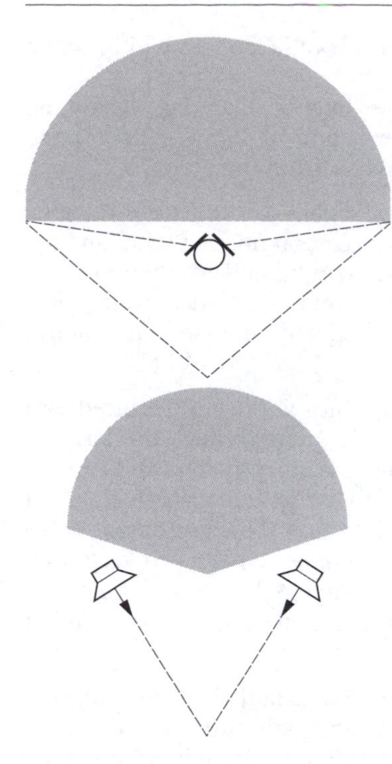

[6.1] **Compression of scale of width**
A sound source that might be spread over 90° or more in real life (*above*) may be compressed on the audio stage to 60° or less (*below*).

Where natural acoustics are being used, the second aim of a good balance is *to place each microphone at a distance and angle that will produce a pleasing degree of reinforcement from the acoustics* – in other words, to find the best ratio of direct to indirect sound.

Here there is another possibility: natural acoustic reinforcement can be enhanced later, usually by artificial reverberation (AR) that is mixed in to the final sound electrically. Taking this further: if the original, chosen balance discriminates against the natural acoustics (or a dead environment is used), AR can be added later. Then, for even greater flexibility, each sound source can have its own treatment. For this reason, and because picking up a sound on a second, more distant microphone will reintroduce the studio acoustics, we have as an alternative aim of microphone balance, *to separate the sound sources so that they can be treated individually.*

For many kinds of broadcast or recorded sound – and especially for popular music – the term 'balance' means even more than all of that: it also includes the treatment, mixing and control functions that are integral parts of the creation of the ultimate composite sound.

Stereo microphone balance

In stereo (and surround sound), microphone balance has an additional objective: *to find a suitable scale of width.* Whereas indirect sound spreads over the entire sound stage, direct sounds should occupy some clearly defined part of it.

Plainly, a large sound source such as an orchestra should occupy the whole stereo audio stage and, to appear realistic, must be set well back by the use of adequate reverberation. On the other hand, a string quartet with the same reverberation should occupy only part of the stage, say the centre 20° within the available 60°; anything more would make the players appear giants. The small group can, however, be 'brought forward' to occupy more of the sound stage, provided that the balance has less reverberation. In either case, the reverberation would occupy the whole available width.

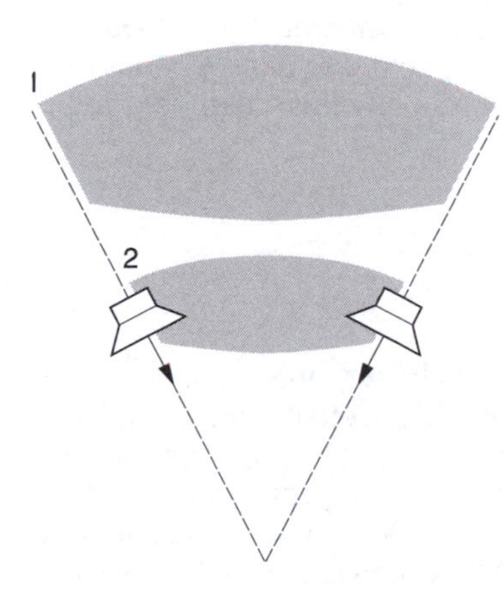

[6.2] **Orchestra in correct scale of width**
A broad subject needs enough reverberation to make it appear a natural size (1); if the reverberation is low the subject appears close and under-size or cramped together (2).

With coincident microphones, image width can be controlled in several ways:

- The microphones may be moved closer to or farther from the sound source. Moving farther back narrows the source and also (as with mono) increases reverberation.
- The polar response of the microphone elements may be changed. Broadening the response from figure-of-eight through hypercardioid out towards cardioid narrows the image. It also reduces the long-path reverberation that the bidirectional patterns pick up from behind and permits a closer balance.
- The ratio of A + B to A − B (the M to S ratio) may be changed electrically. If S (the A − B component) is reduced, the width is

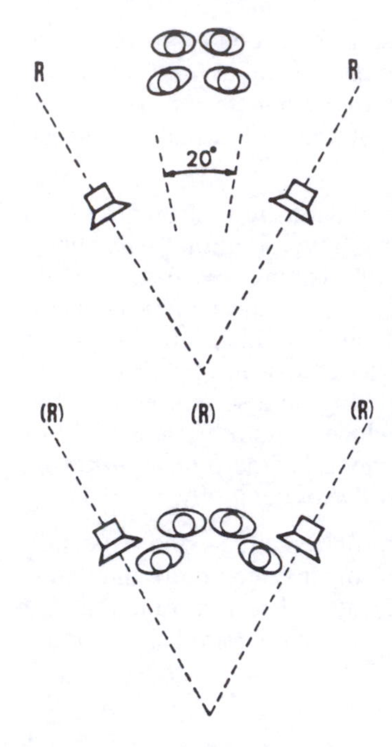

[6.3] **A quartet in stereo**
Above: The quartet uses only a third of the audio stage, but reverberation fills the whole width. *Below*: The quartet occupies the whole of the space between the loudspeakers. Here the reverberation must be light in order to avoid the players seeming gross – such reverberation as there is appears from behind them.

reduced. If M (the A + B component) is reduced, the width can be increased.

- The angle between microphone axes may be changed. Narrowing the angle also brings forward the centre at the expense of the wings, and discriminates against noise from the sides.

With spaced stereo microphones, their separation and the directions in which they point have the strongest effect on image width; and the M/S (A + B/A – B) ratio may again be manipulated. With both kinds of stereo balance, further reverberation may be added from additional distant microphones. This will of course be natural reverberation, in the same acoustic.

Where individual sources for which no width is required are balanced on monophonic (or '*spotting*') microphones, the balance must be close enough (or the studio sufficiently dead) to avoid appreciable reverberation, or there will be a 'tunnel' effect, extra reverberation beamed from the same direction. This closeness is also consistent with application of the '3:1 rule', which proposes that to reduce spill, adjacent microphones should be over three times further from each other than from the sound sources they are set to cover.

A position on the sound stage may be established by *steering* portions of the signal to the A and B channels (this is also called *panning* – see Chapter 9). If a second mono microphone picks up the same source unintentionally, it joins with it to form a spaced pair, which gives a false position; and worse, the amount of displacement also shifts if the level of one of the microphones is changed. Close balance, directional characteristics, layout of sources in the studio and screens may all be used to make sure this does not happen.

Mono spotting techniques cannot be used for subjects that have obvious physical size – such as a group of instruments or singers, a close piano or a pipe organ. For these, two mono microphones may be used to form a stereo pair, or an extra coincident pair set in.

The balance test

To get the best possible balance in each new microphone set-up requires a little experimentation – and it is convenient to call this process 'a balance test'. But first, think about how you want the end-product to compare with the original sound. Do you want it to be a close representation of the original sound? Do you want to adapt it to suit different listening conditions? Do you want to achieve some other effect selected from a broad spectrum of modified sound qualities that might be offered for the listener's pleasure? The end-product of any balance is never an exact copy, but always an interpretation. A vital part of the balancer's job is to exercise choice, making selections from a vast range of possibilities on behalf of the audience.

'Good sound' is not just a matter of applying rules: its quality is subjective. But (paradoxically) the balancer's subjectivity must be allied to a capacity for analytical detachment – which also means that, if possible, the results must be judged in the best listening conditions that any intended audience is likely to use. The surest way of getting a good balance – or at any rate the best for the available studio, microphones and control equipment – is to carry out a balance test.

A balance test is not made on every occasion: with increasing experience of a particular studio and the various combinations of sources within it, a balancer spends less time on the basics and more on detail. A newcomer to microphone balance is likely to see a great deal of attention to detail, and this apparent concentration on barely perceptible changes may obscure the broader objective, and the way that is achieved.

Consider first a single microphone (or stereo pair) picking up a 'natural' acoustic balance. A full comparison test will not be needed very often, but at some stage it is a really good idea to rig at least two microphones in positions that (according to the general principles outlined later) should promise an acceptable result. Compare these two sound balances by using the mixer console and monitoring loudspeakers to listen to each in turn, switching back and forth at intervals of 10 seconds or so, considering one detail of the sound after another until a choice can be made. The poorer balance is then abandoned and its microphone moved to a position that seeks to improve on the better. Ideally, the process should be repeated until no further improvement is possible or time runs out. In practice, this means advance preparation, creating opportunities to move microphones with minimum disturbance and loss of rehearsal time, or to select from a wider range that has been pre-rigged.

For a multimicrophone balance the same process may be carried out for each microphone. Here, what might appear to be a lengthy process is simplified: there is no need for an acoustic balance between many sound sources with a single microphone, as this is achieved in the control desk; nor (in dead music studios) to balance direct against ambient sound. Where multitrack recording is used (as in most music recording studios today) some of the balance decisions can be delayed until the recordings are complete. Mixing down – still using comparison tests – is a part of balance that may be continued long after the musicians themselves have gone.

For speech, in those cases where the speaker works to the microphone (which includes most radio and some television), a similar method may be used: place two microphones before the speaker, and at some stage adjust or replace that which gives the poorer balance, and so on.

A television balance in which the microphone is permitted to appear in vision may be similar to those already described. Static microphones are set in to cover musical instruments or to speakers in a discussion, or a singer may perform to a stand microphone. But where there is movement and the microphone itself should not be

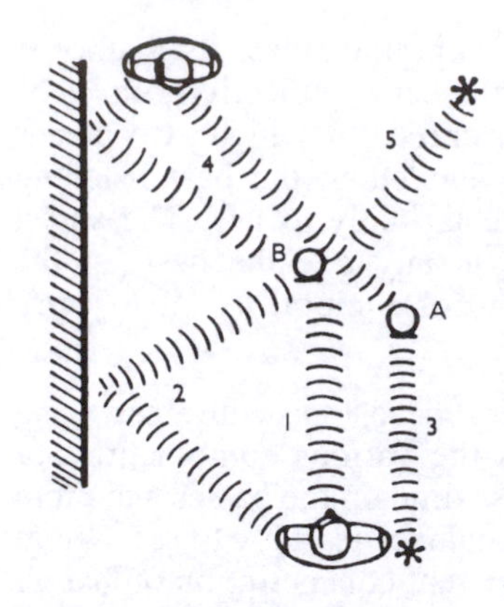

[6.4] **Balance test**
Compare the sound from microphones A and B by switching between the two. The microphones lie in a complex sound field comprising many individual components. For simplicity, the illustration shows monophonic microphones. In a mono balance test listen on each microphone for the contributions from: 1, Direct sound. 2, Reflected sound, reverberation. 3, Unwanted noise from related sources. 4, Other wanted sources and reverberation. 5, Extraneous noise. In stereo, the principles are the same, but the sound paths are more complex. In the stereo balance test, listen also for scale of width and for the location of individual components within the stereo image.

seen, the process may be reversed: ideally, the microphone is made mobile (on a boom), working to the performers. With a moving microphone the balance changes from moment to moment – ideally in harmony with the picture. In either case the way it fits either in or out of vision increases the complexity of any balance problems, to which the simplest answer may be additional microphones.

In any kind of balance, the main things to listen for are the following:

- The clarity and quality of direct sound. In the case of a musical instrument this radiates from part of the instrument where the vibrating element is coupled to the air. It should comprise a full range of tones from bass to treble, including harmonics, plus any transients that characterize and define the instrument.
- Unwanted direct sounds, e.g. hammer (action) noises on keyboard instruments, pedal noises, piano-stool squeaks, floor thumps, page turns, etc. Can these be prevented or discriminated against?
- Wanted indirect sound, reverberation. Does this add body, reduce harshness and blend the sound agreeably, without obscuring detail or, in the case of speech, intelligibility?
- Unwanted indirect sound, severe colorations.
- Unwanted extraneous noises.

When trying out a balance, listen to check that bass, middle and top are all present in their correct proportions, and seem to be in the same perspective. If they do not, it may be the fault of the studio, which may emphasize some frequencies more than others, thereby changing both proportion and apparent perspective. Or it may be more directly the fault of a balance in which a large instrument is partly off-microphone (as can happen with a close piano balance). Remember also that in a live acoustic, a balance that works on one day may not on the next, when the air humidity is different.

Some variables that may be changed during a balance are:

- distance between microphone and sound source;
- height and lateral position;
- angle (in the case of directional microphones);
- directional characteristics (by switching, or by exchanging the microphone for a different type);
- frequency response (by using a different microphone if an extended response is required, or otherwise by equalization).

For greater control over these variables, as few of them should be changed between successive comparisons as is feasible in the time available. Another possibility for improving a poor balance is to set in additional microphones to cover any parts of the original sound that are insufficiently represented.

The balancer working in a studio that has a separate control room to listen to sound (as most do) can use high-quality loudspeakers to judge the results. Working outside the studio, these ideal conditions may not be available, but if artistic quality is both important and possible, balance tests can still be made, by comparing several trial recordings on playback. Since the recordist uses these trial runs to judge how to use the acoustics of the studio to the greatest advantage

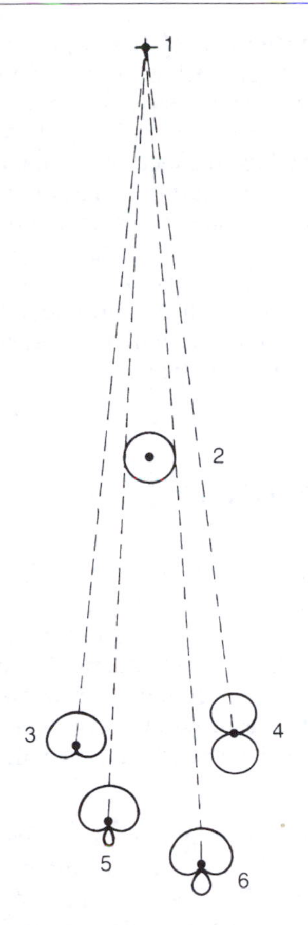

[6.5] **Distance from source**
Their directivity (whereby they pick up less reflected sound) allows some microphones to be placed further away from the source (1). Compared with an omnidirectional microphone (2), the cardioid or bidirectional response (3 and 4) can be set back to 1.7 times that distance; a supercardioid (5) to 1.9 and hypercardioid (6) twice that distance.

(or to least disadvantage, if the reverberation is excessive or highly coloured), high-quality equipment should be used for playback. Headphones can be far better than small loudspeakers.

A recordist working on location may be at a particular disadvantage, one that will not be viewed sympathetically back in the studio. So skill – or in the absence of experience, very great care – is needed.

For reportage, an unchecked balance generally yields results that have an adequate documentary value. Here, experience is invaluable.

Visual conventions for microphones

In film and television the normal criteria for placing microphones for good balance still apply, but in addition there is a second set to think about. These cover the question of what is visually acceptable.

One important convention concerns the match of sound and picture perspectives for speech. Plainly, a close picture should have close sound. Also, if possible, a wide shot should have sound in a more distant acoustic perspective. However, the need for intelligibility may override this: a distant group walking and talking across a crowded room can be made clearly audible as though in a relatively close perspective, while nearer figures are virtually inaudible; again, a reporter in long shot may be heard in close-up. In neither case will the audience complain: they want to hear, and they can.

Other conventions concern whether microphones may appear in vision or not.

For drama, the microphone should always be unseen, as its presence would destroy the illusion of reality.

For television presentations where suspension of disbelief is not required of the audience, microphones may be out of vision or discreetly in vision. In-vision microphones are used where better or much more convenient sound coverage can be obtained in this way: because much of television consists of programmes in which microphones are allowed in vision, they may be seen a great deal without being noticed. However, their very ubiquity is a point against them: there is visual relief in getting rid of them if possible.

For film (including films for television), in cases where the illusion of reality does not have to be fostered, microphones may again appear in vision, but here the reason needs to be stronger than for other television, and the use even more discreet. Most film or location video is shot in places where microphones are not part of the furniture, so to show one can remind the audience of the presence of technical equipment (including the camera). Also, in a film there may be greater variety of pictures or more rapid cutting than is usual in studio-based television, so a microphone that keeps popping up is more intrusive.

When a microphone really has to be in vision, the choice depends on its appearance and sometimes its mobility, as well as on the balance required. Some appropriate microphone types have been introduced in Chapter 5, and their use is described in Chapters 7 and 8. But when microphones must not be seen, the best balance will probably be obtained with studio booms. Here, too, the equipment has already been described; and as it is used mainly for speech balance, boom operation is dealt with primarily under that heading. A television musical performance presents special problems: its balance should compare well with that on radio or records, even though television studios are usually fairly dead. Unless a concert hall is used, this provides an extra reason for multimicrophone balance.

Multimicrophone balance

In a multimicrophone balance a broad range of techniques is drawn together, often to include ways of handling reverberation and equalization, of building up a recording by adding track upon track, and then of mixing down from 16, 24 or even more pre-recorded tracks to a mono, stereo or surround sound finished product. A few brief, introductory notes on some of the techniques used are given below. These go far beyond simple 'microphone placement' but may still be regarded as part of the overall 'balance'.

Equalization. Nominally this word means restoring a level frequency response to a microphone that has inherently or by its positioning an uneven response. But that is just one way it is used: perhaps even more often it is introduced as a second stage in the creative distortion of sound, the response of the microphone in its chosen position being the first. Equalization controls, whether located in an integrated dynamics module or assigned to the microphone channel by computer, include midlift (for adding 'presence') and bass and top cut or boost. Many consoles offer several presence bands or the corresponding dip, or a high or low 'shelf' in response on each channel.

Artificial reverberation (AR). This augments or replaces the natural reverberation offered by conventional acoustics. Greater control is possible, as feeds taken from individual microphones can be set at different levels. On many consoles an auxiliary feed that can be taken from one or more channels is fed to a digital AR program, which may have a great variety of realistic reverberation styles as well as overall times. Older, far less flexible analogue alternatives include acoustic 'echo' chambers. From whichever is used, the signal is then returned to the mixer as a separate and usually stereo source. Sometimes different AR settings are used on different channels. Digital reverberation equipment may also include a range of other effects.

Compression. Reduction in dynamic range may be used to maintain a high overall level (in order to ensure the strongest possible recorded or broadcast signal), or it may have some specialized application. For example, it might be applied to parts of a backing sound that without it would have to be held low in order to avoid sudden peaks blotting out the melody, or to the melody itself, which might otherwise have to be raised overall and as a result become too forward. Limiters as well as compressors are used for this.

Together with further stages in the treatment of sound, such as overdubbing, multitrack recording and then mix-down, whether in stereo or surround sound, these will all be described in more detail in later chapters on studio operations. The idea of 'balance' may then sometimes be taken further to include decisions made when one component of a complex sound is weighed against another, but that is not the primary meaning of the term.

Note that everything about microphone balance before the later stages listed above was still 'analogue'. 'Digital', where appropriate, begins at the console.

Artistic compatibility: mono, stereo and surround balances

If an item recorded or broadcast in stereo is likely to be heard by many people in mono, it is obviously desirable that the mono signal should be artistically satisfying. There are several reasons why this may mean modifying the stereo balance (or signal) a little:

Loudness. Whereas a sound in the centre appears equally loud in both channels, those at the side are split between the M and S channels. As the S information is missing in mono, side sources may be lower in volume than is desirable. Sometimes instruments that have been panned too far to the side sound louder when the balance is heard in mono. This 'centre channel build-up' can be eased by bringing them back in a little in the stereo balance.

Reverberation. The S signal has a higher proportion than M; the mono version may therefore be too dry. One way round this is to make the main stereo balance drier still, and then add more reverberation on the M channel (if this format is employed). In crossed figure-of-eight balances, in particular, there is heavy loss of reverberation due to cancellation of out-of-phase components, and some people would see this as a reason for preferring crossed hypercardioids (in which the effect is less) or crossed cardioids.

Aural selectivity. Individual important sound subjects may not be clearly heard in the mono version, although in stereo they are distinct because of spatial separation. It may be necessary to boost such subjects or so place them (towards the centre of the stage) that in the mono version they are automatically brought forward.

Phase distortion. Spaced microphone techniques may not adapt well to mono, as the subjects to the side have phase-distortion effects: systematic partial addition and cancellation of individual frequencies that no longer vary with position. To check on these, feed A + B to one loudspeaker.

In addition, there are things that are less of a nuisance in mono than stereo. For example:

Extraneous noises. Noises that are hardly noticed in mono may be clearer in stereo because of the spatial separation. Worse, any noise that moves (e.g. footsteps) is particularly distracting unless it is part of the action and balanced and controlled to the proper volume, distance and position. Also, noises from the rear of bidirectional pairs will appear to come from the front, and movement associated with them will be directionally reversed.

When surround sound is heard in stereo in the home, it is often automatically subjected to a process called *downmixing*. In the simplest version, at the receiver the surround left is added to left and surround right to right, with centre divided equally left and right. The subwoofer signal is either blended equally or omitted. If the console has a 'downmix' button, use this to check the effect: the result may be totally unsatisfactory. If so, consider what might be done to improve matters. Because of this, a separate stereo mix is often made and recorded on tracks 7 and 8 of an eight-track mix-down. Super Audio CD (SACD) and DVD-Audio can carry both surround and stereo and will reproduce whichever is required. Some programs downmix more sympathetically (also to stereo headphones).

Chapter 7

Speech balance

Some of the following suggestions may seem almost trivial, but they all help to make a studio (often, here, a radio studio) run smoothly and produce good, clean sound recordings or broadcasts.

Speech balance depends on four factors: the voice, the microphone, and the surrounding acoustic conditions and competing sounds. In a good balance the voice should be clear and sound natural, perhaps reinforced and brightened a little by the acoustics of the room that is used as a studio. There should be no distracting noise, no sound that is not an appropriate and integral part of the acoustic scenery.

A studio designed mainly for speech and discussion may be about the size of a fairly large living room, with carpet on the floor and often treatment on the walls and ceiling to control reverberation. Listen to both the nature and duration of reverberation: a large, open studio, even if dead enough, will sound wrong, and a speaker needs screens set in, not to reduce the studio's reverberation, but to cut its apparent size by introducing more short-path reflections. A warm acoustic might match that of a domestic sitting room, but for film narration audible studio ambience might conflict with picture, so for this and certain other purposes there is more absorber and much less reverberation (often, for convenience of layout, in a smaller space).

Microphones for speech

We begin with one of the most basic set-ups in radio: one or two people at a microphone, and on the other side of the glass, another to balance and control the sound. (This is of course not the very simplest set-up: that would have just one person, solo, at a microphone.)

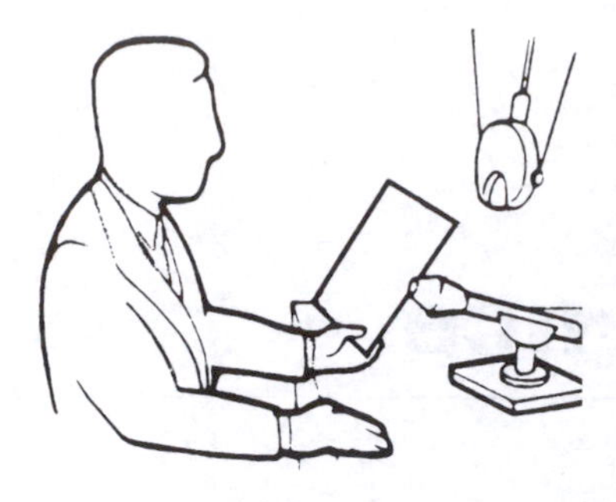

[7.1] **Microphone position**
This shows a good position for speech, with head well up. The script is also held up and to the side of the microphone.

Microphone and acoustics are interrelated. For a given working distance an omnidirectional microphone requires far deader acoustics than a cardioid or bidirectional microphone, which by their directivity discriminate against the studio reverberation and any coloration that goes with it. A hypercardioid, with its greater directivity, reduces these still further. On the other hand, the omnidirectional types (including hypercardioids or cardioids which become omnidirectional at low frequencies) permit closer working.

Consider first a microphone for which, by this reckoning, the studio acoustics are important: the ribbon – which despite its archaic image is still used in some studios because of its excellent figure-of-eight polar response. At its normal working distances, 60–90 cm (2–3 ft), it has a reasonably flat response throughout the frequency range of speech. There will be some roll-off above and below that range, but that is no disadvantage, because these limits discriminate against low- and high-frequency noise. To study the interplay of voice and unfamiliar acoustics, experiment with balance tests as described in Chapter 6. Try different voices: some benefit more by close balance than others. Do this well ahead of any important recording or broadcast, as a last-minute series of microphone adjustments can be disconcerting.

To help get a good signal-to-noise ratio, the speaker can be encouraged to speak clearly and not too softly – if that style is acceptable. On a bidirectional microphone, two voices may be balanced on opposite sides: this is easy enough unless they have radically different qualities. An interviewer can sometimes help by adopting a level that acoustically balances the voices. In an interview where the primary aim is to bring out the views and personality of the interviewee, it often sounds good if the interviewer's voice is a little lower in volume – although the audience may like that more than some interviewers do.

If the relative loudness of two voices does not sound right on the balance test, it may help if the microphone is moved slightly, to favour one or the other. But listen for any tendency for the more distant voice to sound acoustically much brighter than the other; rather than allow this, introduce some control (i.e. adjust the gain) with just a slight dip in level as the louder speaker's mouth opens. Alternatively, balance the voices separately and closer, perhaps on hypercardioid microphones. Random variations in loudness are less tolerable when heard over a loudspeaker than when listening to people in the same room, so if nothing can be done to stop them rocking back and forth, or producing big changes of voice level (without breaking the flow of a good spontaneous discussion), some degree of control is necessary.

It may help if the microphone is less obtrusive, e.g. mounted in a well in a table between interviewer and interviewee. Sight lines are better, but noise from the movement of papers on the table may be worse.

An omnidirectional microphone at a similar distance would give far too full an account of the acoustics, so this field pattern requires an even closer balance. A disadvantage is that the effect of movement in and out on a fixed microphone is more pronounced. However, with a very close balance the importance of the acoustics is diminished, so compensation by fader alone is less objectionable.

In practice, omnidirectional microphones lie somewhat uncomfortably between those designed to use acoustics and those that eliminate them. As a result, microphones that really de-emphasize acoustics are often preferred: it is where the hypercardioids come into their own. They pick up less ambient sound and allow closer balance – partly because of the reduced pressure gradient component (compared with bidirectional types), but perhaps also because they may become somewhat more omnidirectional at low frequencies. This last effect may, however, begin to bring a boomy low-frequency component of the room acoustic back in to the balance, which in turn is de-emphasized by any bass roll-off in the microphone itself.

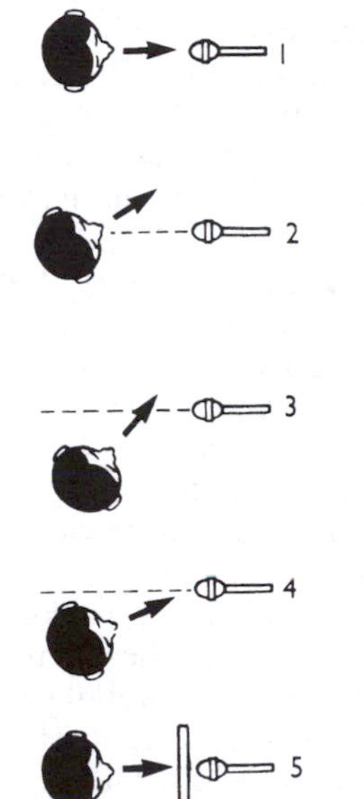

[7.2] **Popping**
Problems with plosive 'p' and 'b' sounds. 1, Plosives spoken (or sung) directly into a microphone may send the diaphragm beyond its normal range. 2, Speaking on-axis but to the side of the microphone may cure this without changing frequency response. 3, Speaking across and past the microphone may cure popping and also change frequency response. 4, Speaking across the microphone but still towards it may still cause popping. 5, A simple screen of silk or nylon mesh over a wire obstructs the puff of air, but does not significantly affect the sound.

Hypercardioids come in a wide range of design characteristics. Variables include angle of acceptance, also the way in which the polar response changes at low frequencies, the degree of intrinsic bass roll-off, and whether additional switched bass cut is provided. At high frequencies the differentially changing response at angles of 30–45° to the axis may be used for fine-tuning the response to particular voices. For increased intelligibility, especially against competing background noise, use a microphone (or equalization, EQ) with a response that is enhanced somewhere between 3 and 10 kHz, but check for sibilance.

Using hypercardioids, voices can be close balanced individually or (for greater numbers) in close pairs, and can sound acoustically dead even in a live room, without sounding harsh. This style is also suited both to voice-over on film and to radio documentary narrative, as well as news reporting and links between segments with different qualities and acoustics. But it does also emphasize mechanical noises from the lips and teeth that sometimes accompany speech, and increases the danger of blasting (popping), particularly on plosive 'b' and 'p' sounds.

There are several ways to deal with popping. To sample the cause, project 'p' sounds at your hand: you will feel the air blast that might push a microphone diaphragm beyond its normal limits. Then move the hand a little to one side. This demonstrates one solution: to speak past the microphone, not directly into it. But a windshield (pop-shield) may also be needed, if not already part of the microphone design. A screen can be made up (as it was originally) from coat-hanger wire and borrowed tights. Set it in about 10 cm (4 in) from the microphone.

Some disc jockeys prefer a close, intimate style with very little of the studio acoustic in the sound. It is possible that the wish to work close to the microphone is partly also psychological: the performer gets a feeling of speaking softly into the audience's ear. A male disc jockey with a soft, low voice only makes it worse by gradually creeping in to an uncorrected directional microphone and accentuating the bass in his voice, particularly when presenting pop records that have what

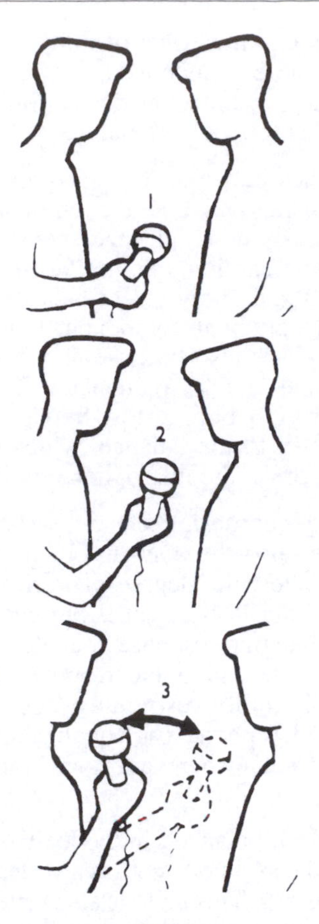

[7.3] **Interviewing techniques**
Position 1: Microphone at waist or chest height – sound quality poor.
Position 2: At shoulder or mouth level – sound quality fair, or good if interviewer and interviewee are close together in satisfactory acoustics and quiet surroundings.
Position 3: Moving the microphone from one speaker to the other – sound quality good. In a fourth variant, the microphone favours the interviewee most of the time.

in other fields of recording would be regarded as an excess of top. Once a good balance has been found, microphone discipline is necessary.

Interviewing with a hand-held microphone

For an interview using a hand-held omnidirectional, cardioid or hypercardioid microphone, experiment with four basic balances:

- Hold the microphone at or just above waist level, so its presence is unobtrusive. This has sometimes been tried, either to keep it out of sight on television or when it is felt that the interviewee might be put off by anything closer. It gives poor quality (very poor, if the microphone is omnidirectional), with too much background sound or too reverberant an acoustic. But a picture can distract conscious attention from sound quality, so provided speech is clearly intelligible this balance may sometimes be permitted. The cone of acceptance should be as narrow as possible.
- Hold it between the two speakers at chest or neck level, with the face of the microphone pointing upward. If it has cardioid to hypercardioid response, both speakers must be within the angle of acceptance, and this may be adjusted during dialogue to favour each in turn. For television, this balance is better.
- Moving the microphone close to each mouth in turn. This balance is needed particularly with omnidirectional microphones, for which the distance is judged according to the acoustics and the noise (unwanted) or 'atmosphere' (wanted) in the background. The interviewee may find the repeated thrust of the microphone distracting, but if the interviewer times the moves well the overall sound quality will be good, and any wind-noise and rumble is less than with directional microphones that are held further away.
- Hold it close to and favouring the interviewee, with only slight angling back during questions, the interviewer relying more on clarity of speech, voice quality and well-judged relative volume.

Where a monophonic balance is heard as a component of stereo, try to discriminate against background sound and reverberation, so those do not seem to come from the same point. With a stereo end-fire interview microphone, check the directional response. This may be clearly indicated on the microphone, but if not, as a reminder, you might mark it with an arrow on tape on the side of the microphone.

Because the interviewer should be close to the subject, this may feel less confrontational if they stand at about 90° or sit side by side. If the loudest extraneous noises come from the direction they are facing, the microphone can be angled to discriminate against it.

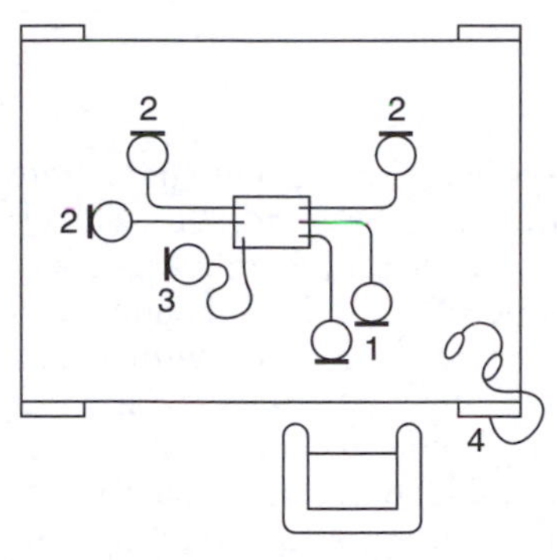

[7.4] **Presenter and guests: I**
All microphones are hypercardioid, arranged to allow individual control of each voice. 1, In this layout for radio, the presenter has a condenser microphone with a moving coil as stand-by. 2, Moving-coil microphones for guests, who take their places at the table while records or pre-recordings are played. 3, For an additional guest, if required.
4, Headphone feed points are under the edge of the table, with switches to select the feed.

Three or more voices

For three or four people in a radio interview or studio discussion, the ribbon microphone has been widely used in the past, and still has its advocates. With its bidirectional response a round table layout is not possible: four speakers must be paired off, two on each side, or three divided two and one. The microphone can be moved or angled a little to favour lighter voices, and two speakers on the same side of the microphone should sit close together so neither gets off-microphone. Participants should be placed so those who are side by side are more likely to want to talk to the others across the table. The change in quality when one turns to a neighbour can be marked, and leaning back at the same time magnifies it.

A more common arrangement has two to six people around a table, each on a separate hypercardioid microphone at about 30 cm (1 ft). The ambient sound pick-up is that of all the open microphones combined, and in total is broader than that of a single cardioid or bidirectional microphone, but the much closer balance more than compensates for this, and in any case allows greater individual volume control. If the order of speakers is known (or can be anticipated by watching the discussion), some faders can be held down a little, favouring those already talking. Different-coloured foam covers on the microphones are used both to guide speakers to the correct place to sit and, by colour coding the fader keys, to match these in turn to the voices. With more people, one microphone may be assigned to each two speakers, who must be close together.

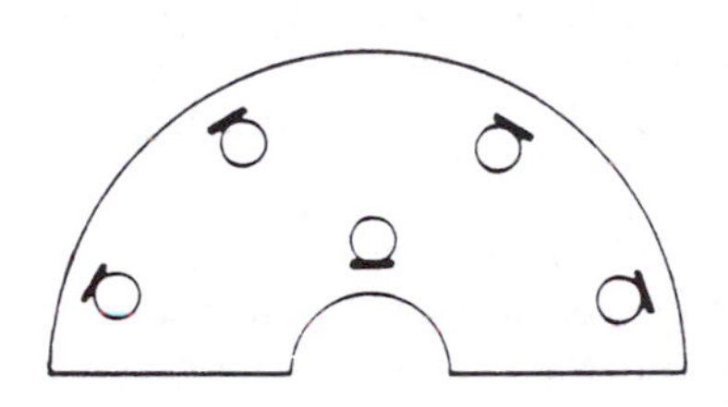

[7.5] **Presenter and guests: II**
Another table layout for radio.

Any nearby reflecting surface affects a balance: the reflected wave interferes with direct sound, producing phase cancellation effects and making the quality harsher. To reduce this, tables in radio studios (and for some television discussions too) can be constructed with a surface of acoustically transparent mesh. If a microphone well is set into this, perhaps with a small platform suspended by springs or rubber straps, it should be open at the sides. For insulation against bumps or table taps, rubber is better when in tension than it is in compression.

In a sufficiently dead acoustic there is one further possibility: an omnidirectional boundary microphone on a hard-topped table.

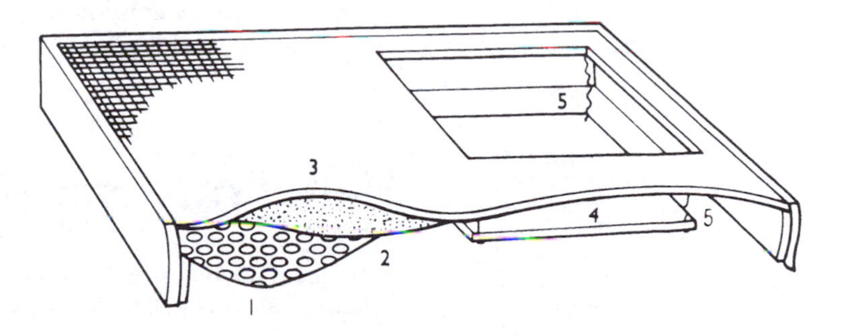

[7.6] **Acoustic table with well**
The table top has three layers: 1, Perforated steel sheet. 2, Felt. 3, Woven acoustic covering (of the kind that is used on the front of loudspeakers). These are not bonded together, but held at the edges between wooden battens. 4, Wooden microphone platform, suspended at the corners by rubber straps (5).

[7.7] **Round table discussion**
Six speakers with a cardioid microphone, either suspended above them (1), or sunk into a well (2). This is not ideal, as all speakers are somewhat off-microphone and are likely to be a little too far back.

Studio noise

For many purposes (e.g. narration, plays, etc.) background noises are undesirable. They are reduced to a minimum by good studio discipline, the elimination of potential problems during rehearsal, and critical monitoring. A check at the time of the initial balance will show if *room atmosphere* is obtrusive when compared with voice. 'Atmosphere' means the ambient noise that pervades even quiet rooms, and if it seems loud and cannot be reduced, then either voice levels might be raised or the balance tightened. Unfortunately, the sort of voice quality that stands out really effectively against background noise is more suitable for a hard-sell commercial than ordinary talk.

Even in radio studios designed to discriminate against external noise, there can be problems: ventilation systems may carry noise from other parts of the building, or the flow of air can sound noisy when compared with quiet speech. Structure-borne noise is almost impossible to eliminate, and rebuilding work is complicated by the need to avoid all noisy work when nearby studios are recording or on air.

Noise may also come from the person speaking. A lightly tapped pencil may resound like rifle fire. A slightly creaky chair or table may sound as if it is about to fall apart. Lighting a cigarette can be alarming; lighting several, a pyrotechnic display. Tracking fainter sounds to their source may prove difficult – embarrassing even. A loose denture may click; a flexed shoe may creak. Some people develop nervous habits at the microphone: for example, a retractable ballpoint pen can become a menace as unseen hands fiddle with it, to produce sharp, unidentified, intermittent clicks.

Several things can make such noises even more irritating. One is that some people listen to broadcast or recorded speech at a level that is louder than real life, so a quiet voice is reproduced at much the same volume as one that shouts. In consequence, extraneous noises become louder too – and, worse, if heard monophonically, appear to come from the same point in space as the voice. In contrast to a visually identified sound that is seen in context and so is accepted, unidentified noises do not. Many experienced broadcasters know this and take it into account. A studio host will notice that a guest's movements are not silent and say, 'Now, coming to join me, is . . .' When explained, a noise is reduced to its proper proportions, so is not consciously noticed by the listener.

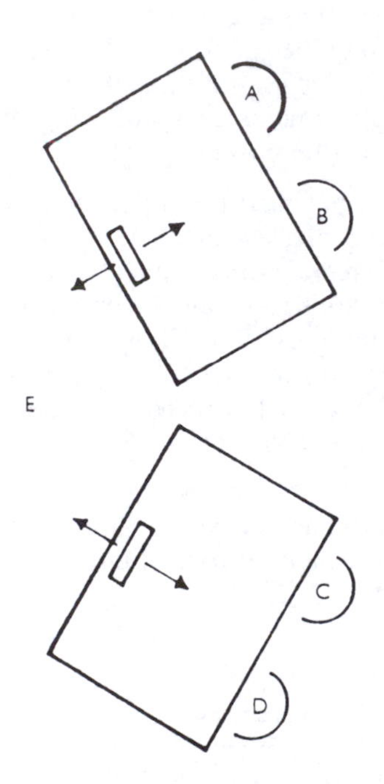

[7.8] **Two bidirectional microphones in mono**
Voices A and B are on the dead side of microphone 2 and vice versa: this is a simple balance for a game show, etc., recorded before an audience. Any voice at E would be picked up on both – provided the two were in phase (as they should be).

Handling a script

A frequent source of extraneous noise at the microphone – and an irritating one – is paper rustle. Ahead of any broadcast or recording

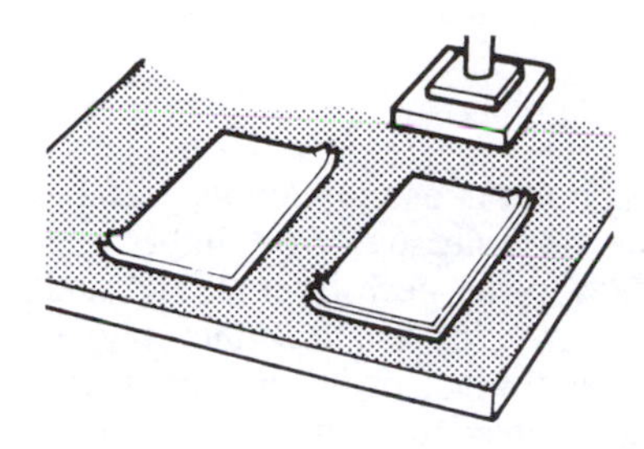

[7.9] **Avoiding script noise**
Corners of script turned up to make it easier to lift pages noiselessly to one side.

[7.10] **Script rack**
No reflections from the script can reach the microphone. The rack may be made of acoustically transparent mesh. The speaker might need to be reminded not to look down when reading from the foot of the page.

from a script or notes it makes sense to give an inexperienced speaker some suggestions on how to handle paper quietly:

- Use a fairly stiff paper (copy paper is good). Avoid using sheets that have been folded or bent in a pocket. Avoid thin paper (e.g. airmail paper or flimsies).
- At a table, take the script apart by removing any paper clip or staple. It is not easy to turn pages silently, so prepare a script by bending up two corners, so each page can be lifted gently aside. (Sliding one page over another will be audible.) Alternatively, hold the script up and to the side of a directional microphone.
- To avoid reflections of sound from the paper that will interfere with direct sound and cause coloration, hold or place the script a little to one side. But do not turn off-microphone or drop the head to read it.
- When standing at a microphone (as actors who need to change perspective must do), leave the script clipped together. Hold it well off-microphone when turning pages – or when using a bidirectional microphone, hold the script up and to the dead side of it.
- When there are several people in a studio (e.g. to perform in radio drama, a sketch or a complex commercial) they should not all turn their pages together. If performers remember to stagger the turnover, they will also remember to turn them quietly.
- If there are only a few sheets of script, try putting them in clear, stiff plastic envelopes.

One device sometimes used is a script rack. Sloping up at an angle to the microphone, it sends reflections that might cause coloration away from it. Unfortunately, it does not prevent a persistent offender from looking down to read the bottom of each page.

Speech in stereo

Whether intended for radio, television or even film, much speech that will form part of a stereo image is, initially, balanced on monophonic microphones. In particular, radio announcements are normally taken mono, and are often placed in the centre, so that listeners can use this to balance two loudspeakers. If the announcement is in a concert hall, a monophonic microphone is again used with a relatively close and dead balance, perhaps with a stereo pair left open at a distance to add stereophonic reverberation. The announcer and microphone are placed where little direct sound can be picked up in stereo; otherwise the position of the announcer might shift (with a change in voice quality) when stereo microphones are faded up or down.

In discussions, individuals are panned to a range of positions on an arc.

One reason for avoiding a stereo pair on speech balances for which no movement is expected is that any slight lateral movement close to the microphone is emphasized and becomes gross on the sound stage.

Boom operation

In contrast to radio, where the microphone is usually static, and the performer works to it, in television there are cameras, lights and action, and the microphone must work to the performer. Radio and television differ most strongly when, in addition, the microphone must be out of the television picture, as the performer moves freely within it. The 'microphone with intelligence' that moves and turns with the action, always seeking the best practicable balance, is the boom microphone. Although its greatest virtue – a capacity to cover long, continuous takes of multi-camera action – is less in demand now that digital editing makes it easier to work in television in shorter sequences or even shot by shot, boom operation is still central to the subject of speech balance in television, and film too. Almost everything a boom operator does is related to or affects balance, so boom microphone coverage is introduced here. How boom and other television studio operations are planned is described in Chapter 20.

From a given position a boom can cover a substantial area, and the platform can be tracked to follow extended action, or repositioned between scenes. In an ideal balance for best match of sound to picture, the microphone is above the line between subject and camera. One factor that affects its main, chosen position is the direction the key light comes from: the boom will generally angle in from the opposite side. Another factor is ease of movement: it is, for example, quicker to swing the arm than to rack in and out when shifting from one subject to another. A boom operator can quickly turn the microphone to favour sound from any chosen direction, or (almost as important) to discriminate against any other. Characteristically, in mono, or for the middle (M) signal in stereo the microphone will be cardioid or near-cardioid, with a diaphragm that is not noticeably affected by rapid movement and is also insensitive to or protected from the airflow this causes, so that the operator is not restricted in speed of response to action. In practice it is best to use a directional condenser or perhaps moving-coil microphone, with a shock mounting and windshield (although the most effective windshield is bulky and can make movement clumsy). For stereo pick-up, MS microphones are often preferred: the 'M' component will generally be located cleanly within the picture.

The boom operator has a script (where one is available) and generally moves the microphone to present its axis to favour whoever is speaking or expected to speak. Although a cardioid has a good response over a wide range of directions, there may be a 2 or 3 dB high-frequency loss at about 45° to the axis, increasing to perhaps 5 dB at 60°, so for the best coverage all these twists and turns really are necessary. The microphone will be dropped as close to performers as the picture allows and this closeness accentuates any angle there may be between speakers. This means that by proper angling there can be not only the best signal-to-noise ratio for the person speaking, but also automatic discrimination against the one who is

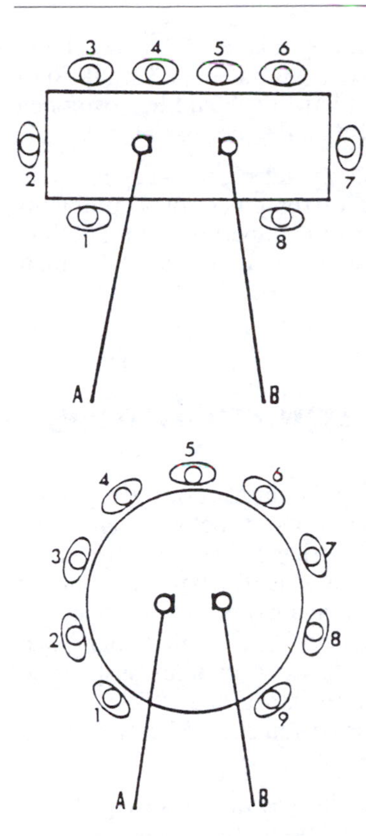

[7.11] **Boom coverage of a large group on TV**
Some groups are difficult to cover on a single boom. In these examples, two cardioids are used. *Above*: Boom A covers speakers 4–8; B covers 1–5. *Below*: Boom A covers speakers 5–9 (with 1 and 4 at low level); B covers 1–5 (with 6 and 9 at low level). For other examples of boom coverage, see Chapter 20.

not. If both speak together, it may be decided to favour the person with the most important lines, or perhaps to cover both equally. With performers of different head heights the microphone should be placed either well in front of them or to the side that favours the shorter.

The microphone position must be a compromise between the positions of all the speakers, and if they move apart, or one turns away, the split can become impossible to cover by a single boom. In this case a second microphone may be set in on a stand out of vision, or hung over the awkward area, or concealed within the picture. Sometimes a second boom is used, but this increases lighting problems.

For 'exterior' scenes that are shot in a studio, the microphone must be kept as close as the limits of the picture allow. To contrast with this, for interiors use a slightly more distant balance or add reverberation.

Because a number of different types of microphone may be used for a single voice, the control console in a studio has equalization associated with each channel that can be used to make the microphones match as closely as possible. If correction is needed, the boom microphone is usually taken as the standard and the others matched to it before any further treatment (such as 'presence') is applied.

Another justification for the boom is that given suitable (rather dead) acoustics, it automatically gives a perspective appropriate to the picture: close shots have close sound; group shots have more distant sound. The boom operator uses vision monitors in the studio to check the edge of frame, dropping the microphone into the picture from time to time during rehearsal. A sight line against the far backing can be used to set and check its position for recording or transmission.

In television the sound supervisor can call the boom operator (and other staff on the studio floor) over his own circuit, which cuts in on the director's circuit to the operator. (This may be automatically preceded by a tone-pip to indicate that the message is from the sound supervisor.) During rehearsal and any time when studio sound is faded out, the boom operator can reply or call the supervisor on a separate reverse talkback circuit. A boom therefore has a multicore cable, permitting communication as well as transmitting programme sound, and returning the mixed sound to the operator's headphones.

In some kinds of television discussion the frame is unpredictable, but this is partly compensated by the influence of an audience, which limits how far performers will turn. For this, the boom may be equipped with a highly directional microphone. Although the inertia of a tube that is nearly half a metre (18 in) long makes it clumsy to handle for big, rapid movements, it can be placed farther back.

Where two people are in a 'face-to-face' confrontation, but far enough apart for a boom operator to have to turn a cardioid microphone sharply to cover each in turn, two hypercardioids may be better, pointing outwards at about 90° to each other for low close working. In a drama this would restrict the use of the boom for other scenes, so in a similar situation – or, for example, for a number of people round a large table – two booms might be brought in.

A boom can cover action on rostra built up to a height of, say, 1.8 m (about 6 ft) provided the performers keep within about 3 m (10 ft) of the front of the high area. The boom platform should be positioned well back, to keep the angle of the arm as low as possible.

The boom operator should occasionally take off headphones to check for talkback breaking through or for other sound problems in the studio and report them back to the supervisor. In addition, being visible to artists, the operator should react positively to their performance.

Slung and hidden microphones

In television balance one aim is to keep the number of microphones to a practicable minimum, and the use of booms helps towards this. But unless set design and action are restricted there may be places that are inaccessible to booms. For these, additional microphones may be slung above the action – ideally, again, just out of frame, and downstage of the performer's planned position, so that sound and picture match. If a slung microphone has two or three suspension wires, its position can be adjusted in rehearsal, and if there is any danger of its shadow falling in shot, provision should be made for it to be flown (raised out of shot) when not in use.

Microphones are sometimes concealed in special props shaped over wire gauze, then painted or otherwise disguised, though an electret is small enough to be hidden more easily. Placed on or very close to a reflecting surface, it acts as a boundary microphone, but check for the effect of hard surfaces further away that may modify the top response. Sometimes equalization can help by reducing high-frequency peaks so the quality is not so obviously different from that of the boom, but it cannot restore cancellation.

Microphones in vision

The presence of a microphone in vision must be natural, intentional and undistracting. A boom microphone that can be seen in the top of frame is disturbing, particularly if it moves to follow dialogue. On video recordings and film this blemish can often be painted out electronically in post-production, which also (conceivably) means that sound coverage might be improved deliberately, with the paint job planned for.

Most microphones used in vision are either *personal*, fixed below the neck or held in the hand, in which case they are moved around by the performer, or they are *static*, on table or floor stands.

For a speaker who is not going to move, a fixed microphone in the picture gives better sound. It will generally be angled to favour the

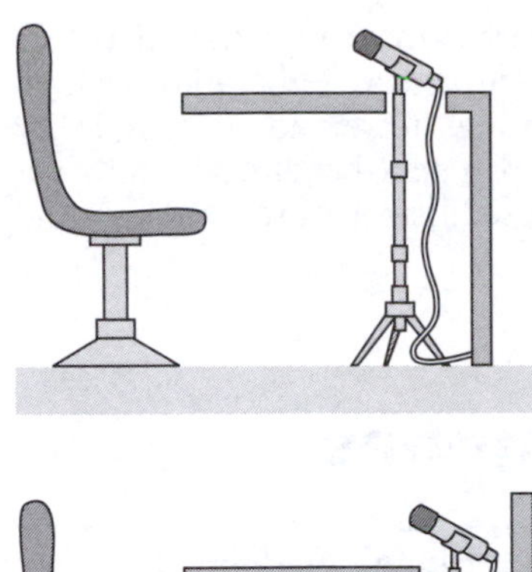

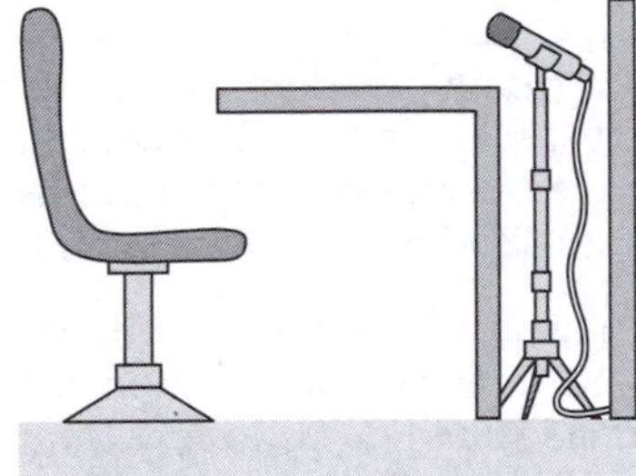

[7.12] **Television desk and microphone**
To avoid bumps, the microphone is on a floor stand, emerging through a hole (*above*), or behind a false front (*below*).

speaker (or split between two): if this chosen angle conflicts with the designer's choice of 'line' within the picture the microphone may be more noticeable, as it will also be in a simple uncluttered design.

The base, for stability and to cushion shocks, needs to be reasonably large, so it can sometimes look more obtrusive than the microphone. If it is recessed into a console, the cavity must be designed with care: if it is fully boxed in it can modify the local acoustics, so one or more sides of the recess should be left open. The microphone capsule itself should stand clear of the surface, so in a shot showing the table it will be in vision, unless it is hidden behind something on or in front of the table – which in turn could be more obtrusive than the microphone. Alternatively, a specially constructed console might have its front continue upward a little beyond the level of the desktop. An acoustically transparent top can be included in the design.

A microphone placed to one side may be neater than one immediately in front; it should go on the side to which the speaker is more likely to turn. However, any turn the other way may go right off-microphone, so it helps if any neighbouring microphone can cover this.

Where many people are spread over a large area of studio, a separate microphone is sometimes used for each individual – but quality demands that as few as possible be open at any one time. So the right ones can be faded up in time, it should be made clear who is to speak next, either from a script or by introduction by name or visual indication. When the output of individual microphones is recorded on multitrack, an improved (or corrected) mix can be left until later.

The use of hand microphones is a widely accepted convention in television reporting. It is convenient for the reporter plus cameraman combination, and also because the microphone can be moved closer if surroundings become noisier. In extreme conditions lip microphones have been used in vision. But hand-held microphones can easily be overused, particularly where they make an interviewer look dominating and aggressive. Alternatives include the use of small personal microphones (often on a lapel) or, for exterior work, gun microphones.

In interviews or discussions, participants using personal microphones can be steered by panpots to suitable positions on the left or right of frame, roughly matching those in the widest shot. In an unscripted discussion it is better to make no attempt to match position in frame: this makes editing difficult and in a television picture a small discrepancy is not likely to be noticed. The pictures establish a 'line' that the camera should not cross, and sound can follow the same convention. Two speakers (as in an interview) can be placed roughly one-third and two-thirds across the stereo image. But note that as two omnidirectional microphones are moved towards each other, they will 'pull' closer still, as each begins to pick up the other subject, and this may need correction by widening the image.

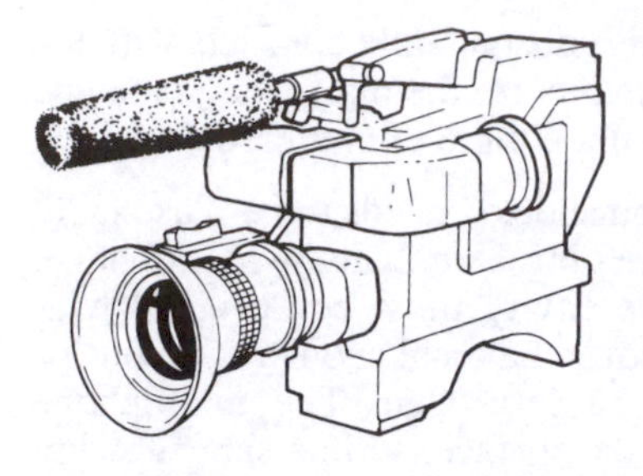

[7.13] **Video camera with short gun microphone**
This ensures that loud effects in the field of view are recorded. For television sports, a full-scale gun microphone is used to capture more distant effects.

The main benefit from stereo output here is that separation of voices from each other and from sound effects and reverberation (which should be full width), as well as from any noise, aids intelligibility and adds realism. A separate stereo pair can be directed into the scene and faded up just enough to establish a stereo background effect.

Using gun microphones

A gun microphone is most useful when filming outdoors, held just out of frame and pointed directly at each speaker in turn. Camera and sound must co-operate to avoid getting the microphone in picture: its big windshield is obtrusive even when still, and more so if it moves.

An advantage of this microphone is that it allows great mobility, though not of course for the speaker to turn right away from the microphone. Accidentally off-microphone sound drops to low level in exterior work, and gives bad quality in interiors. The gun microphone can, however, be used in the relatively dead surroundings of a television studio (in an area well away from strongly reflecting surfaces) for picking up voices at a distance. But in general its directional qualities are not trustworthy in normal interiors or even reverberant exteriors.

The short gun microphone can also be attached to a video camera, pointing along its line. It picks up many sound effects adequately within the field of the average wide-angle lens, but good voice balance only when the camera is close in. One application is to cover events or news when no separate recordist is available.

When a highly directional microphone is used to cover effects such as tap dance, or to follow the action in sports (to some of which it adds great realism), it is often best to filter frequencies below about 300 Hz. A gun microphone with bass-cut has sometimes been used in conjunction with other microphones to cover large press conferences, supplying presence that other distant microphones lack. For this, a parabolic reflector could also be considered: it would be out of frame, or possibly hidden behind a gauze that is lit to look solid. Remember that a reflector concentrates, and so amplifies, the wanted sound.

Another, more specialized microphone for full-frequency pick-up of distant sources is the adapted-array (digitally processed) combination microphone: this has a forward lobe 8° across, switchable to 8° × 30°.

Using personal microphones

For location work, an electret is often fixed to the clothing of each speaker – but take care if two are brought close together, as this can affect the balance. (Older, heavier personal microphones were hung around the neck, and these may still be encountered.) But always remember that a boom would give better sound – often much better.

The electret can often be hidden, perhaps behind a tie – though some synthetics cause crackles, and a woolly tie can produce an equally woolly response. Whether it is concealed or not, it is best to ensure that the means of suspension is also invisible or low-key – for example, by using a clip or cord of the same tone or colour as clothing against which it may otherwise be seen. Some capsules are available in a range of colours, such as black, white, silver or tan (flesh tone). Otherwise a slip-on cover of thin, matching material can help.

The user may have to avoid large or sudden movements, especially those that disturb the drape of clothes (as sitting down or standing up does with a jacket). However, normal movement should not be discouraged unless necessary. A speaker who turns right away from the camera may still be heard clearly – though turning the head to look over the wrong shoulder may cause speech to be off-microphone.

The cable may be concealed beneath a jacket for a tight mid-shot (although a personal microphone may not really be needed for so close a shot). In the more general case it will go to a hidden radio transmitter or continue on its path beneath the clothing, down a trouser leg, or inside a skirt – to a connector not far from the feet, so the wearer can easily be disconnected during breaks. If the user is going to walk it may be best to attach any cable (and connector) to a foot, perhaps by tying it in to a shoelace. But to walk dragging a length of cable is awkward, and perhaps visibly so. It may also cause cable noise – reinforcing the case for a radio link.

Because of its position, the sound on a lapel microphone may be characteristically poorer than for better angles of pick-up. This makes it difficult to match, so once started, it is best to persist with it until there is either a marked change of location (e.g. exterior to interior) or a long pause before the same voice recurs. To a lesser degree, this is also true for many other microphones. The effect of disregarding this is obvious on many television news broadcasts.

A further disadvantage of personal microphones when compared with booms is that there is no change of aural with visual perspective. This is more important in drama than in other productions. But they do usually offer reasonable intelligibility in fairly noisy surroundings, or in acoustics that are brighter than usual for television. The quality gains a little from good reflected sound.

Worn by film or television actors, they will generally be hidden. For ensemble scenes in some films, all major players have been recorded separately, giving the director more options in post-production.

For a hand-held microphone in vision, the sound quality problem goes away, to be replaced by the obvious visual problem, especially if it partly covers the performer's face. Visually, short-stub UHF radio microphones are better than the less common trailing-antenna VHF, and both are usually better than trailing cable. These and microphones on headsets are more often used for music, as described in Chapter 8.

Radio drama

Although virtually extinct in the United States, some American readers may find interest in this account of what they may think of as a lost art. In fact, radio drama is still vigorously alive in countries such as Britain, where both the skills and the facilities for it still exist. The techniques are of course also potentially of use for other complex voice recordings. In some countries (again including Britain) they are also used for educational radio.

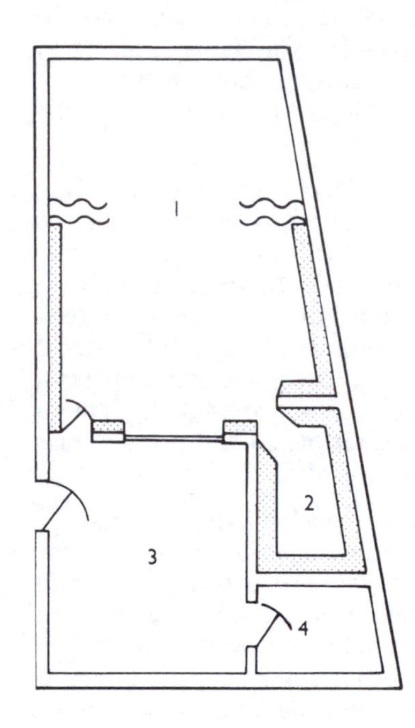

[7.14] **Drama studio suite**
The main areas in a layout of the kind used for broadcast drama: some studios of this kind still exist and are used for their original purpose. In this example a virtue has been made of the irregular shape of a site on which some of the walls are not parallel, so there is less standing wave coloration. 1, Main acting area, with 'live' and 'dead' ends of the studio that can be partially isolated from each other (and their acoustics modified) by drawing thick double curtains across the middle. 2, A 'dead' room, with deep absorbers on all walls. 3, Control room (for acoustic isolation there will be double doors). 4, Machine room.

In radio drama the acoustics of a studio are used creatively, the 'scenery' being changed by altering the acoustic furniture around the microphone. Specialized studio suites are divided into areas with different amounts of sound-absorbent or reflecting treatment; each area provides a recognizably different quality of sound. The live areas offer a realistic simulation of the actual conditions they are supposed to represent; the dead areas, less so. If similar microphone types are employed in the different acoustics, the voice quality does not change but is matched from scene to scene as the action progresses.

Since one factor in the aural perception of studio size is the timing of the first major reflection, delay may sometimes be used to increase apparent size without creating excessive reverberation. But the quality of the reverberation itself also plays an important part. A small room sounds small because first reflections are very early. A combination of this with reverberation time suggests both size and the acoustic furniture of a room: a sound with a reverberation time of 1 second may originate from a small room with little furniture (therefore rather bright acoustically) or from a very large hall with a rather dead acoustic. The two would be readily distinguishable.

For example, a 'bathroom' acoustic would be provided by a small room with strongly sound-reflecting walls, and can be used for any other location that might realistically have the same acoustic qualities. Similarly, a part of the main studio (or a larger room) with a live acoustic can be used to represent a public hall, a courtroom, or a small concert hall – and when not needed for drama can also be used as a small music studio. When going from one type of acoustic to another, balance (in particular, distance from the microphone) can be used to help to differentiate between them; for example, the first lines of a scene set in a large reverberant entrance hall can be played a little farther from the microphone than the action which follows – it would be hard on the ears if the whole scene had heavy reverberation.

For monophonic drama, ribbon microphones have been widely used: their frequency range is well adapted to that of the voice, and with

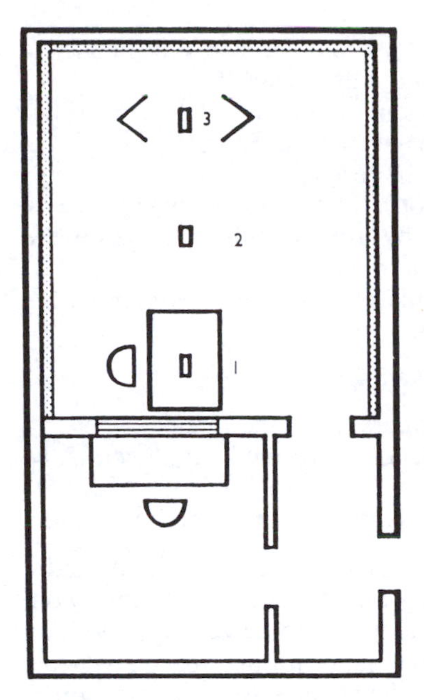

[7.15] **Simple layout for drama (mono)**
A simple set-up using three ribbon microphones for dramatic work. 1, Narrator (close to window). 2, 'Open microphone' – for normal indoor acoustics. 3, 'Tent' – to represent outdoor quality. This is a much less versatile layout than can be obtained in a multiple-acoustic studio, and the results sound more stylized. This might be more acceptable for broadcasting directed to schools, because: (a) school audiences often have poor listening conditions; (b) the number of actors is likely to be small; and (c) educational broadcasting generally has a low budget.

working distances of up to 1 m (about 3 ft) give a generous account of studio acoustics. Actors can work on either live side of the microphone, standing two or three abreast, and can move 'off' either by moving back or by easing round to the dead side. The conditions are comfortable, flexible and easily understood. The actor is working *with* the acoustics.

Still for mono, for a large, open acoustic with many actors, such as a courtroom, public meeting or council chamber scene, an alternative to the ribbon is a cardioid suspended above the action. Working distances and speed of approach or departure are set by listening to the acoustic balance. An additional hypercardioid microphone can be set in for quiet asides that require little reverberation.

For either mono or stereo, if there is a narrator as well as the 'living-room' action, the two must be recognizably different. There will, of course, be differences in the style of acting, and the use of the fader at the start and end of each scene will encapsulate it, and so further distinguish between the two. But acoustics can also emphasize the differences: if the scene is played well out in an open studio the narrator can be placed near one of the walls and at a closer working distance to the narrator's microphone, which may be of a different type – often a hypercardioid. The distance from this to the microphone covering dramatic action must allow for any rapid turn from one to the other (although more often they will be recorded separately). A screen near the narration microphone may help to increase the contrast. Note that even for stereo drama, a narrator will often be mono.

It is sometimes necessary to mix acoustics: to have two voices in acoustically different parts of the same studio, but taking care to avoid spill. There is more danger of spill where the voice in the brighter acoustic is louder. Suitable placing of directional microphones will help, but may not prevent the spill of reverberation from live acoustic to dead. Thick curtains running out from the walls may help to trap the sound – or screens may be used for that – but in extreme cases a separate studio (or the studio lobby) or pre-recording could be used.

Screens are used to add realistic touches. For example, the interior of a car contains a mixture of padded and reflecting surfaces. A 'box' of screens, some bright, some padded, gives a similar quality of sound.

Open-air acoustics

Open-air acoustics, frequently required in drama, are characterized in nature by an almost complete lack of reverberation. Even when there are walls to return some of it, the total reflected sound at the source is low, and is likely to produce only a small change in quality. It is, of course, possible to simulate this realistically by dividing off a section of the studio and padding it deeply to create almost completely dead acoustics, i.e. a dead-room. There are clear advantages:

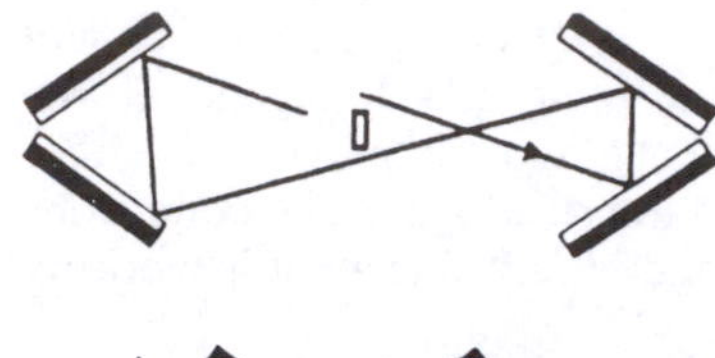

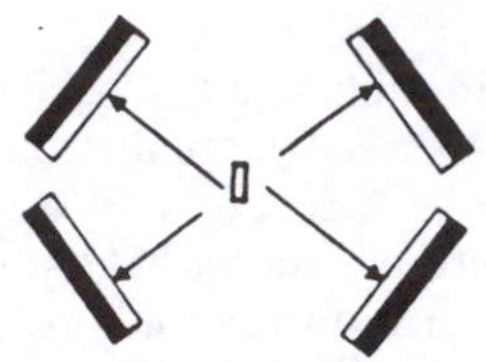

[7.16] **Tent of screens with bidirectional microphone**
In a double-V of screens it is better that sound should be reflected twice on each side (*above*). Standing waves cannot form unless there are parallel surfaces directly opposite each other (*below*).

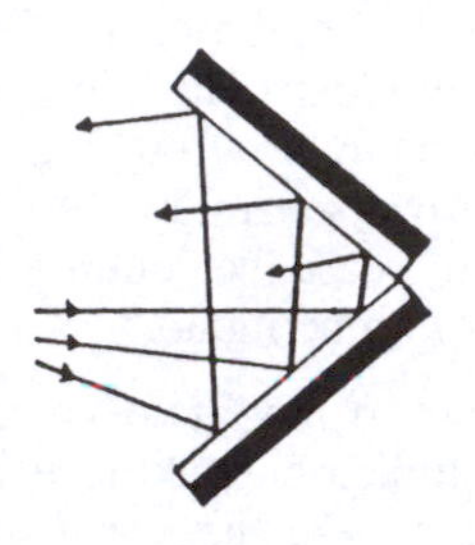

[7.17] **Use of screens**
Double reflection of sound in a V of screens set at an acute angle. For any particular path some frequencies are poorly absorbed, but they are different for the various possible paths. Sound that is reflected back to the region of the microphone has less coloration with this arrangement than from a broad V.

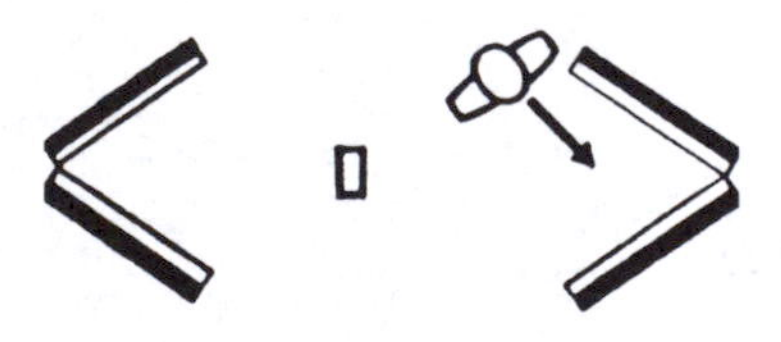

[7.18] **Speaking 'off' a bidirectional microphone**
No performer should go further off than this, or turn further into the screens. More 'distant' effects must be achieved by changes of voice quality or by adjusting the microphone fader.

- it provides the best possible contrast to other acoustics in use, so it makes a wider range of sound quality possible;
- the muffling effect of the treatment causes the performer to use more voice and use the same 'edge' on it as in the open air;
- effects recorded outdoors blend in easily;
- off-microphone voices and spot effects can be blended with on-microphone voices (in mono by using positions round toward the dead side of a directional microphone);
- the amount of space round the microphone is greater than within a tent of screens.

Unfortunately, there are also disadvantages:

- such an acoustic can soon feel oppressive and unreal;
- the transient sounds in speech tend to overmodulate, even if the general level is kept low;
- completely dead sound is unpleasant to listen to for very long.

In fact, the true dead-room is not a practicable proposition, and a compromise has to be found. Where a script is set almost entirely out of doors, it is more convenient to adopt a formalized style with an acoustic that is hardly less reverberant than a normal speech balance. However, if there are only a few isolated lines set outdoors, something close to the dead acoustic is more appropriate. In general, try to set the average acoustic demanded by the script as one that is close to a normal speech balance, and arrange the others relative to this.

Of the various possible sound-deadening systems, try to avoid those that give a padded-cell effect. This is, literally, what a thinly padded dead-room sounds like, and the same thing can all too easily be repeated with a tight cocoon of screens. For monophonic drama, keep it simple: try two Vs of screens, one on each side of a bidirectional microphone. Some points to remember with this layout are:

- Keep the screens as close to the microphone as is conveniently possible; the actors should accept some restriction on their movements. This keeps the sound path lengths short and damps out reverberation (and coloration) as quickly as possible.
- Avoid setting the screens in parallel pairs: it is better to keep the angle in the Vs fairly acute. This helps to reduce bass as well as top. Also, actors should avoid getting too far back into the apex of the V.
- Exit and entry speeches can be awkward. In nature, distant speech is characterized by a higher voice level arriving at a lower final volume. In the studio, a voice moving off is likely to be characterized by a sudden increase in studio acoustic. So stay inside the screens at all times, and avoid directing the voice out into the open studio.
- Off-microphone (i.e. 'distant') speeches may be spoken from the screens, speaking across the V. If this is still too loud to mix with other voices, the actor should change voice quality, giving it less volume and more edge – and should not deaden the voice by turning farther into the angle of the screens: this produces a cotton-wool muffled quality.

The effect of screens is not that of the open air. But it is not a normal live acoustic either. It is different, and acceptable as a convention.

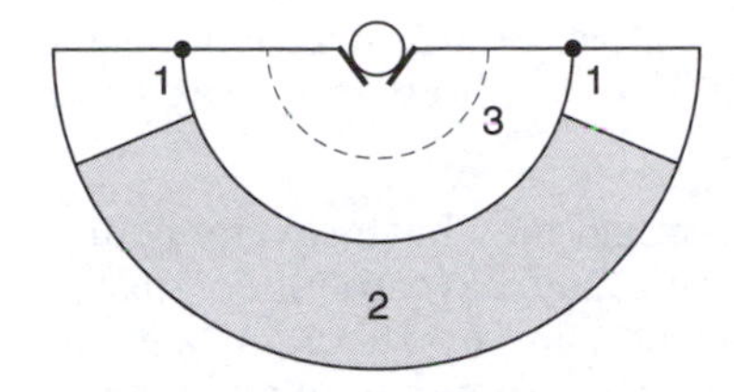

[7.19] **Stage for stereo speech**
1, If crossed hypercardioids are used, voices in these two positions, roughly opposite to each other in the studio, appear to come from the loudspeakers. 2, Most of the action takes place in this area, at a distance of about 2–3 m from the microphones. 3, More intimate speech may be closer.

Stereo drama

For stereo drama, coincident pairs can be used for each main acting area. Crossed cardioids or hypercardioids give the best use of space. Where actors are unfamiliar with the technique, the studio floor may be marked up with strips of white adhesive tape on the carpet showing the main limits of the acting area, together with any special positions or lines of movement. Movement in a straight line on the sound stage is simulated by movement along an arc of the studio: the acting area can therefore be defined by an arc with the microphone at its centre, or by arcs roughly limiting it. Such arcs can be drawn either as having constant radius (for convenience) or at distances that give constant A + B pick-up – which is better for more complex situations, such as group or crowd scenes in which the relative distance of different voices is important. From the microphone polar diagrams it can be seen that for a crossed cardioid this arc is farther away at the centre line and loops in closer at the sides. In addition, the studio itself has an effect: unless the acoustic treatment is evenly balanced, the line may 'pull' closer to the microphone on one side than on the other. Sound stage positions may also be offset laterally from studio positions: the whole studio may seem to 'pull' to the left or right.

Positions 'in the loudspeaker' on the two sides may also be set: for cardioids crossed at 90°, these would in principle be at the 270° limits of the theoretically available stage. In fact, the cardioid response of practical microphones is imperfect at the rear, so the best 'in speaker' positions would usually be found at about 200–220° separation. It is often convenient to reduce this to, say, 180°, by setting microphones with variable polar response one step towards hypercardioid, or using a pair with the narrower response.

In mono, a long approach from a distance can be simulated by a relatively short movement by the performer, who simply walks slowly round from the dead to the live side of a directional microphone. In stereo, long approaches do really have to be long, and must be planned in terms of what space is available: a decision on this in advance will probably determine studio layout.

In stereo, balances that are more distant than those normally used for mono may be necessary, in order to avoid exaggerated movements. For safety, the nearest part of the real stage in which there will be lateral movement should be at least as broad as the sound stage. In consequence, the studio must be somewhat deader than for a comparable scene in mono, and reverberation added if brighter-than-average sound is required.

For this, use either stereo reverberation or a second stereo pair within the same studio. If a coincident pair is chosen for this, it could be about 1.2 m (4 ft) above the first pair. This provides some degree of automatic compensation as a speaker moves in close (but note that a tall person will be closer to the upper microphone than someone shorter would be). An alternative arrangement is to use a spaced pair pointing back to the end of the studio that is away from the stage.

Whereas a small studio for drama in mono may have several acting areas, the same studio used for stereo may have room for just a single stage (this allows for a reduction in reverberation and an occasional need for deep perspective). Therefore there may be good reason to record all scenes that have one basic acoustic before resetting the studio for the next, eventually editing the programme to the correct running order. If very different acoustics are required in the same scene – e.g. a distant open-air crowd scene including studio 'open-air' voices, mixed with close 'interior' conversation – a pre-recording has to be made, the studio reset, and the second part layered in to a playback of the first.

To simulate stereo exterior scenes in a studio, a variation on the same basic studio acoustic may be unavoidable. Cutting bass on the microphone output or at the mixer may help, but the result of that depends on the acoustic coloration of the studio in use. For the effect of a more distant voice in the open air, the actor may help by staying well forward and projecting and thinning tone of voice, or, alternatively, by turning away from the microphone and speaking into a thick screen. Ideally it is bass that needs to be absorbed more.

Stereo drama problems

Dramatic conventions for stereo are well established. One is that movement should occur only during speech from the same source or while other sounds (e.g. footsteps) are produced; certainly, moves should never be made in a short pause between two speeches, except for blind man's buff comic or dramatic effects – easily overused.

Problems that are met in mono drama may be greater in stereo. For example, extraneous noises are worse because:

- more of the studio is open to one microphone or the other;
- with a more distant balance of voices, the signal at the microphone is at a lower level; channel gains are higher and therefore noise is amplified more;
- noises may be spatially separated from other action, and therefore less easily masked.

Special difficulties are raised by a need for artistic compatibility between stereo and mono. Consider, for example, a conversation between two people trying to escape unnoticed through a crowd: they talk unconcernedly as they move gradually across the stage until eventually they are free of the other actors and can exit at the side. The mono listener to the stereo drama needs as many positional dialogue pointers as would be given if the production were not in

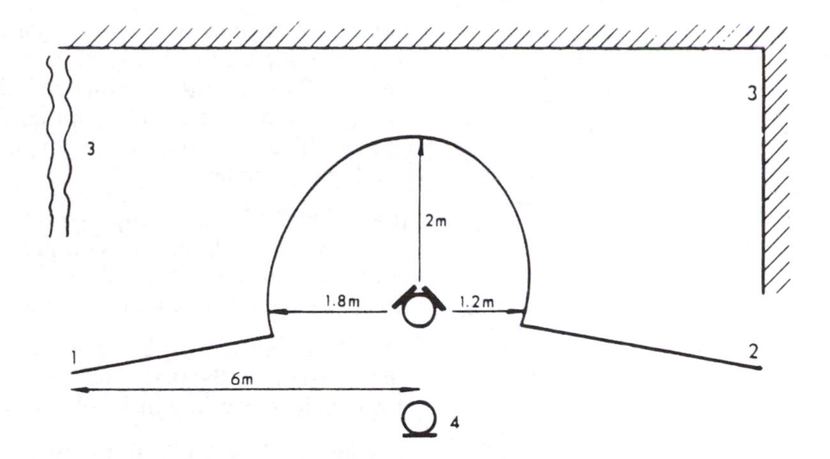

[7.20] **Stereo drama studio layout**
In many studios (as here) there may be asymmetry in the acoustic treatment, distorting the sound stage. In this example, floor markings show minimum working distance around the microphone pair. This is a layout that was designed by its director (and confirmed by aural judgement) to emphasize extreme side positions and allow long approaches from left (1) and right (2). 3, Drapes screening live studio area beyond. 4, Announcer or narrator, centred.

stereo. Indeed, in mono the microphone would probably stay with the main subjects and other near voices would drift in and out of a clear hearing distance; in stereo a convention like this would be more difficult to establish. Imagine or listen to both stereo and mono versions of the chosen convention and decide whether changes could or should be made for the benefit of either the stereo or the mono audience – or whether to disregard potential mono listeners (who may be an important part of a radio audience).

A more extreme case might be a telephone conversation: a theoretical possibility would be to split the stage, putting one actor in to each loudspeaker. This might work well in stereo, but for mono would be meaningless. In this case artistic compatibility requires that the voice of one of the actors must be filtered. The filtered voice could still be put in one of the loudspeakers, but the other, 'near' voice would probably be closer to the centre of the sound stage and in a normal stereo acoustic ambience, but offset to balance the other.

Audience reaction

Audience reaction is a subject that links this chapter with the next. It is a vital element of many comedy presentations; for music broadcasts it provides a sense of occasion that a home audience can share. The two differ in that for the first, the performers' and the audience microphones must be live at the same time, while for music performances they are sequential. For all types of programme, applause (and other forms of audience reaction) must be clear, without being unbalanced – that is, individual pairs of clapping hands should not be loud or assertive.

To achieve this, several possibly conflicting requirements must be met. The audience must be reasonably large – ideally, at least several hundred strong (although a smaller audience can in principle be doubled by using a short delay). The microphone coverage must

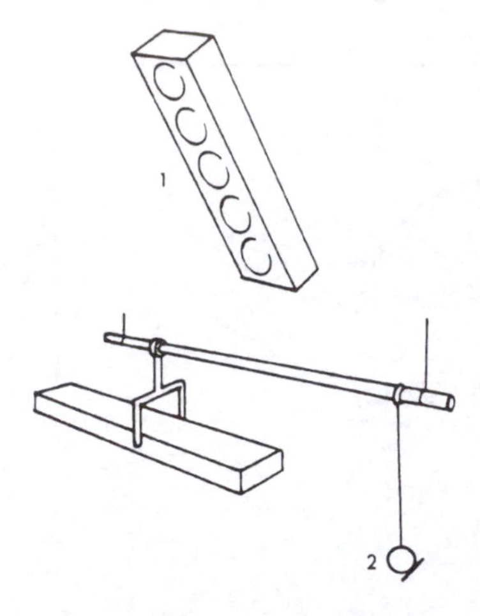

[7.21] **Audience loudspeaker and microphone**
1, Line loudspeaker. Except at low frequencies the elongated source radiates at right angles to its length, and low frequencies are reduced by bass roll-off. 2, The audience microphone is angled with its dead side towards the loudspeaker on the bar, and the line loudspeaker does not radiate sound towards the microphone. Attached to the same bar, their relative positions are fixed.

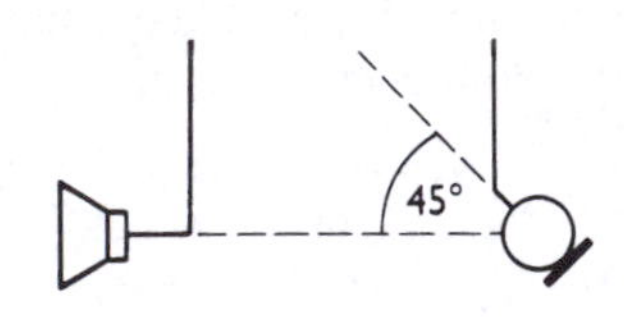

[7.22] **Avoiding spill from public address loudspeakers**
Even when a microphone is nominally dead at the rear it is usually safer to set audience loudspeakers at a rear angle of 45°, where the frequency response is cleaner.

not be so distant as to lose clarity. In addition, there must be adequate separation between the audience and other live microphones. Two or more audience microphones will often be used, and for musical performances these may also act as 'space' microphones. They are pre-mixed on the control console and fed to a single group fader.

It may help to have a compressor or limiter before the main audience fader. A further limiter or compressor can be inserted after mixing audience and dialogue, and used in such a way that speech pushes down the audience reaction as the voice itself is compressed.

In stereo, applause is best spread across the full sound stage. It must be picked up on the main and announcer's microphones only in such a way that controlling its level does not pull the applause to one side.

The audience reaction to a comedy show will obviously be poor if the words are not clearly audible. This requires PA loudspeakers above the heads of the audience, fed by one of the auxiliary send channels available on studio consoles. A line-source loudspeaker radiates most strongly in a plane at right angles to its length, and a microphone placed outside this plane will receive its middle, and high-frequency sound only by reflection. Unless they are well away from these loudspeakers, the audience microphones must be directional and dead-side on to them. Use of the rear axis of a unidirectional microphone is generally avoided, because in many cases the frequency response is most erratic in that direction. Instead, it is safer to place the microphone so that the nearest loudspeaker is at 45° to its rear axis. The PA send is in this case a specially tailored mix of only those channels required for intelligibility. Bass can be cut or rolled off, and equalization introduced to increase clarity and avoid coloration.

In television, for which the performers' microphones may be set well back from the action they are covering, the PA loudspeakers must be brought as close over the audience as is practicable. The loudspeaker volume can be raised by 3–4 dB without increasing the danger of howlround if the frequency of the public address sound is shifted (usually downward) by a few hertz, using a pitch changer.

If the response from some particular individual in the audience becomes obtrusive, fade down the microphones gently, in turn, until that nearest to the problem is located, then reset it accordingly.

The volume of applause at the end of a performance must be held back a little: prolonged clapping at high level is irritating to the listener, who may be impelled to turn it down. On the other hand, laughter within a programme may have to be lifted a little: it could be held down between laughs in order to avoid picking up coloration, raised quickly as the laugh begins, and then eased back under subsequent dialogue just enough to avoid significant loss of clarity in speech. The performers usually adapt rapidly to audience response, and sensitive control should anticipate the progress of this interaction.

Chapter 8

Music balance

There are two main types of music balance. One is 'natural' – that is, music heard in natural acoustics – and is generally preferred for classical music. For a single instrument or for a group having perfect internal balance, one microphone (or stereo pair) may be sufficient. But such perfection is rare: a singer may be overwhelmed by all but the quietest of pianos; an orchestra may be dominated by its brass. So even 'natural' balance often requires additional microphones.

Most music is balanced in stereo, for which the techniques may be further subdivided into those using coincident pairs and others with spaced microphone arrangements. Mono is still used in parts of the world where, for economic or language reasons, stereo broadcasting is restricted. (In television, if several language versions are required, this may take up audio channels that could otherwise be allocated to stereo.) But mono is also used for individual instruments or sections within a stereo balance. Further treatment of individual components of the whole sound is normally used only for purposes of correction.

At the other extreme, natural acoustics are all but eliminated, to be replaced by stereo artificial reverberation. This is a multimicrophone balance that at one time was used for nearly all popular music. It employs many close microphones, predominantly mono, though with some stereo pairs. Each component has its own 'treatment' and its metered dose of artificial reverberation (AR) before being steered to some appropriate place within a composite stereo image. This whole process may be extended over time by multitrack recording before the final mix-down.

Here, too, there are less extreme versions – in part, with the rediscovery of the virtues of natural acoustics for some components of the balance. In fact, many balances are somewhere in the range between the two extremes. Each of the main types of balance will be considered, first in general terms, and then by looking at balances for individual instruments or sections that could be used in either of them.

Music in 5.1 surround sound is generally balanced initially as for stereo, but with reverberation distributed all round, although some mixes may also distribute some of the sources. Sources such as a full orchestra, which may warrant the use of microphones to pick up reverberation, can be balanced for multichannel recording designed directly for 5.1.

'Natural' balance

Here the most important component of balance is that between direct sound and natural reverberation. In a natural balance taken on a single microphone (or coincident stereo pair), its distance from the source depends on both the polar response of the chosen microphone and the acoustics of the studio. A flexible method employs a condenser microphone (or pair) that has a remotely switchable polar response. With this, try various distances for each polar response and listen how clarity (of individual components and of melodic line) compares with blending, or warmth of tone. If possible, set several microphones (or pairs) of the same type at different distances, in order to make direct comparison tests. This is really the only practicable way of determining whether the microphone is within the distance at which reverberation begins to swamp the direct sound and early reflections.

A close balance sounds more dramatic, but may be difficult to live with. Where the acoustics are difficult, it may be necessary to accept

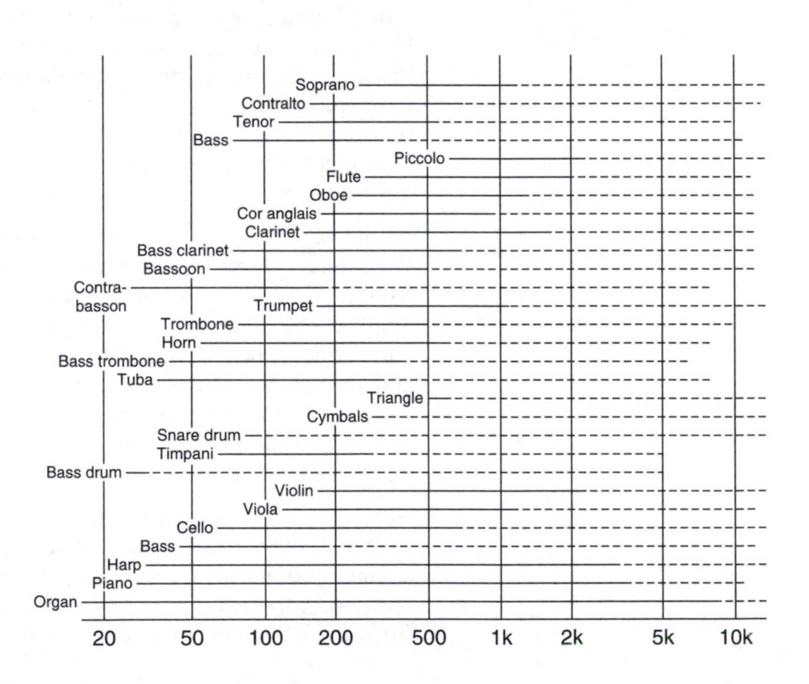

[8.1] **Frequency ranges**
A microphone with a frequency response that matches that of the sound source will not pick up noise and spill from other instruments unnecessarily. The lowest frequency for most instruments is clearly defined, but the highest significant overtones are not. At high frequencies, an even response is usually considered more valuable than a matched one.

a brilliant close balance, with artificial reverberation if desired: this can always be added, whereas a bad acoustic cannot be subtracted.

The spread of orchestra or ensemble must be matched to polar response and distance. With a distant balance a bidirectional response may be possible, but as the microphone is moved in, a broader pick-up is required. In many cases the best mono is achieved by a cardioid with a clean, even response (in which the polar response is independent of frequency). Omnidirectional microphones will have to be nearer still – perhaps noticeably closer to some instruments than to others. They are therefore more suited to spaced-microphone techniques.

As spotting (accent) microphones, smooth cardioids or hypercardioids are again favoured. Bidirectional ribbons, too, can offer an excellent, smooth response in all but the extreme top – which in any case lies beyond the range of the human voice and many instruments.

When they have high-quality loudspeakers, many listeners prefer a more distant and so more blended sound than some balancers enjoy. But for listening on poorer equipment, greater clarity compensates for some of the deficiencies: at least it is possible to hear the main elements of the music. For this reason (and also because it helps to motivate close-ups), a drier balance is often used on television.

The arrival of an audience can change a balance dramatically. Experience will show how much too live the balance must be in rehearsal for it to be correct with a full hall. It will usually be safer to have some separate means of controlling this, either by combinations of microphones or by adding artificial reverberation.

Music studio problems

A 'natural' balance records the characteristics of the studio just as much as those of the musicians, rather as though the studio were an extra member of the orchestra. With stringed instruments, very little of the sound we hear is directly from the strings; almost all is from a sounding board to which the strings are coupled, so the character of each instrument depends on the shape and size of that radiator. In the same way, the studio acts as a sounding board to the entire ensemble: its shape and size give character to the music. But there is one major difference. By uniformity of design, the character of all instruments of any particular type is roughly the same. But music rooms, studios and halls differ widely in character: no two are the same.

In the natural balance directional microphones are angled to pick up not only direct sound, but also the early reflections and reverberation from all three dimensions. In a large auditorium, the volume of direct sound will diminish with distance from the source, while reflected sound is at much the same volume throughout the entire space – so there is a *critical distance* beyond which reverberation dominates the sound heard. Plainly, the best place for a microphone (or microphones) will be inside that critical distance, so that the balance is enriched but not overwhelmed by reverberation. In an excessively live interior (a cathedral, perhaps), an audience seated well beyond the critical distance cannot hear a satisfactory acoustic balance for most music, let alone speech, and this is certainly no place for a microphone.

At the other extreme, in a studio that is unusually dead for its size, a more open directional microphone response may help – a broad cardioid, perhaps. There can even be times when even a distant balance with an omnidirectional microphone does not do enough. Some percussive instruments, such as timpani, continue to radiate sound with a decay characteristic a bit like that of reverberation. If they stand out in too dead a hall, artificial reverberation should be added.

A typical television or film studio is a dead studio on the grand scale. The same problems may arise in microcosm when one or two singers or instruments are recorded on location, say in a domestic environment that is not designed for music making. In this case, taking a microphone to the other side of a small room with heavily damped acoustics will only emphasize their inadequacy. The only effective solution may be furniture removal: if possible get rid of heavily padded furniture such as armchairs. Roll up rugs and draw aside soft drapes. And avoid having too many people in the room.

Theory suggests that the central line should be avoided, because in the middle of a room the even harmonics of the eigentones – the basic resonances of the room – are missing, thereby thinning out its acoustic response. Fortunately, in practice, few studios seem to exhibit serious eigentone problems in music balance. In fact, stereo microphones are often placed on the centre line of concert halls and rarely with ill effect (but note that concert-hall eigentones are very low pitched).

For monophonic balance (which may be a component of stereo) in difficult acoustics, a directional microphone gives the greatest control as it allows an extra dimension for experiment: the angle can be varied, as can the distance from the source and position within the room, and on some microphones the polar response too.

Stereo music balances that are based primarily on coincident pairs of microphones set initially with cardioid elements angled at about 130° may also be modified either by changing the polar diagram, or by changing the relative angle of the elements, or both.

In particular, there are two reasons why the double figure-of-eight system is rarely used. The most important is poor compatibility with mono: there are likely to be some listeners to a stereo broadcast or recording who hear it in mono, and if the A – B signal is large they lose a significant part of the information. In fact they lose much of the reverberation, which is strongly represented in the A – B component of a double figure-of-eight. Another problem is that reverberation is picked up on both back and front of the microphones in roughly equal proportions, so that at low frequencies there is a phase-cancellation effect that takes some of the body out of the stereo reverberation, making it thin and harsh. This is noticeable in a concert hall and marked in the more reverberant surroundings of, say, a cathedral. Switching the polar diagram even a quarter of the way towards cardioid reduces this significantly.

One microphone or many?

The extreme alternative to studio acoustics and the natural blending of sounds is the multimicrophone balance in which an essential, but not necessarily conventionally natural, quality is sought from each instrument or group and blended in the mixer. Neither technique is any more 'right' than the other, but opinion has tended to polarize to the view that most classical music is best balanced on fewer microphones, and that popular music demands many.

There is a rightness about this: the 'simpler' microphone technique for classical music reflects the internal balance of the orchestra and existing acoustics that the composer wrote for, whereas the top 20 hit of the moment is more likely to reflect the search for novelty and opportunities for experiment. But the question 'one microphone or many?' is still open for many types of music that lie between the extremes.

Before choosing a 'natural' balance, consider how the sound source is distributed. Even if the acoustics are suitable, can it be arranged that the sound source lies within the pick-up field of a single microphone position? If it can, then given a satisfactory instrumental layout, careful microphone placing can ensure that the distance between the microphone and each source is exactly right.

Individual instruments and groups

Orchestral players and pop musicians play instruments that have much more in common than in difference. All orchestral instruments are fair game for popular music, but there are others that rarely make the journey in the opposite direction, notably electric guitars and synthesizers, but also some older sounds such as those of the piano accordion or banjo. But it is clear that for all types of music balance we are dealing with the same basic kinds of sound sources, which can be treated in a range of different ways, but for which the various possible treatments required in different types of music overlap.

A natural orchestral balance may require reinforcement of a particular instrument or section, or in a stereo balance mono spotting may be needed to make a position clear: both of these demand close microphone techniques that approach those used in multimicrophone balance.

A question that is still debated is whether it may sometimes be desirable to delay the signal from a spotting microphone to match its apparent distance to the actual distance from the main microphones. To estimate the time involved, roughly measure the distance from the main to the spotting microphone in front of the instruments it covers: as a rule of thumb, this distance in feet (or one-third of the distance in metres) gives the number of milliseconds of delay to go in the spotting channel. A more accurate calculation is: at 20°C, delay in seconds equals number of metres divided by 344. As an example, if a main pair is 10 m further back than the spotting microphones, the delay would be 29 ms. A tiny delay (of, say, 10 ms) would very likely be masked in the mix by sound from the main pair, but nearly 30 ms is too much. This suggests that in a distant orchestral balance, a delay should be inserted, while for most other (closer) balances it may be disregarded.

For what counts as a 'close' balance, there may be differences in terminology. In classical music a distance of a metre (several feet) is 'close' or 'very close', depending on the instrument; for pop music, closeness is measured in centimetres (inches) or replaced by contact. But for all of these, there is no difference in the basic problems and possibilities, and in the following sections they are explored first for individual instruments, sections and combinations, and then for full orchestras and bands.

Most instruments belong to one of three main groups:

- *Strings* – plucked, bowed or struck, and coupled to a sounding board acting as a resonator (or from which vibrations are derived electrically and fed to a loudspeaker).
- *Wind* – where the resonator is an air column excited by edge tone, vibrating reed, etc.
- *Percussion* – where a resonator (often with an inharmonic pattern of overtones) is struck.

Note that while there are general principles that may apply to each of these, some players and some instruments produce different sounds from others of the same apparent style or type (think, for example, of double basses and their players). Note also that any balance that actually works does so only for the particular microphone set-up used, and in all but close balances, only for a particular studio. It may even be affected by the weather: if it is very dry, a good natural balance is more difficult, because high frequencies are absorbed by dry air.

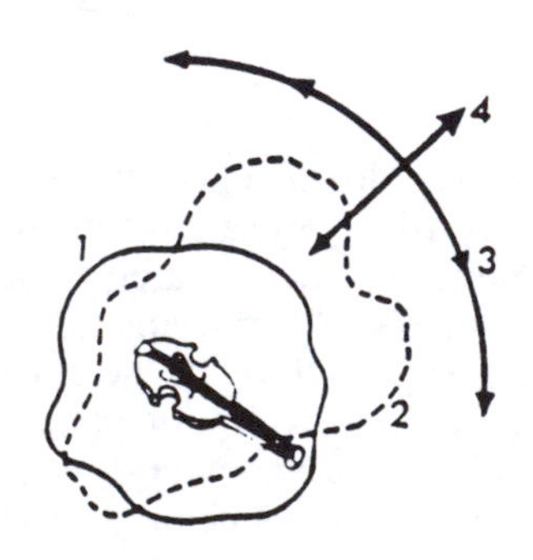

[8.2] **Radiation pattern and microphone position for violin**
1, Low-frequency radiation. 2, High frequencies. 3, Move along this arc for more or less high frequencies. 4, Move closer for clarity, away for greater blending.

Violin, viola

A violin radiates its strongest concentration of upper harmonics in a directional pattern, so that for all but the lower frequencies much of its sound goes into a lobe over the heads of the audience. With that also goes a great deal of the harsh quality, squeak and scrape that even good players produce. The audience normally hears little of the

[8.3] **Close violin balance in pop music**
The player has encircled the microphone stand with the bowing arm. A directional microphone is not subject to much bass tip-up (proximity effect), as the lowest frequency present is a weak 196 Hz. A filter at 220 Hz may be used to discriminate against noise and rumble from other sources. Contact microphones are sometimes used.

high-frequency sound, as even in the reverberation that may be heavily attenuated in the air. At its lowest frequencies the radiation is omnidirectional, but the violin is an inefficient generator of sound at low frequency owing to its size: the lowest notes are heard mainly in their harmonics.

For a concert balance the microphone is placed up and well back from the instrument to add studio reverberation, and frequency content is controlled by movement in an arc around the player. The upper harmonics are heard at their strongest a little off-axis, towards the E string (the top string on the violin). A ribbon microphone can give a good response (though possibly requiring some reduction in top).

For a balance with microphones for each individual instrument or group, separation is more important. In this, the violin microphone will have to be much closer, perhaps at only 60–90 cm (2–3 ft), and angled directly into it. This balance may be harsh, but treatment will be available (and expected) on each channel anyway.

The lowest string of the violin is G (196 Hz), but the fundamental is weak, so even with close directional microphones bass tip-up is hardly noticeable; indeed, a bass filter (high-pass filter), rolling off below 220 Hz, may be used to discriminate against low-frequency sound from other sources. Rather than roll off top, it may be better to lift a mid-frequency band that makes the violin stand out among other instruments: try bands around 300 Hz or 1–1.3 kHz and listen to see what works. The tone quality of the violin will also affect this.

To make a very close violin sound pleasant requires more than the art of the balancer can supply: it demands a modified style of playing in which the scratchier components (normally diminished by distance) are much reduced. When using a stand microphone, very close working may be limited by the athleticism of the player, so a little under 1 m (3 ft) may be the minimum, especially for pizzicato passages in which the bow might hit the microphone. But to get inside that, some players are capable of encircling the microphone stand with the right (bowing) arm. In this rock or jazz balance, the higher overtones ('harmonics' seems too gentle a word) will cut through.

[8.4] **Close balance for three desks of violins**
As an alternative to this, for clean separation it is possible to work with a separate microphone for each desk, but that produces a very strong high-frequency response which must be discriminated against or corrected for.

A contact microphone may also be used, but expect the sound to be different again. If it is attached near the bridge, its output may lack high frequencies: experiment with the position and judge whether or not this is an advantage. Alternatively, try an electret clipped to the player's lapel: this position discriminates against the stridency.

For a close balance on an orchestral violin section, the players sit in their normal arrangement of two at a desk with the desks one behind another, and the microphone is slung directly over and slightly behind the first two players. This should favour the leader enough for solos, but should not lose the rear players as a close balance farther forward would. If this balance is still too distant for good separation in, say, a show band in live acoustics, put a microphone over each desk and, as usual, apply high-frequency roll-off to counter the harsher quality.

Even if used as part of a stereo balance, a mono microphone must be used for close working, as the slightest movement near a coincident pair becomes gross. For stereo, the instrument or group is steered to its appropriate position on the sound stage (for a soloist on his own, in the centre) and full width reverberation is added as required.

For a balance in which stereo microphones are used, they must be more distant. Slight movements of body and instruments then produce an effect that is pleasant, unless they noticeably alter the proportions of sound reaching the microphones. This can happen easily with spaced microphones, but may still do so with a coincident pair: a change in reflected sound can produce spurious apparent movement. Again, reverberation must be of good quality and well distributed.

Violas may be treated in the same way as violins, except that very close working on directional microphones may produce bass tip-up, for which a filter at 220 Hz cannot be used, as this is inside the frequency range of the instrument.

Cello, bass

Cellos and basses differ from violins and violas in that they have a much larger area of resonator to drive the air, so they radiate far more efficiently in the lower register. The high, extended spectrum of resonance that makes the violin shrill and 'edgy' if heard from directly above is scaled down on the cello to the upper middle frequencies. To get the full rich sonority of this instrument's upper harmonics, place the microphone more directly in line with its main radiating lobe. Again, this can be used as an aid to balance. If the objective is just to add depth to a fuller orchestral sound, a balance that is near-sideways-on to the instrument may do enough, but for richness turn the instrument to the microphone, as a cello is turned to the audience in a concerto.

If, in a raked orchestra, the cellos or basses are mounted on light rostra (risers), these may act as sounding boards that add unwanted coloration. Rostra should therefore be solidly braced or acoustically damped. Strategically placed, the mass of the players also helps. If microphone stands are used, avoid placing them on the rostra or pay special attention to their acoustic insulation. In an orchestra, the basses may be helped by a spotting microphone, placed about 1 m (3 ft) from the front desk and angled towards the bridges.

For a cello solo the player faces forward, with the instrument a little off-axis to the player's right. The brightest (and richest) balance will also be on this line, and shifting off it will reduce the upper harmonics. A distance of 1 m (3 ft) is likely to be good, but in a multimicrophone balance may be reduced to as little as 15 cm (6 in), with the usual compensation for close working if a directional microphone is used.

For popular music the bass, or double bass – plucked rather than bowed – is one of the three basic instruments of the rhythm group; the others are piano and drums. Of the three, the bass has the lowest volume, so is the most difficult to separate. This suggests a close balance, directly on to the body of the instrument, but specialists are disinclined to recommend any single balance that will always work, pointing out that there are great differences between one bass or bass player and another. The following alternatives have been suggested:

- a cardioid, hypercardioid or figure-of-eight microphone 30 cm (1 ft) from the bridge, angled down to the strings;
- a similar microphone directed towards the f-hole at the side of the upper strings;
- a miniature ('personal') electret pointing up to face the bridge;
- a miniature electret wrapped in a layer of foam plastic, suspended by its cable *inside* the upper f-hole – this gives virtually total separation;
- so too does a contact microphone somewhere near the bridge.

The first or second of these balances will require the usual bass roll-off to compensate for proximity effects. In general, the first has the most to recommend it, because it picks up a percussive attack quality from the string as it is plucked, and offers fine control over the ratio of this to the resonance of the body of the instrument. In contrast, the fourth balance gives a heavy, hanging quality that must be controlled to a lower level than the others. In stereo pop balances the bass is always taken mono and steered to its appropriate position.

More strings: acoustic balances

Other stringed instruments, such as classical guitar, banjo, the violin's predecessors (the viol family) and so on, all radiate their sound in ways similar to the instruments already described, and the control that the balancer has over the ratio of top to bass is much the same. Groups based on stringed instruments are also balanced in a similar manner.

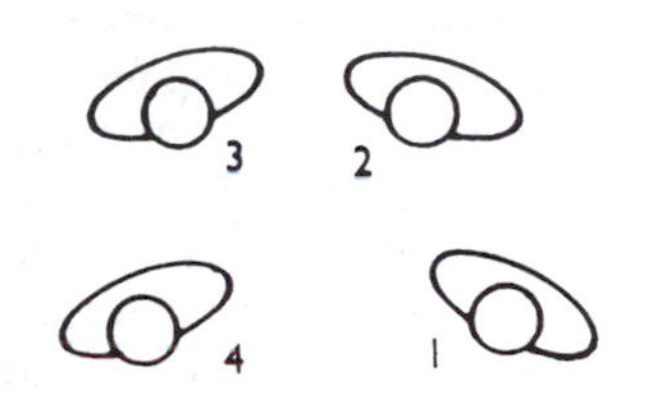

[8.5] **String quartet**
1, Viola. 2, Cello. 3 and 4, Violins. Similar arrangements may be made for wind and larger chamber groups. This mono balance employs a bidirectional ribbon (with a frequency response that offers a good match to the instruments of this group); stereo balances might use coincident pairs of bidirectional, hypercardioid or cardioid microphones, moving closer with each successive pair.

For a string quartet, if the players sit in two rows of two, the microphone can be placed somewhere along a line extending forward and upward at about 30° from the middle of the group. Check that the distance, together with the angle and position within the studio, take full advantage of the acoustics. Listen for each instrument in turn as though it were a soloist, and make sure that the cello can 'see' the microphone. Try to combine the clarity and brilliance of individual instruments with a resonant well-blended sonority for the whole group. For radio or recording, the scale of width should be about one-third of the sound stage; for television it would be wider. As usual, reverberation spreads over the full available width.

As more instruments are added, a similar 'natural' arrangement may be retained, moving the microphone back if necessary in order to keep all the instruments within its field. In this way we progress through the various types of chamber group or wind band to the small orchestra.

For the acoustic guitar as part of a multimicrophone balance, for high output try a position about 15–30 cm (6–12 in) above the sound hole and somewhat off-axis. Use a condenser microphone for its

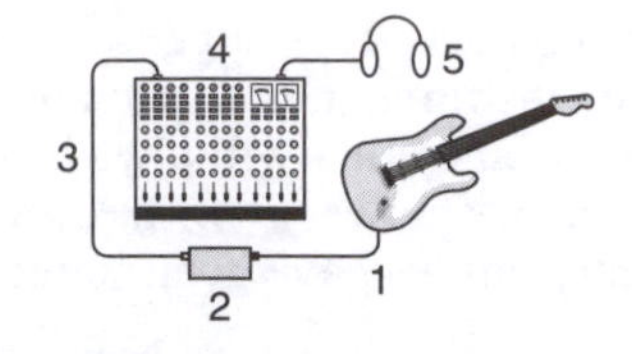

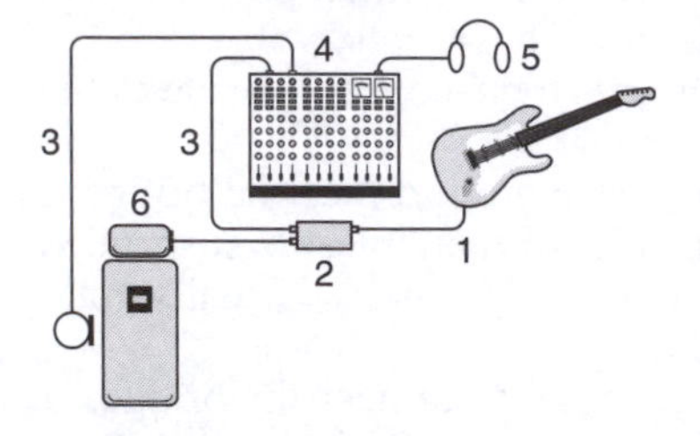

[8.6] **Electric guitar direct**
Above: Direct to console. 1, Guitar cable. 2, DI box. 3, Microphone cable (XLR). 4, Console. 5, Headphones. *Below*: With direct and 'amp' (6) feeds to console.

extended frequency and good transient response. Because air inside the instrument resonates at 80–100 Hz, a microphone near the sound hole will require 5–10 dB roll-off below 100 Hz. If separation problems persist, use an isolation booth or record later to playback.

Electric guitar family

The electric guitar has its own contact microphone near the bridge, from which the output is likely to be an unbalanced, high-level signal, suitable for sending to a nearby 'amp' (amplifier plus loudspeaker cabinet). Some mixers also accept this feed, but in any case the signal can be split by a DI (direct injection) box to send a low-level, electrically isolated and balanced signal like that from other microphones by an ordinary microphone cable to the console. Where effects are added by the performer, their output is already included in this feed. If it is felt that the guitar's built-in pick-up does not have an adequate high-frequency response, an alternative is to add another contact microphone, again near the bridge. The player wears headphones.

A direct feed offers the cleanest guitar sound, and avoids any problems with separation. But if the 'amp' is considered to be an essential part of the musical quality, then in addition (or instead) a microphone may be placed in front of the loudspeaker, about 10–20 cm (4–8 in) from the cone. 'Amps' vary in quality, and may not have an extended high-frequency response: plainly, this is unnecessary if the pick-up is limited, though there may be a compensatory peak below a 5–6 kHz tail-off. In line with the axis of the cone, high frequencies may be too harsh, so try moving the microphone sideways for the best position. If necessary, separate the 'amp' from other instruments with low screens behind the (directional) microphone. If the loudspeaker system is overloaded, reduce the acoustic volume (or the bass), restoring it in the mix. A bass guitar may benefit from lift somewhere between 100 and 400 Hz or in the 1.5–2 kHz overtone range.

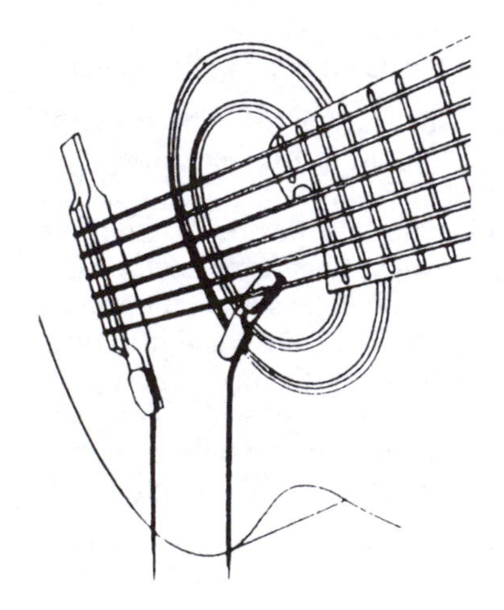

[8.7] **Guitar balance**
For popular music, one possible balance uses both a contact microphone at the bridge and an acoustic microphone clipped over the air-hole.

In principle, a hypercardioid microphone response is appropriate. If the volume of sound results in distortion, check whether there is a switch on the microphone itself to reduce signal level. Some balancers prefer a 'rugged' near-omnidirectional microphone that becomes more directional at high frequencies. This can handle heavy volumes and, if it has a large diaphragm, this may be used to control high frequencies.

In some cases the guitar serves a double purpose as an acoustic and electric instrument. In the simplest balance the loudspeaker is placed on a box or chair opposite the player, with the microphone between loudspeaker and instrument. The player adjusts the overall amplifier level, comparing it with the direct, acoustic sound, and has control of the loudspeaker volume from moment to moment by means of a foot control. A bidirectional microphone can be used for this balance, with any bass equalization that may be necessary at the distances involved.

Other balances include taking a feed from the contact microphone and combining it with the high-frequency response from an acoustic microphone clipped to the edge of the sound hole, hanging inside it directly below the strings and angled to discriminate against noise from the fingerboard. These may be mixed with the loudspeaker balance, taking different frequency ranges, equalization and reverberation from each. Double-check the combination by monitoring it on a small loudspeaker (via AFL, i.e. after the channel fader).

Where a feed of any electrical signal is taken to the console (if not through a DI box then always through an isolating transformer, for safety), a radio-frequency rejection filter may be needed to eliminate unwanted police and taxi messages and, in TV studios, radio talkback.

The combination of contact microphone near the bridge, mixed with acoustic microphone clipped to the rim of the air hole, may also be used for other instruments that have this layout, such as dulcimer or zither. On a banjo, try the contact microphone at the foot of the strings, and clip an acoustic microphone to the bridge for the strings.

Grand piano

The piano generates sound in a manner similar to the smaller stringed instruments: the string vibration is linked to the soundboard. This radiates almost all sound that is wanted, with just a small contribution from the strings at the higher frequencies. If the microphone were to be placed beneath the piano this high-frequency component would be lost.

The pattern of radiation from the soundboard allows some control of treble and bass: in particular, the bass is at its strongest if the microphone is placed at right angles to the length of the piano. In an arc extending from top to tail of the piano, the most powerful bass is heard at the end near the top of the keyboard. As the microphone

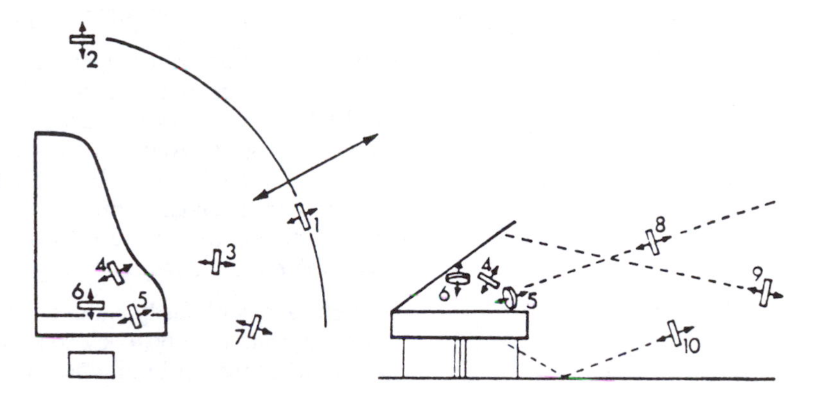

[8.8] **Piano balance**
For comparison of positions, these show mono balances with a bidirectional microphone. The same principles apply to (slightly closer) hypercardioids and, for some of the positions, coincident pairs. The best balance for a grand piano is usually somewhere along the arc from the top strings to the tail. A close balance gives greater clarity; a distant balance increases blending. Of the positions shown: 1, Often gives a good balance. 2, Discriminates against the powerful bass of some concert pianos. 3, Picks up strong crisp bass. A mix of 2 and 3 can be very effective as a spaced pair. 4, With the lid off, close and pointing down toward upper strings, for mixing into multimicrophone balances. 5, Discriminates against piano for pianist/singer. 6, Angled down towards pianist, same effect as 5. 7, One of a variety of other positions that are also possible in a stereo pair: experiment rather than rule of thumb indicates best balance. 8, Concert balance, 'seeing the strings'. 9, By reflection from lid. 10, By reflection from the floor: microphones set for other instruments may inadvertently pick up the piano in this way.

is moved round, the bass is progressively reduced until it is at a minimum near the tail. For very powerful concert pianos the tail position may be best, but with the slight disadvantage of some loss of definition in the bass, together with reduced volume. A point near the middle of the arc should be satisfactory, and is a good starting point for a balance test.

The piano is the first individual instrument that we have considered that may benefit by stereo spread of the direct sound. But even for a solo piano the spread should not be too wide: half the stereo stage is enough, with reverberation occupying the full width.

Microphone distance is governed largely by the desired ratio of direct to ambient sounds. A fast piece may need clarity of detail; a gentler romantic style benefits from greater blending. A closer balance is also required for brilliant or percussive effects, and to some extent these qualities may be varied independently of reverberation by changes of microphone field patterns, as well as by distance. A piano used for percussive effect in an orchestra might need a spotting microphone. Pianos themselves vary considerably: some are 'harder' (more brilliant) than others, so that for a similar result greater distance is required. Conversely, a muddy tone quality may be improved by bringing the microphone closer. A new piano takes time to mature and is better reserved as a practice instrument until the tone settles down.

In a dead acoustic such as that of a television studio, try using two microphones: one is placed close to control brilliance, the other more distant and with added AR, perhaps with a concert-hall duration of delay.

The height of the microphone should allow it to 'see' the strings (or rather, the greater part of the soundboard), so the farther away it is, the higher it must be. If this is inconvenient, try a balance that uses reflections from the lid: this is the balance an audience hears at a concert. For the deepest notes, the piano's pattern of radiation tends to be omnidirectional; but for middle and top, and particularly for the higher harmonics, reflection from the lid ensures clarity of sound.

Depending on the acoustics, cardioid or bidirectional microphones may be used. A bidirectional ribbon can give excellent results: its frequency range is well matched to that of the instrument.

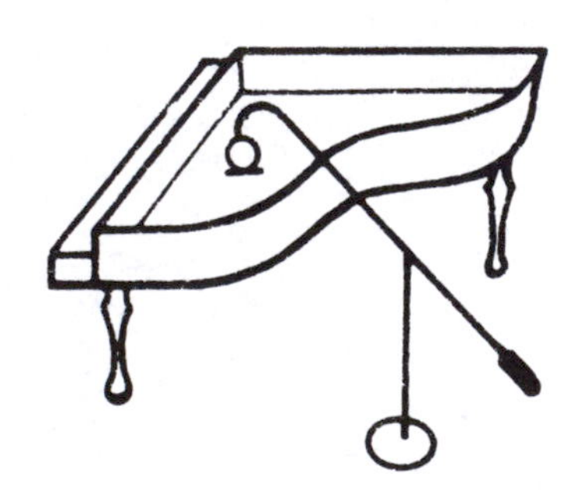

[8.9] **Close piano balance**
Directional microphone suspended over the top strings – in this case, not too close to the action.

As we move closer to an open piano, the transients associated with the strike tone become clearer; at their strongest and closest they may be hard to control without reducing the overall level or risking distortion on peaks. Action noise – the tiny click and thud as keys are lifted and fall back – may be audible, and in balance tests this and the noises from pedal action, etc., should be checked. On a close balance for pop music the lid may be removed, unless this makes separation more difficult.

A bright and still fairly natural balance suitable for the piano in a rhythm group may be as close as 15 cm (6 in) over the top strings. As one possibility, a bidirectional ribbon can still be used: its bass

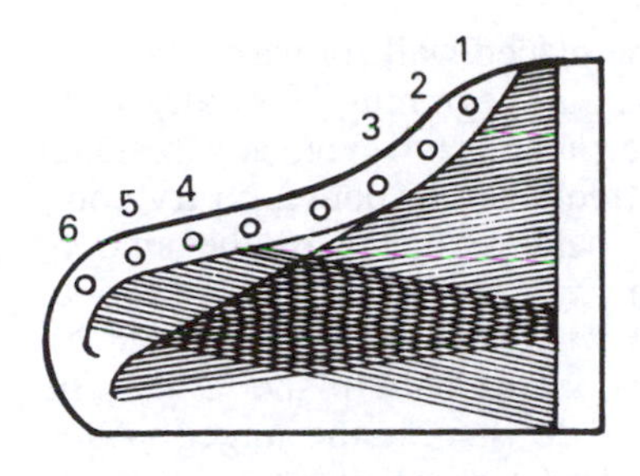

[8.10] **Piano holes**
The focusing effect of holes in the piano frame offer a mellow quality. A microphone 50–75 mm (2 or 3 in) over hole 2 may give a good mono balance, but those on either side may also work; so can pairs mixing the sound at hole 1 with 4, 5 or 6.

tip-up at this distance helps rebalance the sound. The best position depends on the melodic content of the music: one criterion is that the notes should all sound in the same perspective, which can also be affected by the manner of playing. An old trick is to add a microphone baffle, perhaps a piece of cardboard stuck to the back of a ribbon casing, to lift and 'harden' higher frequencies, especially of transients.

Working at such distances, it may be better to use two microphones, one near the hammers in the middle top, with the other over the bass strings, and kept separate to give appropriate breadth to the piano in the stereo picture. For a hard, percussive effect a microphone (or two, or a coincident pair) can be slung 10–20 cm (4–8 in) over the hammers. If the lid has to be kept low (on a short stick) to improve separation, the separate treatment of equalization and AR makes it easier to cope with the resulting less-than-ideal conditions.

Another possibility uses the focusing effect of the holes in the iron frame of some pianos: the full-bodied sound over these can easily be heard by leaning in to them. A microphone 5–8 cm (2–3 in) above the second hole from the top may give the best overall balance. With two microphones, try the top hole and the fourth, fifth or sixth hole.

Overload will cause distortion (particularly of percussive transients), so must be avoided, perhaps by attenuation at the microphone.

For a fully controlled mono contribution to a group, wrap a small electret in foam, place it on the frame at the farthest point and close the lid. Or, for a clear pick-up with good separation, use the lid as the hard surface for a boundary microphone: to retain some bass, tape it on the open lid near the middle of the curve. For greater separation from noisy neighbours, put flats (gobos) between them, or drape a blanket over the lid to cover the opening. If all else fails, move the piano to an isolation room (if there is one) or, for a more natural piano balance, overdub it later.

[8.11] **Piano and soloist on one microphone**
Balance for the piano first, and then balance the soloist to the same microphone. The piano lid may have to be set in its lower position. For stereo, use a coincident pair.

Piano and soloist, two pianos

For a concert balance in a studio with suitable acoustics it is possible to get good separation by turning the soloist's microphone, which may be bidirectional or hypercardioid, dead-side-on to the piano and mixing in the output of a second and possibly a third microphone, arranged to pick up a little of the piano. This time start by getting a good sound from the soloist, then gradually fade in the piano, just enough to bring it into the same perspective. It is of course also possible to employ a backing track, as in popular music. The soloist listens to pre-recorded piano on headphones, and the levels are adjusted at the mixer.

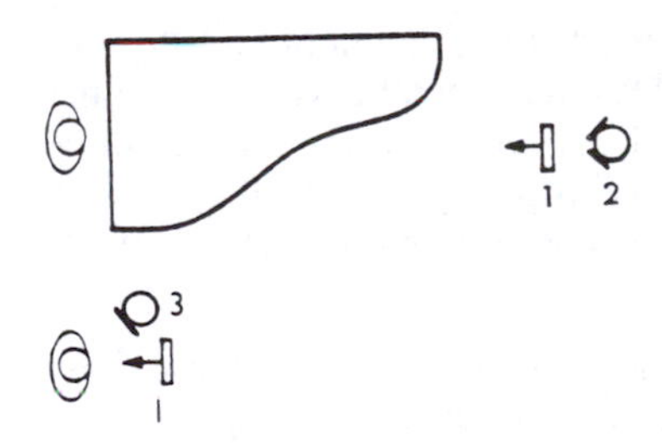

[8.12] **Piano and soloist side by side**
Balance using two ribbons (1) or stereo pair (2) and hypercardioid (3).

A pianist who also speaks or sings may wish to use less voice than would be possible without an additional vocal microphone: this achieves a more intimate effect at the expense of internal balance.

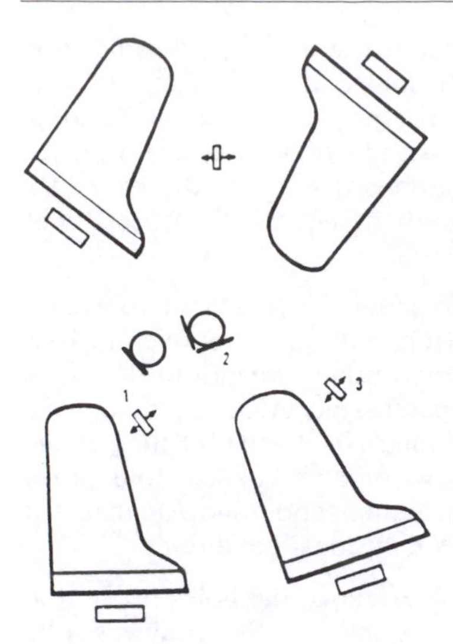

[8.13] **Balance for two pianos**
Above: Mono, with two pianos on a single microphone. *Below*: Two pianos on individual mono microphones (1 and 3) or on a stereo pair (2).

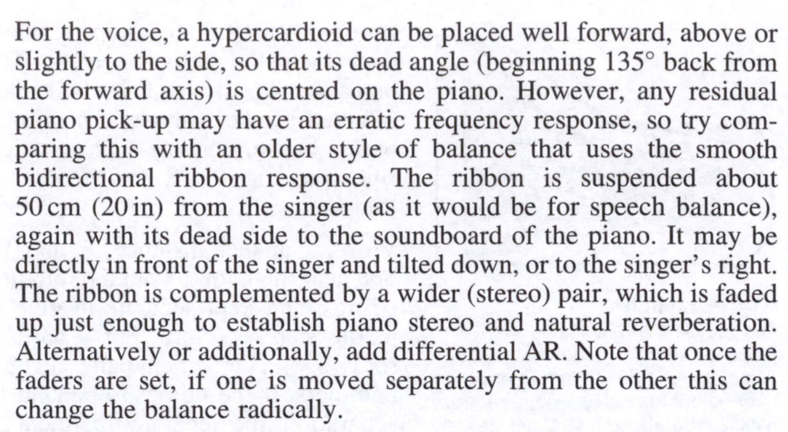

For the voice, a hypercardioid can be placed well forward, above or slightly to the side, so that its dead angle (beginning 135° back from the forward axis) is centred on the piano. However, any residual piano pick-up may have an erratic frequency response, so try comparing this with an older style of balance that uses the smooth bidirectional ribbon response. The ribbon is suspended about 50 cm (20 in) from the singer (as it would be for speech balance), again with its dead side to the soundboard of the piano. It may be directly in front of the singer and tilted down, or to the singer's right. The ribbon is complemented by a wider (stereo) pair, which is faded up just enough to establish piano stereo and natural reverberation. Alternatively or additionally, add differential AR. Note that once the faders are set, if one is moved separately from the other this can change the balance radically.

The best way of balancing two pianos on a single microphone depends on studio layout, on which the wishes of the pianists should be respected if possible. With the pianos side by side, it is possible to use a single ribbon or stereo pair at the tail. A two-microphone balance might use two ribbons, each in the curve of one of the pianos: these can be a shade closer than the single microphone, to give a marginally greater degree of control over the relative volumes and frequency pick-up.

Upright piano

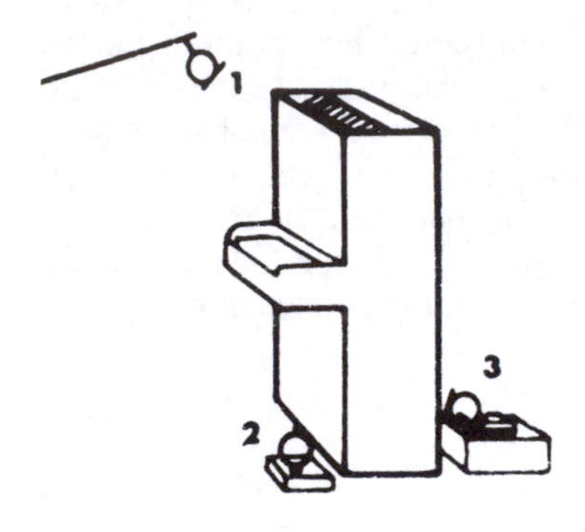

[8.14] **Upright piano**
Position 1: Microphone above and behind pianist's right shoulder. A well-spaced pair might use a microphone over each shoulder, closer in if the front panel is removed. Position 2: To right of pianist below keyboard, or with a well-spaced pair, one on each side. Position 3: At rear of soundboard – avoids action noise.

To avoid resonances at the back of an upright piano, move or angle it away from any wall, then lift the lid and try the microphone somewhere on a line diagonally up from the pianist's right shoulder. Or try two microphones, one on each side. If separation allows, try taking off the front panel, then put microphones about 20 cm (8 in) from the bass and top strings. Or take the lower panel off and again use a pair, in this case spaced well apart to avoid foot-pedal noise. Note also that because it is the soundboard that radiates, it is also possible to stand a microphone at the back (for a close balance), or diagonally up from the soundboard (for a more distant one). Behind the soundboard there is less pedal action noise, but also, of course, less brilliance and clarity.

One of these balances should also work for a janglebox, a type of piano that is specially treated, with leaves of metal between hammer and strings to give a tinny strike action, and with the two or three strings for each note of the middle and top slightly out of tune with each other. Listen and balance for the amount of action noise required.

An upright piano with 'electric' style loudspeakers may have them fed from two very close microphones, that for bass clamped to the frame near the top of G_{ii}; the other somewhere between one or two octaves above middle C (experiment for the best position). Where separation is a problem use a contact microphone on the soundboard

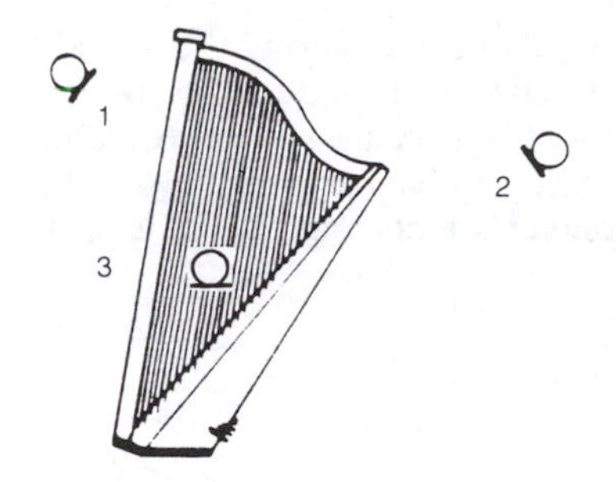

[8.15] **Harp**
1, Frontal coverage, looking down to soundbox. 2, Rear position for microphone favours strings and discriminates against pedal noise, but also loses reinforcement from the soundbox. 3, Close on soundbox, can be used for light or pop music.

for the bass. The output may be taken to the mixer (with a feed back to the two loudspeakers), or sent by DI box to the loudspeakers, which are then balanced as for the guitar. Equalization (EQ) may be used, either to emphasize high-frequency percussive transients or to add weight to the bass.

Harp, harpsichord, celesta

The harp has a soundbox below the strings: in performance this is cradled by the player who reaches round on both sides to pluck (and otherwise excite) the strings. Owing to the relative inefficiency of the soundbox, the vibration of the strings contributes a higher proportion of the sound than in a piano, and ideally any balance should cover both. Foot-pedals control the key of a harp, and if they are operated during quiet passages, their mechanical sounds may be audible. In an orchestra, harps generally benefit by spotting for presence and by locating them clearly, usually towards the side of the stereo picture.

A microphone about 1.5 m (4–5 ft) diagonally upward from the soundbox may be directed to cover both that and the strings. A ribbon gives tight coverage of one or both harps on its relatively narrow forward lobe, while the rear lobe points up and away into the hall. A condenser cardioid also covers the harps well, while a hypercardioid can be placed closer in to the strings, favouring the middle and upper parts of the soundbox, with some discrimination against pedals.

A harpsichord has many of the qualities of the piano, including mechanical action noise, plus the characteristic attack on each note as the string is plucked by its quill. When the instrument is featured, it is useful to have control over the relative values of transients and resonance. For this, set out two microphones, one above and one below the level of the soundboard. The lower microphone takes the body of the sound, perhaps by reflection from the floor; the other, angled to 'see' down into the action, picks up the full sound but with transients and upper harmonics somewhat over-represented. Mix to taste.

[8.16] **Celesta**
A microphone near but behind the soundboard avoids mechanical noise, which can be loud compared with the music from this relatively quiet instrument. Another possible position is in front of the instrument, below the keyboard, on the treble side of the player's feet at the pedals, but keeping well away from them.

In a baroque orchestra (covered by a coincident pair) harpsichord continuo needs to be positioned well back near the basses, with which it co-operates, and also to avoid its penetrating sound appearing too dominant. Use a spotting microphone to fine-tune its contribution.

In other instruments of the same general class, such as virginals, clavichord or spinet, apply the same principles: first identify the soundbox or board, sources of action noise, attack transients and higher frequency components, then place the microphone or microphones accordingly.

The celesta is a rather quiet keyboard instrument. In an orchestra it benefits by spotting. Ideally, a close balance is needed for good

separation, but this has the disadvantage of accentuating the already potentially intrusive action noise. A balance at about 2 m (6–8 ft) would avoid the worst of this, but if a closer balance is unavoidable try a microphone either in the middle at the back or below the keyboard on the treble side of the player's feet – and keeping well away from them.

Piano accordion

In a close balance the piano accordion may require two microphones, one directed towards the treble (keyboard) end, the other favouring the bass. A closer balance on the air-pumping bass end can be achieved by clipping a goose-neck electret to the instrument itself, so that the microphone hangs over some of the sound holes. For greater mobility, attach two, one at each end.

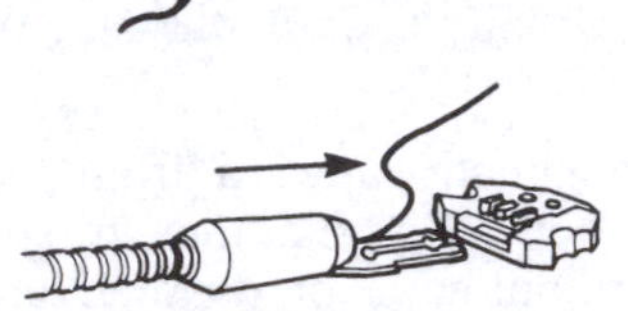

[8.17] **Piano accordion**
A popular music balance can be achieved using close microphones at the bass and treble sides. These may both be attached to the body of the instrument or just one attached over the bass, with a stand microphone for the treble. In this example, a special plate and clip are shown, which can also be used for guitar 'amps' (loudspeakers).

Harmonica

This can be something of an oddity: some players supply their own specialized small microphone, which fits inside the resonant cavity formed by their hands. This avoids the conventional alternative: a simple stand microphone in front of the player.

Woodwind

Each of the wind instruments has a distinctive directional pattern of radiation. In contrast to brass, on woodwind with fingered or keyed holes the bell may make little difference to the sound: for all but the highest frequencies, the main radiation is through the first few open holes. In general, a microphone in front of and above the player and directed towards the middle of the instrument (but not the player's mouth) will prove satisfactory.

For wind, as with other sections of the orchestra, transients (the way notes start) are of great importance in defining the instrument and giving it character. These sounds must be clear but not exaggerated.

[8.18] **Flute**
A microphone behind the player's head discriminates against the windy edge tone, but can 'see' the pipe and finger holes.

On the flute, there is a breathy edge tone. The closer the balance, the stronger are the transients in comparison with the main body of the sound. Listening at an ordinary concert distance, the harsh edges of the notes are rounded off; listening to a flute at 2 feet, it can sound like a wind instrument with a vengeance. For classical music, 'close' can mean 1–2.5 m (3–8 ft); in a multimicrophone balance it may be 15–60 cm (6–24 in). In close balances it is occasionally advisable to resort to devices such as placing the microphone *behind* a flautist's

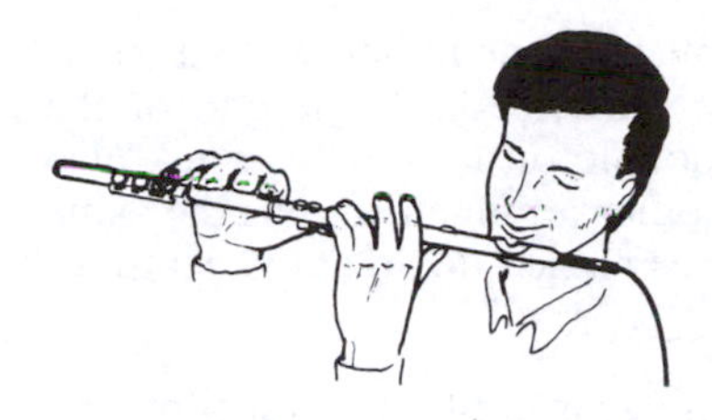

[8.19] **'Electric' flute**
Pick-up within the head of the flute discriminates strongly against breath noise, and offers a distinctively pure, mellow tone.

[8.20] **'Electric' penny whistle**
Pick-up can only be from the side of the body of the instrument.

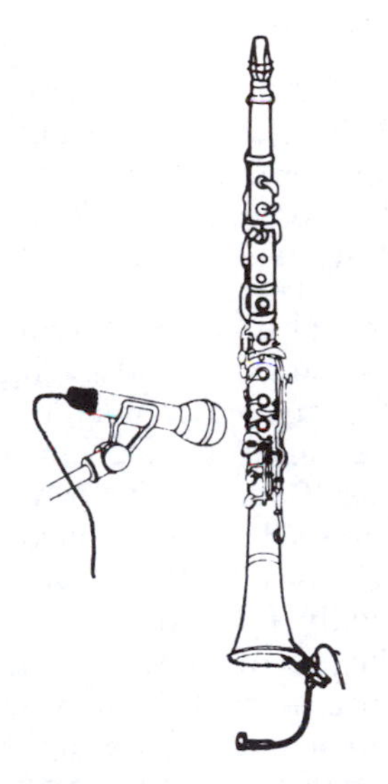

[8.21] **Clarinet**
Popular music balance using two microphones, one of them on a miniature goose-neck for the bell.

head. This reduces the high-frequency mouth noise and edge tones without affecting the tonal sound radiating from the finger holes.

At the other acoustic extreme is a headset carrying a miniature microphone with its axis directed past the embouchure towards the upper holes. Another possibility is an electret on a pad, taped in place near the mouthpiece, but directed away from it. Taking this a stage further, some flutes have been adapted so that the microphone is actually inside the head joint, from where it picks up the sound wave within the length of the flute, but hears no mouth noise or sound radiated from the holes. This gives it the unnaturally pure tonal quality used for the 'Pink Panther' theme. Radiated sound can be added to this as desired, from a stand microphone in front of the player.

An arrangement like that would not work for an 'electric' recorder or penny whistle, as the mouth covers the top end, but a microphone can be fixed in to the side of the instrument above the finger holes.

When used for popular music the clarinet has been balanced using two close microphones. One is on a stand at 15–30 cm (6–12 in), pointing down towards the keys. The other is a lightweight electret on a spring clip attached to the bell, angled by means of a short goose-neck to point back into it: this is mainly for frequencies over 3 kHz.

In a close balance of a full orchestra the woodwind may sound too far back. To correct for this, sling an additional coincident pair or set in three separate microphones on stands at a height of about 2 m (6 ft or so) in front of the section. The extra microphones are for presence, perhaps emphasized by presence in the 2–4 kHz range. They are not intended to provide the main balance, and their signal must be added to it with discretion. But they do also allow precise control of the stereo spread in the central region, perhaps broadening it a little. The width required from a stereo pair on woodwind would still, of course, be much narrower than that of the main pair covering a whole orchestra.

Woodwind orchestral spotting is one of the cases for which a delay may be calculated and added at the console. (Remember: delay in seconds equals number of metres divided by 344.)

Saxophones

For a saxophone try a microphone near the middle keys, angled to include some pick-up from the bell (for higher frequencies), but listen also to check that the breath effects and keypad noises that are also part of its character are present at acceptable levels. For a breathy subtone quality, another pick-up angle might be along the central axis of the bell. As a second microphone, an electret can be attached to the bell, taping it on a soft pad at the rim, or by clipping a goose-neck to the rim with the electret angled into the bell. Saxophones may benefit from presence at about 2.5–3 kHz.

In big-band music a saxophone section has perhaps five musicians. A balance with a bidirectional ribbon can have three players on one

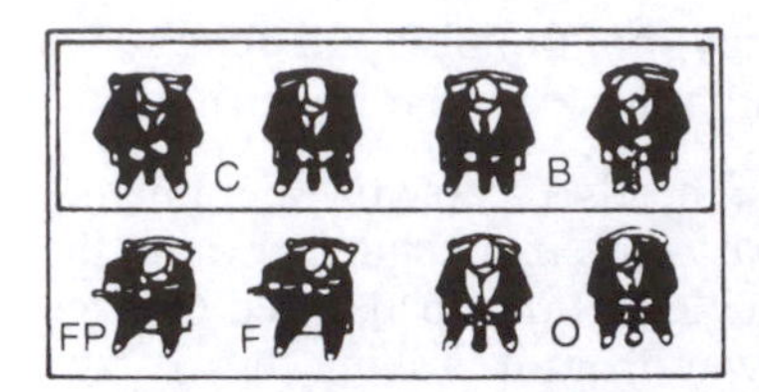

[8.22] **Woodwind orchestral layout**
F, Flute, doubling P, Piccolo. O, Oboe. C, Clarinet. B, Bassoon. If the main orchestral microphone is set for a close balance, it may tend to favour the strings at the expense of woodwind, so that a spotting microphone (or coincident or spaced pair) may be needed.

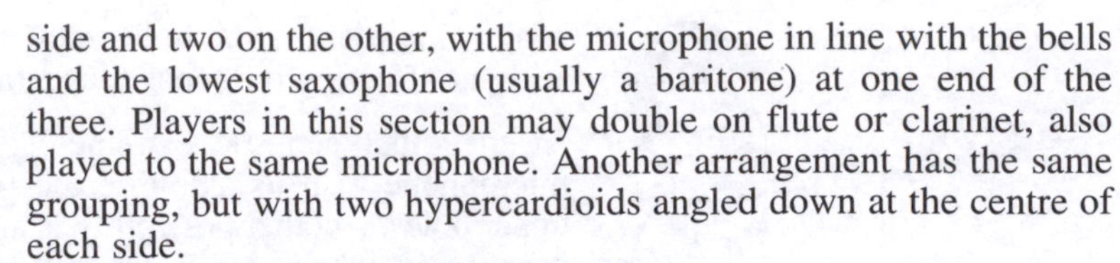

side and two on the other, with the microphone in line with the bells and the lowest saxophone (usually a baritone) at one end of the three. Players in this section may double on flute or clarinet, also played to the same microphone. Another arrangement has the same grouping, but with two hypercardioids angled down at the centre of each side.

An alternative that some balancers prefer is to place this section in a line or arc, with one microphone to each pair of musicians. This helps when saxophone players doubling woodwind fail to move in for a second, quieter instrument, and also allows the height of each microphone to be adjusted more precisely to the needs of the other instrument. A close balance on clarinet, for example, would be lower than for a flute.

Brass

In brass, the axis of the bell of each instrument carries the main stream of high harmonics. Even when the music requires that this be directed at the audience, only a relatively small group receives the full blast at any one moment, the rest hearing the sound off-axis. In consequence, for all but the most exaggeratedly brilliant effects, the main microphone for an orchestral balance is usually quite satisfactory for the brass. It is no disadvantage that its distance from the microphone is greater than that of some other sections, because the added reverberation conventionally suggests the sheer volume produced. Orchestral brass that is balanced well back in a spacious hall is usually easier to control (requiring relatively little manipulation of level) than when, by choice or necessity, closer microphones are used.

[8.23] **Saxophones in dance band**
In a studio layout, up to five players may be grouped around a central microphone. *Above*: A bidirectional microphone set at the height of the instrument bells. *Below*: C, Two hypercardioid microphones set above bell level and angled downwards to reduce pick-up from outside this area. S, Solo microphone for player doubling on a second instrument.

Spotting microphones may have to be set for an unusual reason: in this case because the volume of sound from brass can cause the whole balance to be taken back, making this section seem even more distant unless there is separate control of presence – which will require additional reverberation to match its perspective.

In a close balance – with a trumpet at 30 cm (1 ft) or so, or trombone closer than that – the volume from brass can reach 130 dB, higher than the maximum recommended for the condenser microphones needed to get the full frequency response, unless the signal can be reduced by switching at the microphone itself, for which a pad of up to 20 dB is required. Do not rely on rehearsal levels of unknown brass players: instead, allow for as much as a 10 dB rise in level between rehearsal and recording, especially for a live broadcast, otherwise there is a risk of distortion, reduction of the working range on the faders and loss of separation. It also helps if the players are prepared to co-operate by leaning in to a microphone on quiet passages with mutes, then sit back for louder passages, and also play far enough to the side of it to reduce overload and blasting. The microphone must be near enough for them to reach for close work but far enough back to pick up several players equally. Brass players are well aware of the directional properties of their instruments and will play directly towards the microphone or not as the music requires, though some bands that have done little work in studios and more in dance halls may take this to extremes, perhaps by playing soft passages in to a music stand. They may need to be reminded that a cleaner sound can be achieved if some of the control is left to the balancer.

The microphone should be at a height that is close to the axis of the instruments as they can be comfortably played – although with trombones this may be awkward, as music stands can get in the way of the ideal position. However, the leader of the section will usually state a preference for whether microphones should be above or below the stand: this probably depends on how well the players know the music. In a band where the music played is part of a limited and well-rehearsed repertoire, trombones are often held up, but musicians faced with unfamiliar music need to look down to see it.

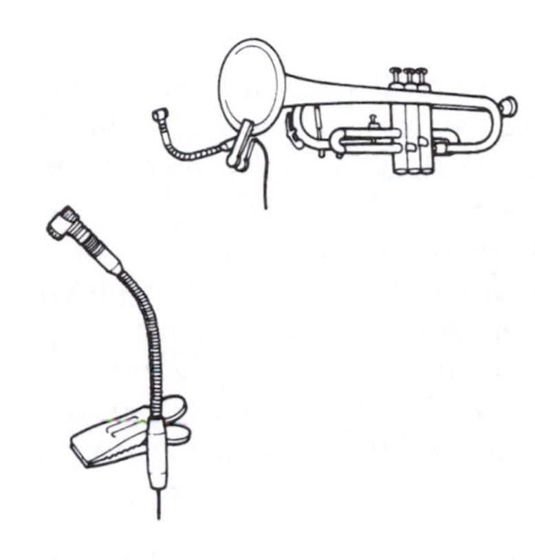

Miniature goose-neck electrets can also be clipped to the rim of the bell of a brass instrument, a technique that gives the player freedom of movement, but loses the effect of movement on performance. Again, try different microphone positions by bending the goose-neck.

Brass may benefit from the presence of 5 dB or even more at about 6–9 kHz. But note that some microphones have peaks in their response in this region and produce the effect without much need for equalization.

In the orchestra, horns can often be placed where their sound reflects from a wall or other surfaces behind them. Such an arrangement – with a microphone looking down from above at the front – is suitable in most circumstances; though exceptionally, for greater separation and a strong, rather uncharacteristic quality, a microphone can be placed behind the player in line with the bell.

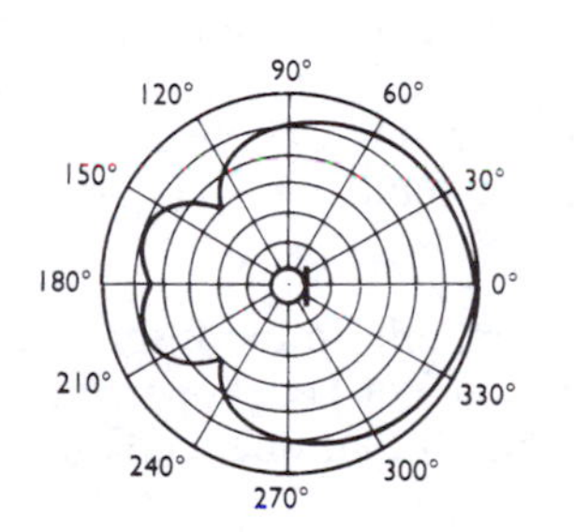

To separate a tuba, try a microphone at 60 cm (2 ft), slightly off-axis.

Several layouts are used for brass or military bands. In one, sections form an arc round the conductor much as in a string orchestra; in another, the performers make three sides of a hollow square, with the cornets and trombones facing each other on opposite sides. In this way the tone of the horns and euphoniums at the back is not blotted out by the full blast of more penetrating instruments. A microphone pair may be placed a little way back from the square, as for an orchestra.

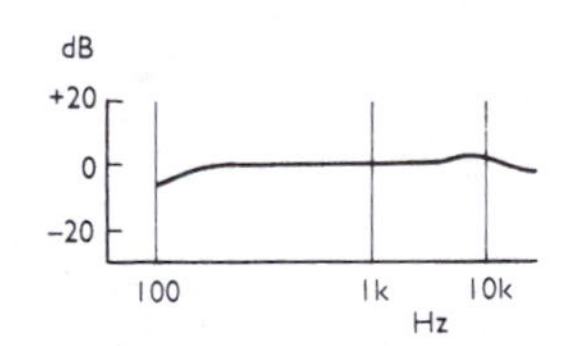

[8.24] **Miniature goose-neck** Electret for brass (trumpet, trombone, tuba and also saxophone) in multimicrophone balances, allowing player freedom of movement.

The main body of euphoniums and bass tubas radiates upward and can sound woolly on the main microphones unless there is a reflecting surface above them to carry their sound forward. Failing this, an additional microphone above them will add presence. A solo euphonium player should come forward to a microphone set high up and angled to separate it from other instruments of similar sound quality. A solo cornet may also have its own microphone, a condenser cardioid.

A military band behaves like a small orchestra from which the strings have been omitted. There is usually plenty of percussion, including such melodic instruments as the xylophone and glockenspiel, and

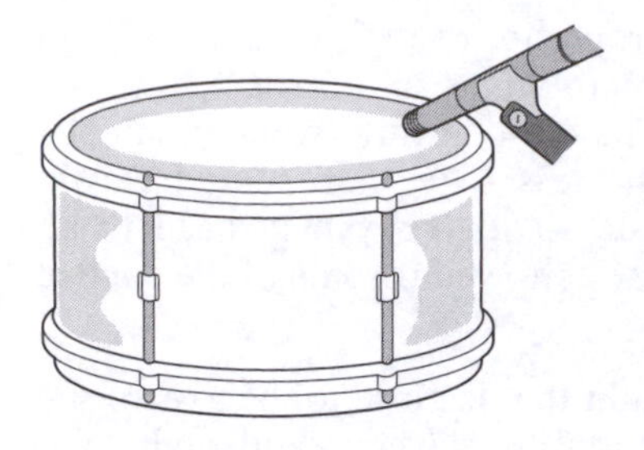

[8.25] **Drum microphone**
A microphone a few centimetres inside the rim is suitable for most of the smaller drums in a kit.

care must be taken to ensure that these are given full value in the balance.

A military band in marching order is balanced for the benefit of the troops behind them; in the open air, woodwind, at the back, will not be reinforced as it would indoors. A microphone balance from the rear helps to correct this, even though for television or film, pictures are more likely to be from the front. A band on the move is even trickier, as the best balance will rarely match preferred wide-angle camera positions. A marching band in an arena can be followed well with a tetrahedral (Soundfield) array: suspended over the action, it can be steered to give an angle and polar response that covers the band, while discriminating against excessive reverberation or crowd noise.

Percussion, drums

Most percussion instruments can be recorded with ease. However, some percussion sounds register more easily than others, and on all but the most distant microphone set-up it is as well to check the balance of each separate item with care. Any rostra (risers) need to be reinforced to prevent coloration. In orchestral percussion a microphone with an extended, smooth high-frequency response is required mainly for triangle, cymbals and gong. Spotting microphones may be set for many of the individual components: for the quieter ones, so that detail is not lost; and for the louder (as for brass), so that presence can be retained when overall volume is reduced. In an orchestral balance delay is often used with spotting microphones.

For instruments such as the glockenspiel and xylophone use a stereo coincident pair above them (though some balancers are happy with a well-separated spaced pair). The characteristic sound of the vibraphone requires a close microphone or microphones beneath the bars, angled upward to the tops of the tubular resonators (the bottoms are closed).

In popular music, percussion shows itself mainly in the form of the drum-kit, including a snare drum, bass drum (foot-pedal operated, as a kick drum), a cymbal (or several), a hi-hat (a foot-operated double cymbal), and toms in different sizes.

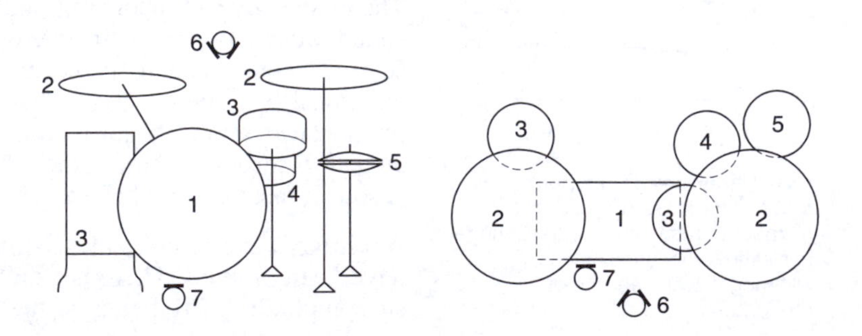

[8.26] **Drum-kit**
1, Kick drum. 2, Cymbals. 3, Toms (and other drums). 4, Snare drum. 5, Hi-hat. 6, Coincident pair with good high-frequency response (a spaced pair may be used instead).
7, Additional microphone (moving coil) for percussive kick-drum effect. Further microphones are usually added, to give each drum and the hi-hat its own close pick-up.

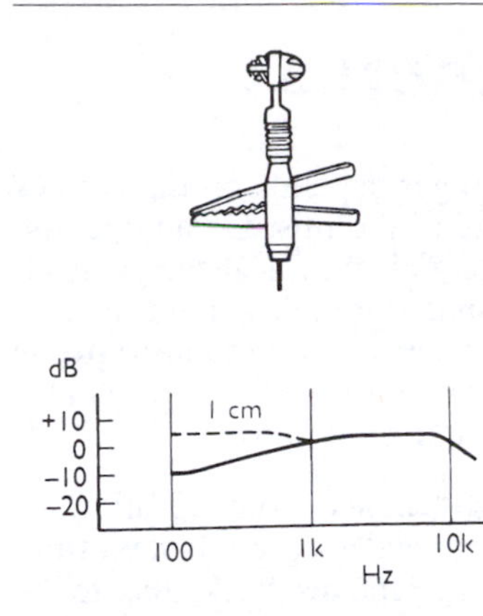

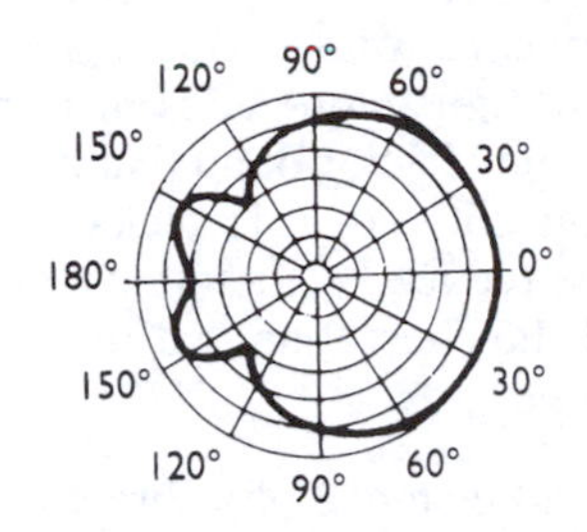

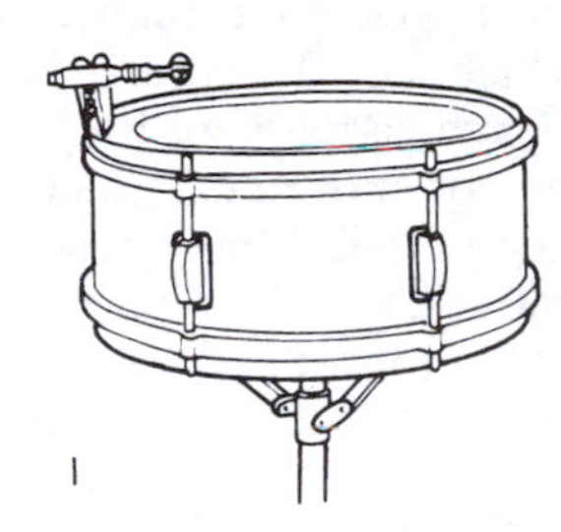

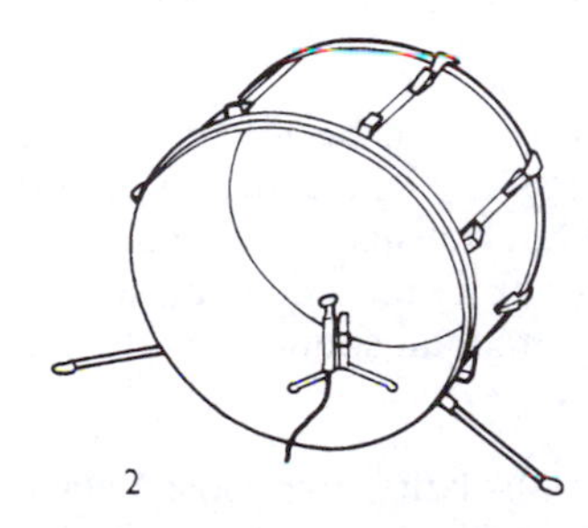

[8.27] **Miniature microphone for drums**
1, Condenser hypercardioid, clamped on top hoop or tensioner: the tone varies with drum balance. 2, The microphone on a small stand inside the kick-drum shell, fairly close to the skin. If a blanket is also put inside to press against the skin, it shortens the vibration to give the drumbeat more attack.

In the simplest balance, a microphone pair (with an extended high-frequency response) points in from above. This will certainly cover the snare drum and cymbals, but any attempt to move down for a closer balance increases their dominance over the hi-hat, which radiates most strongly in a horizontal plane (in contrast to the up-and-down figure-of-eight pattern of the top cymbals), and the kick drum. Avoid a position alongside the hi-hat, as it puffs out sharp gusts of air.

For a tighter sound move one microphone closer to the hi-hat, also favouring the left-hand top cymbal or cymbals and some of the snare and small tom-tom, together inevitably with some bass drum; a second cardioid microphone is then moved in over the other side of the bass drum to find the other cymbal or cymbals, hi-hat, big tom-tom, and again, perhaps, the snare drum. Changes of position, angle, polar diagram or relative level can give reasonable cover of nearly every element except the kick drum.

The distinctive kick-drum sound that is required to fill in the texture of much popular music can be obtained by facing the diaphragm of a moving-coil microphone toward the front skin and close to its edge, where a wider and more balanced range of overtones is present than at the centre. This is one place where one of the older, more solid-looking microphones with a limited top response is as good as a modern high-quality microphone (which might in any case be overloaded). For the dry thud required in pop music put a blanket inside against the beater head to shorten its vibration. Try using EQ to reduce the lower 300–600 Hz range and lift at 2.5–5 kHz to make the sound crisper.

For greater control still (and to increase separation from cymbals) use separate microphones for each part of the kit. For some of the drums (snare, toms, congas, bongos, etc.) they may be clipped to the top edge (tensioning hoop) or pointed down from stands, 3–8 cm (1–3 in) over the head and 5 cm (2 in) in from the rim. A relatively insensitive electret can also go inside a kick drum, but this needs a solid mounting. With a low stand microphone for the hi-hat, an overhead pair for the cymbals and basic stereo coverage, the drum-kit alone can use up far more tracks in a multitrack recording than any other part of a group. Many pop balancers work with the drummer to tailor the sound, and in particular modify the snare drum by tightening snare tension (perhaps also putting some masking tape on it), and may tape a light duster over part of the skin (away from where it is struck). Presence (at 2.8 kHz or more) plus additional bass and extreme top may be added.

The toms are less likely to require attention, but the quality of cymbals varies and the application of masking tape may again help to reduce the blemishes of poorer instruments. With such extreme close balancing it is wise to check the maximum playing volume in advance, and also to recheck the quality of the sound from each drum from time to time: it may change as a session progresses.

Stereo width is introduced by steering the various microphones to slightly separate positions. If a contribution from the studio acoustics is required, angle in two ribbons from perhaps 1.5 m (5 ft) above and away from the kit.

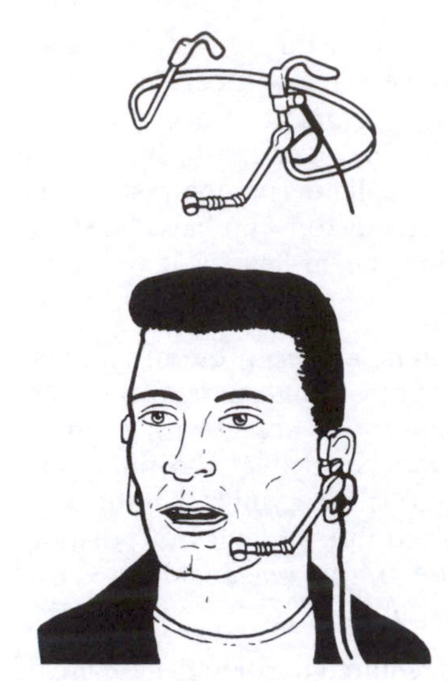

[8.28] **Headset (without headphones)**
For singers, designed for light weight and minimal disturbance of hairstyle. The electret microphone is in a fixed position, and at this distance must be beyond the corner of the mouth in order to avoid popping.

Singers: solo and chorus

What counts as a close balance for a singer depends on the type of music. For a pop singer it may be close to the lips, for an operatic aria at arm's length. For classical music a clean, flat response is all that is necessary: try a hypercardioid condenser microphone at about 1 m (3–4 ft), angled up enough to point over any orchestral instruments behind the singer. High-quality ribbons at the same distance can also work, though with care for what might lie in the opposite lobe.

For lighter styles of music using a close balance, compensation for bass tip-up (proximity effect) is, as ever, needed when using a fully directional microphone, and if the singer moves back, has to be restored. More useful are directional microphones on which the low-frequency response degenerates towards omni. But while a microphone that is hypercardioid only at high frequencies discriminates well against the kind of high-frequency background that might interfere with melodic line and intelligibility, it still picks up a boomy rumble, so again the bass must be reduced – though a singer working very close does produce less of the chest tone that in opera is used to deepen and strengthen the sound.

A skilled performer may use the angle and distance of a directional microphone to precise effect; less reliable singers are certainly safer with an omni at lower frequencies, for which only the volume has to be adjusted (though even this may have some variation of frequency response with angle). For total freedom of movement (and free hands) but constant distance and angle, try a headset (with headband below the hair at the back, for those whose hairstyle is part of the act). The tiny electret must be a little away from the corner of the mouth in order to escape 'popping'. For other microphones, popping is minimized by directing the voice (as for speech) across the microphone – that is, by missing it with the air blast. Windshields, built-in or added, also help.

To improve separation in a studio, a singer can be given a tent of screens, open (for visual contact with any musical director) only on the dead side of the microphone, or with a transparent panel on that line. In some balances a vocal guide track is recorded (on a spare track) at the same time as the main instrumental sound, and the lead and backing vocals overdubbed to replay.

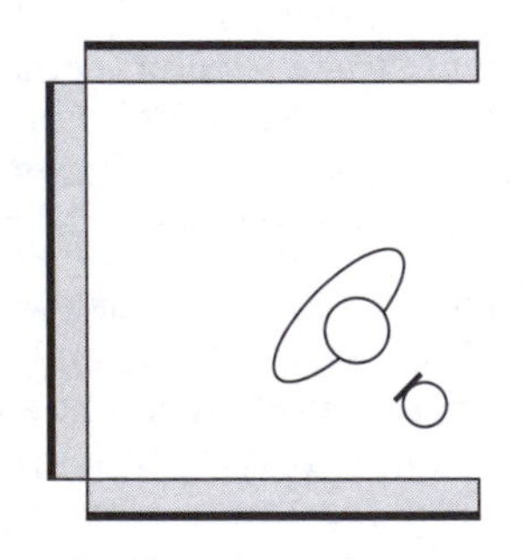

[8.29] **Vocalist**
When the volume of surrounding instruments is high, the singer is screened. The microphone's live side is towards absorbers and its dead side towards the open studio.

Some singers seem subject to a superstitious belief that close balance gives them presence. In fact, if separation (the real reason for close balance) can be achieved in other ways, try setting in a second microphone 30–60 cm (1–2 ft) from the singer's mouth. Not only will popping no longer be a problem, it may also be found that the more distant balance is better, as well as easier to control. Presence can be added electronically, at about 1.5–3 kHz or higher, depending on the voice. Sibilance can be cured by some combination of singing across the diaphragm, careful enunciation, high-frequency roll-off, and choice of microphone with a smooth 3–12 kHz response. Condenser cardioids are preferred for this quality rather than for their extended

[8.30] **Singer with pop-shield**
A gauze on a 10 cm (4 in) hoop protects the microphone from the vocal gusts that accompany a plosive 'p' sound.

high-frequency response, which is not needed. Some balancers are happy to use microphones with multiple high-frequency peaks.

A folk-singer with guitar can in principle be balanced on a single microphone, perhaps favouring the voice, but two microphones mounted on the same stand give greater control of relative level, together with the capacity for separate equalization and artificial reverberation. According to the 3:1 rule, the distance between each microphone and its sound source must be less than a third of the distance that separates them. Another way to avoid phase interference is to delay the vocal signal by about 1 ms, to match its time of arrival at the guitar microphone.

Individual singers are rarely balanced in stereo (except where action is taken from the stage). Mono microphones are used and steered to a suitable position in an overall balance. But a coincident pair is well suited to a backing group or chorus, which can then be given any desired width and position in the final mix.

With choral groups, strive (generally) for clarity of diction. Set the microphone back far enough to get a blended sound, but without losing intelligibility. Spaced bidirectional or cardioid microphones may cover a large chorus. Where a small group can conveniently work to a single microphone position, use a cardioid or, for stereo width, a coincident pair, curving the front line of singers to match the microphone's field pattern.

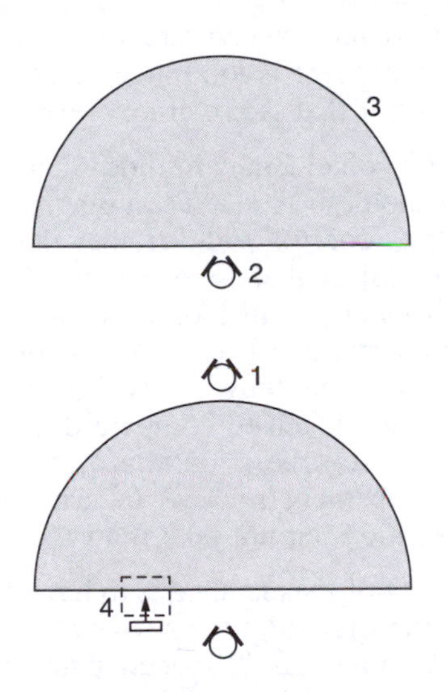

[8.31] **Orchestral balance**
1 and 2, Two stereo pairs with different polar diagrams but giving the same stage width to the orchestra (3). Microphone 1 is a double figure-of-eight, microphone 2 is a double cardioid: a mixture of the two can be used to control and change reverberation where there is such a wide dynamic range that much volume compression is necessary. 4, For a soloist a spotting monophonic microphone is used, steering this to the same position where it appears on the audio stage, as established by the stereo pair.

There is no special virtue in clear separation of sopranos, altos, tenors and basses into separate blocks: the stereo effect may be good, but the musical quality of a layout that blends all sections may be better. As an alternative to the block layout, a chorus may be arranged in rows (and here even more than in the other cases the microphones need to be high up in order to balance all the voices). Occasionally, the singers may all be mixed together. Opera singers must do this, and so can some choirs. Skilled musicianship is required: when a group can sing like this and wishes to do so, the balancer should not discourage them.

Church music, and particularly antiphonal music with spaced choirs of voices, benefits enormously by being brought round to the full spread of stereo: it was here that surround sound and its predecessor, quad sound, came into their own without the need for special composition.

The orchestra

Our next individual 'instrument' is the full orchestra. Indeed, it can really be considered to be one instrument if it is well balanced internally.

The detail of orchestral layout is not absolutely standardized, but the strings are always spread across the full width of the stage, grouped in sections around the conductor, with the woodwind behind them

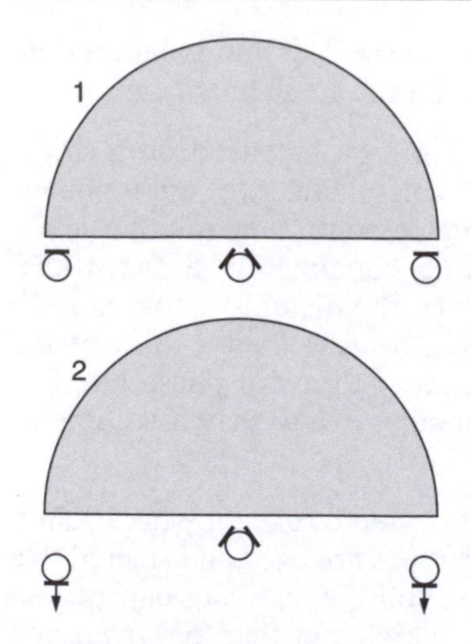

[8.32] **Stereo orchestral layouts**
1, With a spaced pair to 'pin down' the corners of the orchestra (e.g. harp on left and basses on right). 2, With a spaced pair used to increase reverberation.

near the centre. In an older arrangement the violins were divided, with first violins on the left and second on the right, then violas, cellos and double basses in the arc behind them. For stereo, an advantage of this is that the violin sections are clearly separated, and the weight is in the middle. But having the upright instruments (cellos and basses) face forward is not necessary, for either the microphone or the audience, as they are virtually omnidirectional radiators; and there is also some disadvantage in having high-frequency radiation from second violins directed to the back, away from both audience and microphone. The modern layout has the sections placed, from the left: first violins, second violins, violas, cellos; and raised behind the cellos on the right, the basses. Woodwind is always across the centre, and percussion and timpani at the back on one side, with brass around much of the back to the other. That leaves horns, harps and possibly celesta, harpsichord, orchestral piano, etc., all of which may appear in different places.

While the balancer's job is generally to reflect what is offered, in the studio a well-established and experienced balancer may also offer suggestions for improvement, often through a producer. A conductor who is thorough may consult the balancer directly, not only on matters of detail in placing, but might also ask how the internal balance sounds. Many conductors, however, concentrate solely on musical qualities, so that if, for example, there is too much brass it will be left to the balancer to sort it out, if that is possible.

In an orchestral balance that is based on coincident cardioids, theory suggests that the angle between them should be 131°. Treat that as a starting point. Try placing them first to control width, rather than reverberation. This means that for a subject that is to occupy the major part of the audio stage the microphones will look down from the edge of the pick-up area; for an orchestra that is to occupy about two-thirds to three-quarters of the sound stage, they are as high but set back a little from the conductor. Then listen to compare direct sound (enriched by early reflections) with reverberation: this offers a subjective approach to microphone placement relative to critical distance. You can always add reverberation but not take it away.

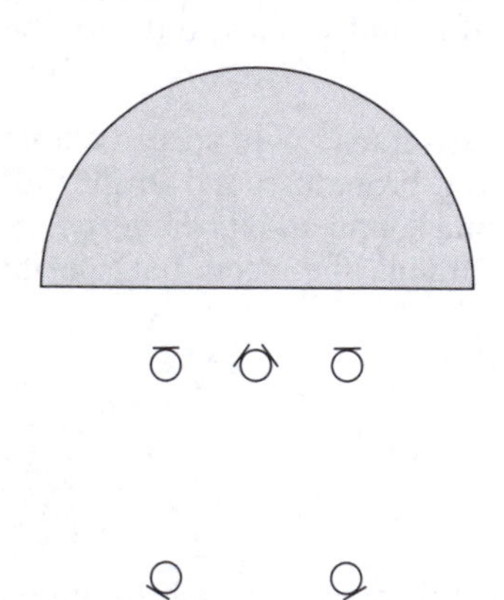

[8.33] **Balance for 5.1 surround sound**
A stereo pair with omnidirectional microphones 1.2 m (4 ft) on either flank and others set back 9–12 m (30–40 ft).

Reverberation may be added by a variety of means. One is to have a second coincident pair set back from the first and facing toward the empty part of the hall or studio. Another has a spaced pair of cardioids at the sides of the hall, well forward again, but looking back along it. In either of these cases the two microphones must be steered to their correct sides of the studio, as some direct sound will reach them, and even more by short-path first reflections. Reverberation microphones should not be more than 12 m (40 ft) back, as this introduces echoes, most noticeable on percussive sounds such as drums. Alternatively, resynchronize by using delay on the closer, main microphones.

Balances using these combinations of forward and reverberation microphones have been used for 5.1 surround sound. A wide range of other successful 5.1 balances have been reported, but some may work best in their original locations or when operated by their originators. The hall's acoustics can easily make more difference

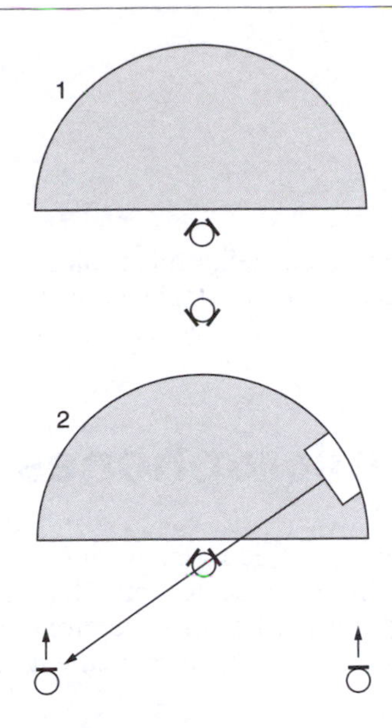

[8.34] **Reverberation in stereo orchestral layouts**
1, Good: behind the main pair is a second coincident pair with reversed output, to control reverberation separately – this can also be used for surround sound. 2, Bad: a forward-facing spaced pair set back into the hall may give false information from the brass. Directed towards the microphone on the opposite side, it may sound as though the brass is in the wrong place.

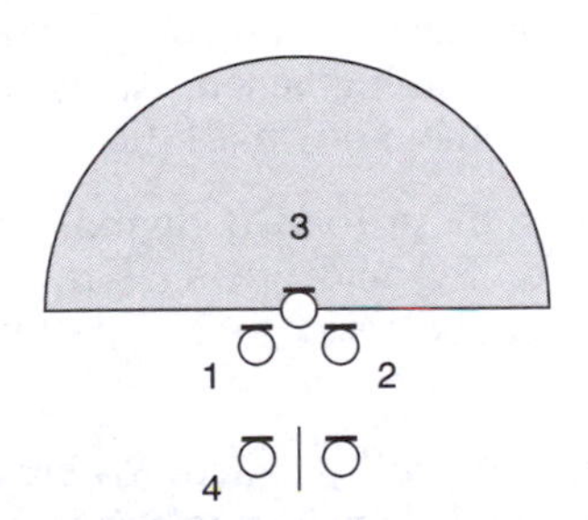

[8.35] **Microphone 'tree'**
1–3, A narrowly spaced pair of microphones, with a third forward from the middle, in a balance originally developed by a recording company. A surround version of this has additional microphones at the sides: these are mainly for reverberation from the hall, but also broaden the orchestral image a little. 4, In yet another balance, a baffled omni pair (similar to the 'dummy head' layout) has been favoured for its spaced effect.

to the result. In practice, any balance that offers good stereo (clean positioning and a well-judged compromise between richness and clarity), together with separate reverberation coverage, is likely to work well in 5.1.

Another stereo layout has a second coincident pair of microphones for reverberation, but this time figures-of-eight looking forward and set back to such a distance that the positions of the various parts of the image roughly coincide. For an orchestra, the second pair may be about 3 m (10 ft) further back. This arrangement has the additional advantages that the balancer can work between the two pairs to handle extreme changes of volume, and that the second pair gives an automatic standby facility in the event of failure or loss of quality of the first pair on a live broadcast. Such failure is rare, but the facility is certainly desirable, and in any case the two pairs of microphones can be used for balance comparison tests. Phase cancellation on the back of the second pair is still a potential problem.

Any microphone that is set up in addition to the main pair and facing toward the orchestra (or a 'bright' reflecting surface) may give false information about the position of the brass, if significant direct or strong first-reflection sound can reach it. The trumpets are likely to be directed over the heads of the strings and may be picked up on any microphone in their line of fire, even on the opposite side of the studio. There is then an extra trumpet signal, pulling the section across the sound stage.

Further mono microphones are often added (for 5.1 too). For stereo, one on each side of the orchestra may be placed to open out the strings. These are faded up just sufficiently to lend presence to string tone, and panned to their proper places in the stereo balance, which is thereby more clearly defined. Woodwind is also often covered: where the main pair give good basic stereo coverage a mono microphone may be sufficient, though a stereo pair is often used here too. It lends presence to woodwind solo passages. Harps and some other instruments may also be given spotting microphones, which again must be used with subtlety to add clarity and positional information.

When close spotting microphones are mixed into a relatively distant overall balance like this, check whether they seem to bring their instrument or section too close. If the added volume is low and the difference in distance not too great, this effect may not be noticeable. If in doubt, it may be safer to insert delay in the spotting channels (the number of seconds delay is distance in metres divided by 344).

Spotting microphones and delay can also be used to restore the balance of sections that are very weak in numbers. Automatic double tracking (ADT; also available on many consoles) adds a copy of the original signal with a delay that can be set somewhere between 4 and 20 ms: this doubles the apparent number of musicians.

With spaced microphones, observe the '3:1' rule for separation. When using pairs of omnidirectional microphones, for each unit of distance between the two microphones, allow no more than a third of that distance between each microphone and its subject. This ensures

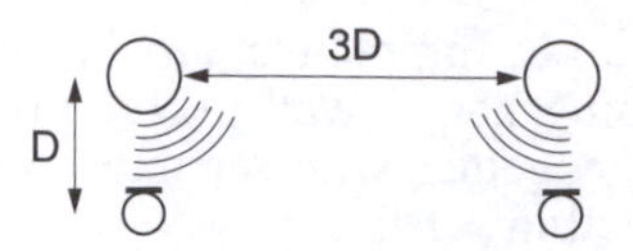

[8.36] **The 3:1 rule**
In a multimicrophone balance the distance between two omnidirectional microphones should be at least three times the distances from each microphone to its sound source. This ensures that separation is greater than 10 dB.

a separation of at least 10 dB, to keep phase-cancellation problems at bay. With directional microphones, this rule can be relaxed a little, but with caution.

Orchestra: spaced microphones

This book recommends coincident pairs for stereo, but accepts that spaced microphones have always had strong advocates, and that the availability of omnidirectional microphones with an exceptionally smooth response reinforces their case. One such arrangement is outlined below.

A 'bar' with two omnidirectional microphones separated by about 40 cm (16 in) gives a rich, spacious sound. Their faders must be locked rigidly together, as the slightest difference swings the orchestra sharply to one side. The axes are angled slightly outwards to allow for high-frequency reinforcement normal to the diaphragm. In a concert hall, their distance may not be easy to change, so a curtain of some five extra microphones may be added, close to the front line of the orchestra, with three more set deeper to favour the woodwind. Their outputs are all panned to their appropriate positions within the stereo picture, and after their relative levels have been set by listening to the effect of each in turn, they are controlled as a group, adding just enough from them for presence and positional information (rather weak from the space-bar) to be strengthened. As usual, fade up until 'too much' is heard, then set them down a little.

Again, spotting microphones (as well as those for woodwind) are added, and considered as candidates for EQ and artificial reverberation. Digital delay of 10–16 ms can be used to avoid time discrepancies.

The virtues of this mass of microphones are still debated. Where a series of concerts is to be covered by several balancers it is best if the hall is rigged in advance with all the main systems currently in contention, so that direct comparison tests can be made. Critics of space-bar and curtain techniques describe them as 'muddy'; critics of coincident pairs call them 'clinical'.

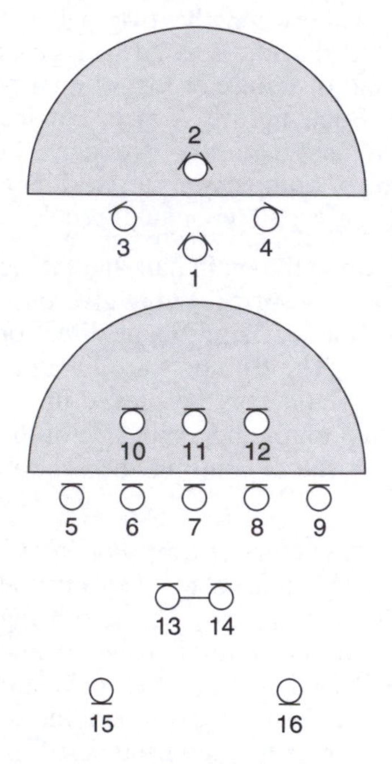

[8.37] **Orchestral multimicrophone balance**
1, Main stereo pair. 2, Woodwind pair 2–3 m (7–10 ft) forward of this section. 3, 4, Spotting microphones for string sections, about 2 m (6 ft) from them, add presence and strengthen positional information. 5–9, A curtain of omnidirectional microphones. Phase and control problems are reduced by a sufficiently close balance. Once faders are set, these are controlled as a group, because changes in the relative level of nearby microphones can affect the apparent position of instruments equidistant from them, including loud, distant sources such as brass. 10–12, Omnidirectional 'drop' microphones, mainly for woodwind. 13, 14, Spaced bar of omnidirectional microphones used as an alternative to the curtain. 15, 16, Spaced reverberation microphones.

Pipe organ with orchestra

If a pipe organ is used with orchestra there is a possibility (in stereo) that its distance will cause its image to be too narrow.

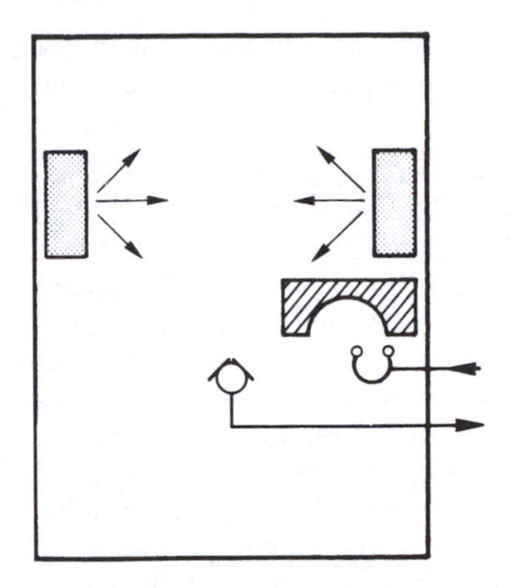

[8.38] **Organist on headphones**
Where a keyboard is badly placed for the organist to hear a good balance, mixer output may be fed back on headphones.

In fact, as a good organ is laid out in a hall, it should have some width (indeed, it is bound to), but not so much that the spatial separation of the various voices is marked: like a choir, organ tone is better blended than broken up into obvious sections. However, in stereo, extreme width (such as that of the organ at the Royal Festival Hall in London) is narrowed by a distant coincident pair, and an average width may even seem narrow. A second pair at a distance that 'sees' the organ as a whole can be used to broaden it on the sound stage. They should be set high and angled to avoid the orchestra, and the main pair set to discriminate against direct sound from the organ.

Orchestra with soloists or chorus

Concert works with solo instruments near the centre can in principle simply use a standard orchestral balance, as the main microphones already favour the soloist. However, in most cases it is usual to set in spotting microphones for additional presence. In contrast, for a piano concerto it may be necessary to angle or place the main microphones to discriminate a little against the volume of the solo instrument.

If there are several soloists, as in a double or triple concerto, or several solo singers, they need to be placed with a thought for the layout of the stereo sound stage.

In public performances a chorus is usually placed behind the orchestra. This creates an imbalance that is greater at the main microphone position than for most of the audience, which is farther back. If the chorus is already weaker than the orchestra, this makes it worse. But in any case, spaced directional microphones are usually added to boost presence and intelligibility, together with extra volume if needed. These are set on very high stands or suspended at the back of the orchestra, to look down at the choir with their dead side to the orchestra. These closer microphones may require delay, to match the choir's distance on the main. Also, the stereo spread can be increased if on the main microphone they appear to be bunched too close together. Weak forces can then be augmented by automatic double tracking (ADT).

For television, to clear sight lines, slung omnidirectional microphones are sometimes used. If so, these may require equalization by adding sufficient lift at 3 kHz to restore intelligibility.

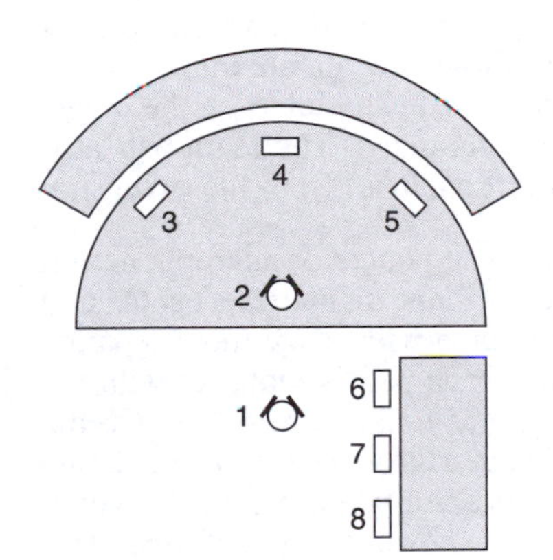

[8.39] **Orchestra with choir**
1, Main pair. 2, Choir pair. 3–5, Directional spotting microphones. 6–8, Alternative layout for choir. Since any spill onto a central coincident pair will be reversed, 6 should be steered to the right and 7 to the left.

Opera

Opera may be recorded or broadcast in concert form; or it may be taken from a public stage performance. A range of methods has been used for each. The simplest employs a standard layout for orchestra,

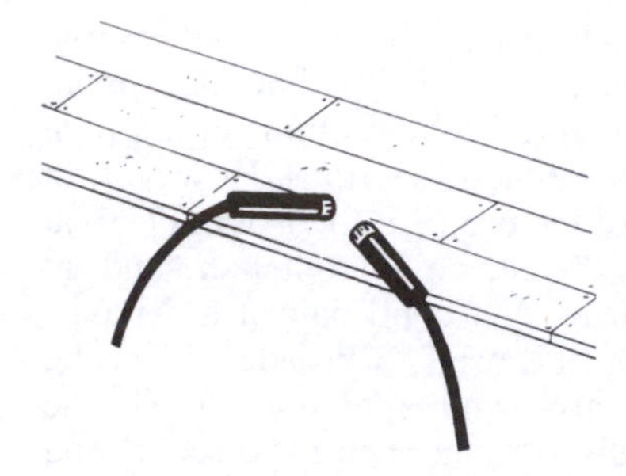

[8.40] **Stage stereo pair, set in orchestra pit**
Capsules on goose-necks are angled slightly down, almost touching the stage. These act as boundary microphones with a directional response parallel to the stage surface, which should be clear of absorbers for at least 1 m (3–4 ft). Additional spaced microphones will be required to cover action out to the sides, or which approaches too close to the central pair.

with chorus behind, and soloists at the front: this works well for oratorio and for concert performances of opera where there is no pretence at dramatic presentation.

Opera on records may have stereo movement, which can be introduced both laterally and in perspective. One layout used for this had a line of five cardioid microphones beyond the orchestra: these covered an acting area marked in a checkered pattern. The square from which to perform was marked on each singer's score; moves were similarly indicated. A disadvantage was that the conductor could not hear the soloists as well as with an oratorio layout. An alternative had singers on either side of the conductor, but facing the orchestra and working to cardioids, which therefore had their dead sides to the orchestra. The singers were recorded on separate tracks, and their perspective and position (with movement provided by panning) were arranged at the mix-down stage. For this, less preparation was required in advance.

In an unusual layout that makes good use of studio space, a standard orchestral layout is combined with a laterally reversed row of singers. With the orchestra at one end of the studio, the chorus is placed at an angle to it along one side of the studio, so that a corner of the orchestra is next to the end of the chorus. The chorus, in rows on rising rostra, works to a line of microphones, as do the soloists in front of them. These microphones are then 'folded' back across the orchestral sound image.

Opera from the stage can be covered by a coincident pair for the orchestra and a row of cardioids along the front of the stage for the singers. If the main pair has to be raised to clear audience sight lines, close microphones are required in the orchestra pit.

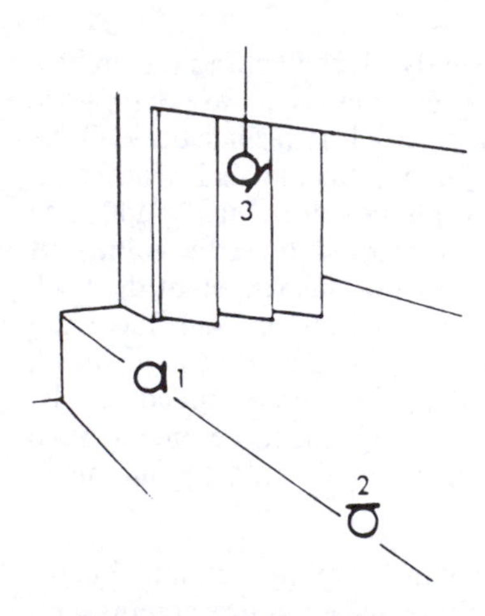

[8.41] **Microphone positions in the theatre (e.g. for opera)**
1 and 2, A pair of (cardioid) microphones in the footlights. Though very close to the orchestra, their response discriminates sufficiently against this. 3, A single microphone or coincident pair placed high in the auditorium – but well forward. At this distance, a cardioid response again favours singers.

To cover the stage, there are further possibilities. One is a coincident pair of cardioids (at 130°), downstage centre, with a spaced pair at the sides, to fill in the corners. If these are all are mounted on brackets, with diaphragms low and close to the edge of the stage, sound reflections from the stage are minimized. The main pair also picks up some sound (though not a full balance) from the orchestra.

A similar arrangement employs directional boundary microphones on the stage. Again, there may be a coincident pair at the centre and single microphones at the corners, but on a wide stage three separate pairs can also work well: the width of each pair is reduced, with the inner microphone of each outer pair steered to the same position as the corresponding element of the central pair. This gives smooth coverage of movement across the stage. If there is a central prompt box to get in the way, two pairs might serve. Note, however, that if boundary microphones are placed on a flexible rather than solid surface, their directional response can change in ways not predicted by simple theory, perhaps by picking up more orchestra than expected.

A deficiency of any footlight microphones is that they can all too easily pick up footsteps on the stage. To counter this, if feet approach any microphone too heavily, fade that one out and rebalance on another to one side. Check the effect on the stereo picture: although displaced, this may be acceptable, or it may be better to pan the remaining microphone to restore the image to its former position.

Important action in an upstage area that is not covered adequately on downstage microphones can perhaps be taken on a 'drop' microphone (i.e. suspended by its cable). Take care not to overdo its use, bringing intentionally distant scenes too far forward. Balance with an ear to the clarity of vocal line, even when television subtitles are added.

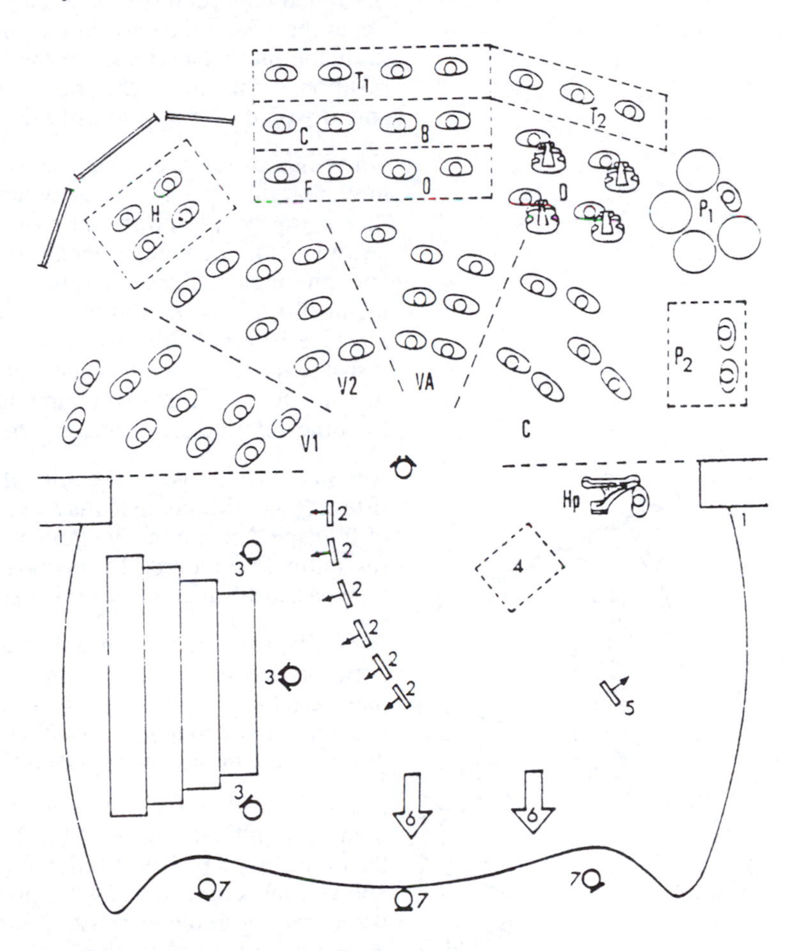

[8.42] **Studio opera layout with chorus and soloists reversed left to right and superimposed on stage** A converted theatre is used in this example. 1, Proscenium arch: the stage is extended into the auditorium. 2, Soloists' microphones. 3, Chorus. 4, Conductor. 5, Announcer. This microphone, dead side to the orchestra, can also be used for spotting effects (e.g. castanets, steered to represent the movement of a dance). 6, An offstage (i.e. distant) chorus can be placed underneath the theatre balcony without a microphone. 7, Audience above. Orchestral layout: V1, 1st violins. V2, 2nd violins. VA, Violas. C, Cellos. D, Basses. H, Horns (with reflecting screens behind). F, Flutes. O, Oboes. C, Clarinets. B, Bassoons. T1, Trumpets. T2, Trombones. P1, Timpani. P2, Other percussion. Hp, Harp. Note that the harp is 'in' or near the right-hand speaker, and that the left-hand side of the orchestra and the side of the chorus and soloists nearest to it will be picked up on each other's microphones, and must therefore be arranged to coincide in the audio stage. For the singers, examples of positioning for particular scenes using this layout are as follows: (i) Part of chorus left, principals centre, off-stage chorus right. (ii) Duet at centre solo microphones, harp figures right. (iii) Quartet at four centre solo microphones.

Classical music in vision

For classical music, a television picture makes only a little difference to the type of balance that is sought. But getting a good balance of any kind can be a little more complicated, because:

- television studios are generally too dead – the halls that are normally used for concerts (or even music sound studios) are therefore to be preferred in cases where they can conveniently be used;

- microphones should not be obtrusive;
- the layout must be visually satisfying, although a good rule is to aim for a layout as close as possible to that of a concert performance.

What is not needed is any attempt to make sound follow picture: when the camera zooms in to a close-up of the oboe, the sound should remain at the concert perspective (note that this is the exact reverse of what may be done with speech). The big close-up in sound would be permitted only for a special effect; if, for instance, someone is talking about the role of the oboe in the ensemble, or alternatively, if this is a dramatic presentation and we must hear as we see the action, from a position within the orchestra. Even for these, the overall orchestral sound would be distorted as little as possible.

One slight difference in balance is desirable for television: it helps to have marginally greater clarity at the expense of the tonal blending that many people prefer on radio and records. The extra presence permits individual instruments or groups to be heard just clearly enough for close shots to be aurally justified. This does not necessarily mean shorter reverberation, but simply that the attack on each note must be heard clearly: this is usually enough to identify individual instruments. A technique that can be used to create this effect is to mix the outputs from two main microphone positions, one closer and the other more distant than for the usual balance.

A balance with this extra clarity also takes into account the different listening conditions that may exist in the home: the sound quality of loudspeakers on many television receivers is lower than that used for radio and records by many people who enjoy music; a more transparent balance serves this better.

[8.43] **Ballet: playback and effects** Dancers need the music to be fairly close, so a loudspeaker may be necessary. Here a series of loudspeakers set in a vertical line radiates high frequencies in a horizontal field. A gun microphone above them receives only low-pitched sound from this array, and this can be filtered out, as the main component of the dance effects is at higher frequencies.

For *ballet* performed in the television studio it is often difficult to get both the dance settings and the orchestra into the same studio: this may be a blessing in disguise, however, as the solution is to take the orchestra away to a music studio (with remote cameras, if required). The music can then be replayed to the dancers over loudspeakers.

Limitations (due to cost) on orchestral rehearsal time can mean that orchestra and dancers must be rehearsed separately. If, in addition, the orchestra is pre-recorded, the tempi are then set, and the dancers know what to expect; ideally, principal dancers will have attended the recording. If piano transcriptions are used for rehearsal, it may be a responsibility of the sound department to make and replay a recording of this. The dancers must have clear audio replay at all places where the action takes them, so the action studio must be liberally supplied with loudspeakers. In some cases in ballet the dancers start before the music; where this happens the replay operator must be as precise and accurate as an orchestral conductor in timing the start of music to match the action.

Effects microphones – usually parabolic or gun microphones, or hypercardioids angled to discriminate against loudspeakers – are needed in the studio to add sufficient noise of movement (the dancers' steps, rustle of costume, etc.) to add conviction to the

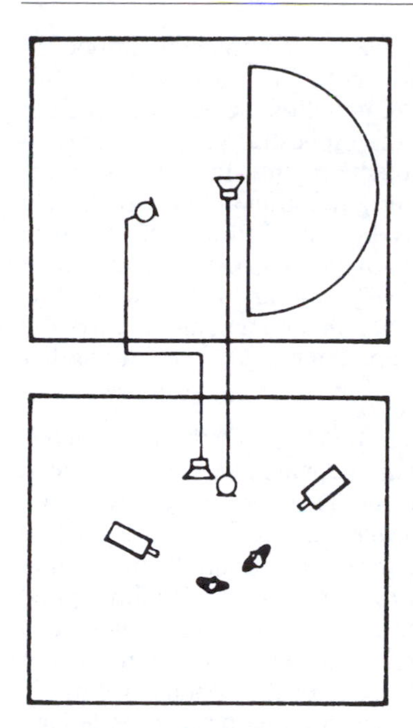

[8.44] **Studio communications**
Links between a television studio and (in this case) orchestral sound. The singers' voices are fed back to the conductor.

pictures of the dance. Consider also an adapted-array (digitally processed) combination microphone if there is one available: its 8° × 30° pick-up lobe should cover the action.

For *singers*, the type of balance depends on whether a microphone may appear in vision; if not, a boom might be used. With the more distant balance, check for separation problems due to orchestral spill on the singer's microphone. Condenser cardioid microphones are preferred.

A marginal advantage of the relative deadness of television general-purpose studios is that when AR is used it can be added differentially – not just in different proportions, but also with different reverberation times for different parts of the sound. Singers usually need less than that required for the music that accompanies them.

Vocal dynamics are an important part of any singer's performance, but when this is broadcast, some compression may be necessary, and in television the problem is made more complex by movement of both singer and microphone, and also by any variations in distance between singer and microphone to accommodate the frame size. Here, even more responsibility for interpretation of the artist's role may fall to the sound balancer, who must therefore think not only about problems of coverage but also – along with the director and singer – about the overall artistic intention.

Where singers and accompaniment have to be physically separated, time-lag problems may need to be rectified. Foldback loudspeakers are needed to relay sound to the singer; the conductor, too, needs to hear how the singer's contribution fits into the ensemble.

For some wide shots, pre-recordings may be used: timing these to match live action requires planning and rehearsal. An additional recorded track can be used for cues that are fed only to the performers. Remember, always, that foldback loudspeakers may be safer when placed off the rear axis of a directional microphone.

Televised opera

There are several ways of presenting opera on television:

- from the stage;
- with actors (perhaps including some of the singers) miming to pre-recorded music;
- with singers in a television studio and orchestra in a sound studio;
- with both singers and orchestra in some hall of suitable (large) size and good musical acoustics.

For stage performances, use the balance already described. If possible, work from a separate sound control vehicle with good loudspeakers, a reasonable volume of space and suitable acoustic treatment.

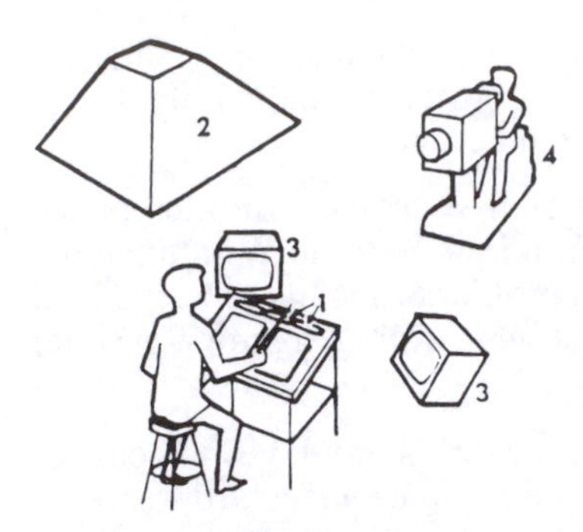

[8.45] **Opera or ballet in the television studio**
Layout for conductor with orchestra in a separate sound studio. 1, Line loudspeaker along top edge of music stand. 2, (Alternative) hood loudspeaker over conductor's head. 3, Monitor for conductor to see action. 4, Camera to relay beat to repetiteur.

Miming opera was once extensively used in continental Europe. In Britain it has been used less, in the belief that the singer's full performance has greater musical conviction. One arrangement employs separate studios. That for the orchestra has good musical acoustics; the action studio has all of the regular facilities to cover sound and pictures – for which the original grand, and deliberately unrealistic, style of stage opera may be modified to that of music drama. The singers can therefore be covered by booms (and, where necessary, other microphones) in a way similar to that for a play. The conductor has a monitor for the action, and a camera relays the beat to the action studio, where a repetiteur takes it from another monitor and passes it to the singers.

Sound communication is a little more complicated. Orchestral sound is relayed to the action studio floor by a directional loudspeaker mounted on the front of the boom dolly. As the boom is moved to follow action, so is the upright line-source loudspeaker that is fixed to it. A line source is directional at middle and high frequencies, radiating mainly in the plane that intersects the line at right angles, so this allows good separation from a high microphone. The voices of the singers can be relayed to the conductor in a variety of ways: one is to have a line-source loudspeaker along the top edge of the conductor's music stand. Three small elliptical speakers set in a line produce little bass and are again directional at high frequencies, helping separation.

Even using this technique, it may be necessary to pre-record or film some scenes in advance. To employ the musicians for a short sequence may be disproportionate, so a system of guide tracks may be used. Here the pre-recorded (or filmed) scenes are accompanied by a soundtrack of bar counts or a piano version, plus any verbal instructions such as warnings of impending change of tempo. This guide is fed to the conductor only (by headphones in this case). Orchestral sound is then added at the main recording.

If singers' voices are included in the pre-recording this can also be used as guide track or, better still, separate tracks can be used. Where singers' voices are pre-recorded in sound only (so that miming can be used, perhaps in order to cover distant action), a guide track can again be included in a multitrack recording.

The fourth arrangement, with orchestra and singers at opposing ends of the same concert hall, employs many of the same communications systems: for example, the acoustic time-lag between the ends of the hall still requires the use of the conductor's music-stand loudspeakers. Separate rehearsal, one of the advantages of twin studios, is lost, but in its place is an enhanced unity of purpose.

Popular music

Having worked up to some of the largest and most complex layouts for classical music, we can now take some of the same (or similar)

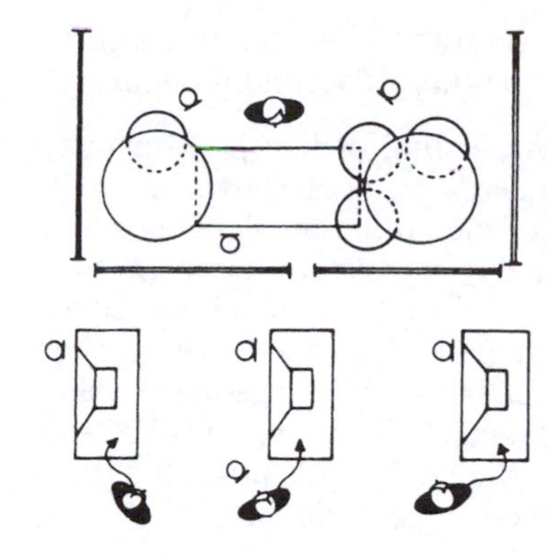

[8.46] **Pop group**
Three electric guitars, drums and vocalist. For maximum separation the drums may have two low screens in front, high screens at the sides, and a further screen bridged across the top. For stereo, microphones are steered to suitable positions on the audio stage, and blended and spread artificial reverberation is added.

components and put them together in a contrasting type of balance that is used for rock, dance, jazz and much other popular music.

A simple pop group may consist of drums, guitars (lead, bass and possibly acoustic), vocalist and perhaps a synthesizer. In the dead acoustics of a big recording studio the various components might be placed in booths around the walls, with tall sound-absorbent screens on either side of the drum-kit (unless it is in a corner) and half-screens in front. To reduce bass transmission, the drums may be raised about 15 cm (6 in) on internally-braced low wooden boxes (rostra, risers), with resonances further damped by laying them on strips of carpet.

Lead and bass guitar loudspeakers can be directed in to half-screened bays, and picked up on cardioid microphones, but an acoustic guitar may need a separate booth. The synthesizer is angled to radiate any acoustic output into an opposite wall, and the amount of screening depends on whether there are any additional, quieter instruments to be balanced: if so, further heavy screening may be required. The guide vocalist is kept physically well away from all of these but in good visual contact with the other players.

Much use is made of tracking (sequential recording) for vocals, vocal backing, additional acoustic guitar, percussion and other effects. Some parts, such as the main drum coverage, keyboard and vocal backing, may appear on stereo pairs of tracks; the rest will be mono. One track may carry time code.

The session may start by recording the basic rhythm plus guide vocal, all on separate tracks, followed by additional acoustic guitar and the synthesizer. Next the lead vocals are laid down, then the backing (as separate stages if complexity warrants this). Finally, the solo guitar, electric guitar riffs, etc., are overdubbed. Alternatively, all or most of the tracks may be recorded at the same time. The mixer will have two faders for each channel, one of which may carry the full audio signal, which is then given nominal (but not irreversible) treatment via the buttons at the top of the strip on its way to the multitrack recorder, MDM (digital modules) or hard disk recording system. The second fader, below the first, controls a separate feed that is used to monitor a trial balance, and this may also be recorded for later reference. In a fully digital console, the faders and all controls mimic their audio counterparts, but the signal is not fed directly through them.

Reduction (or mix-down), in a separate session, begins by returning the tracks to the mixer channels, checking and refining their treatment, starting with rhythm as before, and gradually building up a stereo composite from the separate components. A computer may be used to record and update control data and make the final mix. Multitrack recordings can also be used in this way for classical music.

For radio, where less time and money may be available, the process is often condensed, with the multitrack recordings used mainly for

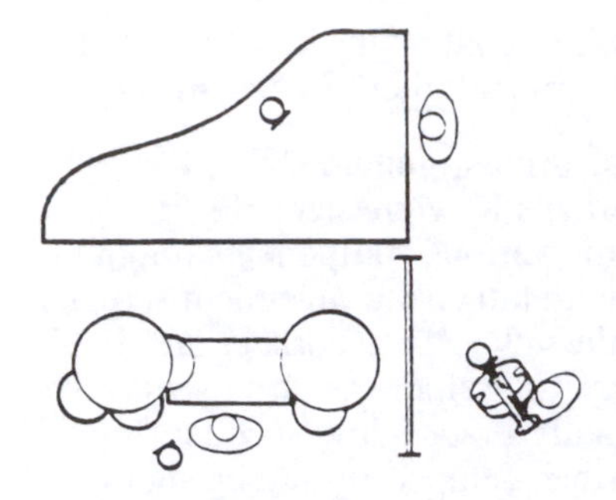

[8.47] **Rhythm group**
The drum-kit and bass are separated by a low acoustic screen. A range of positions is possible for microphones, including several for the drum-kit.

cosmetic changes, to correct errors or for the greater freedom to experiment that is allowed by the possibility of second thoughts.

To regain a sense of occasion and the feeling of live performance, recordings may be made from concerts or road shows. A rock concert can involve a great amount of electronic rebalancing, even before any recording is contemplated. A group of three or four players may have 30–100 channels, linked to a travelling mixer (with a separate mixer for the drummer) to feed the stage speaker system. Studio-style consoles send feeds to audience loudspeakers and make recordings. These are taken back to the studio and generally remixed in the normal way, though records produced from concerts are sometimes proudly marked 'no overdubbings'.

When recording on location, whether from a pop or classical concert, time is limited and it is wise to establish a routine, setting microphones and establishing any split feeds as early as possible. Microphones are identified by scratching them and indicating what instruments they cover. Radio microphones may be used for direct communication between stage and mixer. The recording arrangements are set up and checked out. A mobile sound control room is often used.

Larger groups

A basic component of much light music, including jazz, is the piano trio, i.e. piano, bass and drums. It is also the nucleus in many larger groups. To it are added, stage by stage, guitar (acoustic/electric), brass, woodwind (including saxophones), singers and other instruments for particular effects. These might include celesta, violin, flute, horn, harp, electric organ or any other instrument that interests the arranger. As usual, mono microphones are used for most instruments, but stereo pairs might be tried on trumpets, trombones, woodwind, xylophone, piano and vocal groups. Again, tracking is generally used.

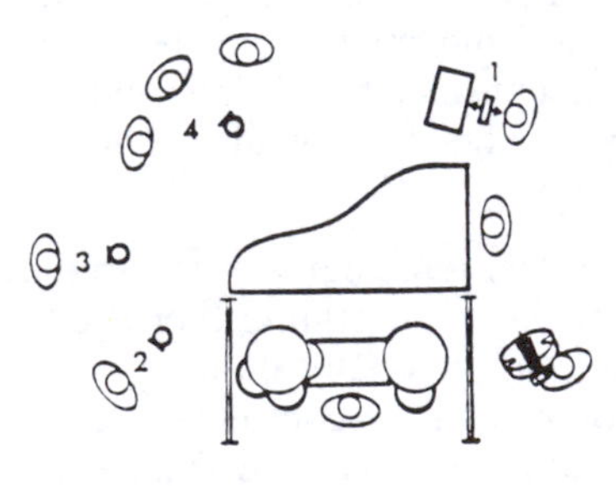

[8.48] **Small band**
Rhythm group with: 1, Guitar (acoustic/electric). 2, Trumpet. 3, Trombone. 4, Woodwind (saxophones). The rhythm group microphones are not shown.

Often the studio is dead, but sometimes only bright surroundings are available. If some spill is inevitable, it might be best to modify the ideal fully separated balance. Even in good conditions there may be enough spill in the studio to dictate some elements of the stereo sound picture. However, certain relationships will always appear in both the real layout and its image: for example, the piano, bass and drums are always close together. These might be steered to centre and left of centre with electric/acoustic guitar near the left speaker, singer and woodwind right centre, and brass on the right. The actual layout in the studio is governed not only by the need for separation (so that this or some other stereo picture can be achieved, and so that all sounds can be individually treated), but also by practical considerations, such as whether the players can hear or see each other.

Multimicrophone layout

When musicians arrive in the studio they expect to sit at positions already arranged for them, with chairs, music stands, screens, piano, boxes for such things as guitar amplifiers and microphones all (at least roughly) laid out. They will have been set according to the balancer's instructions, and the balancer or an assistant will have placed the microphones. Three different situations may apply:

- The group may already have a layout used for stage shows. The balancer finds out in advance what this is and varies it as little as possible (but as much as may be necessary for good separation) with the co-operation of the musical director.
- The group may have made previous recordings or broadcasts in the same studios. The balancer finds out what layout was used before, and if it worked well, uses it as a basis for this one: the group will not expect to be in different positions every time they come to the studio. But there may be changes for the purpose of experiment, or to include ideas suggested by a producer or which the balancer is convinced work better than those used previously: these may require the co-operation of an individual player, section leader or musical director. Bigger changes may be dictated by the use of additional instruments or developments in musical style.
- The group may be created for this session only, in which case the balancer – having found out the composition from the producer or musical director – can start from scratch.

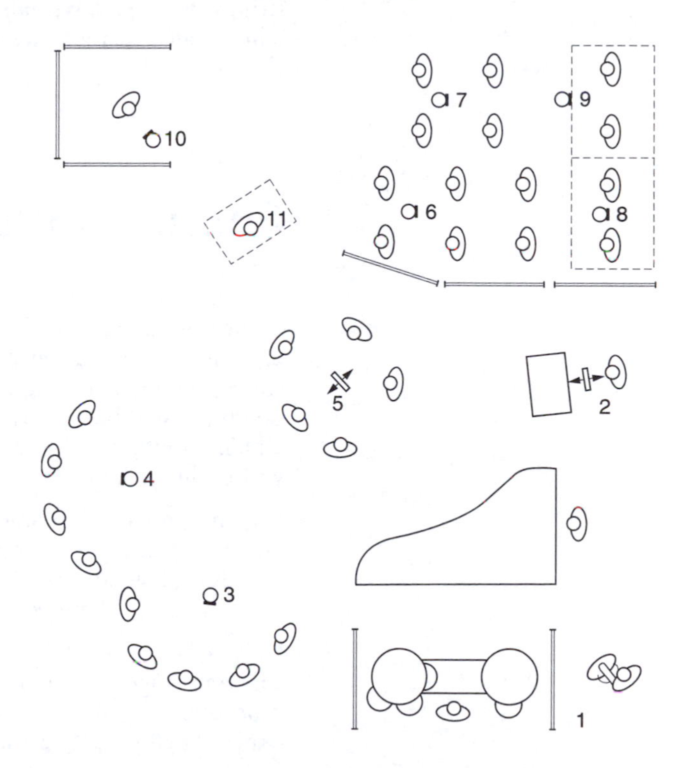

[8.49] **Show band with strings**
1, Rhythm group. 2, Electric/acoustic guitar. 3, Trumpets. 4, Trombones. 5, Woodwind. 6, 1st violins. 7, 2nd violins. 8, Violas. 9, Cellos. 10, Vocalist. 11, Musical director. Variations must be made for acoustic difficulties: in one layout it was deemed necessary to bring the bass forward to the tail of the piano and separate it by acoustic screens from brass, drums and woodwind (the trumpets having been moved farther away). Additional forces complicate the layout further. For example, the pianist may need to reach a second keyboard, for organ or celesta: this could be placed to the player's right and the guitar moved forward.

Even starting fresh, the balancer can apply certain rules that make the musician's job easier. For example:

- The bass player may wish to see the pianist's left hand, to which the bassist's music directly relates. This is important in improvisations.
- The drummer, bass player and pianist form a basic group (the rhythm group) and expect to be together.
- In a piano quartet the fourth player can face the other three in the well of the piano; everybody can see everybody else. But a guitarist may work more with the bass, so could be placed at the top end of the piano keyboard, to see across the pianist to the bass player.
- A bass player with fast rhythmic figures needs to hear the drums, particularly snare drum and hi-hat; these can usually be heard well on the drummer's left.
- Any other two players who are likely to work together on a melodic line or other figures need to be close together.
- When players double, e.g. piano with celesta, the two instruments must be together.
- All players need to see the musical director or conductor.
- The musical director may also be playing an instrument.

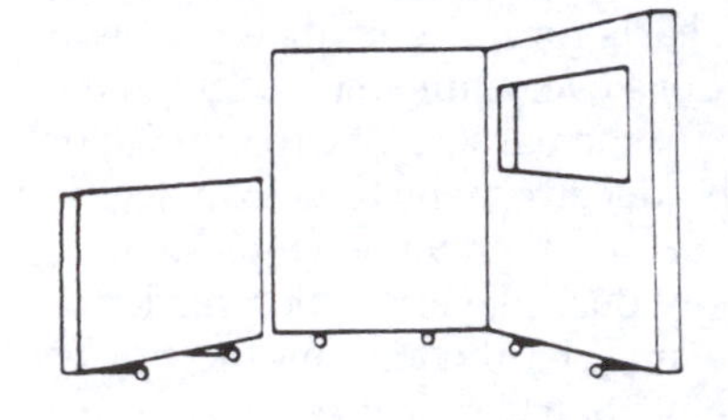

[8.50] **Studio screens**
A dead music studio has a supply of full and low screens with absorbers about 10 cm (4 in) thick that are easily moved on rollers to provide acoustic separation without loss of visual contact. The vocalist's screen may have a window.

This may appear to make the original idea of building up a layout according to good principles of separation less feasible. But although compromises are often necessary, the use of directional microphones and screens should make a good sound possible. Low screens are set between bass and drums, and more to separate drums and brass. Other sections, especially quiet instruments such as strings or celesta, will also require them. A singer may be provided with a tent of screens well away from the players, but should still be able to see them.

Popular music in vision

Plainly the typical recording studio layout for multitrack music cannot be adopted in vision (though a band wholly or largely out of vision will). But apart from visual layout – which may produce some problems of separation – all other elements of the balance and treatment will be as close as possible to that of the recording studio, which inevitably will be the standard by which the television sound will be judged.

Low microphones in among the instruments are not obtrusive: they are readily accepted as part of pop paraphernalia. Large microphone booms and stands are avoided, as are microphones at or just above head height. There are more differences in mountings than in positioning.

The pop singer's hand microphone is accepted by the audience as a prop and not just as something to pick up sound, though some experienced performers also use it to balance themselves with skill.

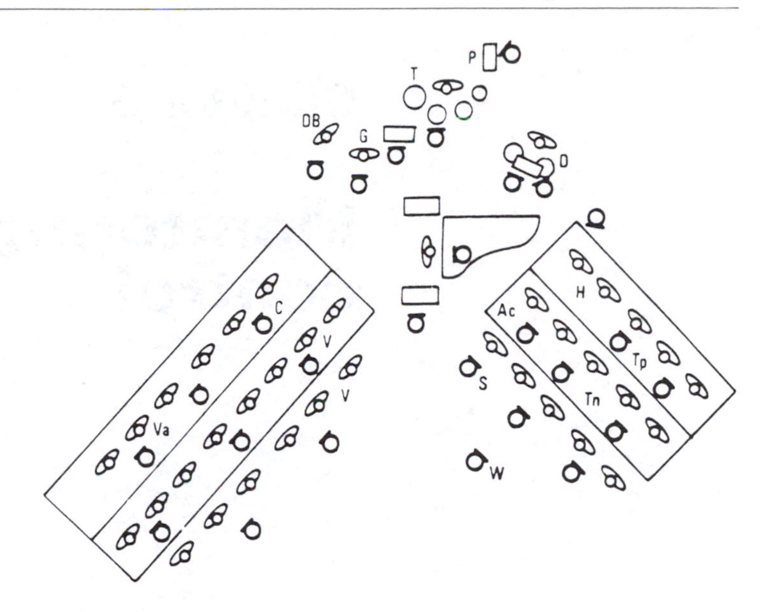

[8.51] **Show band in vision to support singer or featured acts** One of many possible layouts: V, Violins. Va, Violas. C, Cellos. DB, Double bass. G, Acoustic guitar. T, Timpani. P, Percussion. D, Drum-kit. H, Horns. Tp, Trumpets. Tn, Trombones. Ac, Accordion. S, Saxophones, doubling woodwind. W, Overall wind microphone. All microphones may be condenser cardioids except for a moving coil on kick drum.

For the less mobile singer, a stand microphone is an accepted convention.

For any popular music television programme, the control console is likely to have most of the equalization, compression and AR facilities of the recording studio. If several groups are to appear, a separate studio desk may be used to balance each in turn. Digital ‘snapshots’ of the rehearsed mixer settings can then be recalled as each group appears in a live show. It may be necessary to pre-record some tracks even in a live performance if a group is to match its expected sound. Miming to playback of records is cheap and easy – but some would say cheap and nasty: the sense of occasion is prejudiced, and the audience may feel cheated. Such devices should be kept to a minimum.

Television has largely adapted its pictures to the sound requirements of popular music, but sometimes a singer must appear without a microphone in vision. Booms may still be used for this, perhaps with a gun microphone to improve separation. Only where separation cannot be achieved by these means, or in complex song and dance acts, should the music be pre-recorded and vision shot to playback of sound.

Much of this applies also to music on film, although there may be a greater effort to eliminate microphones from the picture.

Although multiple-camera techniques have, at times, been used for music on film and location video, close-ups are usually shot separately, taking sound as a guide to synchronization with a master recording. More is shot than is actually required, so actions before and after cuts will match. Sometimes film shot in sync for one sequence is later synchronized to the sound from another part of the musical work. For this, the beat or action, e.g. drumbeat, is laid down. It can then be matched to a close-up of the action.

Chapter 9

Monitoring and control

Monitoring means using ears, and therefore loudspeakers (or, at a pinch, headphones) to make judgements on sound; *control* is the use of hands to manipulate the sound, mainly in a mixer console. This chapter describes loudspeaker layouts; the path of the analogue or digital signal through typical consoles; and touches on questions of quality, without which the more creative modification of sound is wasted.

During rehearsal, recording or broadcast, the ear should be constantly searching out faults that will mar a production. For example:

- *production faults* – e.g. miscast voices, poorly spoken lines, uncolloquial scripts, bad timing;
- *faulty technique* – poor balance and control, untidy fades and mixes, misuse of acoustics;
- *poor sound quality* – e.g. distortion, resonances, and other irregularities in frequency response, lack of bass and top;
- *faults of equipment or recordings* – wow and flutter, hum, noises due to loose connections, poor screening, or mismatching.

Of these, the faults in the first group are the most important. They are also, perhaps, the easiest to spot. As for techniques, the ability to see faults in those comes with their practice. But the various things that can be wrong with the quality of a sound – supposedly objective, measurable things – appear, surprisingly, to be the most difficult for the ear to judge in other than subjective 'I-know-what-I-like' terms.

Quality and the ear

The quality of sound that the ear will accept – and prefer – depends on what the listener is used to. Few whose ears are untrained can judge quality objectively. Given a choice between loudspeaker qualities, listeners will generally state a preference, but in practice it seems that as many prefer medium quality as want the best. For many, the preference is for a box that sounds like a box: the kind of quality that turns their living-space into a corner of a great concert hall is disturbing. But while such preferences must be respected, it must also be true that quality and enjoyment are often closely related. Higher quality replay of a musical work gives more *information*, so it is often a technically superior rendition that first 'sells' unfamiliar music. And a good loudspeaker allows more of the original intention, as expressed both in content and performance, to reach the listener.

Different levels of quality may be acceptable for different purposes. High quality may be best for brilliant modern music (or for orchestral music heard as a concert); medium quality can be satisfactory for ordinary speech, where the main objective is to hear the words clearly; and 'mellow' quality is often felt to be less obtrusive in music intended as a companionable background noise. On high-quality equipment these different types of programming are all available at the appropriate levels of quality; on poor equipment they are all debased to the same indifferent standard. It is undoubtedly worth persevering with the education of one's ears, and also allowing others the opportunity of listening to sound that is of better quality than they are used to.

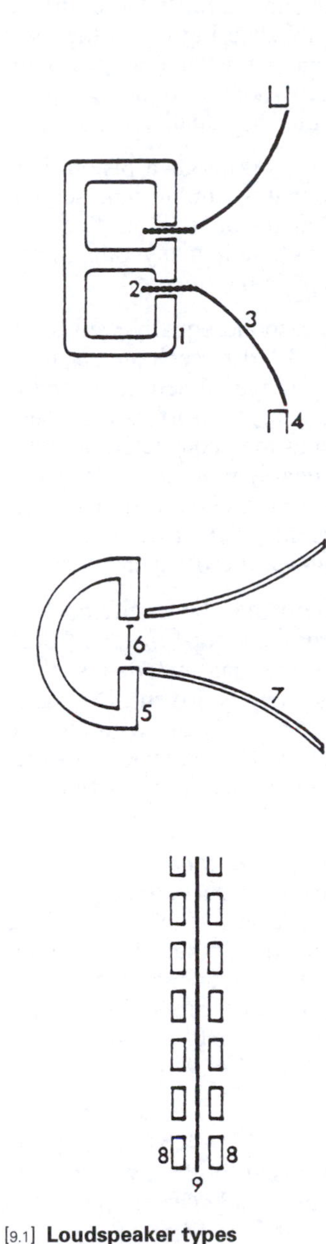

[9.1] **Loudspeaker types**
Above: Moving coil. The most commonly used for all purposes (bass, treble and wide range). 1, Permanent magnet. 2, Moving coil. 3, Cone. 4, Baffle or enclosure. *Centre*: Ribbon unit. Sometimes used for treble. 5, Permanent magnet. 6, Ribbon (seen in plan view). 7, Acoustic horn. *Below*: Electrostatic. 8, Perforated fixed electrodes. 9, Diaphragm.

Nevertheless, any organization that is conscious of cost must decide on a practicable economic upper limit to the audio-frequency range to be offered. In a test by a broadcaster, the range was restricted successively to 7, 10 and 12 kHz. Sounds in the test included tubular bells, brass, cymbals and triangle, snare and military drums, maracas, handclaps, and female speech. The listeners were all less than 40 years of age, both male and female, and included some who were experienced in judging sound quality. The results suggested that:

- only a few critical observers hear restriction to 12 kHz;
- many inexperienced listeners can detect restriction to 10 kHz, while experienced listeners do so readily;
- a filter at 7 kHz was detected even by the inexperienced.

In fact, although aiming to record a full range from 20 Hz to 20 kHz, that broadcaster actually transmits only 30 Hz to 15 kHz on FM radio and NICAM television stereo. For CD and digital tapes the recording industry goes a step further, up to 20 kHz.

Loudspeakers

For technical monitoring, balancers need high-quality sound – which is actually quite difficult to achieve. The main obstacle to good sound has nearly always been at the final stage, where an electrical signal is converted back into waves in air. It is not so difficult to design for a good middle- and high-frequency loudspeaker response:

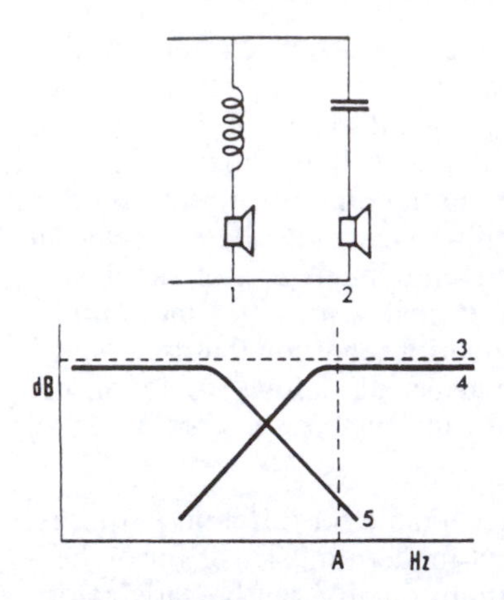

[9.2] **Loudspeaker: simple crossover network**
1, Bass unit. 2, Treble unit. 3, The total power is divided evenly between the two units 4 and 5. For each it falls off at 6 dB/octave beyond the crossover point. This is a very satisfactory type of network provided that there are no severe irregularities in the response of either unit that are close enough to the crossover point to make a significant contribution even though attenuated. For example, a severe fluctuation in a bass cone output at A will still affect the overall response. To counter this, the crossover point might be taken lower or the dividing network might be redesigned to give a steeper cut-off at the crossover.

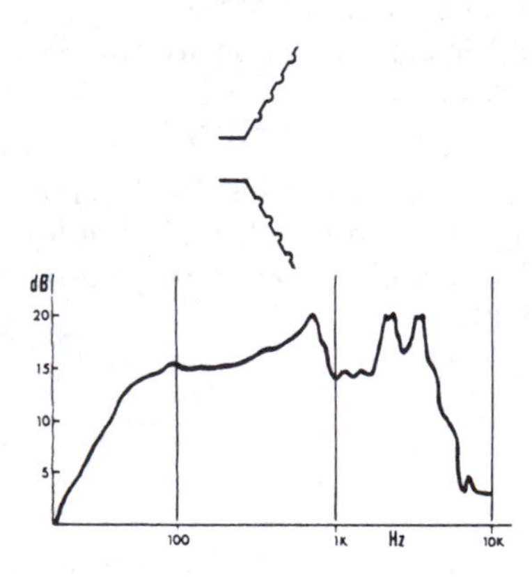

[9.3] **Loudspeaker cone with sides of corrugated paper felt**
The performance diagram shows axial frequency response with output in decibels (arbitrary scale), taken from a particular example.

the hard bit is to combine that with efficient production of the bass. Long-established designs of 'woofer' use either a fairly large surface area to push the air, or (even longer established but now little used except in traditional musical instruments) a horn that gradually flares out to a large aperture. (Separate subwoofers extend the range, but this may be more for effect than to increase quality.)

Most loudspeakers use a stiff diaphragm working as a piston. It is operated by an electromechanical device that is similar to those used in microphones – except, of course, that it converts an electrical signal back into physical movement. Although many other principles are feasible, the most common type is moving coil.

It is not easy for a single diaphragm to respond equally well to all frequencies: usually the signal is split and fed to separate speakers covering different parts of the frequency range. When the sound is divided like this there is no need for a sharp cut-off as one loudspeaker takes over from another, so long as the acoustic output adds up to a full signal. Careful relative positioning of the units is required, to avoid phase cancellation at switchover frequencies. (But the subwoofer in surround sound seems exempt from this rule: its separate existence isolates it from awkward crossover effects.)

For the low-frequency unit the diaphragm may be a stiff cone of paper, corrugated concentrically in order to stop unwanted harmonics from being formed (like a bell, these may include a subharmonic, below the frequency at which the cone is driven). Pistons of other materials are used: again they must be stiff, but with internal damping or they may ring like a wineglass. Of whatever construction, a driver must also be of low mass or its inertia may affect the response.

A feature of loudspeaker design is that as the cone or diaphragm radiates the wanted sound from the front, a second component, opposite in phase, is emitted from the back, and if there is nothing to stop it, some of this back-radiation leaks round the sides and joins that from the front. Long-wavelength low frequencies are most affected by this: as front and rear combine, they are still partly out of phase, so tend to cancel sound that is already weaker than it should be because any cone is too small for effective low-frequency work.

In the design of high-quality loudspeakers there have been many ingenious part-solutions to the problem posed by the size naturally associated with the longer wavelengths of sound. The simplest is to mount the loudspeaker in a much larger baffle (a fixed solid panel). A baffle 75 cm (30 in) square is about the smallest for a modest low-frequency response; a whole wall is better. Another idea is to turn the space behind the cone into a low-frequency resonator with a port that re-radiates some of the power (in this vented box the neck acts as part of a tuned Helmholtz resonator). By changing its phase this can be made to boost the response in the range where it previously fell off. But this 'bass reflex' loudspeaker is bulky, may be subject to cabinet coloration in the bass, and below its resonant frequency cuts off even more sharply than before. It also has a characteristically

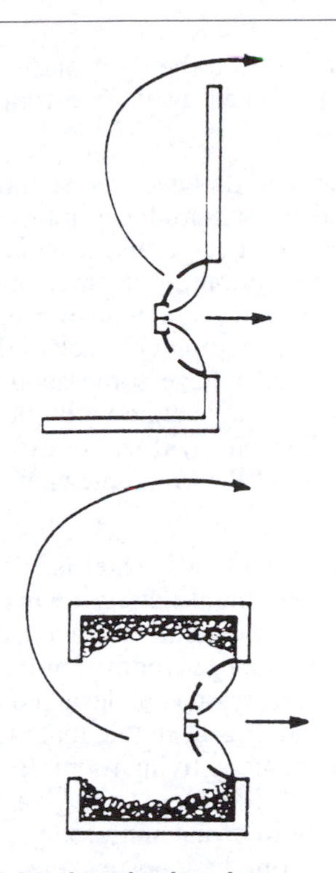

[9.4] **Loudspeaker housings**
Above: Baffle. *Below*: Enclosure with an opening at the rear. Both reduce the frequency at which sound radiated from the back of a cone can recombine with and cancel the direct sound. A box may have internal resonances that need to be damped by lining it with absorbent material.

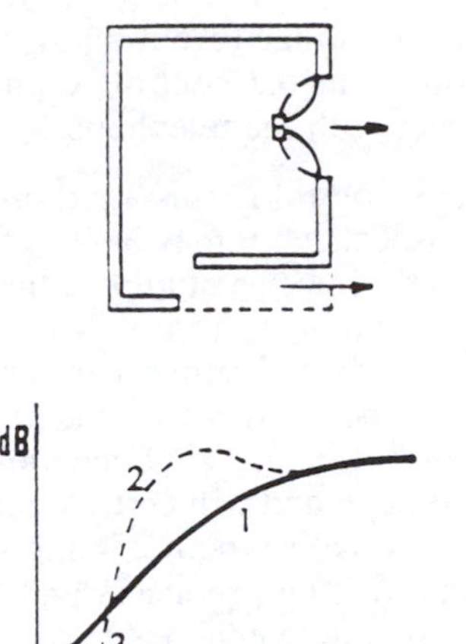

[9.5] **Bass reflex loudspeaker**
This uses the internal cavity and a port at the front as a Helmholtz resonator with a low-frequency resonance peak. Added to the normal bass response of the loudspeaker (1), it can be used to lift the bass for about one octave (2). Below this (3), the cut-off is more severe.

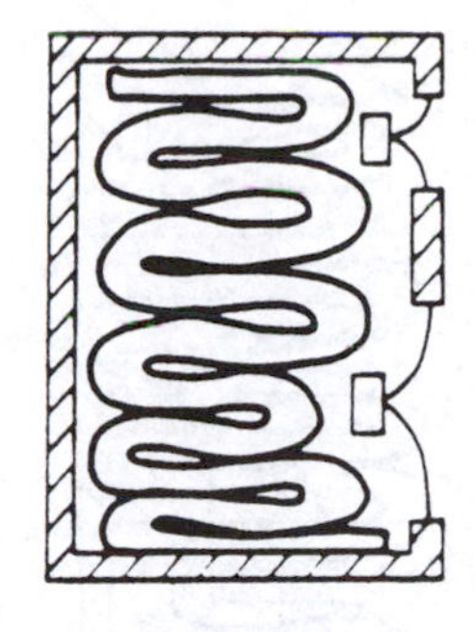

[9.6] **Infinite baffle enclosure**
The sound from the rear of the speaker is trapped and much of it is absorbed. In practice, the stiffness of the air within the box somewhat restricts the movement of the speaker. The cavity contains folded sound-absorbent material.

recognizable quality – not a recommendation for any loudspeaker. Yet another idea is to lose the back-radiation completely by stifling it within the box so the diaphragm is not affected by reflections or by the 'stiffness' of the enclosed air.

Powerful loudspeakers require additional separate amplifiers to drive them. Without these, in the bass particularly, overmodulated sound on which the signal is clipped will create strong harmonic distortion products that are heard on higher-frequency loudspeaker elements.

Monitoring layout

The first question in control-room layout is architectural: whether or not to let listening room acoustics dominate its design. In the more extreme examples (most often found in the music recording industry) the loudspeakers are no longer free-standing boxes (ideally, well away – e.g. a metre – from any wall), but instead are recessed in cavities within the walls. The room is not a rectangular shape, but instead has walls and ceiling that flare out symmetrically and round to the sides, with absorbers and diffusers to break up potential coloration. First reflections continue to disperse, and some are soaked up around the back. And producers (who are paying for this)

[9.7] **Complex loudspeaker enclosure**
The tweeter (HF unit) has an infinite baffle and the woofer (LF unit) backs on to an acoustic labyrinth (also called a transmission line or pipe). This introduces both phase change and attenuation. Again, both cavities are padded.

are discouraged from taking what often seems to be their preferred position in the control room – which is tucked away in a remote corner with the telephones.

Many control rooms are more utilitarian, and do have free-standing loudspeakers. It may be best for the balancer and producer not to see directly into the studio as they face the sound stage, but instead sit where they can turn to see and sort out positioning or other problems. Adapted rather than purpose-built, the room may have irregular shapes and intrusions (often including equipment racks) that disturb a balanced sound field. It will usually have some acoustic treatment, and will certainly be isolated from any studio with microphones. Noise, both coming in and heading out, will be controlled. (And visiting producers and directors will still need somewhere to talk on the telephones.)

A minimum distance from which to listen to a loudspeaker is 1–2 m (3–7 ft). Closer than this, the sound has an unreal quality owing to standing waves near the cabinets; at a greater distance the acoustics of the listening room begin to take over (so larger rooms are made more dead). Except for film dubbing theatres, which are designed for a much bigger (and usually surround) sound, it is desirable that these acoustics should be similar to those of a domestic living room. But it is important to produce both a well-blended signal and to be clear on whether some minor fault is due to the studio or listening room, and the only way to be sure is to sit no farther from the loudspeakers than when listening for pleasure at home.

For monitoring two-speaker stereo the balancer should be at the apex of an equilateral triangle, or along the centre line from a little closer than that to not more than 1.7 times farther back. If the loudspeakers are about 2.5 m (8 ft) apart the balancer will therefore be 1.8 m (6 ft) or more from them. For 5.1 surround sound music balance the layout of the five main speakers is as shown in Chapter 3: these five should all be of the same type and set-up, but the subwoofer should be set for 10 dB higher volume (which is quite hard to judge by ear). For film, the side speakers may be 'bipolar', radiating fore and aft for a more diffuse effect, and set not so far back, at 90° to the balancer.

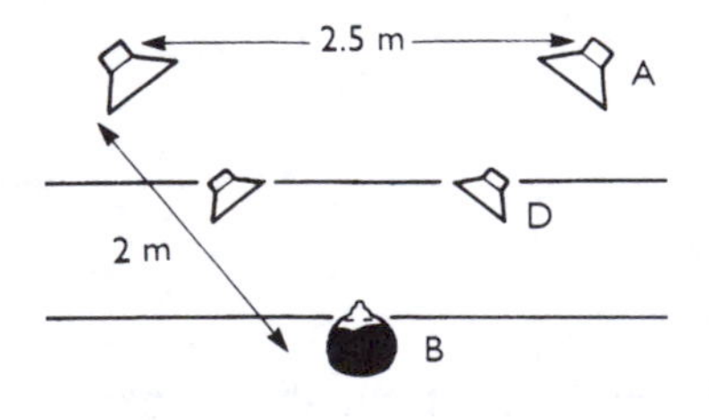

[9.8] **Monitoring stereo**
A, High-quality loudspeakers. B, Balancer. C, Producer, also equidistant from the loudspeakers. D, 'Mini-speakers' offering a distinctively different lower quality.

The producer or director should ideally sit just behind the balancer (but may choose not to). Because monitoring loudspeaker volume is sometimes set rather high, there is an attenuator of, say, 20 dB that can be cut in (by a '*dim switch*') when using telephones, for example. During talkback to the studio it generally switches in automatically.

For stereo an additional pair of small loudspeakers is sometimes mounted closer and in line with the monitoring loudspeakers. These '*mini-speakers*' (or *near-field loudspeakers*), set at lower volume, serve to check what many of the audience will actually hear: all vital elements of the high-quality balance must still be audible when output is switched to them. They are also used for talkback and for listening to channels that have not been faded up (*pre-hear* or *pre-fade listen*, PFL). *After-fade listen* (AFL) takes its feed later, as the name implies. Switching over to these small speakers does not affect output. On some consoles the PFL for any channel can be switched

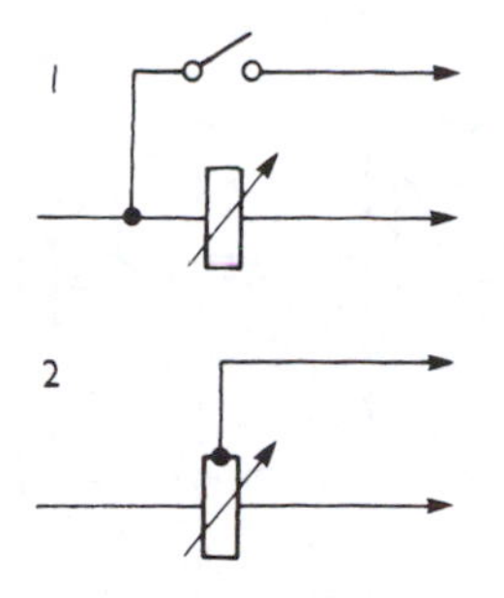

[9.9] **Pre-fade listen (PFL, pre-hear)**
1, Output is taken before channel fader. 2, Symbol for 'overpress' facility: when channel is faded out, pressing the fader knob routes the signal to the pre-fade listen condition. Audio can then be heard on near-field or studio loudspeakers as selected, or on headphones. Other options may be available: for simplicity, these are not shown on the circuit diagrams.

in by pressing on a fader knob when it is fully faded out: this is called *overpressing*.

Most consoles have a wide range of further switches associated with the loudspeakers. '*Solo*' selects just the channel in which the switch is located and feeds it to a main loudspeaker; '*cut*' takes out just that one channel, to hear what the mix is like without it. On some consoles it is possible to solo one channel while retaining another channel or group for background or comparison: for this, press '*solo isolate*' on the second fader at the same time. Some consoles also allow '*solo in place*' – that is, in its stereo position. If these operate before the fader, the individual balance can be checked; if after, that fader's contribution to the overall balance is also assessed. If operated during rehearsal, these switches do affect studio output, so they are linked to the transmission or recording condition that is set up in the console when the *red light* is on: solos, etc., will then still be heard, but on the small loudspeakers, as for PFL. There will also be a headphone feed.

Control consoles and racks

Sound control consoles (otherwise called mixer desks or, sometimes, panels) range in complexity from the simplest that are used in radio for speech and record shows to those for popular music recording, which may have a vast number of channels, each with the facility for a range of treatments and two separate mixing systems so that one can be used to monitor (or broadcast) the music being played while the other feeds a multitrack recording system, for subsequent remixing.

The size (or complexity) of a console reflects the maximum demands expected of the studio. In a large broadcasting centre, radio studios may be more specialized than those for television. A television studio represents a large capital outlay and is used for a greater variety of productions, and the demands of vision add to the complexity of sound coverage, so television consoles may need to be more versatile than their radio counterparts. Consoles used by the record industry may be more complex still. But this size can be deceptive: however big they are, most consoles are designed for use by a single pair of hands. Much of the space on a big desk is occupied by controls that will be pre-set; other adjustments are made one or two at a time, and continuous control will be exercised only over a narrow range near the centre of the console. This, at least, is

[9.10] **Mono channel**
Some of the items that may be found before the channel fader:
1, Microphone (low level) input.
2, Phantom power feed to microphone (48 V). 3, Isolating transformer.
4, Microphone channel amplifier and coarse pre-set. 5, Line (high-level) input: in this example, replay of recordings. 6, Buffer. 7, Fine pre-set gain (typically +15 dB). 8, Polarity reversal control (may be called 'phase'). 9, Equalizer. 10, Equalizer bypass switch. 11, Optional 'insert' routing: allows external module to modify channel signal. 12, Pre-fade listen routing. 13, Remote cut switch, or 'cough key'. 14, Channel fader.

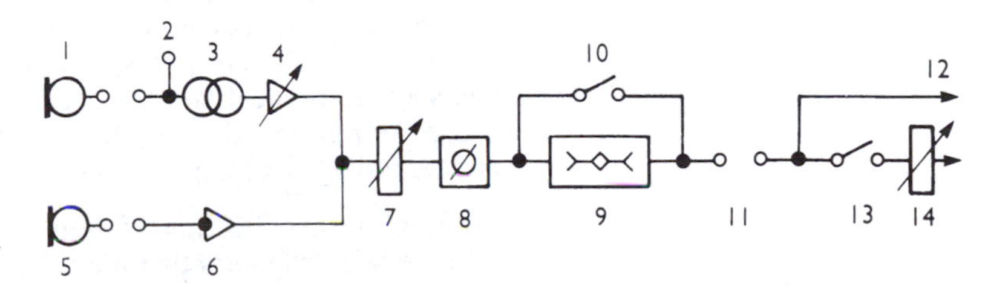

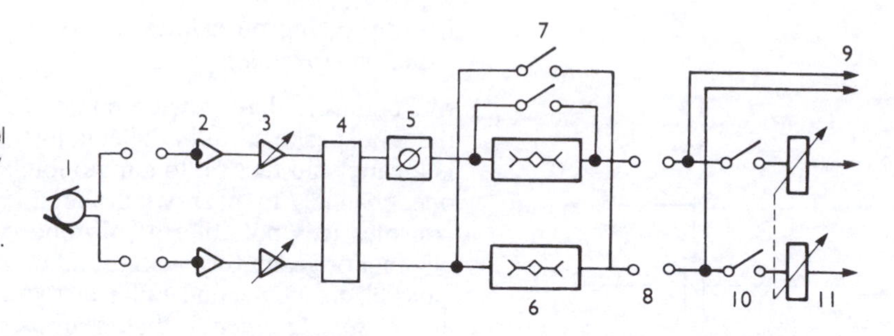

[9.11] **Stereo channel**
Some facilities before channel fader: 1, Stereo microphone. 2, Buffers. 3, Coarse channel amplifiers. 4, Channel input switching (to mono). 5, Polarity reverse switching in left leg only. 6, Equalizer. 7, Equalizer bypass. 8, Optional 'insert' for plugged modules. 9, Pre-fade listen. 10, Remote cut. 11, Ganged channel faders.

the ideal – and as balancers devise increasingly complex sequences of operations, console designers have come up with ways of building these up (or changing them in detail) stage by stage, and of also memorizing and recalling earlier settings.

At one time all consoles carried only analogue sound, and the size of the biggest reflected the great number of individual inputs that must be separately controlled. As the number of things that could be done to each input (or *channel*) increased, the clearest layout had all the buttons, knobs, levers, switches, meters and flashing lights for each channel arranged in an apparently logical strip up the desk (although many of the components they controlled, such as amplifiers, were somewhere else in the desk, or in racks). Even at the first stage of control, with so many of these channels side by side, this already covered a large area on the desk.

The key to understanding how a big console works is, first, to work out what happens in a typical individual channel; second, how channels fit into a hierarchy so that a much smaller number of later channels each allows control over a group of those earlier in the chain, followed by further grouping, to end up with a master output control. Typically, group and master controls should be near the middle of the desk, with larger banks of the earlier individual channels on either side.

A third, smaller group of controls must also be (literally) close at hand. These include: switches, etc., for additional operational devices that may be inserted into selected channels; routing switches and controls for audio feeds to be sent outside the console or beyond its central hierarchy of channels; monitoring controls; and communications. Most of these are in a block that should also be easy to reach, but which can somewhat untidily break up the neat repetitive array of channels.

Racks in the control room or a linked machine room accommodate any equipment that does not fit on the console or which requires easy access or ventilation. (Noisy, fan-cooled or mechanically operated items of equipment need a separate machine room.) Racks may house amplifiers, devices for treating sound, play-in equipment and recording decks, and also cross-plugging bays – patch bays.

A 'rack' is a collection of items of equipment made to fit into a frame with supports that are 20 in (508 mm) apart. Each unit has

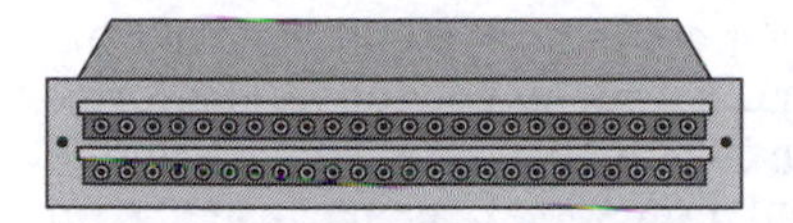

[9.12] **Patch bay (jackfield)**
A standard-sized unit with two rows, each of 24 jacks. There may be several of these, mounted directly in the console, or a whole telephone exchange of them mounted in an equipment bay. Jacks are wired in to intercept signal paths at a variety of points in the desk. Often (but not always) the upper and lower jacks are 'normally' connected together inside the panel – 'normalled', that is, when nothing is plugged in to them. These pairs act as an 'insert point' to a channel (often just after the equalizer) to allow an external device to be plugged in. In a 'half-normalled' pair, the upper jack can be used to plug in headphones without breaking the signal path, but the pair still works as an insert point, diverting its path, if there are plugs in both jacks.

flanges at the front, 19 in (482 mm) wide, pierced by holes used to screw it onto the frame. *Patch bays* are slotted into this at a comfortable access height.

A patch bay is made up from a large number of units, each with 20 or 24 pairs of jacks (or 48 smaller jacks) which, ideally, are connected to as many points as possible throughout the desk. Inside some of the units, the upper and lower jacks in each row are independent ('open'): these serve as inputs to each of the channel lines on the desk. Microphones, replay machines and lines from outside the studio can be plugged in to these. In other units the upper and lower jacks are '*normalled*', which means that unless plugs are inserted, the jacks are linked together. Since the two jacks are connected to two adjacent points in the control console, this allows an external device to be plugged in between them. The top rows are 'sends', the lower rows 'returns'. A '*half-normalled*' variation on this allows the top row (only) to be intercepted without interrupting the internal connection, often so that the circuit can be monitored at this point. Another variation, called '*parallel*', allows both rows to be intercepted without breaking the connection.

Other racks contain 'sends' and 'returns' for all the individual tracks of multichannel recording devices, plus 'in' and 'out' jacks for external devices and for other recorders, and jacks to intercept group and main controls, etc. The bay acts like a highly flexible telephone exchange – one that actually uses traditional telephone plugs and cables.

Microphones are often hard-wired from the studio to racks where it is possible to intercept and cross-plug them. For consoles that require a digital input, the analogue signal may pass though a separate analogue-to-digital (A/D) converter or one that is included in the console.

Digital conversion

Up to this stage in treating sound, analogue electrical signals, in which the waveform mimics that of sound in air, have been unavoidable, if only because microphones and loudspeakers work directly with those waveforms. But as the signal is fed to the console there is an option. While it is possible to continue to work with analogue sound (many excellent consoles are designed to do just that) the alternative is to convert to digital data and operate on that instead.

Here we need to make a short diversion – to see what 'digital sound' is, and then how the analogue version is converted to digital. We might also ask 'why?' There would be no point in introducing two extra stages (including reconversion, taking it back to analogue so that we can hear it) unless there are purposes for which it is better.

First, what is 'digital' audio? Although it has been described as 'pure information', that is simplistic. A sound wave in air also carries information: it is what microphones convert into electrical 'analogue'

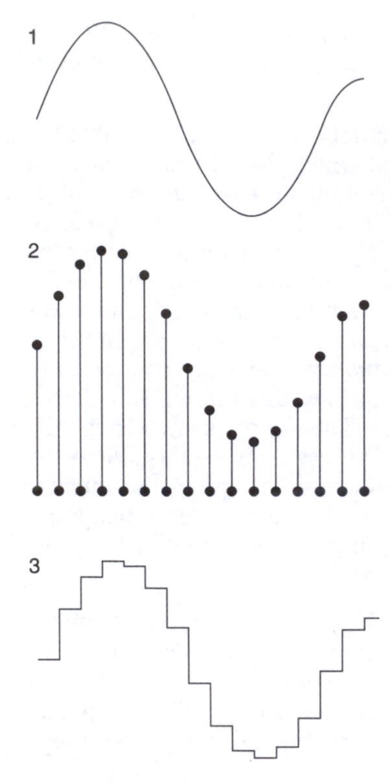

[9.13] **Digitization**
1, Analogue waveform. 2, Digital sampling: for DAT (a digital audio tape recording) this takes place 48 000 times per second. 3, When converted back to analogue form, the data traces out an irregular shape that approximates to the original but with higher harmonics that will be eliminated.

audio. At one time it was argued that computers might reflect what happens in Nature better if they worked directly with analogue forms of information. But measurement uses numbers, numbers are digital and, as it turned out, digital computation soon became faster, so it was possible to do more, at ever increasing rates, and mostly in 'real' time – a big advantage for audio. Ever-increasing amounts of information could be laid down in compact forms, and retrieved with lower losses than with analogue forms of storage. And information could be transmitted more accurately, with enough redundancy of information that damaged data could often be reconstructed.

Next, how is 'analogue' converted to 'digital'? The 'information' in both cases is about how a waveform changes in time. In 'analogue' there is simply a 'picture' of the wave. Digitization means converting that shape into a series of digits (the numbers 0 and 1) by sampling the height of the wave often enough to define its shape, even if it makes 20 000 excursions up and down in a second. There is a mathematical rule about how often we need to sample a waveform in order to be able to reconstruct it perfectly: according to the Nyquist theorem, we must sample it at least twice as often as the highest frequency in it. This is essentially what an analogue-to-digital (A/D) converter does.

A common A/D sampling rate is based on the supposed 20 kHz range of human hearing. The actual sampling rate used for this is not 40 kHz, but 44.1 kHz – the extra margin allows for the sampling to operate at full strength over the full audio range, then tail off in the extra space allowed; any filter should also operate in this marginal range. A 44.1 kHz sampling rate has been adopted by the record industry for CDs.

Digital audio tape (DAT) generally has a more generous 48 kHz sampling rate – on the principle that if professional sound reproduction goes beyond any reasonable limit to human hearing, the response at lower frequencies that people can actually hear is more likely to be smooth. However, in studios where output may be recorded on CD-R, it is simpler to use a common 44.1 kHz for both. Yet another rate, 32 kHz, may be encountered where 15 kHz is accepted as a more practicable limit to transmission of VHF/FM audio and NICAM stereo: the argument here is based on bandwidths (more transmission channels are available than with a wider bandwidth) and on cost. A sound card on a computer might offer all these and other, much higher A/D sampling options.

Many recent recording systems offer a range of sampling rates. For example, a journalist's hand-held recorder can be set for 16, 24, 32 or 48 kHz (mono or stereo). Here the solid-state recording medium is a random access memory Flash card that fits into a PCMCIA slot inside the recorder. A two-channel portable recorder with a removable 2.2 GB hard disk offers a choice of 44.1 or 48 kHz. A four-track digital DAT recorder can sample at 32, 44.1, 48, 88.2 or 96 kHz. Yet another format is DVD-RAM, sampling at up to 192 kHz for four-channel stereo.

Digitization is technically complex, involving many further stages and processes. For CD quality, the height of each sampled point in

AES31 is intended as an industry-wide set of audio interchange standards. The part most likely to be directly encountered by users is AES31-3. It is designed to use EBU 'Broadcast Wave' files (i.e. Wave audio files with added descriptive data) and contains just about everything needed for simple audio project transfer between working studios with deadlines to meet.

the waveform is described by a series of digits (0 or 1): 16 of these make up a 16-bit 'word'. But to calculate with this and finish with an equally accurate set of 16-bit words at the end of the process requires intermediate stages with at least 18- or possibly 20-bit words. But more bits also means a greater signal-to-noise ratio, which allows increased dynamic range or a bigger safety margin ('headroom') before destructive clipping occurs. So at this intermediate stage it is better to work with 24-bit words. There are several processes that are used to minimize, check for and try to correct errors. For CD or broadcast, the end-product of all this computation – with both dynamic range and headroom under tight control – must eventually be scaled back to 16 bits. Paradoxically, the outcome of word-length reduction is improved by introducing noise ('dither'), so that some of the numbers are randomly rounded up and others rounded down. When converted back to analogue, this gives a more realistic waveform.

For sequences of these numbers to be multiplied together they must come together at just the right time, which is not a problem if they all originate from one source, but can be if they come from several (say, a bank of recorders). With multiple sources, one sets the time and its 'clock' is sent (often by a separate cable) to all the others.

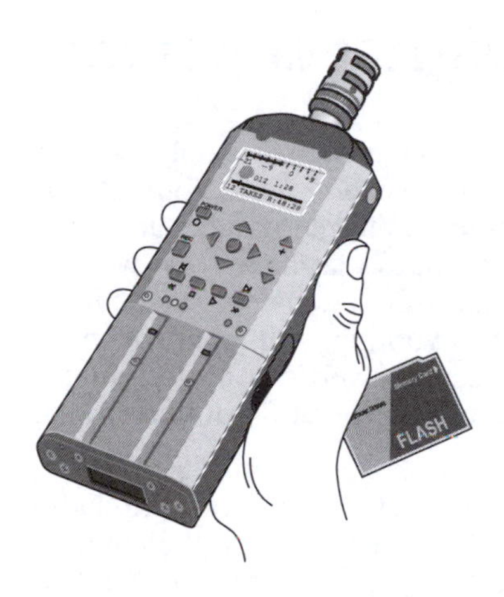

[9.14] **Journalist's solid-state digital recorder**
Digital equipment may be technically complex, but this device is designed for operation by a user who is thinking of something else. The 'Flash card' slots in inside the case and is forgotten until a warning appears. Volume control is by 'low', 'medium' and 'high' signals that interpret a simple bar meter. The microphone may be on a short cable or its capsule plugged directly into an XLR connector on the recorder. The main operational question becomes how far it should be held from the speaker's mouth (see 'interviewing', Chapter 7). The Flash card can be transferred to a PC that has a PCMCIA slot and can be edited non-destructively.

To send a digital signal along wires between components or to greater distances by cable has required the definition of a variety of standards and formats (plus a further vast swathe of the inevitable acronyms). There is AES/EBU, also known as AES3: on a suitable cable with XLR connectors this will carry two channels 60 m (200 ft) – but its word clock wire should be no more than 6 m (20 ft). The semi-professional S/PDIF is for two channels with word clock; TDIF is a commercially originated standard for carrying eight channels each way locally (useful for modular digital multitracks, MDM); and ADAT is good for eight each way by optical cable ('lightpipe'). Wave and AIFF are two professional formats for uncompressed hard disk recordings (typically 16-bit, 44.1 kHz, PCM format). The computer file suffixes are '.wav' and '.aif'.

To reconstruct an audible analogue sound requires digital-to-analogue (D/A) conversion. It reassembles a waveform, initially in the form of a series of stepped blocks. This contains all the frequencies required to define the required waveform, but its jagged, blocky shape adds in a lot of extra harmonics at higher frequencies that must be filtered out.

Digital signal processing (DSP) is superior to the alternatives for both artificial reverberation and delay-related effects (see Chapter 11), and provides a competitive alternative for many other functions in and around the console. It is superior for broadcast and transmission by landline. Also (apart from some 'retro' preferences), it has taken over almost the entire field of recording (see the brief reference later in this chapter). And most professional editing now has the benefit of digital computing power and memory (Chapters 18 and 19).

We may add a fourth question: what can go wrong? One answer to that lies in a cautionary tale from the early days of digitization, when for some equipment that was distributed, the gap between the 40 kHz assumed maximum and the 44.1 kHz sampling rate turned out to be too small. If frequencies in this gap are not filtered cleanly at the Nyquist frequency (half the sampling rate), they are in effect folded back into the top of the audio band. This creates a spurious signal, unrelated to the harmonic structure of the sound, a form of distortion called 'aliasing'. When this fault appeared it was eliminated by replacing the filter with one that had a better cut-off. But the manufacturers had to offer this free of charge for users to retrofit.

In such a rapidly advancing field as digital technology, users may not have to wait very long for the next problem to come along. Bugs in new software are corrected eventually in software updates, but only after a period of frustration by users. Fortunately for the users of analogue equipment, the technology time-lag is longer. Good analogue hardware may be expensive, but it generally has a longer useful life than either the hardware or the software in the ever-changing world of digits.

Automated and digital consoles

In many consoles, whether the audio signal is digital or not, many of the more complex control functions can be exercised by computer. It may be used to set the console to some preferred initial state, or to one of many intermediate 'snapshot' states that have been worked through. In the most extreme case, the entire provisional mix can be reproduced, as though by many hands working with inhuman speed and precision, but still allowing manual override (or just minor 'trimming' changes) to create yet another generation of operational control.

A fully digital console goes one (big) stage further: all the audio signals fed to it are converted into digital data, and all the controls are digital too. A fully digital console can be far more compact than an analogue console of comparable complexity, because sets of controls can be 'layered', set to switch from one set of channels to another, or can be switched to different kinds of operations. Some (designed primarily for the flexibility required in a television studio) have a button switch in each channel fader, to switch the whole channel to operate on a different source, and above the fader four 'wild' controls that can be switched to bring any of a wide range of processors (set on another part of the desk) to operate on the signal in that channel. Digital consoles vary more in layout and operation than analogue consoles, so it may take longer to learn to use a new one quickly and efficiently. But because the vast array of high-quality analogue hardware can be expensive, digital generally has the edge on cost as well as being more flexible. Even so, many users of digitally controlled analogue desks still prefer them to full digital operation.

A radical alternative to this is to dispense with the console completely and allow a computer, with mouse, keyboard and one or more screens, to do the lot. This is also what lies at the heart of the 'virtual

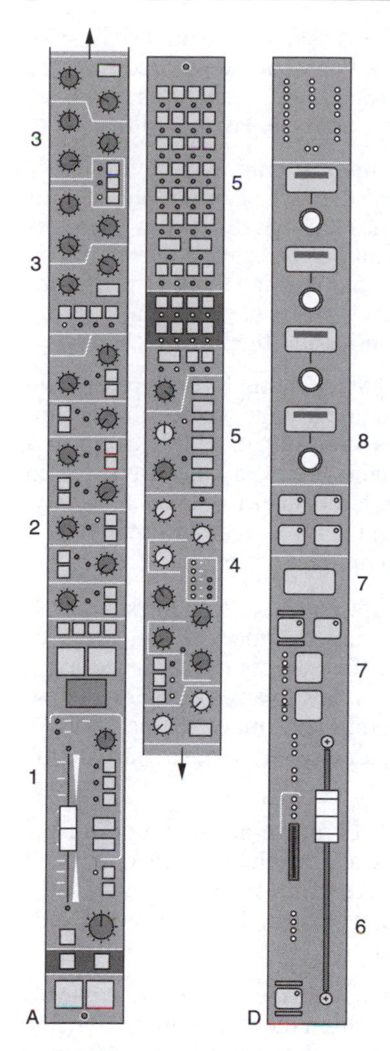

[9.15] **Analogue and digital console channels**
A, Analogue: part of the strip dedicated to a single channel (there may be two faders in line).
1, Monitoring fader and associated controls. 2, Auxiliary sends.
3, Equalization and filters. 4, Dynamics module. 5, Channel assignment.
D, Digital: each whole strip 'flips' between two 'layers'. 6, Fader.
7, Layer selection with window showing channel name above, backlit amber or green to show layer selected. 8, Four 'wild' rotary controls, each with a display above to show its function or setting. Auxiliary send, EQ and filter, dynamics, etc. are assigned to these from modules on other parts of the console.

studio'. In fact, what the virtual-studio computer does is actually very similar to what all the consoles described here are doing: it is just a 'black box' within which the same operations are concealed, then once again revealed as screen operations. In fact, the screens are often visually designed to mimic the manual operations they replace (and as a move in the opposite direction, there is a tendency to reintroduce the hands-on equipment for some operations and plug it in to the computer). But if there is a difference between real and virtual studios it is little to do what the controls do, more a matter of what the studios are used for.

Virtual studios mainly grew up in the world of synthesized music, and in this context are discussed later, in Chapter 16; the use of computers for editing sound is discussed in Chapters 18 and sound-with-video in Chapter 19. But many of the actual functions, which are the same whether analogue or digital, are described here and in the next few chapters.

Channels

The first thing to note about the channels on an analogue console is that they are generally arranged so that they can be worked together in pairs for two-channel stereo. The adjacent channels may be linked (or 'ganged') by a clip attached to the faders, so they are never out of step by more than the fader manufacturer's 'tolerance': a difference that should never be greater than 2 dB. However, at this stage of what may eventually be a stereo or surround sound mix, there will also be many channels that are still operated individually, as mono.

There are two distinct kinds of input. One is from the microphone: this will start at the point where the microphone feed arrives at the input, often hard-wired from a three-pin female XLR socket in the studio. There will often be provision in each microphone channel on the console for a 48 V d.c. supply to feed back to a condenser microphone, using the same three-conductor system that carries the signal (a phantom power supply). The incoming signal passes through a preamplifier that has a coarse rotary gain control, which may be pre-set to lift its level to close to the standard (*zero level* or *line level*) used within the console from this point on.

A second kind of input comes from outside the studio or from other equipment within it, already at 'line' level. Its entry to the console may be through a jack-plug, and the signal will not necessarily be balanced. So this signal is taken first through a small amplifier that balances and feeds it forward to the channel. A mono channel is usually switchable, so it can be used either for microphone or line input. An additional small rotary 'trim' control of about ±15 dB serves either, adjusting its level more precisely, so that the channel fader can be set to the zero marked on its scale when it is making a full contribution to the mix.

Stereo line inputs (including the returns from artificial reverberation devices) may also be fed to dedicated stereo channels, in which the

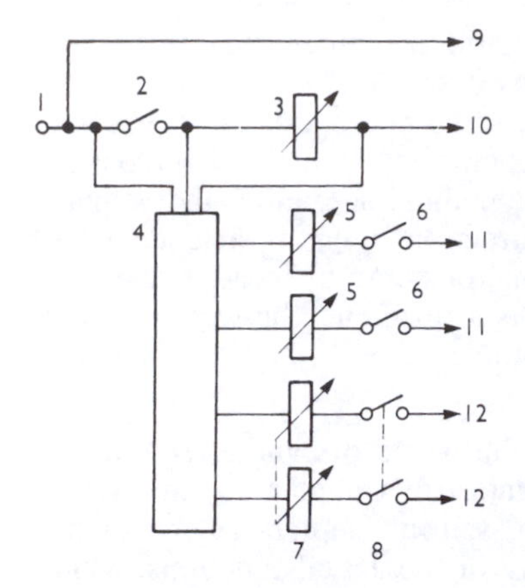

[9.16] **Flexible auxiliary send switching**
Used for public address, studio feedback, AR 'send' or any feed to a remote location that needs a mix that is different from the main studio output. 1, Feed from channel (or group) input. 2, Remote cut. 3, Channel (or group) fader. 4, Auxiliary 'send' switching. 5, Mono 'send' output fader. 6, 'Send' output switch. 7, Stereo ganged output faders. 8, Stereo ganged output switches. 9, Pre-fade listen. 10, Channel out to group bus or main control. 11, Output to 'mono send' buses. 12, Output to 'stereo send' buses.

faders operate equally on both legs. This saves space on the console, and treatments on the strip above the fader can be tailored more to the demands of stereo. On some consoles, these fader units also have a 'remote-start' button that can be set up to replay recordings.

On both mono and stereo channels there should also be a polarity control switch, often designated by the Greek letter phi, ϕ (for 'phase'). In mono, this simply switches the signal between the wires that carry it, so that a stereo signal can be corrected if the wiring on one leg is wrong (or it can be used to put the signal out of phase, for some specially distorted effect). On a stereo channel, the polarity control reverses one leg only, conventionally the left, or 'A' leg.

Progressing up the strip, the signal passes through (or bypasses) an equalizer and filters (described later) and a stage where other equipment may be plugged in. A 'PFL' switch diverts a side chain to a small speaker. Back on the main chain, there may then be a switch which, if activated, permits the whole channel to be cut remotely: this allows a '*cough key*' to be operated by a studio presenter, or a '*prompt-cut*' when a performer is given instructions.

Still before the fader, there should be provision to split the signal and send to *auxiliary* circuits – which may include grouped feeds of selected channels to AR devices, public address (to audience loudspeakers), foldback (to the studio) or feeds to anywhere else. These may all require different combinations and mixes of the sources, which must then be controlled separately, in another part of the console.

After all this, the signal reappears back down at the bottom of the strip, at the most important control of all, the channel fader. With all the faders in a row along the near edge of the console, the balancer can easily see and change their relative settings. In this position, they can also be faded smoothly in and out.

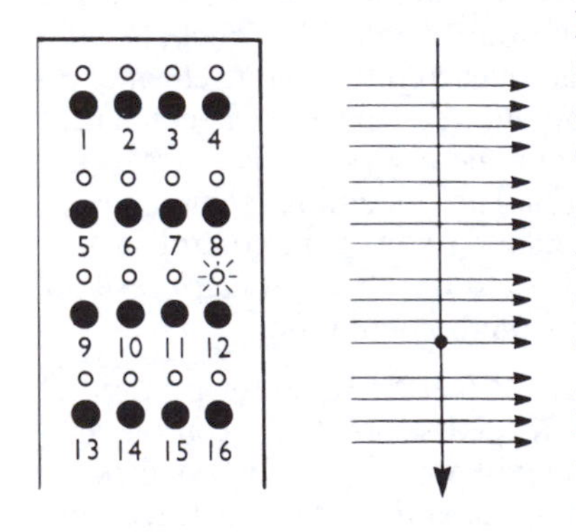

[9.17] **Routing matrix**
A small panel at the top of each channel strip on some mixers designed for linking a separate channel feed to a multichannel recorder. The mass of press-buttons looks more complex than it is: press any button to send the channel output to the selected track.

Back up the strip, following the fader there will be controls (described later) that place the signal within the stereo image. Finally, a switch assigns the channel output to a group fader (usually stereo), or via a subgroup fader (which may be mono or stereo).

Consoles for use with multitrack recorders may have an additional row of small modules, each with tiny buttons (for selection of send or return) and also dedicated 'dynamics' (compressor, limiter, expander – see next chapter) associated with each channel return. At this level of complexity each channel is also likely to have some kind of meter and this will be right at the top of the strip. Although they are a long way up from their associated faders, this row of meters, in line along the top of the console, gives an instant indication of where the action is and a good feel for its significance, as well as all the relative levels.

The strip for each channel is often built up from a series of plug-in modules. This strip layout does not necessarily reflect their true sequence: it is arranged for operational convenience, as determined by the console manufacturer, and in any case the actual wiring makes repeated excursions to the patch bay. But it does make visual

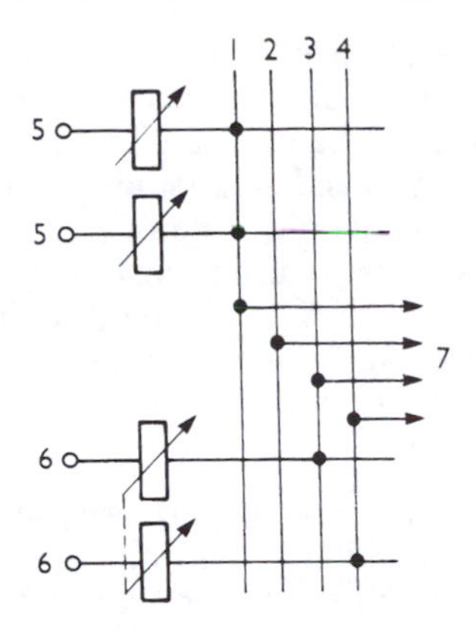

[9.18] **Groups**
1–4, Bus-bars: these are actually wires that cross all channels. In this example two mono channels (5) are both connected to bus 1, and the stereo channel (6) is connected to bus-bars 3 and 4, while nothing is fed to bus 2. The output of each bus (or stereo pair) is fed via group controls to a main control. Auxiliary 'sends' also have their own buses and output controls. 7, Group outputs.

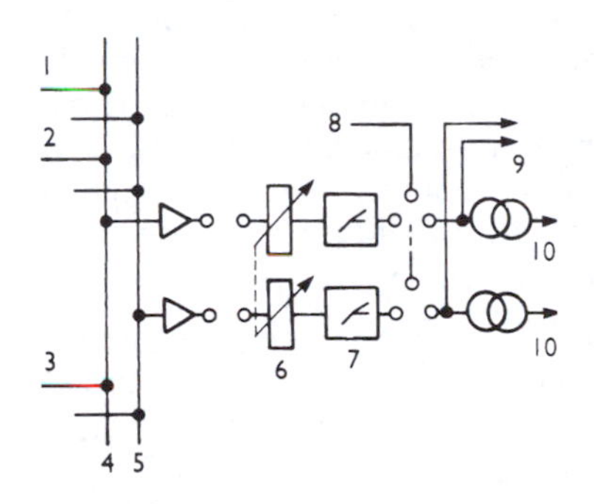

[9.19] **Main or master control**
1–3, Stereo inputs from group faders. 4 and 5, Left and right main buses. 6, Ganged main control. 7, Limiter and make-up amplifiers, also linked for identical action on left and right channels. 8, Zero-level line-up tone input. 9, Feed to monitoring meters and loudspeakers. 10, Stereo output.

sense, and when a source has been selected and plugged to a channel, it is convenient to write its name or a microphone number with a wax pencil or on a label in a line along the desk just above the faders.

Group and master controls

For effective manual control, each of the mono and stereo channels is assigned to a group, which is usually given a number. Colours can also be used to identify groups. Letters, too, may be used, but at the risk of confusion with 'A' and 'B' stereo legs.

The switches for group selection may be at the top of the channel strips, or down near the faders; in some cases a display by the fader shows which group it goes to. The selector may actually route the audio feed to the group fader strip, or alternatively the group fader may be part of a voltage-control amplifier (VCA), which acts as an additional control on the individual channels. Whether or not the audio signal passes through the group faders (or any fader on the console), it is convenient to imagine that it does. Even though on automated consoles many operations are remotely controlled, operationally the result is much the same.

Group faders require fewer elements on the strips above them – mainly stereo balance controls, routing and feed switches, and some means of checking the signal at this stage: PFL and possibly, again, a meter.

Analogue signals are added together at each stage by feeding them to a common, low-impedance bus, from which they cannot return back down through the other channels feeding the bus, because these present a high impedance to the signal. The mixture is taken from the bus by an amplifier that feeds it to the next stage.

The groups are in turn mixed (on a stereo pair of buses) to the *master fader*, also known as the main gain control, then through a final compressor/limiter to the studio output, where a monitoring side chain feeds the main loudspeakers and meters. Meters, compressors and limiters, and their role in volume control, are described in Chapter 10.

The console will also usually have elaborate provision for treating and routing the various auxiliary feeds or 'independent main outputs' (taken from before or after the faders at each stage and passed through their own *submixers*). Some of the uses for these are described later. Alongside the main output there may be a second *independent* or *mix-minus* (also called *clean feed*) output, taken off before a final channel (perhaps for narrator or announcer) is added.

Plainly, the capacity of its console largely determines the limits of what can be achieved in a studio. Individual balancers may become proficient in operating and understanding the quirks of one console to the almost total exclusion of all others, and conversion to another

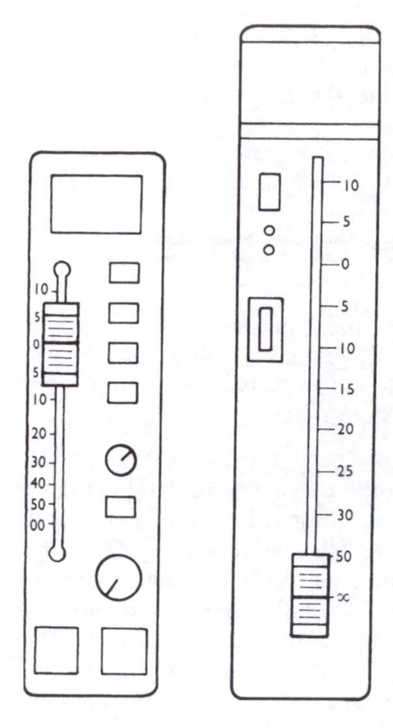

[9.20] **Faders**
Short- and long-throw conductive plastic faders. These may also have channel monitoring and group selection controls on the fader strip. Some faders used for line input may also have remote-start buttons on them.

may be accompanied by temporary loss of speed. This is an extra reason why innovation is often disguised within relatively conventional layouts: 'intuitive' layouts make for greater interchangeability. This is at a premium when working outside the studio using equipment hired by the day. The balancer would usually prefer a known console, even if it is a bit too big, but careful costing may require that something smaller be selected. Brief catalogue descriptions will include the number of mono/line, stereo, group and independent channels that are available, plus details of equalization, auxiliary feeds, communications, and so on.

Remember, big is not necessarily beautiful: for some purposes, such as modest film or video recordings on location, a small mixer with just a few simple faders and no in-line facilities may be all that is needed.

Faders

A fader is essentially just a variable resistance or potentiometer, sometimes called a 'pot' for short. As a contact moves from one end of its travel to the other, the resistance in the circuit increases or decreases. At one end of the scale the resistance is zero; at the other it is infinite (it moves right off the resistive surface or otherwise interrupts the signal). The 'law' of the fader is near-logarithmic over much of its range, which means that a scale of decibels can be made linear (or close to it) over a working range of perhaps 60 dB. If the resistance were to increase according to the same law beyond this, it would be twice as long before reaching a point where the signal is negligible. But the range below −50 dB has little use, so here the rate of fade increases rapidly to the final cut-off.

Faders on most professional equipment are flat and linear (as distinct from older rotary or quadrant designs). The biggest 'long-throw' faders have an evenly spaced main working range (in decibels); where less space is available, 'short-throw' faders have the same distance between 0 and −10 dB, but progressively less beyond that, and with a more rapid fade-out below −40 dB. Music consoles designed to work with two complete, parallel fader banks may have long-throw faders on the nearest to the operator, with short-throw faders above. On automated consoles, they have hidden motorized drives, so they can move under the control of a computer. (On the most advanced consoles other settings on the control strips are also automated.)

Each fader may be associated with several amplifiers. One of them compensates for the action of the fader itself, so that the scale can be marked with a zero (for 0 dB gain or loss) at a comfortable distance from the top of the stroke – in fact, with some 10 dB available above the zero. The layout can be the same whether it directly controls an audio signal that passes through it or, as a VCA fader, controls the audio level from a distance. A minor refinement is that the knobs may be in a range of colours, so that white might be used for channel faders, blue for groups and red for the master.

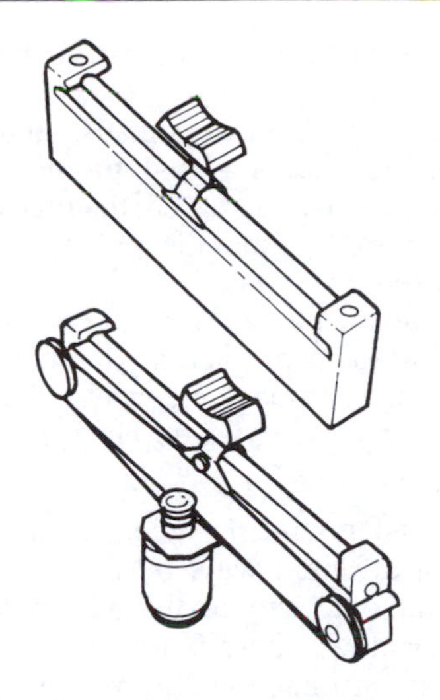

[9.21] **Manual and motorized faders**
Above: Manual operation only.
Below: This can be operated either manually or by the wire from the computer-controlled motor.

The faders in most common use employ a strip of conductive plastics. Carbon granules (in a complex layout, distributed to produce a 'law' that is scaled in decibels) are printed on a plastic base and heat-cured to render the surface smooth and hard-wearing. For ease of operation and to reduce wear further, the slide makes only light, multi-fingered contact with the resistive surface. High-quality individual specimens may vary by up to a decibel from the average (a range that is within the typical limits to human awareness). One measure of performance is evenness of movement, the lack of any 'stick/slip' effect a user might detect when making light, fingertip movements of the fader.

Simpler faders based on carbon-track or wire-wound resistors tend to wear more rapidly and then make unpredictable crackles as the slide contact is moved (as the owners of most domestic equipment eventually discover). At one time, the only satisfactory alternative was a fader with separate studs that the sliders engaged in turn, with a fixed network of resistors between each successive pair of studs. But it was difficult to make studs and sliders that always moved in perfect contact, unaffected by dust, moisture and other atmospheric contaminants, so these are now used only for coarse pre-sets.

In the days when they were used throughout a mixer, a rare crackle or two was the price paid for being able to set levels precisely, literally stop by stop in increments of 1.5 or 2 dB (for which the term 'stop' is still used). With the advent of stereo, these stops provided exact matching between the two signals, though with some danger of getting momentarily out of step when moving from one pair of fader studs to the next: this could result in stereo image *wiggle* or *flicker*.

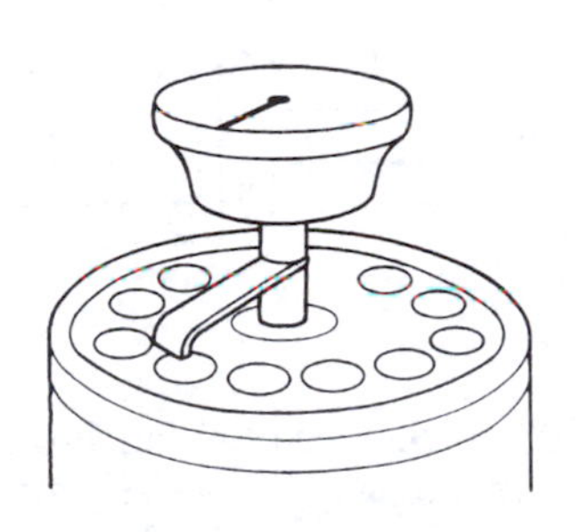

[9.22] **Stops**
This term originally referred to the discrete jumps in level associated with the spacing between studs on older, stepped faders. These (with many more studs than shown here) were arranged at intervals of approximately 2 dB, close to the limit of human perception. The term 'stops' is still used, and the scale markings on many of today's faders correspond to them over their main working range. Rotary faders are still used for coarse pre-sets, or for 'trimming' over a narrower range, but individually wired studs are now rarely seen.

Some of the simplest designs of fader have been used in low-level mixers, i.e. those in which there is no preamplifier, or internal amplifiers compensating for the loss in the faders. But these are not suited to mixing, where there is much use of faders, because as each additional channel is opened it provides an alternative path for signals from the others. In consequence, the faders are not independent in operation: as fader 2 is brought up, there is a slight loss in signal 1, so that its fader must be adjusted to compensate. Such problems are encountered only in simple or non-professional mixers. A simple mixer for which it might be suitable is that used for a subgroup of audience microphones from which the combined output is fed to one stereo fader, so freeing several channels with full facilities for other uses.

Stereo image controls

Rotary potentiometers are used at many points on the console where space must be conserved. Several may be used (after the channel faders) to construct or modify the stereo image:

- *Panpot*, short for panoramic potentiometer. This steers or pans a monophonic element to its proper position in the stereo image.

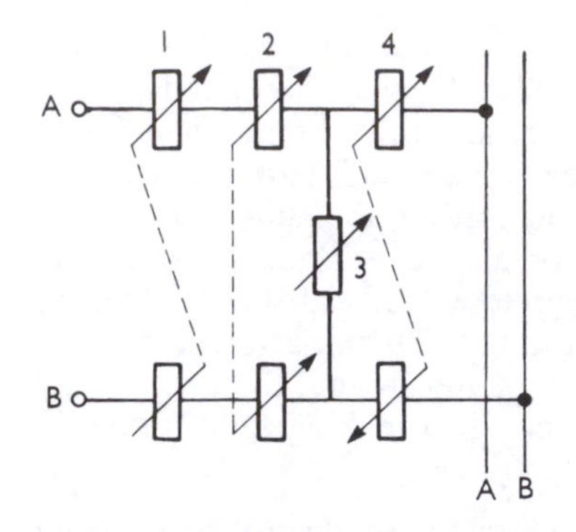

[9.23] **Stereo controls**
1, Pre-set control to balance output between stereo microphone A and B legs. 2, Ganged channel fader. 3, Image width control. 4, Channel offset control (to displace the whole spread image).

The input, fed through the moving contact, is divided in two and fed to the A and B legs.

- *Image width control.* This cross-links the A and B signals, feeding controlled and equal amounts of the A signal to the B channel, and vice versa. As this cross-feed is increased, the image width narrows. However, if at the same time the polarity is reversed, the image width is increased.
- *Displacement control.* This is a ganged pair of faders, one on each path, and working in opposition to each other, so that turning the knob has the effect of moving the image sideways. This control can also give apparent movement to a static but spread source.

In a stereo channel the main pair of faders and the width control might also be put into the circuit in a different way: by first converting A and B into M and S signals and operating on those instead. When M and S signals are faded up or down there is no 'wiggle'; instead, there might be minor variations in image width, but this is less objectionable. In addition, image width can then be controlled

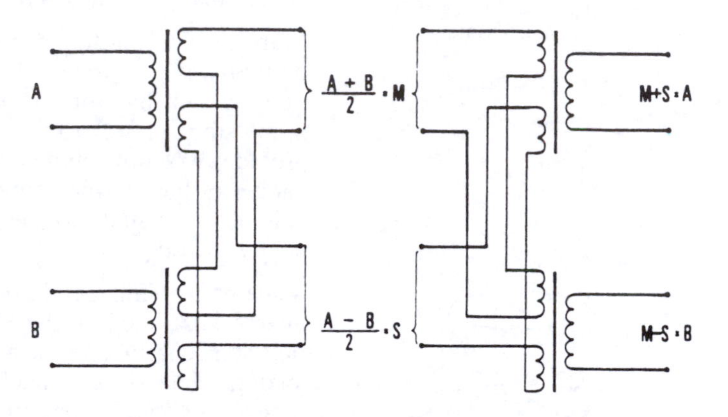

[9.24] **Stereo signals**
An electrical method for converting A and B signals to M and S, and vice versa. 'M' is equivalent to the signal that would be obtained by directing a single suitable monophonic microphone towards the centre of the audio stage. It therefore gives a mono output that can be used by those without stereo loudspeakers, although the balance between centre and side subjects, and between direct and reverberant sound, will not necessarily be as good as for a normal mono balance. 'S' contains some of the information from the sides of the audio stage.

simply by changing the proportion of M to S. In particular, image width can be increased without change of polarity. However, after passing through its channel fader and width control, the M and S signals must be converted back to A and B before passing through the displacement faders.

Using microphones that produce the M and S signals directly can eliminate one stage in this process. However, this implies a radical change in working practices and the use of mixers that are at present laid out primarily for AB operation. So it is usual to convert MS microphone signals to AB and work with them in the usual way.

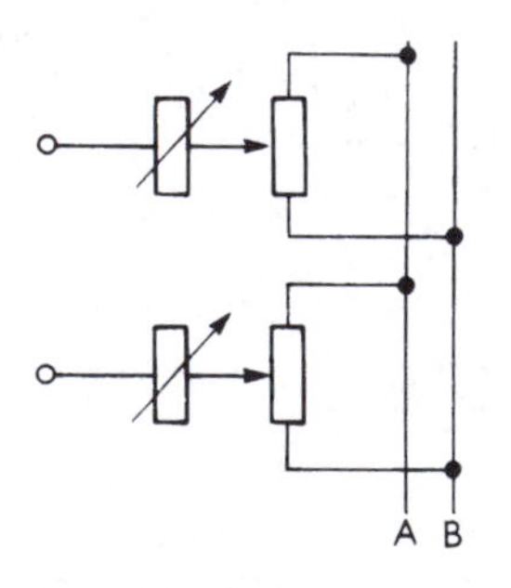

[9.25] **Mono channels in stereo**
A small rotary panpot in each channel is used to place individual mono components within the stereo image.

Where two monophonic channels with panpots are linked together to form a stereo channel, image displacement and width are controlled by the relative settings of the two panpots. If they both have the same setting, the source has zero width, displaced to whatever position the panpots give them. As the panpots are moved in opposite directions

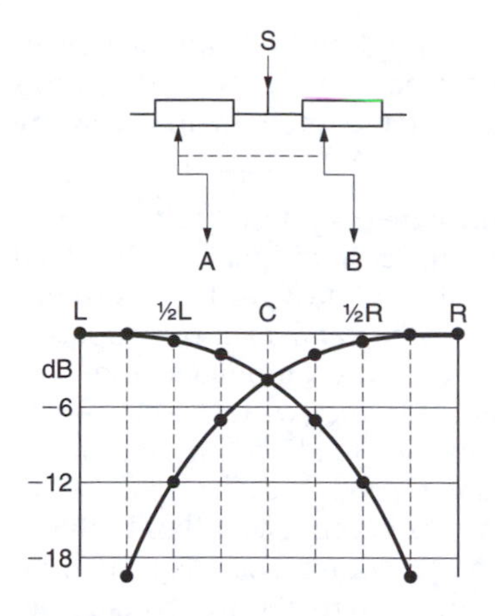

[9.26] **Panpot attenuation**
A signal from a monophonic source is split and fed through a pair of ganged potentiometers: step by step, the attenuation in the two channels is matched (dotted lines). The apparent volume will then be the same in all available positions across the stage. L, Left. $\frac{1}{2}$L, Half left. C, Centre, etc.

from this point, the image increases in width. Full width is given by steering one of the signals all the way left, to the A side, and the other all the way right, to B in the mixed signal; this can, of course, be done without a panpot at all.

A stage in the creation of surround sound employs two stereo panpots, one for side-to-side operation and the other for forward-to-rear. By using the two together a monophonic source may be steered anywhere within the four corners. For most purposes it is sufficient to do this with the separate controls. But occasionally, for more complex, continuously smooth movements, say of a character in a play or an opera, or for special effects such as an aggressive insect or a circling plane, signal pairs might also be fed through a joystick control, or possibly controlled in two dimensions by a mouse or a trackball.

Tracklaying, multitrack recording and mix-down

In music recording studios, popular music is generally built up in stages – by *tracklaying* and *overdubbing*. An instrumental backing might be recorded first on several tracks, followed by a soloist who listens to a replay on headphones and adds the vocal, then a further instrumental or vocal track might be provided by a musician who played earlier. Treatment of various kinds may be tried, but not yet confirmed.

The recordings are laid down using one of several multitrack recording systems. The longest established is the analogue multitrack recorder: here the tracks are laid down side by side on broad tape (24 tracks on 2-inch/52-mm tape or eight tracks on 1-inch/26-mm tape) with separate stacks of heads to record and replay them. Next came digital multitrack systems (24 or 48 tracks on half-inch/13-mm tape). A cheaper competitor is the multitrack digital module (MDM): each MDM has eight tracks on an inexpensive domestic-style videotape, and the modules are stacked to provide anything up to 128 recording tracks. Eight is a good number for 5.1 surround sound mixdown: 5 plus 1, with 2 for a separate stereo mix.

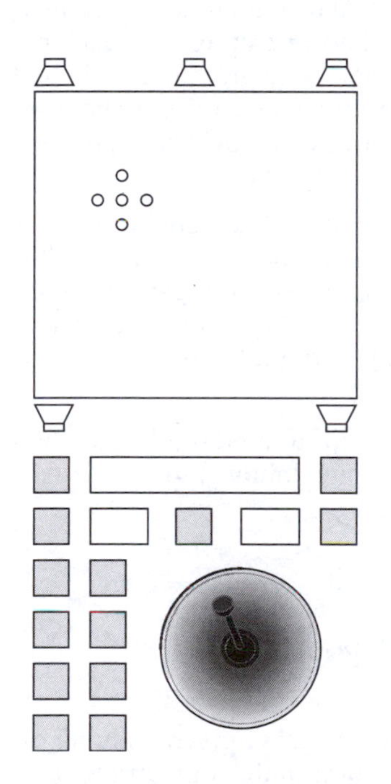

[9.27] **Joystick panpot**
On some consoles, in addition to the rotary panpots in each channel, any selected signal can be fed via a small joystick control. The display above shows its panned position in 5.1.

Any of these recording systems can be traded in for (or backed up by) hard disk on a PC or Mac computer, which costs even less, and also removes the obligation to think in multiples of eight. Alternatively, a removable hard drive is housed in a small rectangular box with SCSI connector pins at the back. A recording made in one place can be taken away and plugged directly to the input of an audio editing system (which can be anywhere) and later returned to another studio. This whole pluggable SCSI box is usually called a 'scuzzy' for short.

On many consoles each channel has an array of numbered selector buttons, each of which sends a feed from it to a particular track (or stereo to pairs of tracks). A trial mix is obtained during the session (so producers and others can get a feel for what is being achieved) and this may also be recorded for reference: the levels and settings set for all channels are noted – or better, stored in the desk by computer.

A console that can store information generally has ‘write’, ‘read’ and ‘update’ modes, and may also have additional ‘write’ and ‘update’ buttons associated with each fader (moving fader systems normally have touch-sensitive fader knobs) and in more sophisticated consoles, the EQ and panpots, etc., are automated too. The initial balance is made with the mixer set to the ‘write’ mode. When the tape is replayed with the mixer in the ‘read’ condition, all the earlier fader settings and changes are reproduced (irrespective of their current positions). If while the mix is being read a ‘hold’ button is pressed, replay continues but with the controls unchanged from that point on: this could be used (for example) to get rid of a fade-out on the end of the initial mix.

Except in the case where the computer also moves the controls, the mix is updated as follows: first, all faders are set to zero or to some nominal setting which becomes the ‘null’ level. Then, after the relevant channels have been set to ‘update’, the tape is replayed and any changes from the null position (and their variation in time) are fed to the data store, changing the selected parts of the mix by the amounts indicated. Alternatively, the whole of the previous control sequence for a channel can be rejected and written afresh by using the ‘write’ button instead of ‘update’ and operating the fader normally. Also associated with each fader is a grouping switch. If several faders are dialled to the same number to form a subgroup, the ‘write’ and ‘update’ instructions for any one of them operate on the control data for all members of the subgroup.

This *mix-down* generally takes place, after the departure of the musicians, in a post-balancing session, perhaps even on another day. At this stage the recording systems of choice (again keeping a backup) might include DAT, magneto-optical (MO) or hard disk. If it is destined for CD, it will go through a digital audio workstation (DAW) anyway.

In radio and television, where time is more limited, multitracking is more often used just to allow for additions, minor improvements or second thoughts.

Control console checks

Operating a console is like driving a car: easy, given training and experience – and provided that the car starts and is in working order. With hundreds of faders, switches and routing controls to choose from, if you cannot hear what you think you should be hearing it is as well to have a systematic strategy for recovery, which should include the apparently obvious. Consoles vary, but a basic trouble-shooting checklist might start like this:

- *Power*. Is the console switched on? Is power reaching it? Has the supply failed?
- *Microphone or line source*. Is it working? Is it fully plugged in? If in doubt, try another (but not one that takes power from the console).
- *Jackfield*. Is the routing correct? Is the path intercepted (and so cut)? Are the jacks fully in? Has moisture condensed into it? (This is among the hazards for equipment taken away from the studio.)
- *Channel input selector*. Is it correctly set to microphone or line?
- *Routing within console*. Is it going to an audio group or VCA or as an independent direct to the main fader?
- *Faders*. Are all on the chosen routing open? Open them all to '0 dB'.
- *Solo, cut and AFL switches*. Has the signal been interrupted by a switch on another channel?

Make up your own checklist, or use one that others have written for that studio. This might also contain notes on performance limitations and how to make the best of them. Keep a fault, maintenance and repair log, and in a studio used by several operators a suggestion book.

With the growing complexity of studio equipment, maintenance is increasingly a matter of module replacement, of dealing with 'black boxes' rather than understanding the detail of their internal circuits, and how to investigate and repair them on the spot. But it is essential to be aware of operational problems, such as any switch that makes a click on the line, so that it can be avoided when on-air or recording. The user should know how and be prepared to route round any module that fails, especially on a live broadcast. It helps if operators understand signal pathways through the console, are familiar with common deficiencies in quality, and know how to localize and describe them.

Monitoring sound quality: noise

Before we can begin to use the console and its monitoring system creatively, we must first know how to judge sound quality – if only to distinguish between artistic and technical faults. When there is a fault, there is liable to be a frustrating and potentially expensive delay until it is tracked down to its source and cured or isolated. Some of the 'traditional' problems associated with analogue audio are described below. Digital audio bypasses some of these, but introduces others of its own: some digital problems are described in Chapter 16.

Sound degradation problem number one is *noise*, which for our present purpose may be regarded as random sound of indefinite pitch (there are many other forms of 'unwanted sound', but they can each be treated separately). Irregularities in the structure of matter cause noise: the particles that are magnetized (or missing) on tape or disk, roughness in the wall of an old record, the granular structure of a resistor, the random dance of individual electrons in semiconductor or metal; all produce some noise in a recording or broadcast. They are heard as crackling, buzzing or hissing, and must all be kept at low level compared with the audio signal. We need a good *signal-to-noise ratio*.

Noise becomes serious when components before the first stage of amplification begin to deteriorate; at a later stage, when there are more drastic faults (e.g. a 'dirty' fader or a 'dry' joint), where vibration reaches a transducer, where there are long lines, or where radio transmissions are received in unfavourable conditions. To combat noise, in recording or transmission systems where noise levels increase with frequency in a predictable manner, the high-frequency signal is increased (*pre-emphasized*) before the process and reduced (*de-emphasized*) after it. For radio or television, these systems require national or international standards, so that professional and domestic equipment produce complementary effects. One advance in noise reduction came with the switch from amplitude to frequency modulation of broadcast audio signals, another when signals could be passed along the audio chain in digital rather than analogue form. Even so, a great amount of analogue equipment remains in use.

For the recording industry a range of commercial noise reduction systems is widely used, with different standards applied to domestic recordings, professional audio, or optical or magnetic recordings on film. Most divide the sound into a number of frequency ranges and apply treatments to each, together with a control signal that allows the reproducer to decode it.

In one noise reduction system that is used in professional studios for analogue recordings, the signal is split by filters into four frequency bands: 80 Hz low-pass; 80 Hz to 3 kHz; 3 kHz high-pass; and 9 kHz high-pass. Each band is separately compressed, so that very-low-level signals (40 dB down) are raised to 10 dB, while signals from 0 to 20 dB down are virtually unaffected. The recovery time (the time during which the level returns substantially to its original state) is 1–100 ms, depending on the signal dynamics. There are also crude volume limiters in each channel: these are simple peak choppers. A matching network is used to reconstitute the original signal on playback.

Such systems rely on the idea that noise that is similar in frequency to the louder components of the signal is masked by it, while noise that is widely separated in frequency from the louder parts can be reduced in level: the improvement is therefore partly real and partly apparent. As the system produces a coded signal that must then be decoded, it cannot be used to improve an already noisy signal. It is widely used to improve the quality of tape recordings, where an increase in the signal-to-noise ratio is used partly to reduce hiss, hum and rumble, and also by lowering recording levels a little, to produce less distortion and to lower the danger of print-through. It is particularly valuable for multitrack recordings, but can in principle be used around any noisy component or link. The signal is, of course, unusable in its coded form: characteristically it sounds bright, breathy and larger than life. (The noise reduction systems used on domestic equipment are much simpler. For further information on these and other analogue noise reduction systems, refer to specialist literature.)

Digital recording offers a 90 dB signal-to-noise ratio without further treatment.

Hum

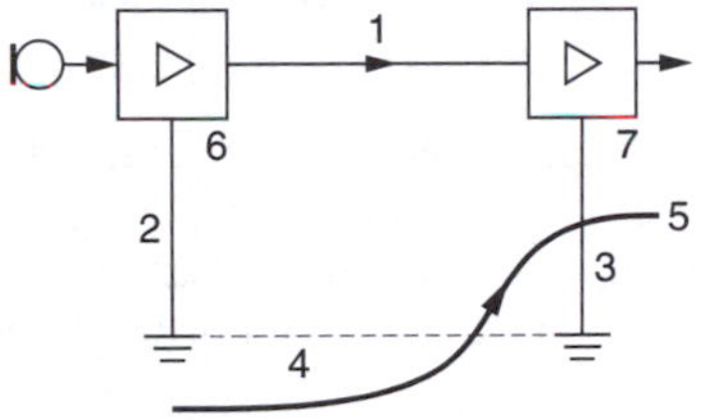

[9.28] **Earth loop (ground loop)** In this example: 1, Microphone channel. 2 and 3, To earth by two separate paths. 4, The loop is completed invisibly, through the common earth. 5, An unrelated mains cable, crossing through the loop, creates an electromotive force that generates a matching alternating current in the loop. 6, In this worst case, a preamplifier selectively boosts only the forward half of each cycle. 7, This amplifies the hum and generates rich harmonics, of which the first overtone (second harmonic) is particularly strong and audible. If it is difficult to find and eliminate the additional path to earth, it may be necessary to filter the signal, giving particular attention to the second harmonic (at 100 or 120 Hz).

A noise that does have definite pitch is *hum* – often due to the mains frequency and its harmonics getting into the audio signal. Mains hum may be caused by inadequate smoothing of an electrical power supply when rectified from alternating to direct current, or by inadequate screening on wires carrying audio signals, particularly those at low level, or in high-impedance circuits, such as output from condenser microphones before the first amplifier. Magnetic microphones can be directly affected by nearby electrical equipment, such as mains transformers or electric motors. Hum may also be caused by making connections by separate paths to a common earth, which creates an earth loop that picks up signals from power lines that pass through it.

The second harmonic of mains hum is generally the most audible, so most hum filters just get rid of that (at 100 or 120 Hz). More elaborate 'notch' filters take out narrow slices of a wider range of harmonics without substantially affecting music. Some 'plug-in' digital signal processors have more sophisticated programs that include this; they also tackle ventilation noise, guitar amp buzz, hiss from archive recordings and have extra programs for clicks and pops. (But note that if applied too aggressively, they introduce artifacts of their own.)

Hum can be a problem when recording popular music from public performances, where feeds are split from the system set up for the stage foldback loudspeakers. To identify the source of hum quickly, fit the earth (ground) leads for each circuit with a switch so that each can be individually checked. Where sound is combined with vision, lighting (or their dimmers) is the usual suspect. A full rehearsal will show up danger points – or ask for a quick run through the lighting plot to hear what it does to open microphone circuits. Portable hum-detection devices are also useful when setting up any temporary rig that involves a lot of audio and lighting circuits.

Distortion

Distortion should not be encountered in modern systems that are working well, but was a serious problem in the days when successive components were less well protected. If at any point in a broadcast

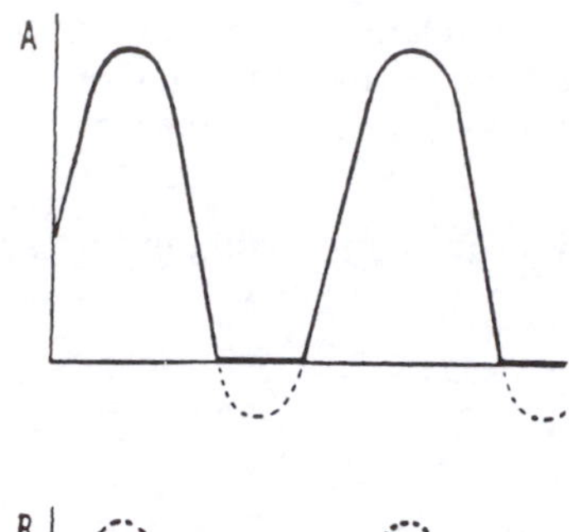

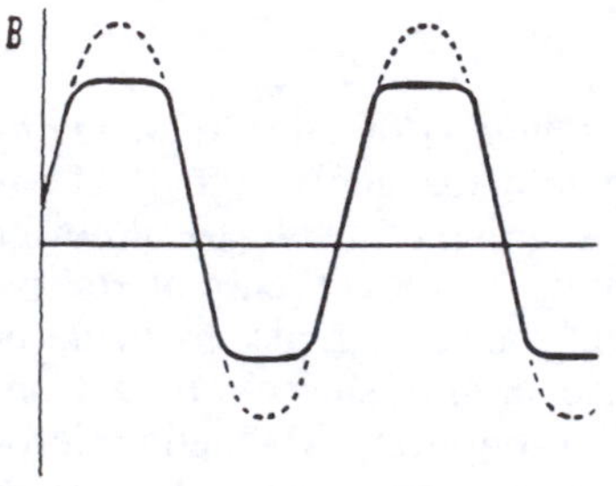

[9.29] **Harmonic distortion**
The dotted portion of the signals is chopped off, flattened.
A, 'Bottoming' causes part of each cycle to be clipped. B, Both peaks are flattened, as excessive swings in each direction cannot be followed by the signal in some item of equipment. In both cases the distortion produces many overtones, at exact multiples of the frequency of clipping.

or recording chain the volume of an analogue signal is set too high for any stage that follows, the waveform is changed – the peaks are flattened, creating new harmonics. However, most are added at frequencies that are already present in much sound, and most music. The introduction of this *harmonic distortion* is therefore partly masked by existing sound, and even where it is not, it does at least have a pseudo-musical relationship to the wanted sound. Speech sibilants, which do not have a harmonic structure, bring it out strongly. Note, however, that human hearing is itself subject to distortion at high acoustic intensities.

Small amounts of distortion may be more acceptable than a high noise level: 2% harmonic distortion is generally noticeable and 1% represents a reasonable limit for high quality, but with up to 3% on some peaks.

Unduly high distortion may be due to faults in equipment, such as where the moving part of a transducer has worked loose (e.g. a ribbon or moving coil hanging out of its magnetic field), or errors in their use, such as if the wrong bias is used when recording on magnetic tape.

Intermodulation distortion is far worse than harmonic distortion: it occurs when two frequencies interfere with each other to produce sum or difference products. It should also be rare: good sound equipment should be designed to avoid intermodulation effects like a plague.

Distortion may arise at the start of the chain, at points that are, in theory, under the control of the balancer. Some air pressures (perhaps caused by wind, or by plosive consonants in close speech) may move a microphone diaphragm outside its normal working range. Or microphone output may be too high for the first preamplifier. Amplifiers have a linear response to changes in volume (i.e. output is proportional to input) only over a limited range. The sensitivity of professional microphones is related to the volume of expected sound sources in such a way that they mostly feed signals to the mixer at about the same range of levels. Exceptions are bass drum, brass and also electric guitar loudspeakers that are set very high. If following preamplifiers are of fixed gain, a 'pad' (a fixed-loss attenuator of, say, 12 dB) may be required between the microphone and amplifier. Alternatively, use a microphone of lower sensitivity or one with a head amplifier that can be switched to accept different input levels.

Another example of this form of distortion may occur in very close piano balance. In some popular music the piano becomes a percussion instrument, with powerful attack transients. The brief, high volume of these transients may not be obvious from any meter that takes time to respond, but this too is liable to distort.

Note that if a digital signal is clipped, this (unlike analogue clipping) immediately results in unacceptable loss of data. To avoid this, digital recordings therefore generally require either 10 dB of clear, extra 'headroom' or consummate care if higher levels are permitted.

Crosstalk

On stereo, *crosstalk* is a defect to listen out for. It is measured as the level at which a tone at a particular frequency in one stereo leg can be heard on the other. But what is regarded as 'unacceptable' seems to vary. For example, for broadcast tape recordings the limit might be set at −38 dB at 1 kHz, rising to −30 dB at 50 Hz and 10 kHz. These figures compare favourably with the −25 dB (at middle frequencies) available in the groove of a stereo vinyl disc, but are extremely poor compared with digital recordings. In a good mixer, the specification will be much higher, perhaps requiring a separation of more than 60 dB – and one console offers better than 75 dB at 15 kHz.

Subjective quality checks

Checking the technical quality of work in progress should be automatic and continuous, so that any fault (and with experience an immediate diagnosis of its probable cause) simply seems to jump out at you. But there are other circumstances, in principle just as important, in which faults may be there for all to hear, but none of those responsible seem to notice. This may happen when the product is no longer just a single work, but a whole body of output, such as a broadcast channel.

[9.30] **Subjective scales for quality assessment**

Scale 1	*Sound degradation*
1.1	Imperceptible
1.2	Just perceptible
1.3	Audible but not disturbing
1.4	Somewhat objectionable
1.5	Definitely objectionable
1.6	Unusable

Scale 2	*Overall quality*
2.1	Excellent
2.2	Good
2.3	Fair
2.4	Fairly poor
2.5	Poor
2.6	Very poor

Judgements of sound quality are personal and subjective, but in such cases it may still be helpful to try to apply standards that are common to any group of people who work together. A broadcaster may attempt to do this by setting up a system of observations that might cover general sound quality (including background hum, noise or hiss), change of sound level at programme junctions, and interference. One of the items in this last category is the talkback induction sometimes heard on radio, but even more often on television because of the continuous stream of instructions and information fed from the gallery to the studio floor. And – this may seem a lost cause, but add it in anyway – most of the public would like a less intrusive match between the sound levels of programmes and commercials.

In one example of such an assessment system, sampling was done for periods of 5 minutes on a wide range of material, including productions as they were transmitted (both before the transmitter and off-air), studio output (but taking into account conditions that may exist in rehearsal before all adjustments have been made), material incoming from other areas (by landline, radio link or satellite) and all recordings.

Because they cover such a wide range of possible defects and because others might appear unexpectedly, the most useful scales are likely to seem highly subjective. The divisions in the scales to the left may seem coarse, but they are readily assessed and easily understood.

Chapter 10

Volume and dynamics

There are three kinds of volume control. The first uses meters to control overall level. For the second, to control internal dynamics, we need loudspeakers and our ears. The third employs compressors and limiters and is fully automatic but may still require monitoring by ear.

The first application of volume control is purely technical. Meters are used to keep the signal within limits that allow it to pass with minimal unwanted changes through equipment, along lines and links, to reach broadcast transmitters or recording media. If an analogue level is too high, preamplifiers will overload and distort, tape recording media will be saturated, and (even worse) any attempt to convert analogue to digital will cause signal clipping, in which the data is severely damaged. If the signal is too low, it has to compete with noise, and even normally robust digital signals can suffer. For checking maximum volume, loudspeakers are no help at all, as their settings are arbitrary, but they are essential for listening to changes in volume (dynamics).

Ensuring that what the audience should hear lies between acceptable maximum and minimum volumes may also require reduction of dynamic range, by either manual or automatic compression. Fortunately, many people listening at home prefer a reduced range. For this, it is the meter that gives little help: aesthetically satisfying comparative volumes can be assessed only by ear. Meters are a guide, but ears must be the judge.

Automatic compression can be used to control some kinds of sound reasonably well – but not all. Automatic control lacks intelligent anticipation: it would, for example, ruin Ravel's *Bolero*. In most music, subtle dynamic control by the performers is important to the artistic effect and no compressor will do an adequate job unless it can be set for the specific item. Another kind of problem with relative volumes is that different control characteristics are required where several types of output are heard in succession.

Specialized uses for compressors and limiters include short-wave radio, where a sustained high-signal level is necessary to overcome static and other noise; also certain other types of high-intelligibility radio, where a limited range of output is broadcast to a maximum service area for a given transmitter strength. In addition, they serve particular purposes within programmes, e.g. as an element of pop music balance.

Meters

Although a meter cannot be used to judge relative levels, such as that of one voice against another or of speech against music, it is nevertheless an essential item of studio equipment. Its uses are:

- to check that there is no loss or gain in level between studio and recorder or transmitter;
- to avoid levels that would result in under- or overmodulation at the recorder or transmitter;
- to compare relative levels between one performance and another;
- to check that levels meet prescribed values for good listening;
- to provide rough visual checks on the relative levels (and correct polarity) of stereo channels and mono compatibility.

There are two main kinds of meter for analogue audio, operating on completely different principles. One responds to the audio signal logarithmically (as our hearing does); the other does not.

A *peak programme meter* (PPM, or 'peak meter'), well established in European practice, is basically a voltmeter that reads logarithmically over its main working range. In order to give prominence to the all-important 'peaks' that may be of short duration but vulnerable to overload and therefore distortion, it has a rapid rise characteristic ('time constant') and slow die-away that makes it easy to read. The PPM originally developed for BBC use has a rise time constant of 2.5 ms, giving 80% of full deflection in 4 ms (our ears cannot detect distortion of less than this duration), and a die-away time constant of 1 s, a fall of 8.7 dB/s. Other European meters may have different rise and fall times and face-scales from those favoured in Britain.

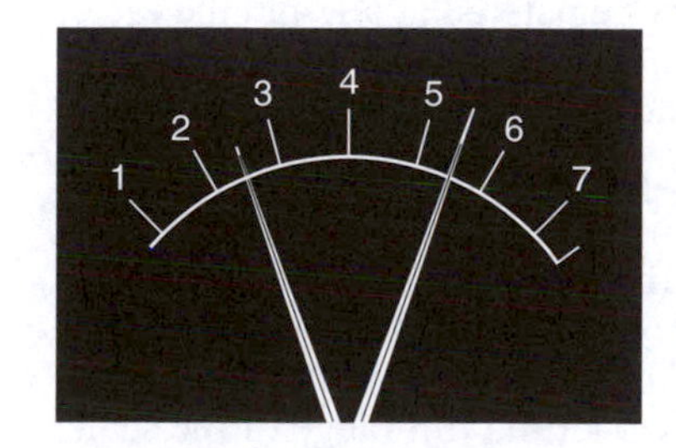

[10.1] **Peak programme meter (PPM)**
The BBC standard PPM is widely used for analogue metering in Europe. Markings are white on black. Each division is 4 dB; '4' on the meter is a standard 'zero level' and '6' is full modulation, above which distortion is to be expected. The needles may be set up to show either A and B or M and S signals.

Placed prominently on the middle of the console, the original mono PPM had a single needle that, for visual clarity, moved over a black face with white markings. For stereo this was replaced by two meters, each with two concentric needles. One meter is for the A and B signals, in colours that follow the nautical 'red-port-left' and 'green-starboard-right' convention. The second shows M and S signals, conventionally with white and yellow needles (and not always easy to distinguish). Because $M = A + B$ it normally reads higher than either: for a source in the middle it is 3 dB higher. Because $S = A - B$, it is normally less than M. Unless there is phase cancellation, S is to the left of M, so if S swings to the right it suggests that a leg is wired with its polarity reversed: this is an immediate visual check for the fault. (But beware: in some consoles the functions of the two PPMs can be swapped over.)

PPMs are used, as the name suggests, to protect against analogue peaks that may cause distortion. However, although the rise is rapid, peaks of very short duration will not be fully registered: the true peak may be up to 4 dB higher. Also, if digital recordings exceed the '100%' level, the damage due to loss of data is so severe that this 'maximum coding level' is taken as an absolute limit. To guard against excessive unpredictable peak levels and allow for any under-reading, an alignment with peaks 10 dB below that – 10 dB '*headroom*' – is adopted for digital recordings. On a DAT recorder, for example, peaks should be this much below the '0' that marks maximum coding level. The loss in signal-to-noise ratio is a fair trade against the danger of digital clipping.

An exception is made for CDs, which are used more for the distribution of digital recordings than as the original recording medium. CD mastering is controlled very precisely. Digital editing systems allow very accurate measurement of the highest peaks, so when the recording is 'burned' to CD-R this can be used to ensure that even momentary peaks do not exceed the maximum coding level. By using this '*normalization*' process, a CD-R disc can have a level similar to that of a commercial CD. If the level is read on a PPM its apparent headroom will be 4 dB (to allow for the meter's slower response).

The alternative (again designed for analogue sound) is the *VU meter*. Widely used in American equipment, this is marked with two scales: one shows 'percentage modulation' (used more in broadcasting) and the other reads in decibels. The VU meter is fed through a dry rectifier and ballast resistance, and (unlike the more complex PPM) may draw from the programme circuit all the power needed to operate it. For cheaper models the ballistic operation may allow a needle to peak substantially higher on music or speech than on steady tone. In contrast, a more expensive, studio-quality VU meter overshoots only by a little when a pulse signal is applied (in which it compares favourably with a PPM). But a time constant of 300 ms prevents transient signals from being registered at all, and the meter under-reads on both percussive sounds and speech, so at its '100% modulation' level these may be distorted.

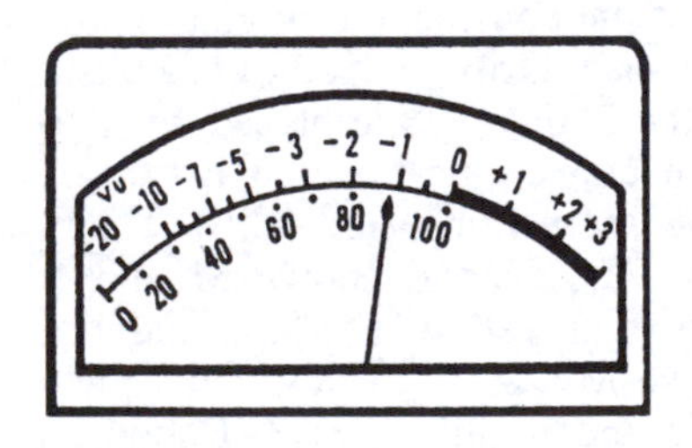

[10.2] **VU meter**
Widely used for analogue sound metering in America. The lower scale indicates 'percentage modulation' and the upper scale is decibels relative to full scale.

Though it may have a linear calibration in 'percentage modulation' (the rectifier characteristic should ensure this), more than half of that scale is taken up with a range of 3 dB on either side of the nominal '100%'. Little programme material remains consistently within such a narrow range, so the needle is likely to register either only small deflections or flickers bewilderingly over the full range of the scale. For maximum volumes, continuous sounds (including most music) may read '100% modulation' on the VU meter; however, for staccato sounds such as speech or drum taps a true maximum is given by an indicated 50%, because the meter does not have time to register a full swing to the 100% mark on the louder individual components of speech.

Some balancers who have always used a VU meter say they like the way its reading seems to relate more to average loudness than a PPM does. For a signal that is supposed to stay close to full modulation they try to keep the needle close to the nominal zero in a range somewhere above −5 and not above +3. For some kinds of music the roughening of tone caused by a slight overrun can even be regarded as beneficial: the term ‘gutsy’ has been used for this effect. A drum-kit that may already be overmodulating at ‘VU 0’ sounds even more dramatic (different, certainly) when the needle is ‘pinned’ over to the right. VU meters seem better suited to the record industry, which has more time for experimentation, than they are to radio and television, where excessive peaks are more of a menace.

Note that only those VU meters that conform to the relevant American standard can be used to compare programme material. Many meters do not meet this standard. Some bargraph meters can be switched between peak and VU meter calibration, or (for the best of all possible worlds) have both, side by side.

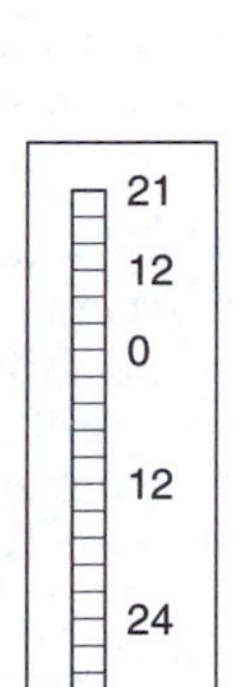

[10.3] **Simple LED bargraph**
As a rough guide to level (or overload) in an individual channel or group, a simple scale with LED segments may be adequate. Here each division below zero (maximum volume) is 3 dB.

A meter designed (or set) primarily for digital signals has the zero (to mark maximum coding level, above which ‘clipping’ will occur) near the right (or top) of the scale. The main working range is well to the left of (or below) that, 10 dB below if recommended ‘headroom’ is allowed for safety. Unlike the PPM, this meter shows instantaneous peaks, but still with a decay that is slow enough for the eye to register easily.

On a radio control console, the meter (of whatever type) is likely to be either centrally placed or roughly in line with the main control fader; a television sound output meter is mounted in line with the transmission monitor and, to avoid eye fatigue, in much the same visual plane.

Line-up tone

Meters are used to establish common levels at all stages in the chain by using a pure tone, usually at 1000 Hz and of known volume, as a reference. Organizations (such as the BBC) with a central control centre for signal routing will have a reference tone source at that point, and most consoles and some other items of equipment such as recording machines also have their own. A studio will therefore often have the choice of using the central source or its own local tone – which should, of course, be the same. This always registers ‘4’ on a monophonic PPM (or as the M reading in stereo, with nothing at all on S). Equal stereo A and B signals should each read 3¼, i.e. 3 dB below ‘4’. Tones from other sources will be faded up until they reach this same reading, to establish the incoming zero level; any losses on the incoming line should be made good at the input to the console.

The standard used is 1 mW into 600 Ω, and for an analogue signal is equivalent to 40% modulation at the transmitter (so that 100% is about 8 dB above this). At all stages in the chain the tone should give the same steady reading on all meters, and when a recording is made (other than for public distribution), reference tone is recorded on it at the appropriate level for the equipment and recording medium, so that the process can be repeated when it is replayed.

Line-up tone is also used to set a common level on recordings that are to be used in broadcasting. Different standards are set for analogue recordings, digital recordings such as DAT that require more headroom in order keep clear of the digital maximum coding level, and the anomalous CD-R, for which the (known) recorded level may be higher, in order to match that on a commercial CD. Documentation kept with recordings should be clear on what alignment has been used.

As a practical guide to consistent standards, the following are maximum programme and line-up levels – in general, following AES/EBU recommendations – as adopted by the BBC and its suppliers for the most common media. As usual, the meter is a PPM on which a reading of '6' represents maximum programme level ('100% modulation') and '4' (the standard A + B = M zero level tone) is 8 dB below that.

Open-reel analogue tape recordings should have 20 s of 1 kHz line-up tone at the start. For full-track mono this will be at zero level. For twin-track stereo this corresponds to a level 3 dB lower on each leg, but M (the sum of A + B) will still read '4'. Follow this with 10 s of recorded silence, then 4 s of yellow leader immediately before the programme. At the end put 3 s of red or red-and-white trailer. If any additional material (such as retakes) is to go on the end, leave 2 minutes of recorded silence, then mark the extras with yellow spacers only – not red. Note that even though a standard alignment has been used for line-up, its level should still be noted in the documentation (and on the box).

Analogue multitrack tape should have, on all tracks in parallel, 1 minute of 1 kHz tone at 4 dB below peak level, then 1 minute each of 10 kHz and 100 Hz, both at 14 dB below peak level. If a noise reduction system is used, this requires more tone (check the instructions for the system used). Follow this with 30 s of recorded silence, then the programme.

On *DAT*, for which the maximum A + B programme level is taken to be 10 dB below the full scale, stereo (A + B) line-up tone will be set at 18 dB below full scale (or the individual A and B legs will be 21 dB below full scale). On replay these will be matched to '4' (or '3¼'). If the sampling frequency is set at 44.1 kHz (16-bit) to match that for CD, it should also be possible to disengage 'copy prohibit' in order to allow for an unlimited number of copies to be made. To allow quick-start replay, time code should be recorded continuously from the start of the tape.

Unless otherwise required, it is good working practice for a DAT tape to begin with 30 s of recorded silence, 30 s 1 kHz tone (set at the above level), another 30 s recorded silence, then a start ID: this will be the first and ideally only ID on the tape. The programme should follow half a second after the ID: this allows the ID to be used to set it up (unambiguously) for replay, and can also be used for quick-start or memory-start operation. Note that the programme (especially music) may begin with a second or two of atmosphere.

If further items must go on the same DAT tape, allow 15–30 s gaps, followed by consecutively numbered start IDs, but without any more line-up tone. If there is an 'auto-ID' function, switch it off, to avoid any confusing double IDs. Finally, set the cassette 'record protect' tab.

Digital multitrack tape also follows the rule of allowing 10 dB headroom on the actual recordings – but start the tape with 1 minute of 1 kHz tone at zero level (i.e. 8 dB below the standard peak level), then 30 s of recorded silence before the programme.

On *CD*, for which the maximum A + B level on a PPM needs to be 4 dB below full scale, stereo (A + B) line-up is set at 12 below full scale (or the A and B legs at 15 below full scale). On replay these match up to '4' (or '3¼'). CD-R is recorded with 30 s of 1 kHz line-up tone after 2 minutes of recorded silence *at the end* of the disc (so an 80-minute disc can hold only 77 minutes of usable material). Documentation should make it clear whether normalization has been used to lift the level to the maximum available.

With *MiniDisc* (MD), before recording, select for analogue or digital input and follow instructions in the manual for other pre-sets. There is a mode on the machine to name the item (but if it is fed by computer, it is quicker to use its keyboard). Check the level of the material to be recorded: this should have 10 dB clear headroom above peaks, as for DAT. No additional line-up is necessary. Set the 'record protect' tab.

If copying to MD from a digital edit (for example, in one common use for MD, of a broadcast trailer) run the play-out, then about a second before the required material crosses the play cursor, switch MD to record. For each successive track, press 'record' again to insert a new track number. At the end, switch back to 'standby' and wait a few seconds for the machine to complete writing additional information on the disc. There are procedures for trimming recordings ('topping' and 'tailing') if required. Machines vary, so check the instructions.

Note that, in addition to line-up, broadcasters generally ask for (or by contract, require) recordings to conform to rigorous technical standards with specified documentation, and will usually supply information on this on request. It is plainly in everyone's interest to ensure the best available quality where this is possible. For some suggestions (or requirements) about documentation, see Chapter 20.

Multitrack meters

In music recording studios, multitrack techniques ideally require separate metering of each channel that is being recorded. For this, there will be in-line displays on all channels, and meters may also be supplied for auxiliary 'sends'. Mechanical PPMs are unnecessarily complex, expensive and also difficult to follow in such numbers. If needle-pointer displays are used, the simpler VU meters are adequate, indicating which channels are not only faded up but also passing a signal. In America, these have sometimes been arranged in pairs, one above the other for each stereo channel, all across a large mixer.

A clearer option is *bargraphs*. These are vertical strips ranged side by side, presenting a graphic display that is much easier to follow. With processor control, they can be switched to show a variety of VU or PPM recordings, with stereo in AB or MS pairs, and may also have overload or 'hold' options. In one example, they can be switched over to show a frequency spectrum, analysed as one-third octave bands.

A bargraph display may take one of several forms. A strip of either LED or liquid crystal segments is perhaps the simplest. More flexible, in that its brightness can be controlled, is the plasma bargraph: this becomes brighter in the 'overload' range. But it also takes more power and costs more. A third type, displayed on a small screen, just simulates the bar.

For some channels that do not justify the cost of any kind of meter, an 'overload' light may be all that is needed – to give the simplest possible warning that the upper limit has been breached, so 'back off!'

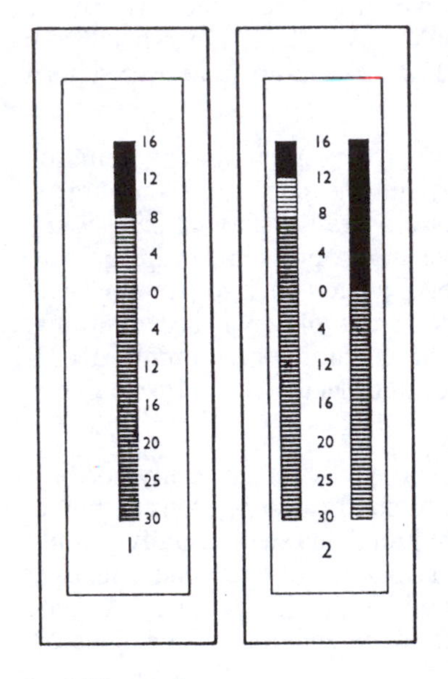

[10.4] **Bargraphs**
1, Mono. 2, Stereo (or any pair of channels). The simplest has LED segments. The more complex, bright plasma display can be programmed to offer a variety of calibrations, including the standard PPM and VU response. Above maximum volume (here +8 on the PPM), the plasma glow is brighter still, to signal overload.

Controlling stereo

Where the stereo broadcast is on FM, control is easy. Since the deviation (size) of the signal does not exceed the greater of the A and B channels, all that is necessary to avoid overmodulation is that these do not exceed their normal limits. The same is true for analogue recordings where the A and B channels are kept separate. For systems like these, it is necessary only to keep the separate A and B signals below the standard 'full modulation' mark.

In broadcasting, however, several problems may arise:

- in a large network some of the transmitters may be in mono;
- some of the listeners to the stereo broadcast have mono receivers;
- some items on the stereo service may be broadcast in mono only.

For practical purposes it is assumed that the average level for A + B is 3 dB higher than A or B on its own; accordingly, the stereo and mono parts of the system are lined up with a 3 dB difference in level. However, what happens to a dynamic signal is a little more complicated. There would be no difficulty if the signals added simply, so the A + B maximum volume really were never more than 3 dB above the separate maximum volumes for A + B, but in fact this happens only for identical signals, such as tone or a source in the centre of the sound stage. If A and B are both at the same volume but are actually different sounds, their sum varies from +0 to +6 dB. So where A and B are at maximum volume the M signal may be overmodulating by 3 dB. This should therefore be controlled for the benefit of mono systems.

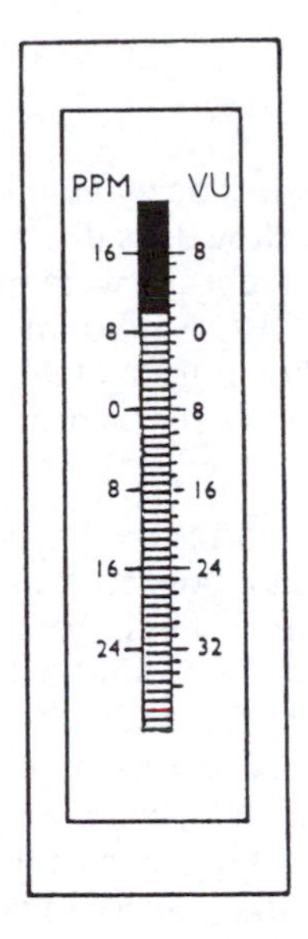

[10.5] **Switchable bargraph**
Here, zero level on the PPM scale is indicated by 0 not 4, as on the older, electromechanical PPM. Maximum volume is therefore +8, which corresponds to zero on the VU.

On the other hand, if the signal is being controlled primarily on the M signal, the full stereo signal can overmodulate. This happens when A + B is at its maximum and the whole of the signal is in the A or B channel, which is then 3 dB over the top.

A visual comparison of the M and S signal levels is less crucial, but can give useful information about the width of the stereo image. If S is close to M, the stereo is wide and a significant amount of information may be lost to any mono listener. If S is 3 dB lower, the stereo content is modest, and the width of the image is relatively narrow.

Programme volume: relative levels

A peak programme meter can be used to suggest or check levels. For example, if ordinary talk or discussion is allowed a normal dynamic range with occasional peaks up to the maximum, then newsreaders (who generally speak very clearly and evenly) usually sound loud enough with peaks averaging about 6 dB less.

A whole system of such reference levels can be built up; the table shows specimen PPM peak levels for speech and music that experience indicates are about right for listeners hearing radio in reasonably good conditions. Some more unusual instruments are listed as examples of how far some sound sources can depart from 'average'.

[10.6] **Suggested PPM peak levels** (dB relative to 100% modulation)

Talk, discussion	0
News and weather	−6
Drama: narration	−8
Drama: action	0 to −16
Light music	0 to −16
Classical music	0 to 22 or lower
Harpsichords and bagpipes	−8
Clavichords and virginals	−16
Announcer between musical items	−4 to −8*

*Depending on type of music
These recommendations predate the commercial era in which promotional material is habitually given maximum volume, but they do still offer guidance on what listeners might really prefer.

Even if such a list is adopted as a general guide, it does not solve all the problems posed by junctions between speech and music or, on the larger scale, between one programme and another. For radio stations with a restricted range of programming, there is a limited number of types of junction, and it is fairly easy to link them without apparent jumps in level. But for a service that includes news, comedy, magazines and discussion programmes, religious services and various kinds of music, problems of matching between successive items can be acute – and there is no rule-of-thumb answer. It is a question of judgement, of trying to imagine what the audience will hear when listening to your programme for the first and (probably) only time. Remember that your own perceptions are coloured by long acquaintance with the material: familiarity breeds acceptance. Remember, too, that the audience may not agree that your masterpiece should be displayed more prominently than anything around it.

A further slight complication is suggested by the results of a study of listeners' preferences for relative levels of speech and music. Taking a large number of pieces in different styles (but excluding modern pop music) and with different linking voices, the results were:

- speech following music should be (on average) 4 dB down;
- music following speech should be (on average) 2 dB up.

This apparent contradiction is easy enough to resolve: it just means that announcements linking music items should be edged in a little. The aesthetic use of faders for links like these is discussed in Chapter 14.

Maximum volumes

[10.7] **Preferred maximum sound levels** (dB, referred to 2×10^{-2} Pa)

	Public *M*	*F*	*Musi-cians*	*Studio mngrs* *M*	*F**
Symph. mus.	78	78	88	90	87
Light mus.	75	74	79	89	84
Speech	71	71	74	84	77

*M, male; F, female.

For a broadcast or to make a recording, the balancer monitors at a fairly loud listening level with full attention. How does this compare with most of the audience? In one experiment, checks were made on the maximum sound levels preferred by BBC studio managers, musicians and members of the public. As shown in the table it was found that studio managers preferred louder levels than the musicians, and very much louder than the public.

For classical music, the figures for the public correspond reasonably well to what they would hear at a distance in a concert hall. Musicians, normally much closer to the sound sources, might be expected to choose higher levels. But the studio managers (sound balancers) chose unrealistically high levels. One reason for such diversity is that those professionally concerned with sound extract much more information from what they hear, and the greater volume helps them hear fainter wanted and unwanted sounds, checking technical quality as well as the finer points of operations like fades and mixes. Musicians, however, draw on their experience of instrumental quality and musical form, and listen for performance, usually with less concern for technical quality.

Those preferred maximum levels refer specifically to the case where the listeners, like those concerned in creating the programme, do want to listen attentively and have reasonable conditions for doing that. But this is often not the case: for example, many people's early evening listening may be while driving home or preparing food. Where listeners' attention is limited and background noise may be high, speech may sound too quiet unless it peaks 8 dB higher than music. But if, when listening to a programme controlled in this way, you decide that a particular item of music appeals to you and turn it up, you will find that the announcement following it seems unduly loud. As a general rule, listeners can do little to combat the levels as transmitted.

Television sound in the home is often set at levels comparable to those appropriate to background radio listening. This may be satisfactory for some material, but if it is followed by a play with a wide dynamic range the average level drops, and may not be corrected by the viewer. The television sound supervisor (monitoring at high level) may have set relative levels, of speech to effects or singer to accompaniment, that are not appropriate to such quiet listening. Alternatively, the viewer may raise the volume, and then subsequent announcements and programmes (and especially commercials) are much too loud: indeed, they seem all the louder because it is irritating to have to turn it down. For the broadcaster, all this requires a fine judgement of compromises. Start this process by listening to part of a rehearsal at realistic domestic levels on small loudspeakers.

Linking together speech items of varying intelligibility presents a special problem, and particularly in the case where an item containing noise and distortion has to be matched to studio quality. Here, the least unsatisfactory solution is found if the noisy item is matched for loudness at the beginning and end, and lifted for intelligibility in between: the item might be slightly faded in and out.

Sadly, some readers will find that these suggestions about relative and maximum volumes are impractical in today's commercial environment, in which an overriding demand is for promotional material to be as loud as possible, however much this may annoy the audience. This inevitably distorts any ideal based on what the listener would prefer, and leads to a general flattening of levels. It also means that any attempts to apply aesthetically satisfying standards must be rather more subtle than we might wish. Try not to lose them completely!

In more practical terms, there is a further way in which listening conditions may vary. At the one extreme, there are the noisy listening conditions (in-car radio, for example) that demand high intelligibility and a fully modulated broadcast signal: this contrasts sharply with the quiet listening-room conditions and attentive audience (perhaps not as common as we might wish) that requires 'high-fidelity' sound with a wide dynamic range. The two sets of requirements are largely incompatible, and may best be served by different radio stations.

Dynamic range and compression

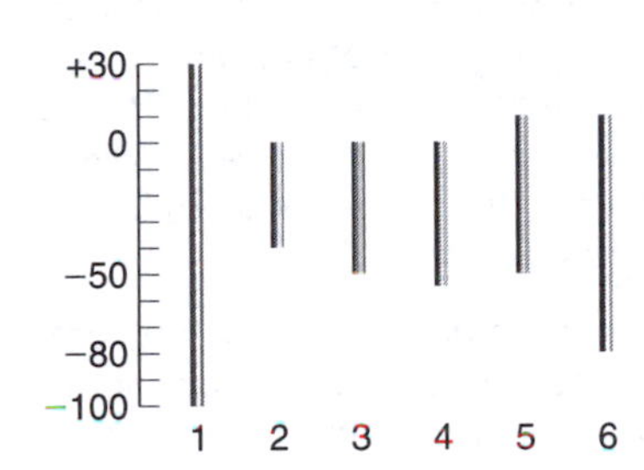

[10.8] **Dynamic range and the need for compression**
1, The dynamic range of music can fill the entire range of human hearing. The maximum ranges for some media are: 2, Videotape. 3, Vinyl record. 4, FM radio. 5, Analogue tape. 6, CD, DAT, etc.

The ear can accept a great range of volumes of sound. At 1 kHz the threshold of pain is 110 dB or more above the threshold of hearing – that is, a sound that is just bearable may be over a hundred thousand million times as powerful as one that is just audible. Studios or microphones with noise limits that are 10–20 dB above the ear's threshold level are good for a range of perhaps 100 dB or even more. Analogue recordings are limited to a signal-to-noise ratio of 50 or 70 dB, but digital recordings raise this to a maximum of 80–100 dB (which demands a noise reduction system). FM broadcasting, back down at 50 dB, was actually a huge advance on AM at 30 dB. Even that may be more than some audience preferences or listening conditions warrant.

Plainly, some compression of the signal is usually necessary. For this, the old concept of 'stops' remains useful. 'Taking it down a stop', a reduction of 1.5–2 dB on the fader, amounts to a drop in level that will be just perceptible to the operator, but which the audience will not associate with a hand on the control. A script or score can be used during rehearsal to note settings and other information that will be useful on the take. The standard procedure is roughly as follows:

Set the gain at a suitable average level for the type of programme, and adjust the loudspeakers to a comfortable volume. Judge divergences from the average level by listening to both relative loudness and intelligibility (remember that at times these two factors are at odds). Glance at the meter occasionally, mainly to watch for very quiet or loud passages. After recording, the effects of any high peaks can be rechecked, to discover whether there is appreciable distortion. For broadcasts, an occasional peak may be better than reducing the effective service area of the transmitter by undermodulating a whole programme; compressors or limiters deal with peaks big enough to cause technical problems. On speech programmes also check the meter occasionally in order to maintain the same average volume. The ear is not reliable for this: it tires.

At the end of rehearsal of a programme with recorded inserts, it may be worth checking the levels of the first few inserts again, to see if one's judgement of the right level for them has changed. It often has.

Compression of music

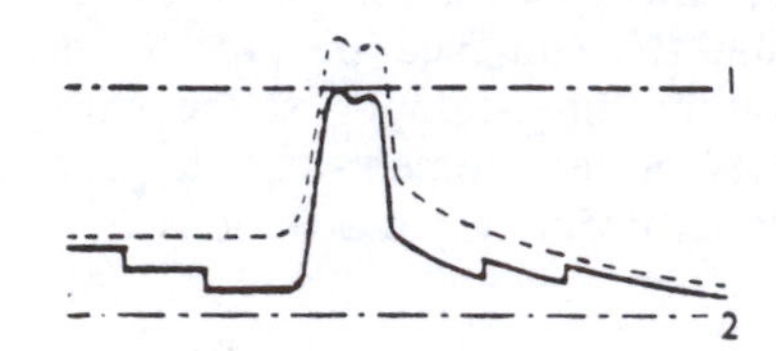

[10.9] **Manual control of volume**
1, Highest permissible level.
2, Lowest acceptable level. This method of compressing a heavy peak retains the original dramatic effect but ensures that the signal falls between the maximum and minimum levels. Steps of about 1.5–2 dB are made at intervals of, say, 10 s.

To reduce the extreme dynamic range of classical music, we should again use manual control – and also ears (both ears, for stereo), eyes (for meter and cue-sheet or perhaps a musical score and to observe actions that may affect sound) and brain, to anticipate events.

Avoid overmodulation on music that has loud and quiet passages, not by 'riding the gain' but instead by careful preparation, gradually pulling back on the gain control for half a minute or so before a peak is due, with subsequently a slow return to the average level. Each operation should consist of steps of about 2 dB. During rehearsal, appropriate settings – the start point, total compression and end – can be noted on a score. A similar procedure may be used to lift long periods of low level. It may also be necessary to vary the level repeatedly throughout a work, so that all the elements of light and shade fall within the acceptable dynamic range.

Compression of this kind is most likely to be necessary when recording orchestral works but, depending on the nature of the music and the closeness of the balance, control may also be required for smaller combinations or groups, or even a solo instrument such as a piano.

There is one other way of dealing with the control of orchestral music: simply leave it in the hands of the conductor. It has been known for a conductor to have a meter to check during rehearsal and recordings, but this is unusual. On the whole, it seems preferable that a conductor should concentrate on the performance, bearing in mind the limitations of the medium if that is possible, but leaving control to the balancer.

However, music specifically designed for background listening may have some measure of compression designed in to it by an arranger; and also, of course, for that, smaller musical forces will be employed.

The further the microphone is from sources, the easier it is to control music: studio or concert hall acoustics smooth out peaks. The balancer who adopts a bright, 'tight' balance has more work to do without necessarily earning the thanks of listeners – who might prefer a well-blended sound for classical music.

Control of music affects the reverberation just as much as the direct sound. The effect may be disconcerting: the microphone apparently moves closer, but its perspective remains the same. The two may be controlled more effectively by using two microphones (pairs or arrays), one close and one distant, and working between them on changes of music volume. Alternatively, reverberation microphones may point away from the orchestra: they are controlled independently to achieve the best effect as the acoustic volume of the music varies.

Compression: speech and effects

In controlling speech the problems are different, as are the methods employed for dealing with them. Studio interviews and discussions on radio and television also benefit by careful control, particularly if one voice is much louder than the others, or habitually starts off with a bang and then trails away. Here, control is a matter of anticipation: whenever the needle-bender's mouth opens, drop the fader back – sharply if in a pause, but more gently if someone else is speaking. A slight dip in level is more acceptable than the risk of a sudden loud voice. Laughter may be treated similarly if the general level is held well up. But edge the fader up again soon after the laugh.

Careful control should not be noticeable: do not miss the first high peak and only then haul back on the fader. No control at all is better than bad control.

Do not discourage speakers from raising their voices, unless they actually ask for advice; even then the best advice is 'be natural'. A natural but controlled programme is better than one in which voices are never raised and the meter never has any bad moments.

A good programme depends on variety of pace and attack. It should not be necessary to hold back overall level in order to accommodate possible overmodulation. But if editing is expected it is better to deal with the more violent fluctuations of level afterwards, or there may be difficulty even in a studio where extraneous noise can be kept low.

When recordings are made on location and subsequent editing is envisaged, it is better that a level should be set such that background atmosphere can remain constant throughout a sequence. Sometimes, gentle control can be exercised by slight movements of a hand-held directional microphone.

For scripted drama something like 16 dB may be an acceptable range for peak values of speech, the average level being about 8 dB below the maximum permissible. Even so, scenes involving shouting have to be held back, possibly making them seem distant, which in turn may not match a television picture. In cases where close shouting must be mixed in with more normal levels of speech, the actors should hold back on volume and project their voices.

This upper limit makes even more difference on loud, percussive sound effects such as a gunshot or a car crash. Even a sharply closed door is too loud. It is therefore fortunate that radio and television drama have established a convention in which such effects are suggested more by character than by volume. Peaks must be evened out and the overall volume held back. This explains why unlikely-looking methods for making sound effects sometimes produce surprisingly good results.

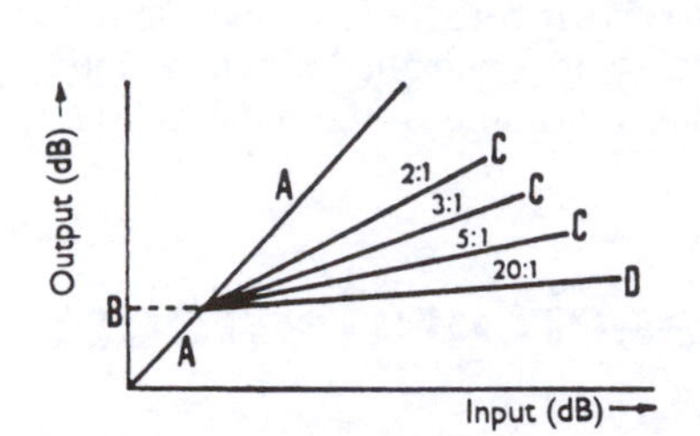

[10.10] **Compression ratios**
A, Linear operation: input and output levels are the same. B, Compression threshold. C, Various compression ratios reduce the output level.
D, Extreme compression: here it acts as a limiter, holding the maximum output volume close to the threshold level, but still permitting a small proportion of any high peak to pass.

Compressors and limiters

The full use of any recording or transmission medium requires that some passages are close to the maximum permissible level, and that they may momentarily or unpredictably exceed it. In such cases, the danger of analogue sound distortion (or, far worse, digital clipping) is replaced by a controlled (and usually less noticeable) distortion of dynamics. After most corrections have been made manually, the few that remain are 'cleaned up' automatically by a *compressor* or its more extreme form, a *limiter*. In popular music, these can also be used to raise the level of individual channels in a mix, or to make the whole recording denser or 'hotter'. Essentially, a compressor works as outlined below.

Below a predetermined level (the threshold or onset point) the volume of a signal is unchanged. Above this threshold the additional volume is reduced in a set proportion, e.g. 2:1, 3:1 or 5:1. For example, if a threshold is set at 8 dB below the maximum and 2:1 compression is selected, signals that previously overmodulated by 8 dB would now only just reach full modulation. Similarly if 5:1 had been chosen, signals that previously would have over-peaked by 32 dB now only just reach the maximum. In practice this means that the overall level can be raised by 8 dB in the first case or 32 dB in the second. Compression allows relatively quiet components of a complex sound to make a much bigger contribution than would otherwise be possible.

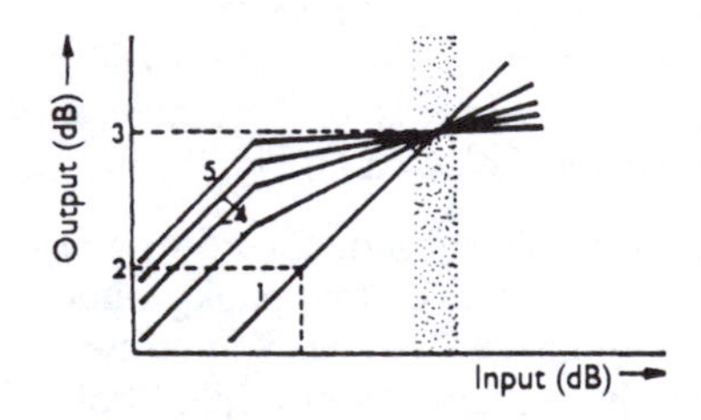

[10.11] **Lifting low levels with a compressor**
1, Original signal. 2 and 3, Lower and upper limits of required dynamic range. With the compressor out of circuit, only a small proportion of the input goes into this range. 4, With the signal lifted, increasing degrees of compression allow more to be accommodated. 5, The effect of a limiter. The greater the compression, the less the effect of overmodulation (above 3).

If the compression ratio is made large, say 20:1, and the threshold is raised close to the maximum permissible, the compressor acts as a limiter. It can then be used *either* to hold back unpredictable high peaks *or* to lift the signals from some lower level into the usable working range of the equipment. Note, however, that high-level signals may be excessively compressed and that background noise will be lifted as well. Reduction of noise in gaps in useful material (noise 'gating') is described at the end of this chapter.

The effect of a 2:1 compressor with a threshold at 8 dB below peak maximum is to compress only the top 16 dB of signals while leaving those at a lower level to drop away at a 1:1 ratio. An interesting variation on this would be to place two limiters together, working on a signal in parallel but with the threshold set at different levels. If one is set for 2 dB below 100% and the other for (say) 24 dB below, but raised in level, this introduces a variable rate of compression that is greater at the lower end of the range than near the top, where

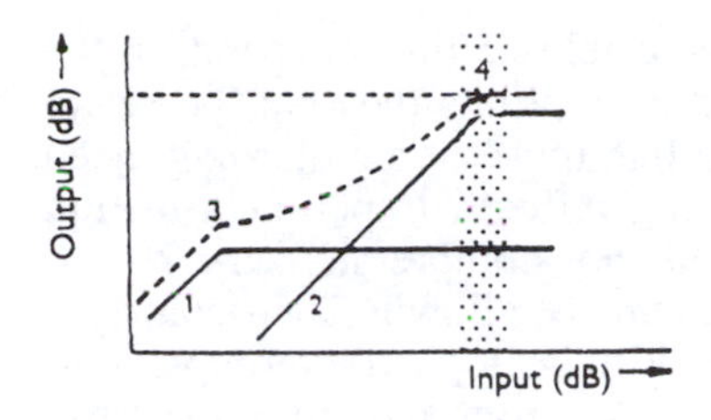

[10.12] **Compression by two limiters**
1, Effect of a low-level limiter.
2, Effect of a high-level limiter.
When the two signals are combined, the compression characteristic is different from that of a single-stage compressor, which would give a straight line from 3 to 4 and beyond.

something approaching the normal dynamic range is used until the upper limiter takes over. Another option is for the lower limiter to have a weighting network to arrange that quiet high- or low-frequency signals are lifted more than those whose power is mostly in the middle (500–4000 Hz) of the audio range.

Expanders operate as compressors in reverse, allowing further fine control of dynamics. Used on music, they thin the sound and selectively emphasize intermittent peaks. More often they are set to operate only below a selected threshold, so that useful sound above that level is unchanged but quieter sound in the gaps, which may include unwanted noise, is reduced. Expanders can also be used for more technical reasons, for example as part of the *compander* of a radio microphone, where the signal may be compressed before its local transmission, then expanded in the receiver to restore the original dynamics.

In large consoles (especially those used for popular music), a range of these controls is included in a dynamics module in line with each channel fader (or, in 'assignable' consoles, are available for programming it). In simpler mixers they may be offered only on output and, if needed for individual (usually mono) channels, are inserted into the signal routing.

Compression using digital circuitry is very versatile. One device has controls that can be set for expansion, compression or dynamic inversion. The expansion ratio starts at 1:10 then proceeds by steps to 1:1, after which it becomes normal compression. The scale continues to infinite compression (with a ratio of 1:0, a 'perfect' limiter) and then the ratio becomes negative, so that quiet sounds are made louder and vice versa. This carries the compressor principle into new realms, as an instrument of creative sound treatment and synthesis.

Compressors and limiters are also available as digital 'plug-ins'. Some programmes offer versatile high-quality effects, while others use the versatility to recreate the peculiar characteristics of early valve (vacuum tube) designs that were used in old pop music recordings.

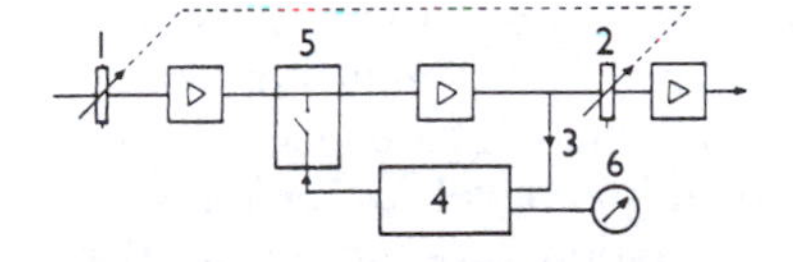

[10.13] **Simple compressor/limiter**
1 and 2, Threshold level control (ganged). 3, Sample signal fed from the main programme chain.
4, *Feedback* side chain with compression ratio and recovery time controls. 5, Control element operating on the main programme chain.
6, Meter showing the degree of attenuation. Alternatively, if a small delay is introduced into the main chain, a side chain taken off earlier can intercept peaks in the delayed signal before the first transient arrives: it operates in *feedforward* mode. Although this can be made to work on an analogue signal, it is easier when the signal is digital.

Compressor design and use

Compressors and limiters work by sampling the signal at a particular point in the chain, and if it is above a designated level, deriving a control signal that is fed back through a side chain to reduce the overall volume in the main programme chain. It is usually arranged that this control signal operates with a fast attack and slow decay (also called release or recovery) – though both 'fast' and 'slow' may have a wide range of actual durations, with some controlled by the balancer. A typical (compressor) attack may vary from 0.1 to 30 ms, and the decay may take from 0.1 to 3 s. Other controls include threshold and compression ratio, and there may also be a meter to show the degree of attenuation.

In popular music, the ideal attack and release times depend on the instrument. For staccato sounds such as drums they can be set to operate quickly; for sustained tones like those of a guitar they need to be much longer, to avoid 'pumping' effects. Experiment to find the best settings for each instrument. For compression, try 2:1 or, say, 4:1 on individual microphone channels, allowing their contributions to the mix to be boosted. It can also be applied to groups, or to the master control to make the final mix denser and more powerful. But beware: even for popular music, use automatic compression with caution. What can add life can also take life away.

In some compressors the attack time can be reduced to a few tens of microseconds. However, such a very sharp change of level is liable to introduce an audible click or, short of that, to modify the transient of a musical sound enough to change its character. The alternative is to delay the whole signal for about half a millisecond, then make a gradual change of level over the same period, in advance of any sudden heavy peak. The transient problem is largely eliminated, but any audible background still changes level abruptly, which is likely to be noticeable and therefore objectionable. The quality of its unpleasantness is also affected by the rate of decay of the control signal. Heavy limiting, or strong compression with a moderate recovery time, again produces mechanical and ugly pumping effects.

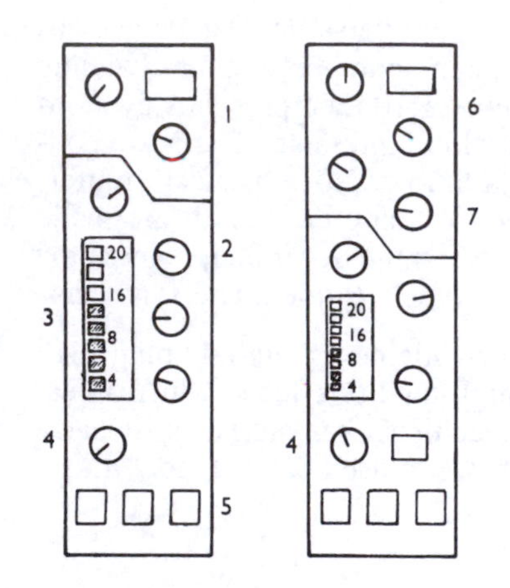

[10.14] **Compressor/limiter and dynamics modules**
Examples of in-line channel controls. *Left*: Mono module. 1, Limiter threshold and release (decay time) settings and 'gate' key. 2, Compressor threshold, compression ratio, and attack and release settings. 3, Meter showing gain reduction due to compressor or limiter. 4, 'Make-up gain' amplifier setting. 5, Switches for operation as compressor, limiter, or under external control. *Right*, Stereo module. 6, Expander controls. 7, Compressor controls. Compressor/expander modules may also be available for fine control of popular music dynamics.

Early limiters had gain-recovery or control-signal decay rates of the order of seconds: this allowed a reasonable time for the overload condition to pass so that it would not be pumped by too rapid a train of high-level signals. The more sophisticated limiters (described above) have a selection of decay times available, typically 100, 200, 400 and 800 ms and 1.6 and 3.2 s. The fastest of these can hold isolated brief excessive peaks without having a severe effect on the background, but they are not so satisfactory for a rapid series of short peaks: any background is given an unpleasant vibrato-like flutter.

If there are only occasional peaks, and they are not too big, say not more than 3 dB or so, the recovery time is probably not critical: something like half a second may be adopted. The exact choice of recovery time becomes important only at high levels of compression. An alternative is to allow automatic variation between about 30 ms and 10 s, varying with the level and duration of the overload signal.

'Gating' background noise

A major problem with limiters and compressors working on high gain is that background noise is lifted too, unless something is done to stop that. In particular, when there is no foreground signal to regulate the overall level, that is, during a pause, the noise is liable to rise by the full permitted gain. Such noise will include ventilation hum and studio clocks, movement and other personal noises, traffic, electronic or line noise, and so on. Even where such sounds are not raised to an objectionable level, certain consonants at the start of subsequent

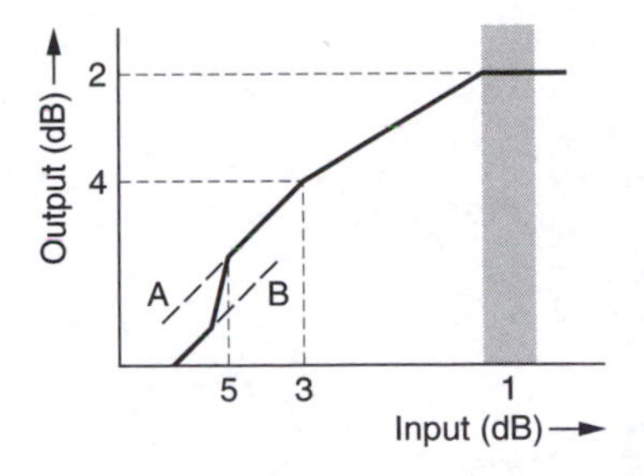

[10.15] **Noise gate**
1, Range of maximum expected input level. 2, Maximum output level.
3, Lowest level of useful input.
4, Lower end of desired dynamic range. 5, Gating level: if level of input signal drops below this, the output falls from level A to level B, so that noise in gaps between useful material is reduced.

speech may not be powerful enough to trigger the compressor. The most unpleasant result is excessive sibilance at the start of some words, though this can be avoided by high-frequency pre-emphasis in the input to the limiter: this is sometimes called 'de-essing'. It can also be used to reduce sibilance on voices that naturally have too much, or where a microphone with an inappropriate response has added it.

One device that can be used to discriminate against noise is a *noise gate*, a specialized use of the expander in a dynamics module. In a gate, when the sound level falls below another, lower threshold setting, the gain is allowed to fall to a parallel lower input/output characteristic: this might, for example, be 8 dB below that which would apply without a gate. At this setting, a device that could not previously have been used for more than 8 dB compression on a particular signal without the background noise becoming excessively loud in pauses could now be used for 16 dB compression.

Noise gates are often made to work very quickly, so that even in tiny pauses between the syllables of a word noise is sharply reduced. For this to be successful, the settings must be chosen with care: the gate must be *below* all significant elements of speech, but *above* all the underlying noise. If these conditions cannot both be met, the result can be unpleasant, with either speech or the noise wobbling up and down on the edge of the gate. In any case, either extraneous staccato noises at a higher volume or a rise in the background to an unexpectedly high level will fool the gate, as also (in the opposite sense) would a very quiet passage at the start of a piece of music.

In the 'automatic level control' available on some high-quality portable recorders used for professional radio and film location work, yet another type of limiter is used: one that avoids most of the rise and fall of background noise that is heard with other limiters. A short delay is introduced before the control signal begins to decay. The duration of this delay can be adjusted to allow for different lengths of pause between syllables or words as spoken by different people or even in different languages, but is normally 3 s, after which the 'memory' is erased and a new level of adjustment selected. This same recorder has a separate circuit for the control of very brief transients, which do not therefore affect the basic level. Obviously, such a device still has limitations: there would undoubtedly be many cases where a human operator would adjust the level, for example on a change of voice, so it will probably be used only where it is inconvenient to have an operator. Like the fast-operating gate, the device can all too easily be fooled by unexpected events.

Chapter 11

Filters and equalization

The word 'equalization' reflects its historical origin, from when its main function was to restore inadvertent changes to frequency response. These were among the biggest obstacles to be overcome as audio engineering has developed. Microphones still often introduce marked changes, differentially responding to the frequencies present. Acoustics, too, may require correction.

For television drama, a range of different microphones may have to be used to pick up a single voice as an actor moves around a set, and the varied responses must be matched to some standard, perhaps of a main boom microphone, or the transitions will be obtrusive. This demands a capacity for changing each microphone's frequency response in detail. So, when linking a film documentary, does the jump between a personal microphone on location and a studio microphone for commentary. 'Equalization' rediscovers its original meaning when applied to the correction of unintended changes like these.

Today, equalization (EQ) is used just as much for the creative manipulation of frequency response. In general, it is pop music balance that makes the greatest use of EQ, with bass and treble emphasized or rolled away, and selected bands of mid-range frequencies lifted to lend presence to particular components of the mix. If frequency bands containing the characteristic elements of musical quality of each instrument are raised, the result is a kind of musical caricature or cartoon. In fact, that is precisely the object intended – and when it succeeds, the selected essentials do, indeed, seem to press themselves upon the listener. Note that effects of this type may also involve the use of compressors, as described in Chapter 10. At its best, selective EQ, sometimes allied with compression, allows contrasting sounds to blend together better, without loss of individual

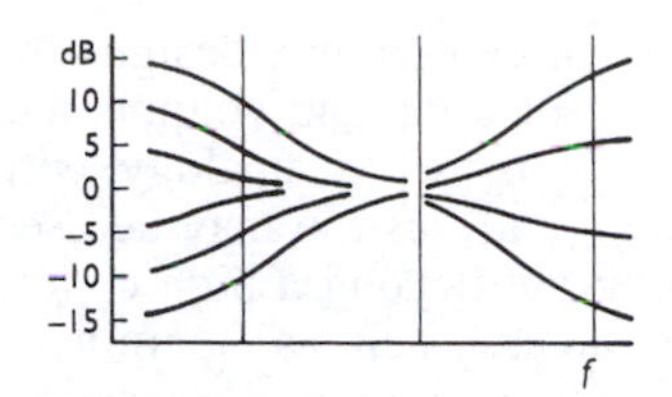

[11.1] **Simple bass and treble control**
The effect of some domestic 'tone' controls.

character. At its worst, everything gets pushed up and the result is no better than simply raising the main gain control – which is pointless.

In this chapter we are concerned with equalization, not just as a means of correction, but also as a creative technique. Many examples of the use of EQ in music have been given in Chapter 8, where EQ was included as a component of music balance. Here many of the examples are of the effects of EQ on speech, and particularly its use in drama.

Filters

The simplest kind of filter – a passive filter using one resistor and one capacitor – cuts just the bass or top, rolling off at 6 dB/octave above or below a given frequency. At the nominal cut-off frequency the signal is 3 dB down. A band-pass filter has two filters and the designated band lies between points that are 3 dB down at each end. Steeper roll-off at, say, 12 or 24 dB/octave requires more complicated networks. Active filters have amplifiers built in, to lift peaks to higher levels or make up for losses. Boosting very low frequencies, say below 100 Hz (below 60 Hz for subwoofers in larger auditoria), adds the power sometimes felt in surround sound. For some instruments, boosting lower frequencies can add warmth. Boosting very high frequencies, above 5 kHz, can add brilliance – or, more likely, harshness.

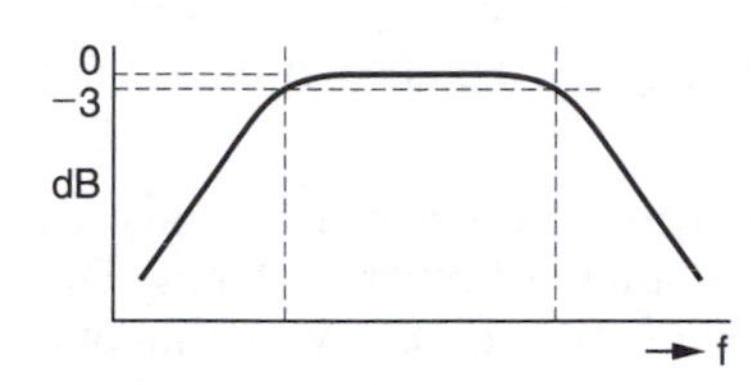

[11.2] **Band-pass filter**
This is a combination of high- and low-pass filters. The band is defined as the range out to −3 dB down at either end.

Variations on these effects require greater complexity. An active *midlift* filter with a broad or narrow peak is used for *presence* and is a vital part of pop music balance. The middle of the audio range may be taken to be about 256–440 Hz, but presence, more often applied a few octaves above that, acts on the harmonics of the notes produced, strengthening and extending them. As different notes are played by an instrument, the presence frequency remains constant, so it behaves rather like a physical part of that instrument, creating a formant – or, more likely, enhancing a formant already present that helps define its character. Midlift in the 2–5 kHz range enhances the clarity or definition of many instruments such as a snare drum or woodwind.

Vocal formants in the middle of the audio range shift continuously in the process of speech, but the higher formants are caused by resonance in cavities of less flexible or fixed size. For presence on vocals it is usually the upper range that is enhanced, perhaps by 6 dB somewhere in the 2–5 kHz range, and this can also add intelligibility.

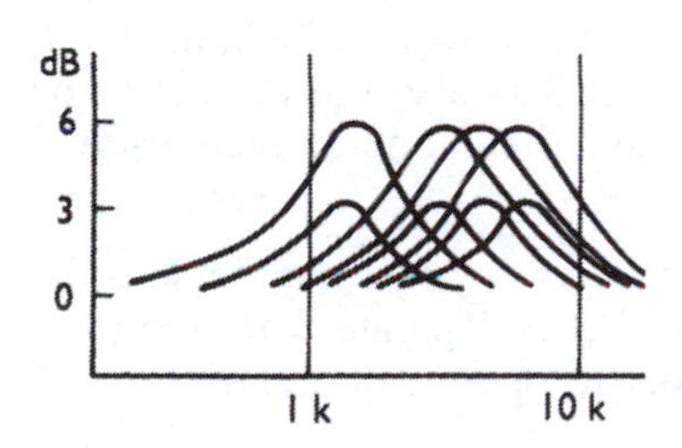

[11.3] **Simple 'presence' boosting**
The effect of an early two-stage, four-band midlift system.

Note, by the way, that the recommendations here and elsewhere in this book are necessarily imprecise, since the final effect must always be selected by ear in the light of particular qualities of each instrument, the relationship of one to another, and the overall melodic or harmonic effect desired. Also note that EQ can also all too easily be overdone.

Early presence devices had a limited range of possibilities, offering perhaps four or five nominal frequencies separated by intervals of rather less than an octave, together with several degrees of lift, switched at increments of 2 or 3 dB. In more complex designs the range is more flexible in both frequency selection and volume, and will also allow for a dip in place of a peak: by analogy with 'presence' the effect would be '*absence*', and this too may be used in pop music mixes. *Shelving*, flattening out beyond a high or low filter, can be used to hide problem areas, such as boominess, scratchiness or sibilance. It can also help to improve separation, reducing an unwanted range picked up on the wrong channel. For this kind of filter, the signal may be diverted into a 'shift register', in which there is a delay of the order of microseconds that is different for every frequency present: the longer the wavelength, the greater the delay. When recombined in different strengths with the original, the changing phase differences produce the desired 'shelf' effect.

Top-cut
Bass-cut
Top lift or cut with shelving
Bass lift or cut with shelving
Presence lift or dip
Presence lift with variable Q

[11.4] **Some commonly used EQ symbols**

A variable in more flexible systems is the narrowness (known as the 'Q') of the peak or dip. In most consoles today, a group of controls for bass and treble cut (or shelving) and several degrees of presence is provided on all of the individual channels, mono and stereo, and also in group fader channels. If 'Q' is marked on a control or quoted in a specification, this is calculated as the centre frequency divided by the bandwidth (between frequencies where it is 3 dB down). A Q of 1 covers more than an octave; a Q of 10 picks out individual tones.

On a complex analogue console, each mono channel may have as many as a dozen rotary controls and a variety of other switches. The high- and low-pass filters remain relatively simple, with (in one example) an 18 dB/octave cut at a choice of ten turnover points from 3 kHz up and 12 dB/octave cut at another ten from 350 Hz down. The equalizers operate in bands of 1.5–10 kHz, 0.6–7 kHz, 0.2–2.5 kHz and 30–450 Hz, each with a range of peak and shelf settings. In addition, two in the middle have variable Q. There is an overload indicator light, monitoring for this at three points in the EQ module.

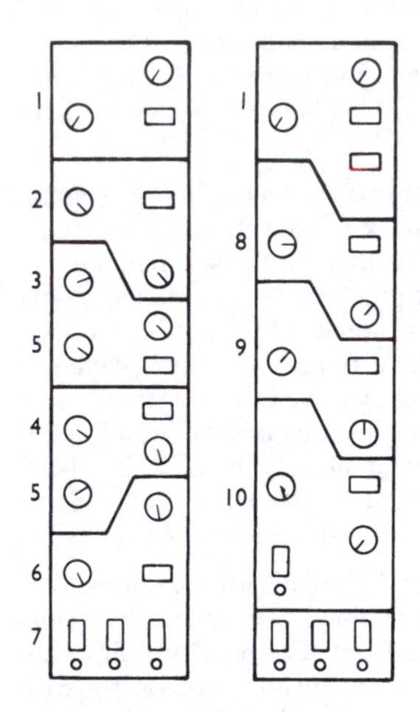

[11.5] **Equalization modules**
In-line channel controls on a console designed for mixing popular music. *Left*: Mono channel. 1, High- and low-pass filters. 2, HF equalization. 3 and 4, High and low mid-band equalization. 5, Variable Q (bandwidth). 6, LF equalization. 7, Routing switches. *Right*: Stereo channel. 8–10, HF, mid-band and LF equalization. All equalizers have a wide choice of centre frequencies and boost or dip levels.

The stereo modules have fewer controls, but each is available over a wider range; and overload is indicated separately in the two channels.

If the signal passes through a filter network with everything set to zero, an analogue-controlled setting can never be quite as smooth as the original, so it is usual to include a bypass switch which, in effect, switches the whole module on and off. It is also convenient to be able to switch any filter in or out, so that direct and corrected sound can be quickly compared.

In some computer-controlled systems, the settings may be selected and assigned from a central module that is separate from the individual channels, but stored in memory to operate on them.

'Notch' filters (with very high but usually undefined Q) are used to eliminate sharply defined whines or whistles such as the television line frequency at about 15 kHz. A dedicated hum eliminator takes out mains hum at 50 or 60 Hz, together with its main overtones. Such specialized filters are not needed on every channel, so are patched in at an insert point when required. Increasingly, digital signal processors are used; these have a wider and more flexible range of functions and can be tuned far more precisely than their analogue predecessors.

When used for EQ, digital signal processing programs often emulate the naturally smooth curves of analogue EQ (and often of existing analogue EQ characteristics), but with the potential for greater versatility.

Graphic equalizers

[11.6] **Graphic filter**
In one design the slide faders are at one-third-octave intervals, and control the gain or attenuation in a band of about that width. In principle, the overall effect should match the line shown on the face of the module, but this may not be achieved in practice. The result should be judged by ear.

Even in combination with bass and treble filters there is a limit to the flexibility of equalization based on individual presence controls. In a television studio or when filming on location, the degree of mismatch between microphones can be more complex. One approach has been to take the principle of midlift further and split the whole spectrum into bands, typically about 30, each of one-third of an octave, and apply correction separately to each. If all controls are set at zero, to make a horizontal line across the face of the panel, the response should also be level throughout the audio range; in practice, the actual line may be distinctly bumpy and there is a surprising amount of difference in performance between units from different manufacturers. In any case, the unit should be switched out of circuit when not in use.

If individual sliders are raised or lowered, their associated circuits form peaks or troughs that overlap, each blending into the next. The selection of the desired effect is made by setting the slides for the various component bands to form what appears to be a graph of the new response. In fact, the actual response achieved may be different: it depends on the design of the filter and the size of the fluctuations set. The true arbiter – once again – must be subjective judgement.

Equalizers of this kind are plugged (patched) into the channels where they are needed. So, too, would be the digital alternative. One 'plug-in' stores 'models' of many present-day and older microphones, together with the effects of windscreens, polar response, proximity effects and switchable filters on their frequency response, so that by entering source and target microphones, one can be 'converted' to the other.

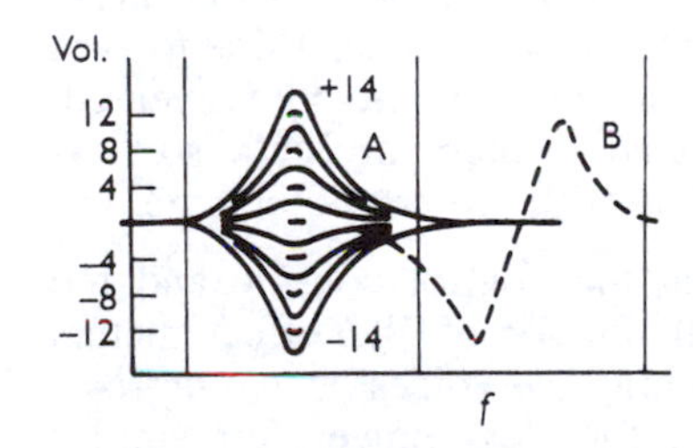

[11.7] **Audio response control**
Ideal effect (A) of a single fader and (B) two adjacent faders set at opposite extremes. In practice, sharp jumps in the response can only be achieved at the expense of a more erratic response when the sliders are set for a smooth curve. Apparently similar units may differ in their capacity to compensate for this.

Graphic filters have been used in film dubbing theatres, on television control consoles and more generally for any creative sound application. When speech is put through one, it is possible to make severe changes, for example to introduce several dips of 30 dB without much reducing intelligibility. If this is intended not just to restore the deficiencies of a balance made in adverse conditions, but instead to achieve a positive and marked dramatic effect, the imposed distortion has to be drastic. However, the selective emphasis of particular frequencies is noticeable where a dip might not be. The brain reconstructs defective harmonic structures, perceiving sounds as experience says they should be, but finds it difficult to disregard what should not be there, but plainly is. Given that peaks are more noticeable than dips, a useful strategy for locating and reducing distortion or noise that affects a limited but unknown frequency range is to lift the faders successively, until the problem sounds worse, then reduce its setting to below the median.

Telephone effects

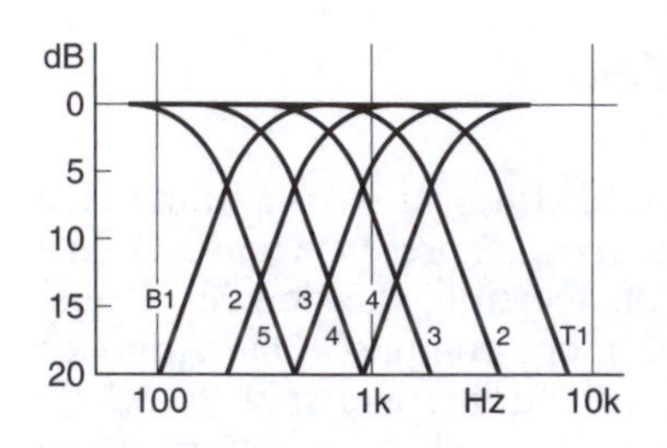

[11.8] **Simple telephone effects unit**
The example has four settings for bass-cut (B1–4) and five for treble-cut (T1–5).

Telephone circuits often have a bandwidth of some 3000 Hz with a lower cut-off at about 300 Hz, but this should not necessarily be copied when simulating telephone quality. Settings for bass and treble filters should, as always, be made by ear. How much is cut may vary even between different simulated telephone effects in the same presentation, just as, in life, the amount of the lower frequencies can change with how closely a telephone is held to the ear.

A band-pass filter is used for telephone effects, but there is no set range for this. It depends on differences between male and female voices, and between different examples of each. If the fundamental of a male voice is at 110 Hz, then bass-cut at 220 Hz will give a small but noticeable change of quality. For a female voice an octave higher, 440 Hz bass-cut is more suitable. In practice, 440 and 880 Hz may be better still: the loss of bass is stressed for dramatic effect. The top-cut will be closer to that of the telephone circuit. A microphone used like this may be called a *filter microphone* or perhaps a *distort*.

In a full-scale studio telephone conversation set-up it is usual to have two microphones, one normal and the other with the filter. In radio, these two can be placed in different parts of the same studio, but not separated so much that the actors cannot hear each other direct. This arrangement avoids them having to wear headphones.

It is important to avoid acoustic spill, which in this case means pick-up of the supposedly filtered voice on the normal microphone: otherwise the effect of telephone quality will be lost. If the 'near' microphone is directional, and it is placed with its dead side toward the 'far' voice, the pick-up of direct sound is effectively prevented. Discrimination against reflected sound is more difficult, so it is better to use a dead part of the studio for the filter microphone.

The two microphones are mixed at the control console and the relative levels are judged by ear. With this sort of filter effect, meters are even less use than usual in matching the voices, as the narrow-band input has less bass, and therefore less power, for similar loudness. At normal listening levels there is little difference in intelligibility between full-range speech and the narrow band used to simulate telephone quality, so a fairly big reduction in the loudness of the filtered speech may be tolerated in the interests of creating a feeling of actuality.

In a television studio the two microphones should be placed well apart. Then, as the picture cuts from one end to the other, the normal filter arrangement can be reversed without spill either way. If the

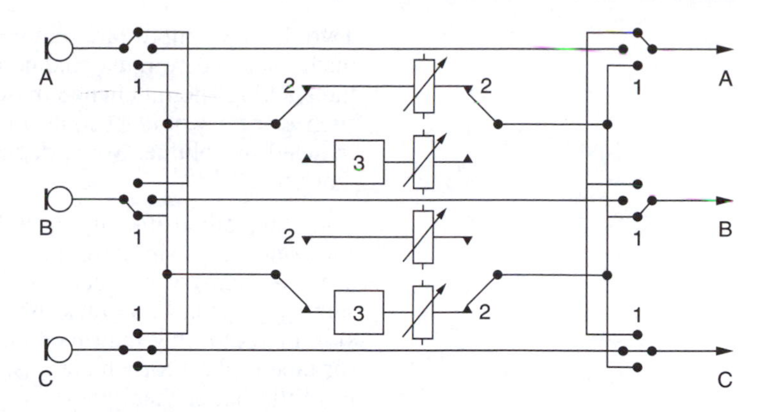

[11.9] **Telephone effects switching unit for television**
1, Pre-selector switches. 2, Switch linked to video-cut buttons. Any sound routed through this is automatically switched in and out of the filter condition when cutting between pictures. 3, Bass and treble filters. In this example, microphone A has full sound only while B is filtered and vice versa, and microphone C remains the same whatever the picture.

selection is linked to the vision cutting buttons the sound and picture changes can be synchronized. But it is best not to cut in the middle of a word as this sounds unpleasant and draws attention to the technique.

Note that as distorted sounds must be held at a lower level than 'clean' sound, care must be taken not only in balance, but also before that, in studio layout. If the two performers cannot be acoustically completely separated, because both must appear in vision and must therefore be in the same studio, it should be arranged for the two settings to be as far apart as possible. If they are not, coloration can affect both voices, as distant, reverberant but otherwise undistorted sound is superimposed on the telephone quality speech. Also, if one performer speaks louder than the other, the loud voice is likely to be heard on the quiet speaker's microphone. It may help for the actors to face each other in the studio layout, so that directional microphones point away from each other. But note that the use of wide shots makes it worse, because in those the microphones are farther away, so it more likely that the distant voice will be heard as the channel gain is raised.

Background effects added to a telephone conversation can make a big difference to the way voices are matched. The degree of filter and its volume relative to the clean speech must depend on both the loudness and the quality of the effects. In particular, 'toppy' background sound tends to blot out quiet speech, but if the backing is woolly or boomy the shriller telephone effect cuts through it. A complicated scene may require varying intelligibility, as for example in this radio script:

EFFECTS	(*Loud party chatter background; phone dialling close, ringing heard on line: it stops.*)
GUEST	Hey, is that you, Charlie?
CHARLIE	(*On filter, barely audible against party chatter.*) Yes. This is Charlie.
GUEST	Say, this is a great party, why don't you come on over?
CHARLIE	What are you at, a football game or something?
GUEST	What am I what?
CHARLIE	I'm in bed.
GUEST	Hang on, I'll shut this door. (*Door shuts; chatter down.*) S'better! Now wha's all that about a football game? It's three in the morning.
CHARLIE	(*Now more distinct.*) Yeah. Three in the morning. I'm in bed.

Two different mixes are required for this, and the balance and pitch of the near voice must change on the cue 'door shuts'. In addition, the backing should change in quality too: heard through a real door, such effects would become muffled and muddy, as well as being reduced in volume. Some degree of top-cut might be introduced at that point.

There are other dramatic uses for filters besides telephone quality: for example, to simulate station or airport announcements (adding a touch of echo) or intercom (at the same time simulating a poor radio link by playing in a crackling, hissing background). A film, television or radio play may include the effect of a radio announcement (or tape replay) for which a useful convention is that the reproducer has little better than telephone quality. This again is a caricature in which one type of distortion, reproduction of sound through a single loudspeaker, is represented by another, limitation of frequency range.

Chapter 12

Reverberation and delay effects

This could be a very short chapter. You want reverberation? Just buy an artificial reverberation (AR) device, put it in a channel and power it up. Do you want really long reverb? Turn the knob to, say, '5.5 seconds'. Maybe also scan through a seemingly endless list of acoustic environments to reach 'cathedral'. Plug it across the channel you want to reverberate – and you are in business. You are talking, walking or playing your trumpet – in a space like a cathedral. And, just like it says on the wrapper, the quality is amazingly realistic. The same goes for other 'delay' effects. Just plug in the 'reverb' device, check the instructions and off you go!

Digital AR and delay-related effects are the ultimate black box success story – particularly when compared with some of the alternatives.

The longer version of this chapter does just that. It explains what AR tries to emulate and describes some of the earlier more-or-less successful attempts to solve a seemingly intractable problem. In the past, various forms of artificial reverberation have been used, including echo chambers, plates and springs. These all differed in quality from one another, and from the real thing. Interestingly, despite daunting limitations, some really do work well, and others that you might find gathering dust in a storeroom have on occasion been disinterred and brought back to life to show off their strange 'retro' character. Rather more easily you could run down to 'plate', just one of a hundred descriptions offered on the AR list, retained for the benefit of those who recall and can still find a use for its distinctive response. But now that artificial reverberation can be constructed to mimic the characteristics of natural acoustics so much more accurately, most users, for most purposes, will now prefer to head for a rather more alluring environment – if only as their starting point.

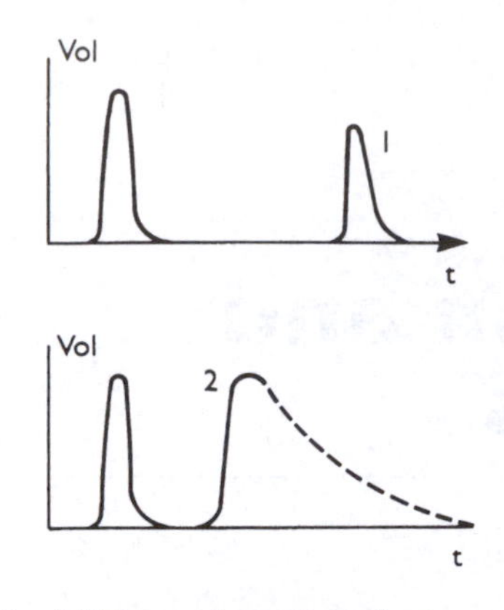

[12.1] **Echo and reverberation**
1, An echo is a discrete repetition of a sound after a delay of 20 ms or more. 2, Delays shorter than 20 ms (which are associated with the dimensions of all but the largest halls) are perceived as components of the reverberation.

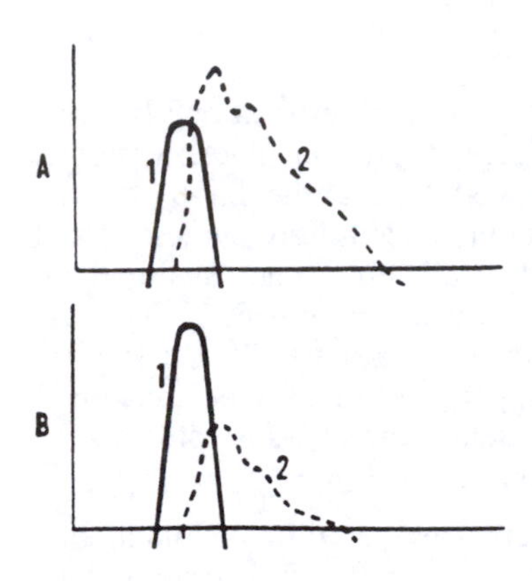

[12.2] **Echo mixtures**
1, Original signal. 2, Reverberation. With separate control of direct sound and artificial reverberation, a wide range of effects is possible.
A, Reverberation dominates the mix.
B, Reverberation tails gently behind.

This chapter covers the techniques for using AR (or even real reverberation) and also a related group of *delay* effects, in a wide range of applications. Artificial reverberation, too, can be an effect in its own right, but is more often used when more (and perhaps also better) reverberation is wanted than the built-in acoustics of a studio can supply. For music it is used to enrich a dry environment, to rebalance the contribution of spotting microphones, or in a close multimicrophone balance to supply a vital component of the mix that would otherwise be missing.

The term 'artificial reverberation' has at times been shortened to the more convenient '*echo*', but that is something of a misnomer for a studio technique that extends reverberation without, it is to be hoped, introducing distinct repetitions. A true echo may be heard only when reflected sound is delayed by a twentieth of a second or more. (How short the delay can be depends on how sharply defined the sound is.) As separation is reduced, there comes a point when even a staccato original and its repetition are no longer perceived as separate entities, but become a single extended sound, a composite that has changed in character. There is, in fact, partial cancellation at some frequencies.

Delay effects may fall into either category: they may be short, to make use of the phase cancellations; or long, producing perceptibly separate echoes, perhaps with further repetitions. They can also be used simply as a time-shift device, for example, to move a close-balanced component farther back, so it does not stick out in front of a widely balanced whole. In this case, delay will be combined with AR. Delay-related effects also include pitch and/or duration changers.

AR feeds in the console

Because AR and delay-related effects devices are nearly always digital, when used with analogue consoles they need A/D conversion before the process can begin, and D/A conversion to feed the result back into the signal. This is all quite routine: the converters are part of the 'black box' that contains the device. But the box in question generally lives outside the console and is plugged in to the channel (or group of channels) as required. A great variety of these devices is now available, offering a range of AR quality and an even greater range of other delay-related effects that may come with the AR or separately packaged. Inevitably, this kind of device is liable to be more expensive than a purely digital program in a digital console (or for the computer at the heart of a 'virtual studio'). This is because of the cost of all the hardware (the box) plus the additional A/D and D/A stages.

In fact, the device is just a computer in disguise – together with its built-in sound card (which does the converting). AR and delay require a processor and some random access memory (RAM). Digital samples of the data (the sound, now converted into pure information) are fed to memory, then after a pause for as long as the sound would take

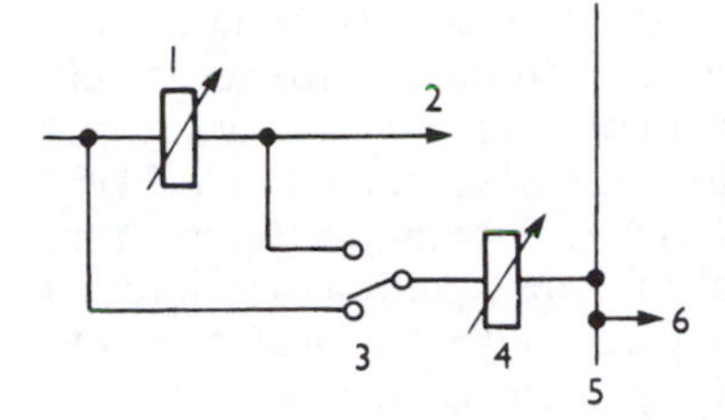

[12.3] **Flexible 'echo send'**
1, Channel fader. 2, Channel output. 3, Pre- or post-fader switch. 4, AR send fader. 5, Bus-bar allocated to AR. 6, Feed from console to AR device. On many consoles, the feed to AR passes through auxiliary 'send' circuits that may also be used for other purposes. In the simplest case these are taken from a point after the channel fader, so that when a channel is faded out the reverberation goes with it. In a complex balance, several AR devices may be used, each fed from a separate bus-bar.

to travel some given distance it is read out again, perhaps slightly reduced. For a computer, this is a simple operation, although for good AR, it can take a very large number of operations like that, in many different distinctive combinations. In principle, any other delay-related effect is much simpler: all it takes is a separate set of instructions (the program) and a channel and some switches to feed it out.

Most consoles have provision to split the feed from each source channel fader, diverting part to a range of 'auxiliaries', one or more of which can be sent to the AR and effects devices located outside of the console. In many consoles, auxiliaries can be fed from a choice of points, either before or after the fader. For AR, the signal should pass through the channel fader first. (If it were to go to the auxiliary feed before reaching the fader, the AR return level might have to be adjusted every time the source channel was faded up or down.)

If the whole of the sound from a particular single source is to be passed through a delay device, this may be cross-plugged (patched) directly into the channel, and in this case it can be before the fader.

For stereo echo, A and B signals are often combined before being fed in, but A and B reverberation is extracted separately. This also means that mono sources taken as part of a stereo mix automatically have the stereo reverberation that is essential to a realistic sound. The reverberation from each device is returned to the console as a line source (usually stereo) in a new channel, so it has its own fader (or faders) and equalization.

Since the console and the AR devices are completely separate, there is also freedom to use any other means of creating reverberation. It does not even have to be artificial. It can be real. After all, real (at least in principle) is the standard by which any alternative is judged.

Real reverberation

Natural reverberation begins with character-defining *first reflections* from the roof and walls – in the centre of a concert hall, after a substantial delay – followed at progressively shorter intervals by many more, smaller products of multiple reflections until, vast in number, they trail away below the threshold of hearing. This is the process that most Western instruments and ensembles were designed to exploit: it is an unwritten part of the music. When considering alternatives to natural reverberation we must therefore ask how their decay effects compare with the natural form and its many variations, not all of which are benign. An alternative could magnify the acoustic infelicities, or introduce new ones. At the other extreme, it might eliminate all the natural rough edges, and sound not just smooth but lifeless.

In the music room or concert hall, some frequencies are absorbed more readily than others at each reflection; if after many reflections some are sustained while others diminish, the result is *coloration*.

Coloration occurs in both natural and artificial reverberation, but note that the build-up and blending of natural reverberant sound depends on a three-dimensional system of reflections and decay that greatly enriches the quality at just the stage when coloration is likely to be most severe. When the artificial reverberation differs from natural acoustics, its use is justified if the new pattern of coloration and density of decay products is still perceived as musical in quality – and if its origins are not obtrusively apparent, as they were in some of the older systems. When real reverberation is added from a source that is remote from the original studio, this must also be judged by the same standards. How good, how natural, does it really sound?

Historically, the first way of enhancing reverberation was to use a separate room, often called an *echo chamber*. (Properly, it should be called a reverberation room, of course, but that is a lot less snappy.) This 'chamber' certainly had the advantage of real acoustics, although often in far too small a volume. But real acoustics are still a real option, so this needs to be considered not just as a historical curiosity, or even as a baseline for comparison, but as a practical possibility.

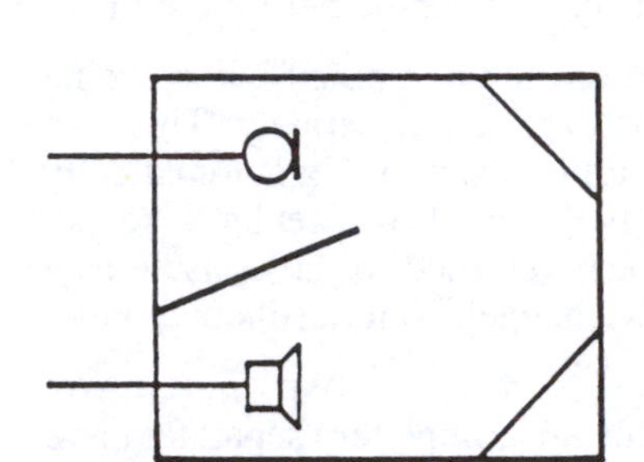

[12.4] **Echo chamber: size**
A U-shaped room may be used to increase the distance the sounds must travel from loudspeaker to microphone. Here the interior wall is angled to reduce standing-wave coloration.

A typical echo chamber has reflecting walls, perhaps with junk littered about at random to break up the reflections and mop up excess mid-range reverberation. If space is limited it might have a wall down the middle, so that sound has to follow a U-shaped path in order to delay the first reflection. The loudspeaker itself can be made directional so that no direct sound is radiated towards the microphones, which in turn can be directional, discriminating against both sound from the region of the loudspeaker and coloration due to structural resonances (between parallel roof and floor, for example).

A humid atmosphere gives a strong high-frequency response; a dry one absorbs top. An echo chamber linked to the outside atmosphere therefore varies with the weather.

When a large hall is simulated, there should actually be a reduced high-frequency response, corresponding to the long air-path. In some chambers, the bass response may be too heavy; or alternatively the original sound may already have sufficient bass reverberation, so that none need be added. Equalization (EQ) and bass-cut controls in the return channel allow the necessary adjustments. Note, incidentally, that if less bass is fed to the chamber, its loudspeaker can be used more efficiently, but if bass is reduced for this reason it needs subsequent equalization.

Because it takes up valuable space within a building, an echo chamber may be rather small, perhaps 110 m^3 (4000 ft^3) or less. This will have broadly spaced coloration, due to natural room resonances in the lower middle frequency range. Like radio studios, it needs to be isolated from structure-borne noise, and can be put out of commission by building or engineering works unless noisy work is stopped while it is in use.

Another disadvantage is that, once laid out, its reverberation time is fixed. If this is about 2 seconds it may be suitable for music but little use for speech. If, however, only a touch of echo of this duration is

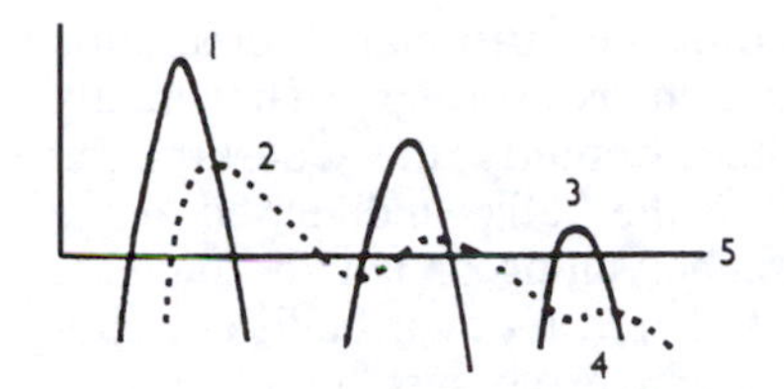

[12.5] **AR: the effect of reduced level**
1, Loud direct sound. 2, Audible reverberation. 3, Quiet direct sound. 4, Reverberation is now below threshold of hearing (5).

used it is noticeable only on the louder sounds, a situation that corresponds well with what we sometimes hear in real life. In this example an overlong reverberation time is apparently reduced.

For music the (fixed) reverberation time may be about right, but with too early a first reflection (because the shortest path length between loudspeaker and microphone is far less than in a bigger hall). To restore both clarity of attack and a sense of spaciousness, a delay should be introduced into the echo return feed.

Far better results have been achieved by using a spare empty studio or full-sized hall as the echo chamber. This is rarely practicable, but if it is, it is well worth considering for works that might really benefit from the added spaciousness. The real luxury is to start out afresh and lay out a new, dedicated space for this purpose. It is then also possible to fit movable doors and absorbers to make the acoustics variable.

Analogue AR

The first reasonably successful AR device – more versatile than most echo chambers because reverberation time could be varied – was the reverberation plate. In principle, this was rather like the sheet of metal once used in theatres to create a thunder effect, except that, instead of being shaken by hand and delivering energy directly to the air, it had transducers. One vibrated the plate, working much like a moving-coil loudspeaker; two others (for stereo) acted as contact microphones, to pick up the vibrations. The plate was 2 m^2 of tinned steel sheet stretched tight inside a steel frame. Unlike the echo chamber, its natural resonances at low frequencies did not thin out to sharp peaks in the bass, but were spread fairly evenly throughout the audio range.

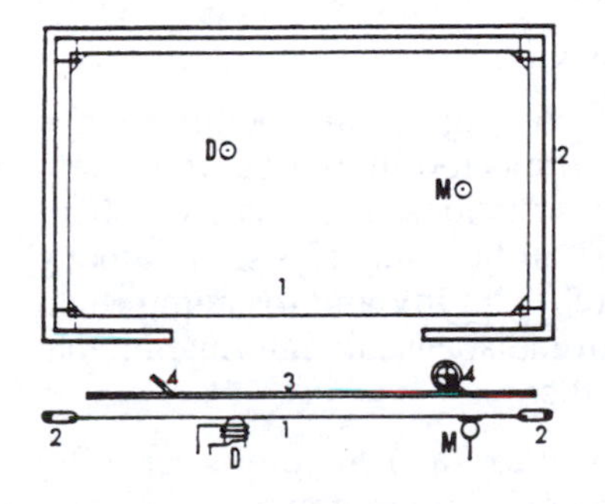

[12.6] **Reverberation plate**
1, Metal sheet. 2, Steel mounting frame. 3, Damping plate. 4, Pivots on damping plate used to change the spacing and so change the reverberation time. D, Drive unit. M, Contact transducer. The unit shown is mono; stereo has two contact transducers, asymmetrically placed.

The (asymmetric) layout of the transducers, together with the frequency characteristics of the plate, its variable damping and the response of the transducers and their amplifiers, were all arranged to give a maximum duration of reverberation at mid-frequencies. It had some roll-off in the bass but more at higher frequencies (15 dB at 10 kHz) to simulate the high-frequency absorption of air in a room of moderate size. With the broadest separation of damping and plate, it had reverberation of 5.3 s at 500 Hz, dropping to 1.5 s at 10 kHz, as in a cathedral. With increased damping this could be reduced to under half a second, although in practice it was better when used for reverberation times suited to music rather than for speech. A short direct path between the transducers was unavoidable so, again, to simulate a large hall, delay was required.

The 2-D structure of the plate allowed the wave to radiate outward and break up, producing reflections that arrived in increasing numbers as the reverberation decayed. This was a fairly 'natural' effect, although distinctively different from true 3-D, especially on string tone.

A third, and rather more weird, medium for artificial reverberation employed what were usually described as 'springs'. They really looked like springs, but worked quite differently: the acoustic input was applied torsionally to one end of the helix, and sensed at the other. The metal was etched and notched along its length, to create internal reflections, each of which fathered its own family of subsequent echoes, and so on. Early, inexpensive systems of springs were developed for use in electric organs. With improved engineering, they produced a quality of reverberation comparable with chamber and plate. With reverberation times of 2.0–4.5 s they were not suited to speech, but they were better than their early rivals for string tone.

Plates and springs may still be encountered and used for 'retro' special effects, but are hardly worth the effort or expense of maintaining them. Echo chambers, too, are pretty much an expensive, acquired taste. Digital techniques now cost less and are far more versatile.

Digital AR

A typical good digital AR device has a very wide range of reverberation characteristics, many of which are pre-set, but with an additional capacity for customized programs, which can be saved and reused. As with other advances in computer technology, this has depended in part on vast increases in random access memory (RAM), together with high-speed processing and improved software. Each AR program will have several stages. First, there is a group of delays that are characteristic of some particular shape and size. These are followed by a second generation of delays with feedback loops set for gradual decay, and the more stages that are added, the richer and more flexible it becomes. There may be separate programs for specific delay effects.

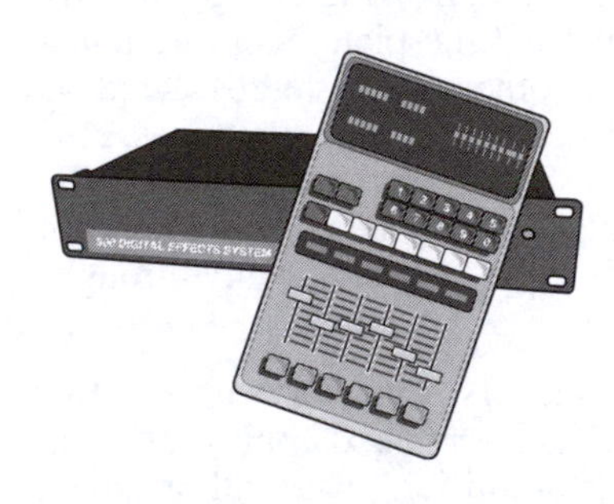

[12.7] **Digital AR control unit**
The deceptively compact controls for one 'digital signal processor' that offers a wide variety of AR and other effects. Features and controls include: LED display showing a descriptive name of the selected program, its main characteristics and the volume (including overload warning); program or register keys; display showing number of program or register selected; function keys; sliders and keys associated with the currently selected control 'page'. Operation is not intuitively obvious: new users should follow the step-by-step instructions, guided initially by the supposedly 'user-friendly' program names.

Few operators use more than a limited range of the facilities made available by any computer system – life is too short – but it is worth experimenting. The manuals and instructions are often so full of jargon that they are likely to put off all but enthusiasts, but among their wild swings between incomprehensibility and oversimplification, there are likely to be nuggets of indispensable information that apply to the system in use and no other.

To learn to operate a relatively complex (and therefore versatile) device, first run through a few of the existing programs in memory. As well as displaying strings of numbers, these may have been assigned 'user-friendly' names, but it is best to take these purely as a general indication of content and as a starting point in the search, rather than expect a programmer's idea of a particular acoustic to match your own. Check how to select from what is on offer, and also how to store your own modified (or originated) acoustic programs. The simplest strategy is to use only those already available, of which there is usually a range that is good enough for most purposes, but you may soon want to start 'tuning' them. A simple way of learning to use the controls (press buttons, knobs and sliders) is to convert one existing program until it matches another already in memory: the match is confirmed by the series of numbers – 'parameters' – which define the acoustic effect. Then add the original and the intervening stages to a short, dry recording in order to monitor the changes.

The next stage is to alter the parameters to something completely new. Here there can be an inhibitingly large range to choose from, and you may suspect that some offer different routes to the same objective. One device offers 22 parameters that can be altered to modify reverberation, with another 13 to develop 'effects', all within a stereo image. The aim is to control each stage of the simulated acoustic reinforcement of a sound, beginning with the 'pre-delay' (i.e. arrival time of the first reflection, in ms). Controls for the character of early reflections may be described vaguely as its 'pattern' (of coarseness and irregularity).

A suitable setting for the initial 'pre-delay' not only creates necessary effects of distance and spatial volume, but also allows the transients of the original sound to be heard clearly before the arrival of any reflected sound, of which the initial volume – its 'attack' – may be another separately controlled parameter. Other controls may be for relative levels of high and low frequencies. ('Room size' – area in m^2 – may also be offered.)

The most important single measure of an enhanced, natural-sounding acoustic is its reverberation time. With digital AR this can usually be set for different values in several frequency ranges. Depending on its period and style, classical music generally sounds best in a hall with a mid-range decay of 1.8–2.5 s. However, low frequencies may require longer, say 2.5–3.0 s, and it will be necessary to set a crossover frequency at perhaps 600–800 Hz. Treble decay may need to be shorter, with a crossover at 6 kHz or above.

In addition, the length of decay in one or more bands might be made to depend on the presence or absence of gaps in the music, so that for vocal or instrumental clarity it is relatively short when the sound is continuous, but longer (and so richer) in pauses. This is a modification that will be used cautiously: on lead vocals it helps to define phrasing, but if it is also applied to supporting lines that have gaps in different places, it may muddy and confuse the overall effect. In a reversed or 'stooped' version of this, a component can be given rich reverberation as it runs continuously, but rapid decay to clear any pauses in order to allow intervening figures to stand out in them.

Advances in computer technology have been such that extremes of complexity (often including a wide range of other delay-related effects) are now available for far less than the price of a plate – versions of which can in any case be mimicked by most AR programs, so that balancers who had learned to make constructive use of its severe limitations need not feel deprived: they can have a known (if simulated) reference point to move on from.

Less complex devices can now, in principle, be provided at much lower cost, but will still do far more and with better quality simulations than in the systems they have replaced. But while the choice of AR and delay-related effects has grown so much, the microphone techniques for joining them with live sound have remained much the same.

Using AR on speech

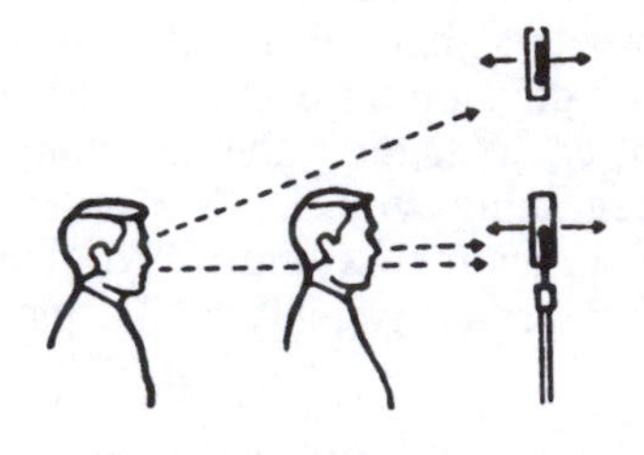

[12.8] **AR for perspective**
A second directional microphone is suspended above the main microphone. If the upper microphone has a stronger echo feed this adds depth to crowd scenes. Here bidirectional mono microphones are shown; for stereo, try two crossed pairs. This layout is for radio, but the same principle can be used for television.

Because echo chambers and early devices such as the plate were so poor at providing short reverberation times, one of the greatest improvements offered by the use of digital reverberation is found in its application to speech – usually for dramatic effect. The main limitation is now due to the studio itself, and the performer's working distance. An actor who works close to the microphone and whispers will, for the same setting of the echo fader, produce just as much reverberation as when standing 3 m (10 ft) away and shouting. So compensate for perspective effects: as an actor moves in toward the microphone, fade down the AR.

If there are other voices or sound effects that have to be picked up in a different perspective at the same time, this requires a more complex set-up. One useful arrangement has a second directional microphone (or stereo pair) with a strong AR feed suspended above a main microphone that has little or none. An actor who moves in close to the lower microphone is no longer in direct line with the upper one. In this way, perspectives can be emphasized without constant resort to the AR fader. This is more likely to be of use in radio than in the visual media, where there are restrictions on the ideal placing of microphones, but the same principle (of direct and AR microphones) can still be used.

Delay-related effects

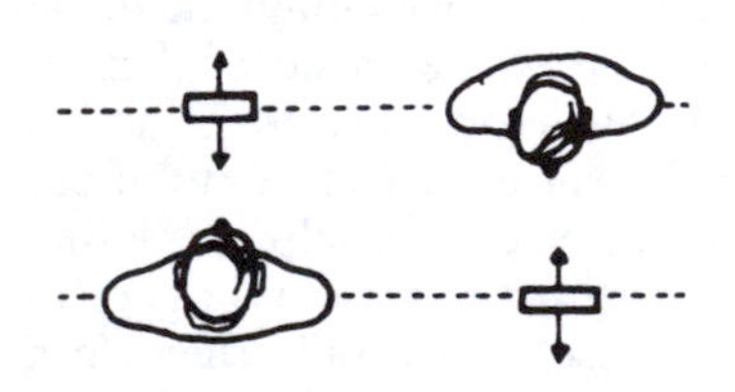

[12.9] **Separate AR on two voices**
If differing amounts of echo are needed on two voices taking part in a conversation, it helps if opposing microphones and voices are placed dead-side-on to each other. For stereo, the two can be steered to suitable positions within the image. This layout might be adopted in a radio studio; in a television studio there is normally more space available, making separation easier.

A word of warning here: any of the effects described below can easily be overused; for example, echo effects (typically with a delay of about 35 ms) are often most effective when applied to a single phrase of music, repeated only when that phrase recurs.

Some of the less natural effects of feedback and delay had already been explored and developed by more laborious techniques long before the present digital systems became available. For example, a simple echo could be achieved just by allowing tape to travel from the recording head to the replay on a tape recorder, though for any flexibility, the machine needed a variable-speed drive. If the output was mixed back into the input, still with a delay of 35 ms or more, the echo circulated, dying away if the feedback was attenuated, or becoming a howlround if it was amplified. This effect is sometimes called *spin, recirculation* or *slap echo*, and has a mechanically repetitive quality. The effect varies with the length of the delay: at a fifth of a second, for example, there is a marked flutter effect. All of these, and many more, can now be selected from programs of digital delay effects and control parameters.

For example, it is possible to start with a staccato effect such as a tap on a glass and let it spin indefinitely by controlling the feedback volume manually. Before the advent of digital technology, any slight unevenness in the frequency response of the circuit was quickly exposed: using tape, even a high-quality recorder soon revealed its deficiencies and within half a minute the coloration characteristic of the system completely swallowed the original sound. This effect (or defect?) can now be eliminated.

It was also found that a signal delayed by about 15–35 ms and added to the original could be used to double the apparent number of instruments. For a while, 'Automatic Double Tracking' (ADT) was provided by a separate special digital device; now more often it is just one of the many options on a digital signal processor (DSP).

Doubling (or tripling) works well with sounds that are sustained and have a broad frequency spectrum: typical applications include guitar, strings, horns, cymbals or vocal-backing chords. In popular music it can reinforce thin backing or enrich foreground. Other interesting effects are found when it is applied with a 1–5 ms delay to transients. It can also be used to resynchronize sounds that have been produced a little out of step owing to the time a sound cue takes to travel through air.

The short-delay effects in which phase is important were also originally explored in pop music by tape techniques before pure electronic devices were introduced. Two recordings were played, initially in synchronization, and then one was slowed by gently pressing a hand against the flange of the tape spool – which is how the resulting effect got the name of *flanging*. The recordings were separated by a progressively increasing interval, causing phase cancellation to sweep through the audio range. This effect is now achieved by variable digital delay, and is most effective when the interval between the signals lies in the range from 50 μs to 5 ms.

Recombined sounds with delays in the 5–15 ms range have a different quality, but are still dominated by phase cancellation that is heard most clearly when allowed to sweep through different frequencies. *Phasing* effects were originally explored by using analogue

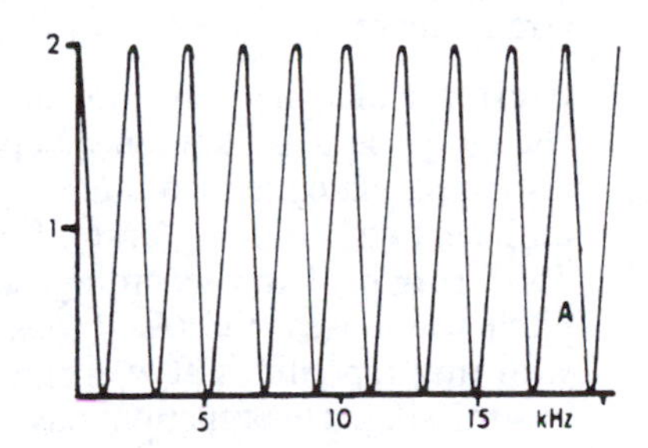

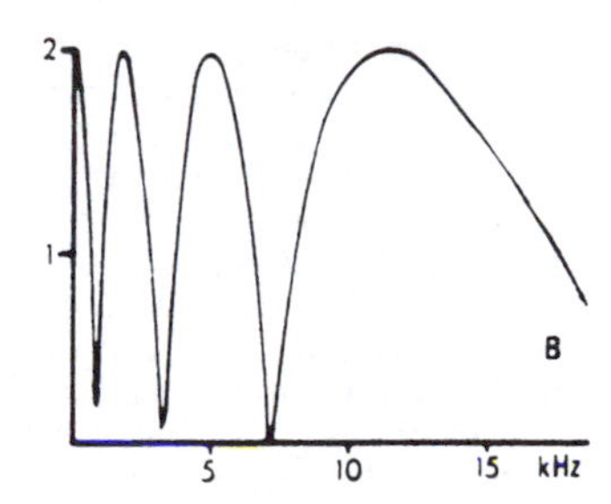

[12.10] **Flanging and phasing**
The change in amplitude (vertical scale) with frequency. A, The effect of flanging – here with a delay of 0.5 ms – is evenly distributed throughout the frequency range, but is not related to the musical scale, which is logarithmic. B, Cancellation and enhancement produced by the use of a phase-changing network: the dips are distributed evenly over the musical scale.

phase-shifting networks, but these were distinctively different from the 'flange' effect in that cancellation was distributed throughout the audio range in an exponential manner (in the ratios 1:2:4:8:16, etc.) compared with flanging's arithmetical array (1:2:3:4:5, etc.). The analogue phasing effect was therefore directly related to the harmonic tone structure, unlike flanging's denser overtone structure, much of which is not conventionally musical in quality. With the introduction of techniques based on the use of delay circuits, 'phasing' reverted to the arithmetical pattern of cancellation originally found in flanging, which many users still preferred.

In a further variation on this, called *chorusing*, a signal in the doubling range is also very slightly changed in pitch. In the mid-frequency range this lends a feeling of depth to the sound; at high frequencies it enriches the harmonic structure.

Other effects on offer might include differential delays for variously defined bands or tones within some of the more complex DSP programs, and move on from those to go even further beyond the realm of natural sound modification.

Changing pitch

It is easy enough to change the pitch of a sound: simply make an analogue recording (say, on tape) and replay it at a different speed. But this immediately creates two problems. The first is that as pitch increases, duration falls, and vice versa. The second is a radical change in what the beginning and end of every component sounds like. For example, if replay is speeded up to increase the pitch by two octaves, duration comes down to a quarter and the characteristics of both attack and decay change so that the sound becomes both far more percussive and drier than before.

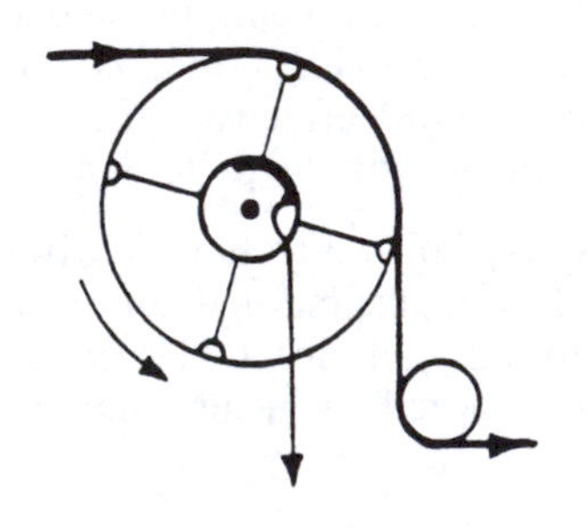

[12.11] **Tempophon**
This was a mechanical precursor to the digital pitch changer. It changed duration and pitch independently. The head assembly could rotate either way: *if the sense was opposite to that of the tape, pitch increased; if the heads chased the tape, it decreased.* Signal was fed from the head in contact with tape, via a commutator and brushes to a second recorder.

If this is acceptable, it can be used for simple effects. An early example was a double speed piano composition. On a first recording the piano was played with deliberation, to avoid the appearance of inhuman virtuosity when speeded up. This was then played back at double speed, and conventional accompaniment from the same piano was added. The double speed piano sound was not just a minor variation on ordinary piano quality: it sounded like a completely new instrument, and had to be treated as such in the musical arrangement.

If effects like these are not the objective, it will be necessary to change pitch and duration independently. Early attempts to do this also used analogue recordings: the tape was read by a series of pick-up heads on a rotating drum. Each moving head in turn sampled a short length of the recording, and these were remixed. When the pitch was raised, the result was composed of a series of segments with tiny repetitions at every junction, or if lowered it would have bits missing. This function may now be included as one of the many on a digital signal processor. For this effect it samples overlapping or gapped segments from random access memory, and joins them together. This process can never be perfectly natural: there are always either bits of sound missing or bits repeated, however small, so the quality achieved depends on the skill of the digital programmer. A good digital program can make the end-product much smoother.

Whether it is done by the archaic analogue method or by a digital pitch changer, the result also depends on the nature of the samples. There is no problem with sustained unchanging tones: they just get longer or shorter. But at the other extreme, transients may be clipped or completely lost, or they may develop a stutter, all of which are bad. There is a tendency for a gritty, grainy quality of sound to appear. It is sometimes heard on commercials where the advertiser wants to put in too many words (or to shorten a required product warning), and then tries to cut its duration by more than it will comfortably take.

Devices vary both in the smoothness with which they blend sound and also in their capacity to avoid the worst defects. Experiment will show the limits within which any particular change is acceptable: it may be ±20% in duration, pitch or combination of the two.

Some problems can be reduced by pre-emptive action. For example, if the plan is to record and then change the pitch of an actuality sound – such as a foghorn in a busy seaway – any background recorded with it that cannot be filtered out will be treated as well. Because it is natural, the background may not be noticed when heard with the ordinary sound, but treatment can make it vividly apparent. Plainly, for such applications, close microphone techniques on the original recordings (to give good separation) are desirable. The foghorn might then, for example, be used to construct a melody from a sequence of pitch-transposed components.

There are all sorts of possibilities for special effects. A hollow-sounding alien voice might be obtained by adding a delayed, pitch-changed repeat, while adding subharmonics creates a monster (such as *Star Wars*' Jabba the Hutt).

At this point it is necessary to add a few caveats. One is that when a singer's voice is subjected to a change that affects its duration, the result can sound odd. This is due to vibrato, the cyclic variation in pitch that gives colour and warmth to a singer's sustained notes – a technique that is copied on orchestral instruments such as the violin. Analysis of voices that are generally considered beautiful suggests that pitch variations of as much as 8% may not be excessive (a semitone is about 6%). But the really critical quality of vibrato is the number of these wobbles per second. The acceptable range is narrow: for an attractive sound the vibrato rate must be between 5 and 8 Hz, and is independent of the pitch of the note sung. Outside of 5–8 Hz the effect resembles either wow or flutter: the vibrato range can be taken as separating the two. The direct consequence of this is that only small duration (or speed) changes are possible without taking vibrato outside its normal range. At worst, the result is to make the vibrato both obvious and ugly. At best, halving duration (and doubling speed) produces the effect of a small, vibrant-voiced animal.

Tremolo, a similar variation on the volume of a note, is also dramatically affected by such changes; so too is portamento, an initial slide into a note or from one note to another, which if slowed down changes its character, or if speeded up almost disappears.

Another unwanted side-effect might be on the quality of male and female voices. While these may differ in fundamental by an octave, the *formant* ranges (i.e. the parts of the vocal sound that are emphasized by throat, mouth and nasal cavities and used to give the main characteristics of speech) are, on average, only a semitone or so apart, and vibrato ranges are also little separated. Small changes of pitch therefore tend not to change the apparent sex of a voice, but rather to de-sex it.

A more benign use of pitch change is offered by a program that slightly spreads the nominal pitch as an anti-feedback measure. If the treated signal is returned through a studio by acoustic feedback or public address, the loudspeakers can be louder. Unlike the frequency-shifted feedback sometimes used for public address, this can be fed to musicians, too, as the spread is centred on the original frequency.

Some of the (potential) problems associated with pitch change (while leaving duration unchanged) can be sidestepped by proceeding directly to a pitch-correction program. This should be able to take a sample of sound and shift it up or down as required, matching it precisely to a chosen note on any of a wide variety of western, ethnic, historical or invented-for-the-occasion scales. It will also preserve intonation by defining a tonal centre, retuning that and preserving the intonations around it. Variations on this allow the creation of harmonies from an original (for example, to accompany a lead vocal or 'thicken' its quality) or to add vibrato.

The most sophisticated professional pitch and duration changers are now extremely flexible, offering independently changing tempo and pitch that can be mapped over an extended waveform or several waveforms in parallel – without loss of timing accuracy, stereo phase or surround sound data (all possible dangers in digital processing).

This kind of process bridges a link between the range of DSP functions that apply to all audio control operations as described in this chapter and the much broader range that will be outlined in Chapter 16, which introduces the virtual studio along with 'plug-ins' such as those offered on the Pro Tools platform and the MIDI operations used in conjunction with them.

Sampling

Sampling and looping are further possible products of digital signal processing, in this case bridging between delay effects and digital audio editing (which is described in Chapter 19).

Originally, sampling was achieved by physically cutting audiotape and putting spacers on either side of it; to loop the sample, its two ends were joined so that it would replay repeatedly for as long as required. It could be played backwards or at a different speed. It could be copied, introducing changes in volume or EQ. All sampling and looping effects mimic this. Some MiniDisc and CD players offer means of marking in and out points so that short samples can be selected and replayed as loops, originally in order to help find an exact point in the replay.

Digital developments from this are more versatile. Essentially they at least start by doing the same things, but use digital recording, editing and signal storage. Keyboards can be used for replay (and to change pitch): they are widely used for percussion. And drum machines use high-quality samples to build up repetitive sequences that can be used either as a minimally-skilled foundation for other-wise live popular music or as the rhythmic underpinnings of a synthesized composition.

Chapter 13

Recorded and remote sources

All studios have a variety of additional sources that are mixed in with the output from the studio microphones. These include disc players – for CDs, MiniDiscs and vinyl – analogue and digital tape decks (plus, for television, videotape and film sound), hard disk sources including Zip drive and the output from digital editing systems, and lines from other studios and remote locations, including ISDN and regular phone lines. Some of the same sources are used in film and video dubbing theatres – in fact, in any kind of recording studio – but their use is central to much of the output in both radio and television production. This chapter deals with the operational techniques for using them.

These are all programme sources comparable to studio microphones, but differ from them in that they usually reach the console at or near the much higher 'line' level (zero level). Volumes may need adjustment, but no great preamplification. Most go straight to stereo channels.

In radio, and to a lesser extent in television, each small segment of this type of programme is collected, selected or pre-recorded individually. The creative process is spread out in time, and within the limits of human fallibility and commercial pressures, each element may have lavished upon it all the care it requires for perfection. There is often a stage when most pre-recordings are complete or require only last-minute editing, and all this 'insert' material may be taken to a studio to be broadcast with live links or re-recorded into a complete programme. Another possibility is that it is scheduled into an automated play-out.

In order to match levels, mix in effects or music, introduce fades and avoid abrupt changes of background atmosphere, recordings are mixed in to the programme, not just cut in 'square'. Ideally, each insert must be controlled afresh, so it is fully integrated into its new setting, with special care for the relative levels at the beginning and end of each insert (see Chapter 14). In the operation of replay equipment, the most vital single element is good timing, because where successive sounds are linked by mixing it is more difficult for errors of timing to be corrected by editing. On live radio there is no calling back miscues.

Digital automated broadcasting uses time code to control the output from playlists and 'jukeboxes' (which may just be different aspects of the same play-out systems). These are capable of cueing the routine transmission of many items in a random order, and can even have overlaps programmed in. But manual techniques are still used for much creative production work – in fact, to build many of the programmes or items that will later be broadcast automatically – and they also employ the techniques that, in their most developed form, automatic systems must simulate. The fully digital alternative is to assemble the material on hard disk using a digital editing system: this can achieve much the same results but by more analytical techniques. If a studio is still used at the end of this, it will likely be for fine-tuning and review. Whatever the method employed, the objective is the same, and is described here in terms of the studio techniques. For audio editing, whether as part of the studio operations or carried out elsewhere, see Chapter 18.

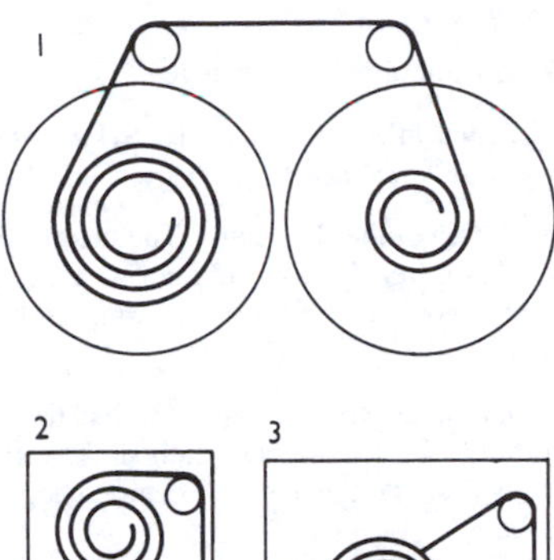

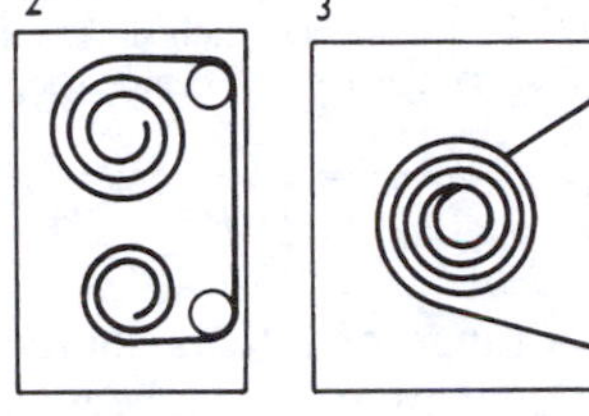

[13.1] **Tape transport systems**
1, Open reel (reel-to-reel). 2, Cassette. 3, Cartridge, a continuous loop. Pressing the cartridge record button automatically puts a tone (below the audible frequency range) on the tape. On replay, when this point on the tape returns to the head, the tape stops and is re-cued automatically, ready for immediate replay. Tape cue tracks may also have time code recorded on them.

Live-and-insert programmes

The linked actuality programme is typical of what might be termed 'live-and-insert' work: location recordings of effects and interviews arranged in a sequence that helps tell a story. Unless each collected item cues directly to the next, linking narration will almost certainly be needed. Very similar in construction is the illustrated report, in which summary, comment and studio discussion are mixed with extracts from recordings of some event. Disc jockey shows, whether of popular or classical music, may also be regarded as live-and-insert shows. Here we can be sure that exquisite care has been exercised in recording the 'inserts': their arrangement within a programme deserves care too.

In the following example, where the raw material is assumed to be on tape, the script of a feature that combines recordings with live narration might look like this:

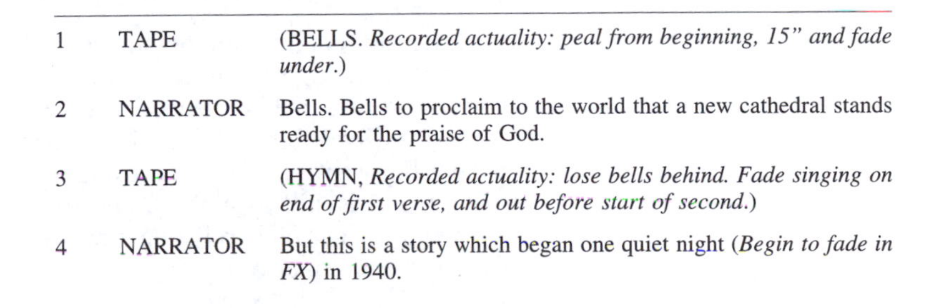

1	TAPE	(BELLS. *Recorded actuality: peal from beginning, 15" and fade under.*)
2	NARRATOR	Bells. Bells to proclaim to the world that a new cathedral stands ready for the praise of God.
3	TAPE	(HYMN, *Recorded actuality: lose bells behind. Fade singing on end of first verse, and out before start of second.*)
4	NARRATOR	But this is a story which began one quiet night (*Begin to fade in FX*) in 1940.

5	FX	(NIGHTINGALE. *Fade in drone of* BOMBERS *until nightingale is drowned out. Add distant* ANTI-AIRCRAFT FIRE. *Hold bombers behind speech.*)
6	NARRATOR	The bombers swept across the quiet countryside to the city. In the city centre, the old cathedral . . .
7	FX	(BOMB *screams down and explodes.*)
8	NARRATOR	. . . a glory of Gothic architecture. Windows of richly stained glass, set in delicate traceries . . .
9	FX	(BOMB *explodes, tinkle of* BROKEN GLASS.)
10	NARRATOR	. . . carvings and statuary that were the life-work of countless medieval craftsmen . . .
11	TAPE	(*Pre-mixed FX sequence*: 2 BOMBS, *begin to fade in* FIRE *and hold, gradually fade out bombers.*)
12	NARRATOR	. . . the hammer beam roof a triumph of human ingenuity . . .
13	TAPE	(*FX continue: Several* FALLING TIMBERS *then* CRASH *as roof falls.* FIRE *up to full, then fade to silence.*)
14	NARRATOR	With the cathedral had gone many other buildings. The centre of the city was largely a wasteland. Rebuilding could not be started until after the war. But nothing could stop people dreaming of the future . . . and planning for it.
15	TAPE	(*Edited Interview: quick fade-in on atmosphere.*) At first there was quite a lot of opposition to the idea of a new cathedral: well, of course, people said houses first and the new city centre, shops; the cathedral could come later . . .

This requires two tape replay machines, a range of effects sources and a microphone. During the preparation ('prep', or rehearsal) stage, many of the cues on the console and play-in operators' scripts will be overlaid with all the notes and numbers required to complete it.

In fact, this example is at (or perhaps even a little beyond) the limit of what is now put together without editing. An alternative and perhaps more likely approach would be to supply the inserts as components of a digital edit, to which studio narration is added by recording it directly on another track. The editing is completed at the back of the studio while preparation of other items continues.

Preparing recorded inserts

Each insert must first be considered as though it were a programme in its own right: it must go through all the processes of rough and fine editing that are needed to give it point and clarity. When carried out 'as live' this can often be as drastic as is possible when digitally editing a programme. Anything that does not work well can be cut out (but is still available), and the essence of confused or muddled actuality can generally be interlinked with script to make it both lucid and concise.

Heavy actuality background effects should be recorded for long enough before and after foreground speech to allow for fades (although if the effects, as recorded, are too short or sound wrong, a loop or effects recording can often be used to overlap and provide material to fade). Try to allow a few seconds of atmosphere for fades even where the background is not heavy. An intake of breath before a speaker's first word may be left on the recording. At normal listening levels, neither atmosphere nor breath stands out, but they do sound natural. Take particular care with any background atmosphere on speech inserted between music recordings on which studio atmosphere is low.

If manual replay is envisaged, pre-recorded segments are assembled to replay them in sequence, ensuring that any to be mixed or cross-faded are replayed from separate sources. At this stage a script may give narration in full but inserts in cue form, with their durations, like this:

NARRATOR	. . . spoke on his return to London Airport.
INSERT	*Item 10 Dur: 27* In: What we all want to know . . . Out: . . . or so they tell me (laughter)
NARRATOR	The minister stressed the need for unity between the two countries and then said:
INSERT	*Item 11 Dur: 70* In: It's all very well to be critical . . . Out: . . . there's no immediate danger of that.

In this example there are two types of cue from narration. The first needs about the length of a sentence pause, but the link ends with a 'suspended cue' that demands to be tied closely to the sentence that follows. In the first case the atmosphere fade-up may follow the cue; in the other it should go under the last words of the cue. In the second case the timing also depends on the pace of the recording that follows. For instance, 'It's (pause) all very well . . . ' sounds odd if tied too closely to the cue ' . . . and then said:'. Mark the peculiarities of each cue in some bold private shorthand on the script. Again, these suggestions are equally valid for play-in or for the digitally edited alternative.

Other more complex overlaps between narration and insert might be needed for artistic effect: how such fades and mixes can best be arranged is described in more detail in Chapter 14. The object of 'prep' (or rehearsal) is to try these out, and unless the programme is assembled by trial-and-error, or in short sections, cues may come up so fast on the take that some quick visual reminder is needed for each of them in all but the simplest sequences, and will be used as a musician uses a score. There is no virtue in trying to play it from memory.

The main things to mark on the script are:

- the format and location of each source: which play-in source it comes from (although on the console it is helpful to group inputs from each format, so that a late change in which of a pair of machines is used is not missed);
- any oddity, peculiarities of timing or hazards to listen out for;
- volume control settings and type of fade-in;
- duration (or other notes on when the 'out' cue is coming up);
- type of fade-out.

Where the replay system allows precise, manual cueing, rehearse this to get the feel for the timing. After cueing a replay there must often be a tiny pause before fading up, to allow a mechanical device to run up to speed: on analogue tape for example, this might otherwise be heard as a wow, if only on atmosphere. Do not rely entirely on remembered timings: different devices may have different run-up times, although digitally controlled (time-coded) replay can be very fast.

Speech recordings with different background effects present a typical case for a mix. If the changeover point is internal to both recordings, there are several ways of performing the mix, depending on the length of pauses between words. In the simplest case, with a full sentence pause on both recordings, the pause can be retained at its natural length. Analogue tape is a medium that allows visual control of the process – by putting marks on both tapes to show where any cross-fade should begin and end. Even if one of the originals has no pause, it should still be possible to arrange a tidy mix: for example, if there is no pause at the 'out' point of the first recording, then the second can be cued and faded up early. Where there is no pause on either side, matching atmosphere can be filled in from somewhere else on the recording.

In all cases match the levels with care: use the fader to 'edge' one side of the join in or out a few decibels to avoid a doorstep effect on the voice. The same technique can be used where heavy atmosphere ends too close to a last or first word.

Even if digital editing is used in the studio, for live transmissions some operators still prefer to transfer the output to analogue tape before going on air. Computers crash more often than tape replay fails – and the edited version still exists as backup. As a further justification, play-out to tape offers a final chance to check the edit, a potentially vital safeguard. And last-minute cuts can be made quickly on tape. However, the progress of digital play-out seems inexorable – and is acceptable provided it is rigorously separated from other uses of the same platform, and provides backup both of material and processors.

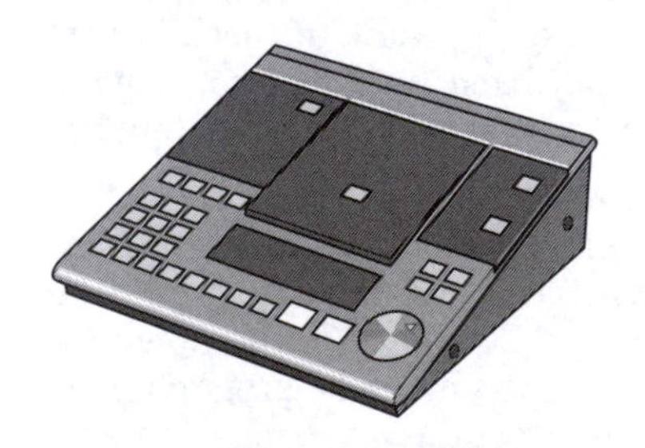

[13.2] **Programmable CD player**
For broadcasting studio use, a CD player has a wide variety of functions, including precision set-up at any point – with memory for a sequence of many 'start points' on different discs – and also looping and variable speed.

Cueing play-in

A general-purpose radio studio will have a wide variety of play-in sources, including perhaps two digital editing systems, three open-reel tape decks (for backup or for on-air play-in), two DAT record/

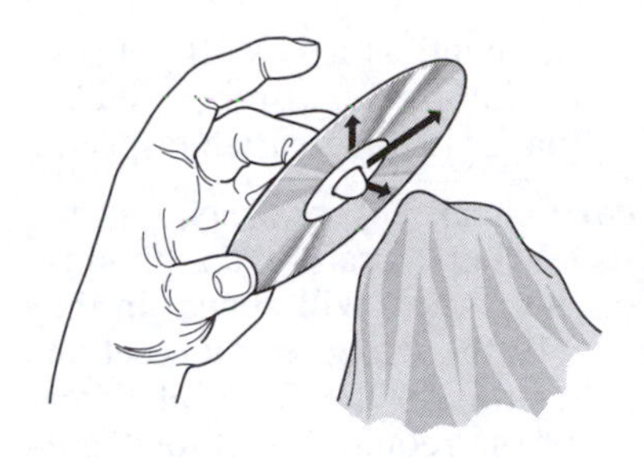

[13.3] **Care of compact discs**
CDs should be handled with as much care as vinyl discs. Treat them like precision optics (which they are). To avoid fingerprinting the tracks, hold only by the edges or by rim and centre. Remove any dust with a soft air-brush. Avoid cleaning, but if a soft cloth (or lens cloth) must be used, brush very gently from centre to edge, and not along the line of the tracks. The only solvent that may be used is dilute isopropyl alcohol (and even this can lift the label).

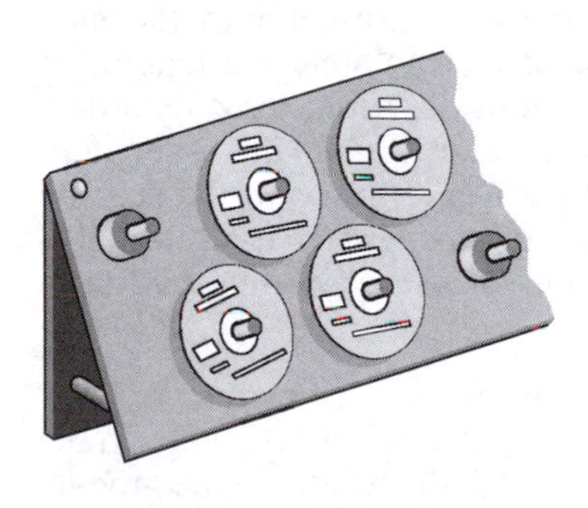

[13.4] **Rapid-access CD rack**
A home-made record rack for CDs. This has a separate numbered spigot for each disc. Each spigot consists of two slices of wooden rod with a screw through the centre to fix them to a board that leans back a little. The outer rod is about 14 mm (5 in) in diameter, so that the hole in a CD slides smoothly over it and rests with its centre against the larger disc. The spigots are about 14–15 cm (5–6 in) apart. The side of the CD with printing on is at the front and the recorded side, touching nothing, is at the back. It is easy to lift the disc by its edges into the player.

replay machines, two cassette players, three programmable CD players, a MiniDisc cart, two 'gram' (gramophone or phonograph) players, and perhaps a Zip drive record and replay cart.

For analogue open-reel tape, run the recorded sound backward and forward over the replay head to identify the 'in' point, mark that at the head with a wax pencil, and then if there is atmosphere to fade up, also mark the place from which the fade up will start. Ideally, insert a 4-second yellow leader to end at this point and set this back from the replay head far enough for the tape machine to run up to a steady speed. Some professional machines do this quickly, in about 1 cm (½ in), but in any case, if time permits, rehearse to check this. On cue, replay and fade up from the first mark (or the end of the spacer). The play-in machine fades up to full, with the console pre-set for the required (rehearsed) play-in level. Do not fade it in at the console unless replay is cued by a remote start from there.

For *DAT* machines with 'quick start' or 'memory start', for near-instant replay from the beginning of a recording, set up at the end of the start ID: programme should follow in half a second. Any subsequent items on the same tape should also have their own start ID (after a clear gap on the tape). As usual, check for any fade-up on atmosphere.

A *tape cassette* also offers rapid replay: to set up at the end of a leader, turn the spools by fingertip or a thick pencil to move the tape.

Commercial *Compact discs* (CDs) are designed to be run on domestic machines only from the points programmed on to them, as listed in the notes in the box. When setting up a CD on a standard replay machine, allow time for these cues to be read automatically into memory, then the track can be selected and held on 'pause'. The delay after pressing 'play' is checked, and the track reset on 'pause' until required. On cue, it is started early enough to allow for any delay.

Studio CD replay machines can also be set to replay CDs and CD-Rs from the start of a track in the same way. If 'auto-cue' is selected, the disc will set up for instant replay from the start of sound on the track (but beware very quiet starts, as these have sometimes been clipped by auto-start devices).

For more flexible CD operations, begin by choosing the required track, then press the appropriate buttons to select start and stop times in the display window, followed by 'auto pause' to activate the stop time. If no stop time is selected, it should play to the end of the track. There should be instant replay buttons to play several cues from the same disc – and the machine should also store a large number of cues from different discs, and recall them when any disc is put in the machine.

For precise location of a start point on the CD, it is useful to have a search wheel. Play up to the required cue, and set the machine to 'edit': the search wheel then allows the sound to be run backward and forward just like running tape over a replay head. When a suitable point is found, press 'edit' again to store it (or enter the time, as before). The search wheel may also be operated with a

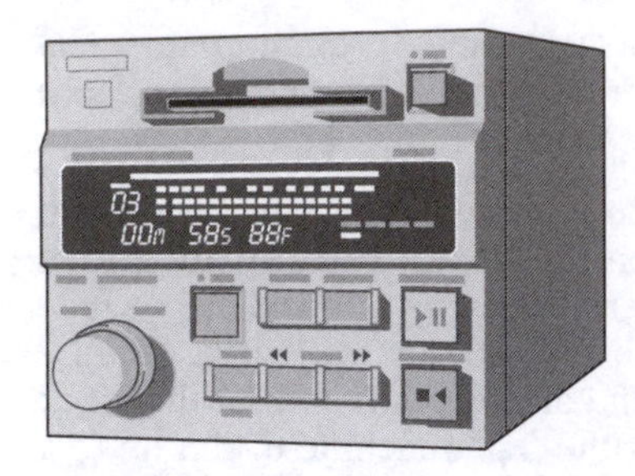

[13.5] **MiniDisc cart**
An MD player designed to act as a cart machine – to play spot announcements such as station ID, jingles, trailers, commercials and sports, news or other (usually fairly short) recordings.

'loop' that repeatedly replays a tiny fragment, as it is shifted back and forth until the exact point is found. 'Loop' can also be used as a sampler. Other functions of this CD player include variable speed.

Before playing a *MiniDisc (MD) cart*, check any pre-sets, e.g. to 'single' replay. Insert the disc with its label up (an arrow on it points 'forward'). The name of the disc (or 'no name') will appear in the display window for a moment, then it will show that replay is set on the first track; turning a knob switches this to the starts of other tracks. Press 'play' to check the start of the required track or 'end/monitor' to check the last 5 seconds; at any stage 'standby/cue' resets it to the same start ready to go again, and after playing the whole track it will again reset to the selected start.

To start from a point within the MD track, play to somewhere near it, then use the 'search' keys: these repeat a short loop to find the precise 'in' point. Again, 'play/pause' (from the chosen point) can be used to check it, and 'standby/cue' resets to it. The display window shows elapsed time, or can be switched to give time to end of track.

A MiniDisc cart machine can be used for spot announcements such as station identification, jingles and commercials, as well as trailers.

A Zip drive cart, a single replay slot used for the same purpose, may have 10 cues ready for instant replay, each on a numbered button, with others (in double figures) almost as easy to find and run.

A *tape cartridge* (if one should still be encountered) is made ready to start from the beginning by simply pushing it into the slot. Cartridges contain loops of lubricated tape in standard lengths of say 10, 20 or 30 seconds, to allow various durations to be recorded. They switch off if there is a pause of more than, say, 2 seconds, after which the tape carries on until it has returned to the starting point, to await re-cue. Cartridges were once widely used in multi-slot carts where their automatic reset was useful in run-and-forget operations.

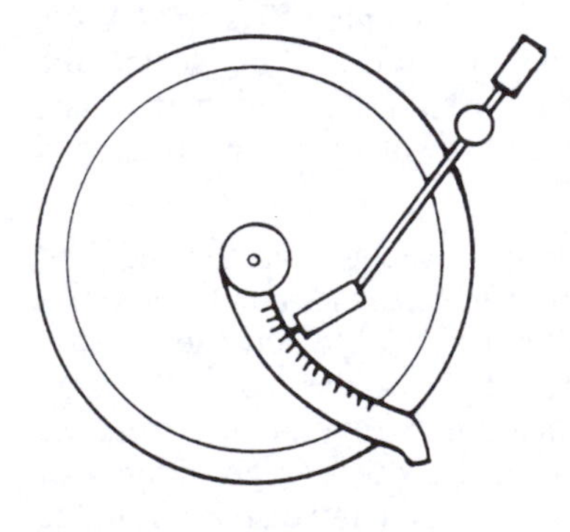

[13.6] **Simple arc scale locator**
This can be used to help find a particular groove on a vinyl or archive disc. The scale rests over the turntable spindle at one end and in the hand or on a block at the other. It can be made very simply from card. Some record players have optical systems to display a scale in a nearby window.

'*Grams*' (phonograph players) are used for vinyl or archive discs. A fairly brutal way of setting up is to run the disc a little beyond the 'in' point, stop and run back manually with the needle in the groove: this was the method actually used in the years before it was adapted (less destructively) to running tape back and forth over a replay head. Some turntables have groove locators that show the position of the arm on a scale in an adjacent window. A strip of plastic or card with a scale extending across the grooves can also be used – though both methods can only give a rough indication. On vinyl discs, successive excerpts can be found reasonably quickly by lifting and relocating the head. It helps that on a grooved disc, variations in volume are clearly visible.

If several discs are to be played, they can be numbered with a wax pencil on the smooth part just outside the label. In the days of 78s it was even possible to mark the actual groove with a wax pencil sharpened to a knife-edge. If it is driven from the centre, the turntable acts as a flywheel, to even out speed variations: 'flanging' effects were originally created manually by resting a finger against the edge of one of a pair of identical discs that had been running in sync. During replay, there is always a danger that a pick-up will run forwards (a groove run) or, worse, back. If a repeating groove occurs during a radio transmission, one cure is as violent as the effect: fade out, jump the stylus on a little, and fade back up. Alternatively, as the stylus may require little persuasion to return to its original path, or at worst to jump forward, simply rest a pencil very gently across the pick-up arm and 'squeeze' it towards the centre of the record.

Archive disc recordings in outdated formats are transferred to a more suitable medium before use. Use vintage replay equipment or a modern replica to convert the recording into an electrical signal without unduly wearing the original, and follow this by any treatment that helps to clean up the sound: there are DSP programs for this (but prized 'vinyl quality' may also be diminished by digital treatment).

Returning to more modern techniques, *hard disk replay* employs a recording stored as a digital file in computer memory – often in a digital editing system. This allows instant replay, and can be set to run a series of overlapping files. For digital editing, see Chapter 18.

Digital audio files can also be cued in arbitrary order from a *MIDI keyboard* controller. In principle, MIDI devices can readily be combined with other digital operations such as sampling (re-recording between pre-set cues) and pitch changing. They could be used, for example, to create a tune out of a single toot on a car-horn, or to comply with a director's request to 'make the footsteps a bit deeper'. They are therefore particularly useful for sound effects (see Chapter 15).

Remote sources

Another type of source that may appear on a console at a radio or television station is a line from some other studio or remote location (in Britain these are called *outside sources*). A local radio station might have lines to the regional weather bureau and recreational areas, local government offices, traffic monitoring points and vehicles, sports arenas and concert halls, together with other information centres for news, business or entertainment reports. International links, network lines, other local or distant studios and the telephone system may all be used. Contact between the central studio and the remote source may take various forms. Often there are two lines: the programme circuit, which may be one- or two-way; plus a control line, which is effectively a telephone link between control points at the two ends.

For many links of this kind, the programme material is on a broadband digital telephone line (ISDN). Both remote source and the studio have matching codecs. When ISDN is used for broadcasting, the greatest potential hazard is the quality of the headset and codec at the far end, but in this application, the broadcaster can usually ensure that they match, and any minor loss of quality due to data compression is acceptable. The studio usually has a calling unit with a keypad, display window and a few extra buttons. Calling is similar to off-the-hook phoning: key in the number, then press 'dial'. In the window, 'ISDN line' appears, followed by confirmation (or not) that the codecs match, which means that the line is working both ways. It is then ready for diversion to the console as an outside source. In case error message numbers appear, it is best to know in advance what they mean: mostly they refer to the usual telephone dialling problems.

Note that a codec with greater bandwidth (but still with 'acceptable' data compression) is sometimes used for music. If the studio does not have the necessary unit, it might be available in a switching control area outside the studio: the number can be dialled from there and the line fed to the studio, again as an outside source.

Generally, any kind of link is set up and checked before it is diverted to the console. In the case of a one-way *music line* (i.e. broadcast-quality line) plus control circuit, in order to prove the line, contact is established first on the control circuit, and the distant contributor is asked to identify the broadcast line by speaking over it and also to send line-up tone. If the programme is not yet on air, these can both be checked with the channel fader open, so that the level of the incoming material can be accurately set and its quality established.

If the line is not available until the programme of which it is to form part is already on air, the line can be fully tested to the last switching point prior to the active part of the control console. On its assigned channel within the mixer it is checked on *pre-hear* (also called *pre-fade* or *PFL*) using a small loudspeaker or headphones.

A control line may be double-purpose: as well as being used for discussion between the two ends before, during and after the contribution, the return circuit may have cue programme fed on it. Also, the producer's talkback circuit might be superimposed on the cue programme. The cue feed is checked by the contributor as part of the line-up procedure.

Sometimes the outside source is a 'self-drive' studio. If there is a separate, manned control point the cue programme may terminate there and the contributor may be given a cue to start by the operator. Alternatively, it may be fed to the studio for the contributor to hear on headphones or (for television) a concealed earpiece – better than a loudspeaker, which might cause coloration. The feed might be for the beginning and end of the contribution or, in an interview, will continue throughout. If there is no alternative to a loudspeaker being on while the microphone is open, the contributor could be asked to use the directional properties of both to minimize spill, and then reduce the loudspeaker level until it is just clearly audible at that end. When the contributor speaks, it is possible to test the circuit by raising the incoming level until coloration is heard: this establishes a level to keep well below. During the link, try reducing the volume of cue feed while the contributor is speaking, but take care to give no reason to fear that it has been cut off. Anti-feedback pitch spreading can also help.

Another arrangement is to have broadcast-quality or broadband lines both ways. This is necessary if the material is to be broadcast or recorded at the same time at both ends.

In practice it may be unwise to go directly on air with members of the public whose language or manner are unpredictable. A digital editing system can be used 'on the fly' to mark and delete unacceptable language and select a good out point for a 'near-live' contribution.

Clean feed and multi-way working

The simplest form of *cue programme* is the output of the mixer studio. This is satisfactory if just a cue to start is needed. The remote source then switches to hear local output until the contribution ends.

In the more general case, participants will hear the originating studio and contributors from other sources throughout. Some speakers like to hear their own voices coming back – it may help them judge their delivery and effectiveness – but most do not; at worst it may seriously inhibit their performance. Ideally, all speakers who hear cue on headphones should have the choice of hearing themselves or not.

The choice is not always possible: if, say, speakers at the New York end of a transatlantic discussion were fed the full programme from a mixer in London they would hear their own voices coming back, delayed by the time it takes for the signal to travel there and back. By cable, at the speed of light, this is a thirtieth of a second, bad enough to make speech decidedly uncomfortable, and if the return signal is loud enough, it becomes impossible to carry on without stuttering. Digital conversion can also introduce delays that do this. Satellite links are worse, with quarter-second delays in each direction. For television, links with delays may be used even for contributors in the same town.

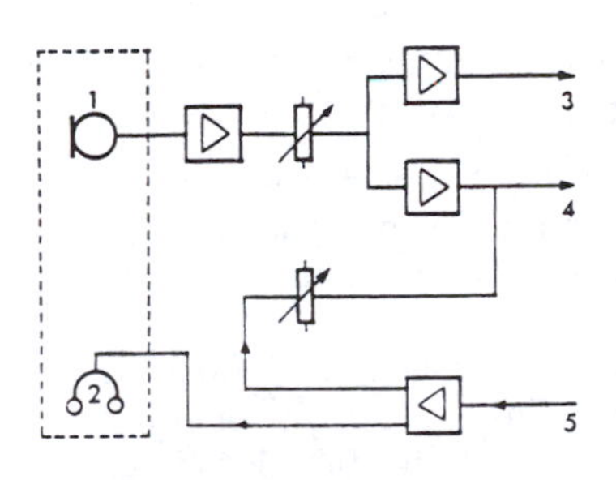

[13.7] **Clean feed circuit**
For multi-studio discussion, speakers in the studio can hear a distant source on headphones without hearing their own voices. 'Clean' cue is also provided. 1, Microphone and 2, Headphones (or concealed hearing device for use on television) for a speaker in the studio. 3, Clean feed to distant studio. 4, Combined studio output. 5, Incoming line. This arrangement is essential for both radio and television when studios are linked by satellite, which introduces a long delay to any sound that is returned to its point of origin.

For these it is essential to use *clean feed* (also called *mix-minus*), in which participants hear everything except their own contribution. Most consoles provide circuits for this. For two-way working where the outside source line is added to the studio output after the main gain control, a clean feed can be taken off just prior to the main control and fed via a parallel fader: in effect, there are then two main controls, one with and the other without the remote contribution.

A typical use of clean feed is to provide an actuality feed of an event to which various broadcasters add their own commentary. The actuality sound is fully mixed at the originating console, where commentary for the home audience is added through an independent channel fed in to the parallel output. This is similar to the provision of mixed music and effects tracks in film dubbing: there, too, a user may add commentary separately, perhaps in a different language.

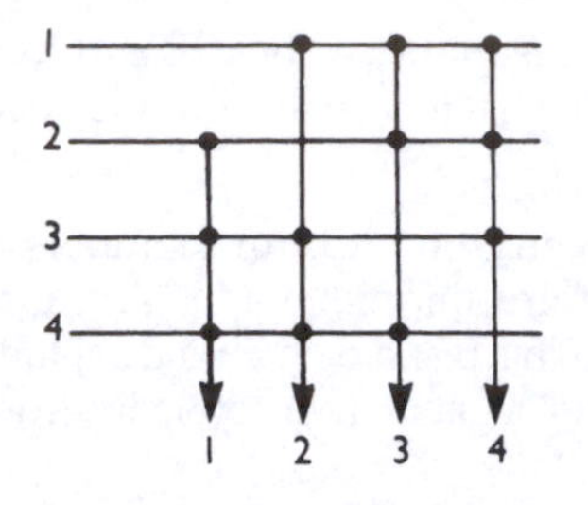

[13.8] **Multi-way working**
In this example, four remote sources each receive a broadcast-quality mix without their own contribution. This matrix can be achieved by the use by 'auxiliary sends' or 'independent main output' facilities in the central mixer console.

Another possible arrangement uses auxiliary circuits. Each channel in the mixer has provision for a feed to be tapped before the fader and fed to an auxiliary bus, where the output of all channels selected is combined in a rough mix that can go to any local or remote point. Many consoles have two or more auxiliary 'sends', to allow *multi-way working*.

An application of this is the 'independent evidence' type of programme where a presenter questions two people in separate studios who cannot hear each other (e.g. to examine the reactions of various partners who know each other well: brother and sister, secretary and boss, or husband and wife). Three studios are required: the host, listening on headphones, hears the full programme output, while in two remote studios the 'victims' are fed nothing of the other person's answers, and only such questioning as is directly addressed to them.

A complex form of clean feed is used when recording sporting and other events of international interest. Picture and sound are recorded on videotape and a synchronized multitrack sound tape. Commentaries in different languages can be recorded at the same time and recombined with the actuality sound when the tapes are replayed.

Telephone circuits in radio and television programmes

Prior to the introduction of ISDN links, telephone circuits were not routinely used in all broadcast programmes for several reasons:

- Speech quality from the handset is good enough for communication but far below that normally required in broadcasting. Distortion and noise levels are high, with an uneven frequency response.
- The standard telephone bandwidth is narrow, 300–3000 Hz in the USA and 300–3300 Hz in the UK. This is enough for full intelligibility, but, again, less than is desirable for broadcasting.
- The telephone system cannot be expected to guarantee the freedom from interruption needed on lines used for broadcasting.
- Connection by automatic equipment at a given time cannot be guaranteed: junctions, overseas circuits or the number may be busy.

Some of these conditions still apply where ISDN broadband is available. Plainly, since broadcasting is geared to different technical standards, telephone circuits are used only for good reason. Valid reasons are:

- *Extreme topicality*. There is no time to arrange higher quality lines or to get a recording or the reporter to the studio, and the same programme value could not be achieved in any other way.

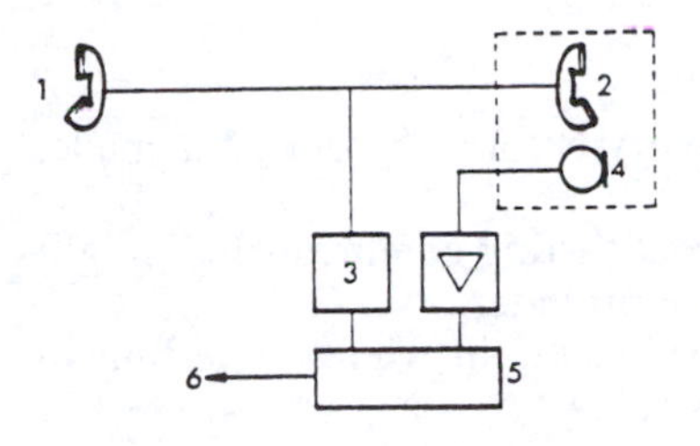

[13.9] **Broadcasting live telephone conversations**
1, Remote telephone. 2, Studio telephone. 3, Constant-volume amplifier. 4, Studio microphone. 5, Console. 6, Studio sound output.

- *Extreme remoteness.* Here, physical inaccessibility is the problem – a distant country from which better quality lines are not available. The speaker may be in an area that has been struck by natural disaster. (But satellite links can still often be set up in such circumstances.)
- *Audience participation.* The ordinary citizen at home, without elaborate preplanning, is invited to take a direct part in proceedings to which the audience is normally only a passive witness.
- *The phone-in.* This is an extreme example of audience participation, in which telephone quality is an intrinsic part of the show.

In practice, after the circuit is established on an ordinary telephone – often one of several set up in the studio – a feed is diverted through a 'telephone balance unit' to the studio console, where it is mixed with studio output to add the full frequency range of the near voice.

The use of a constant-volume amplifier or (essentially the same thing) a combination of amplifier and limiter will ensure that incoming speech is as loud as that originating locally. Note that the gain of a simple limiter should not be set too high – say, to a maximum of 8 dB – or line noise may be boosted noticeably during pauses in the conversation. However, if there is a noise gate in the circuit (and assuming there is adequate separation between speech and noise levels) the gain of the limiter can be substantially increased. Also, when the call is checked for quality before transmission, it is essential to establish that no crosstalk is intelligible.

The call remains connected as long as it is routed through to the studio, even if the studio telephone line is not open, and should not be disengaged at the end of the conversation when the studio line is closed. It is finally disconnected after switching back to the control phone or by the remote caller replacing that receiver.

For the broadcaster, special services may be arranged: fixed time calls may be booked, or timing pips that would normally be superimposed on a call may be suppressed. From the point of view of the engineers of the telephone service, there are two rules that must be observed:

- no equipment should be connected that might feed a dangerous high-voltage signal to the line;
- no unusually loud signals should be transmitted as these could inconvenience other users, inducing abnormally high crosstalk.

The way in which calls are used in programmes may be subject to a code of practice agreed with the telephone company. Whether the following are seen as rules or just as matters of common courtesy depends on the local arrangements and on regulations imposed by the licensing authority. These may vary from country to country and also with the type of use (recording or live broadcast) and to a lesser extent the kind of programme. Some are common sense; others are designed to avoid complaints to co-operating organizations whose service is inevitably seen at a technical disadvantage; yet others are intended to limit intrusion.

The broadcaster should:

- Obtain the consent of the other party before a broadcast is made.
- Avoid using party lines.
- In the context of a call, avoid broadcasting comments that might sound like criticism of the service provider.
- If calls become unintelligible or subject to interference, terminate their broadcast as quickly as possible.
- Immediately abandon calls that are interrupted by an operator or crossed line. Someone should then call the other party as soon as possible to explain what happened.
- Broadcast telephone numbers, or permit them to be broadcast, only by prior arrangement. Measures may be needed to avoid exchange congestion and prevent emergency calls getting blocked – this may prove expensive.
- Avoid involving operators or other telephone service staff in the broadcast.

Where a conversation is recorded, the same result must be achieved as if the item were live, except that editing may be used to achieve it.

In order to ensure that quality is as high as possible, speakers at both ends may be asked to speak clearly but without shouting, not too fast, and with the mouthpiece by the mouth, not under the chin. Broadcast sound should not be loud near the telephone. If there are parallel extensions at the caller's end, these should not be used by others to listen to the call, as this too could reduce the quality.

Broadband digital links may be used if the speaker at the far end has a suitable headset and codec. A public or company office, for example, may have the equipment – though it may turn out that the quality is not as good as when both ends are controlled by the broadcaster, in which case ordinary telephone quality may be better than some peculiar compromise. If ISDN is used, it is set up as described earlier.

Chapter 14

Fades and mixes

This chapter covers what may seem to be just a few tiny details of one of the simplest operations that any console (or virtual studio) user can face. But it is more than that. In fact, as much as anything else, it is the way fades and mixes are carried out and timed that characterizes a polished, professional product. Not every fade should be just a slow, steady slide in or out. Everything from disc jockey shows to high drama on film or television depends on a sensitive hand: a good fade is like a well-played piece of music. A good operator does not wait for a cue and then act on it, but rather, anticipates it so that cause and effect appear to coincide – which is, after all, the purpose of rehearsal.

What constitutes a smooth fade? The ear detects changes of volume on a logarithmic (decibel) scale, so all faders are logarithmic over much of their 30–40 dB working range. Below that, the rate increases – again, roughly in line with our expectations. At the lower end of a fade, as the listener's interest is withdrawn from the sound, the ear accepts a more rapid rounding off. The fade is, in fact, a device for establishing a process of disengagement between listener and scene: once the process is well established, it remains only to quickly wipe away the tail end. An even movement on a standard fader does most of this, but the rate of fade in or out can often be increased at the lower end, and reduced to allow fine control in the final few decibels at the upper end.

A more complex use of faders overlaps their operation, as a mix. (Music mix-down, a separate subject, is described in Chapter 9.)

The fade in radio drama

Consider first the fade in radio drama. For many users, this must seem an almost forgotten application. If so, think of it as a model for the more general use of operational techniques – and not just in all kinds of radio production, but throughout all uses of sound in all media.

Radio drama often adopts a convention that each scene starts with a fade in and ends with a fade out; narration, on the other hand, is cut off almost square. This ensures that there is never any difficulty in distinguishing between story and teller. Within the story, the type and rate of each fade and the length of pause that follows it all carry information for the listener. The example of radio (and particularly how dramatic effects are achieved in radio) is useful because in creating a sound picture the resources available are much more slender than those for film or television pictures, which can employ cuts, fades, dissolves and wipes, plus all the elements of camera technique. A great deal has to be conveyed in radio by subtle variations in a fade.

For a scene change, the gentlest form is a fade to silence over 5 or 6 seconds, a pause of a second or two, and an equally slow fade-in. This implies a complete discontinuity of time and action. Faster fades imply shorter lapses of time, smaller changes of location, or greater urgency in storytelling. For instance, for 'Maybe it's down in the cellar, let's go and see . . . well, here you are, have a look round', the fades can be quickly out and back in, just enough to establish a tiny time lapse.

Moving off, then back on microphone, perhaps helped by small fades, is used for still smaller changes of location, say from one room to the next. It is convenient to assume that a microphone is static within the field of action that may be moving, e.g. a scene in a car. Any departure from this convention must be signposted by dialogue or effects, or the audience will be confused. A scene starting by a waiting car that then moves off may create a moment of confusion if it is not clear whether the microphone goes with the car or not. Dialogue pointers are often used to clarify this, and their deliberate misuse can raise a laugh:

FUNNY MAN	Quick, drive to Tunbridge Wells!
FX	*(High-powered car departs and fades out.)* *(Pause.)*
FUNNY MAN	He might have waited for me to get in.

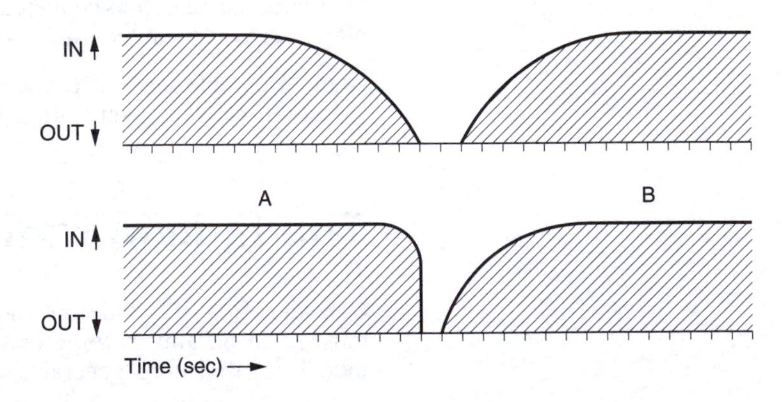

[14.1] **Fades in drama**
Above: Simple fades between scenes, with smooth fade out then smooth fade in. *Below*: Fades between narration and scene. Slight edge out at end of narration (A) and smooth fade in (B). In these diagrams, the baseline represents silence and the upper limit is normal volume.

To some extent, moves on and off microphone are interchangeable with fades. In one survey, listeners found that a fade gave as good an impression of moving off as did actual movement. It only works that way round: if a move is used to represent a fade, the shift in acoustic perspective makes it sound odd – unless it is deliberately exaggerated, perhaps by combining a fade with increasing AR to suggest a move off in reverberant surroundings.

How deep should a fade be, before it is cut off? Plainly, a scene that has a strong curtain line cannot be faded much, and a new scene with a first line that contains important information needs a fast fade-in. For both of these the adjacent fade must be conventionally slow, or the intended scene change will sound more like a pause in conversation.

Difficulties arise when a scene begins with what on stage would be an actor walking in: in radio, this is reinterpreted using dialogue, acoustics and effects to establish a scene, together with deliberate fades for exits and entrances within it. This may mean breaking a fade into several parts: it is no use fading on a pause. So, if an actor has two separate lines while approaching the microphone, the move in might stop for a moment during any intervening line. Similarly, it is better to hold up a fade-in momentarily if there is a gap in speech. It may even be necessary to reset, jumping up or down in level between lines from different speakers to make the overall effect of the fade smoother.

'Door opens' is another instruction that may head a scene. For such an effect to be clear, it has to be at close to normal level, and the fader is then taken back a little for the fade-in proper. In general, a short spot effect of this kind is not recommended for the start of a scene.

Educational radio may employ shallower fades, partly because of the listening conditions in classrooms: a poor receiver or boomy room acoustic. Producers who visit schools to listen with their audiences recommend high intelligibility, low dynamic range and background effects held well down so they do not interfere with speech.

Listening conditions in some parts of the world are poor because crowded short or medium wave transmissions are heard at the limit of their range, or on small, cheap radios. For these, too, a form of presentation that uses shallower fades is necessary.

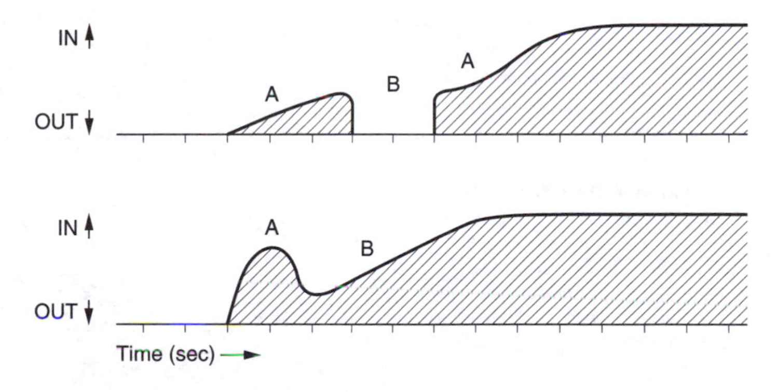

[14.2] **Complex fades**
Above: A, Speech. B, Silence. Do not fade in on silence. *Below*: Complex fade to accommodate a quiet effect at the start. At A, fader is open for 'Door Opens', then taken back for fade-in (B) on speech.

Mixing in effects

If the convention is that each scene opens and closes with a fade, plainly there must be sound to fade on. Using unimportant dialogue for this is dramatically weak. To fade out on laughter (where appropriate) is better, but using background effects is of more general use. As the scene comes to a close, speech is faded down and effects lifted to swamp the line; then they too can be slowly faded out. Similarly, a continuous background of sound can be used for a fade-in – effects are good for establishing new localities – and here again the sound may be peaked for a moment, then faded gently as dialogue is brought up.

For another kind of fade-in, bring up effects a little before speech – just enough for it to be clear what they represent, but not enough to draw attention away from the first words. The more complicated the background effects, the more they will need to be established first. In either case, they must nearly always be dropped to a lower level behind the main body of the scene than at the beginning and end. The level then may have to be unrealistically low, and lower in mono than in stereo, where the spread background interferes less with intelligibility.

Varying the level of effects behind a long scene helps to stop the sound getting dull: it should be lowest behind speech that is important to the plot, and louder when dialogue refers to the backing sound, or when characters have been directed to pitch their voices against it. Performers should be encouraged to sharpen their tone if in similar natural circumstances they would do so.

Effects can often be peaked as a bridge between two parts of the same scene: different conversations at a party can merge into and emerge from a swell of background chatter. In film terms, this is like either a slow camera move from one group to another or like one group moving away and others coming closer to a static camera. When you construct a scene for radio it is all too easy to imagine that you are being more precise in depicting movement than is actually the case. Fortunately it is not usually important if listeners see it differently from you. Nevertheless, it may be possible to be more precise, perhaps by giving the impression of a tracking shot by peaking effects and then gradually changing their content.

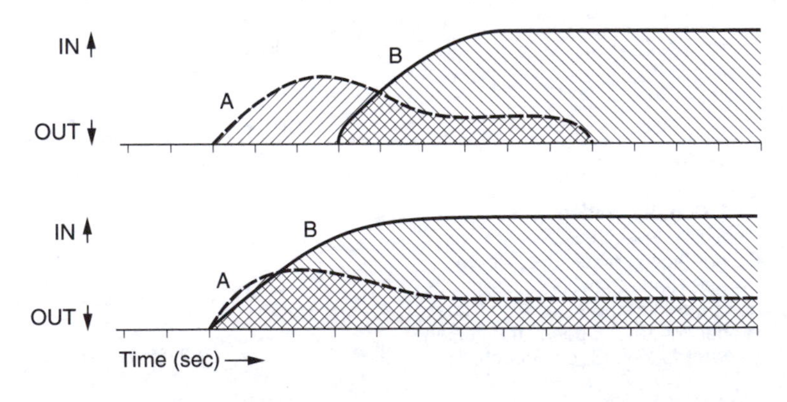

[14.3] **Fades with effects**
Above: Establishing effects (A) before speech (B) on separate faders. *Below*: Fading up effects (A) with speech (B) and then holding them back to avoid an excess of distracting backing sound.

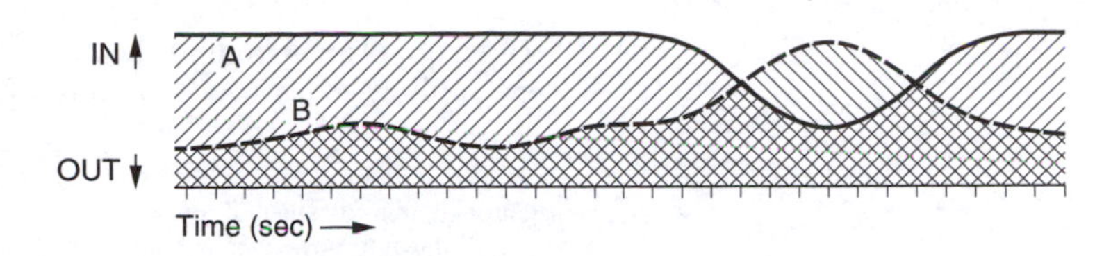

[14.4] **Mixing effects**
Background effects (B) can often be varied continuously and can sometimes be peaked to bridge scenes, here as the fader for speech (A) is dipped.

Beware the overuse of background effects. Used intermittently, they enliven a production; used for every scene, they become tedious, and more so if they are too loud.

Fades in television and film

For television and film the rate of fade and timing of any pause is normally set by picture: in the simplest case a fade of picture is matched exactly by fade of sound, but when there is low but significant sound during the fade that should be emphasized. Similarly, an effect of little significance that sounds too loud can be reduced.

There are plenty of obvious uses for effects and music during picture fades. For example, to bridge a time lapse in a party scene, peak the chatter background rather than fade it out: this firmly implies that action and location are unchanged. If the new scene then turns out to be a different party, a deliberate incongruity is introduced. A change of scene from one noisy location to another might be made by peaking effects against the fade and mixing while the picture is faded out. This works best if the two sounds are characteristically different. Such techniques should be used only with calculated and deliberate intent. It is self-conscious in that it draws attention to technique, at least for as long as it takes to establish a convention. The dramatic reason must be strong, with sound that is important to the action.

The less obvious use of effects, as described for fades in radio, also has its uses. When dissolving between pictures it is normal to mix sound backgrounds (or fade the louder of them in or out) to match picture. Here a peak in effects to bridge the mix may be accepted.

More interesting, because it is more open to the creative use of sound, is the picture cut from one location to another. The simplest case uses a straight cut with dialogue (or strong action) to carry it. For example:

Vision:	*Sound*:
MS oilman	*FX: Burning oil rig*
... turns to see distant figure striding towards burning rig	OILMAN (V/O): Sure it's easy. When you've stayed alive as long as this boy has, the hardest job is working out what to charge the client.
Paper in manager's hands, fast tilt up to CS face.	MANAGER: Nine and a half million!!

Here the effects behind the first scene must fade out fast. The next example uses a variety of techniques on three shots:

Vision:	*Sound*:
Through arch to cathedral tower ... tilt down to boys approaching under arch and past cam	*Distant choir sings* *Chatter, footsteps*
	NARRATOR: In the shadow of the medieval cathedral this group's interest in electronics complements traditional activities like choral singing ...
	RUGGER PLAYER: Break!
WS playing field, cathedral and school buildings b/g, pan L as ball and players cross diagonally towards cam ... when centred on science block, z/i past players to windows	NARRATOR: ... and Rugby football. Perhaps rather surprisingly, the work we're going to see started not with physics, but here in the chemistry lab.
Wide HS in lab, 2 boys and glassware f/g	BOY: Oh no! If we've got to do this, it's set all wrong ...

The picture has been shot with the commentary in mind. The music was recorded separately; but chatter and footsteps, football effects and laboratory sound were all recorded in sync with the picture. Before mixing, the chatter and footsteps track is displaced a few frames because the approach then matches better with the tilt down. The music is held low from the start and chatter and effects fade in with the tilt down to dominate it. The shout 'break' cues the cut to the football field and kills the music. The football effects that follow on the same track are therefore brought up to full volume before the picture cut. With the first zoom, the effects dip sharply to a level matching that of the laboratory, so that a quick mix on the cut is enough.

The key device is that of sound anticipating picture. Here the result seems natural, rather like hearing a noise in a street and turning to see where and what it is. This can also be used for strongly dramatic cuts. More noticeably to the viewer, the anticipation of sound can be used stylistically, as when a football fan who is talking about some big match lapses into a daydream during which we hear the sounds of the crowd distinctly as though in his thoughts, followed by a cut to the crowd scene at the match and the roar of the crowd at full volume.

Dialogue overlapping picture is now a widely used convention, but needs care as it can call attention to technique or, worse, cause confusion.

A whip pan generally implies a spatial relationship, which may offer interesting possibilities for the use of sound, as here:

Vision:	*Sound*:
LA through rigging to flag on mast	*Music: military band starts playing*
LA figurehead of Nelson, band crossing R–L	NARRATOR: ... and the martial music, the uniforms, the discipline and the impressive reminders of a glorious past,

MS Nelson figurehead ... whip pan R and down	all tell us that this school trains its boys for the sea...
Cut in to matching whip pan to boys turning capstan	*FX: capstan turns, ropes creak, marching*
Closer, pairs of boys push spokes of capstan through shot	Their project, too, is suitably nautical.
MS boy, boat in b/g hauled up slipway	BOY: With two of our boats at school we had a problem....

In this case the whip pan is not a single shot of the two objects: the figurehead and the capstan are half a mile apart. That they are all part of the same general location is suggested by a series of atmospheric detail shots with sound overlaps. The sync music from the second shot has been laid back over the first (mute) picture and also continued through the whip pan, but drops sharply during the camera movement. On the capstan shot, the associated effect (recorded separately) is allowed to dominate, and the music is eventually faded out under the boy's speech. The noise, relevant to his words and to subsequent shots, is held low behind his voice.

The use of sound in all these ways is suitable to both television and film. So, too, is the use of music links between scenes. Indeed, the music bridge is a device common to radio, television and film.

Theme and background music

Few things about broadcast drama and documentaries attract so much controversy as the use of music. This is because, to be effective, and perhaps also to justify having specially composed music, it is often loud. In contrast, in the home, many people like speech to be clear, with less and quieter music (cinema audiences seem less likely to complain). Most people agree that, when it succeeds, music helps create works of high quality; but at worst it may just be used to smudge a supposed elegance over a scene that would be as effective with natural sound.

When television shows works that are the product of, or derived in style from, the older traditions of film, music is still 'in'. But to meet the taste of home audiences, simpler styles have been developed, artistically justified and found to be effective. This is linked to a greater use of well-recorded effects: these either increase realism or enhance emotional response, just as music does. It follows, more than ever, that music must be selected with great care (or rejected if there is not a perfect fit) and integrated into programmes with delicacy.

The ideal is either specially composed music or well-chosen selections from commercially released records. Sometimes suitable background music can be found in 'mood' record catalogues. But use this with care: it is easier to lose tension and coarsen texture by adding music than it is to enhance it. When in doubt, avoid it.

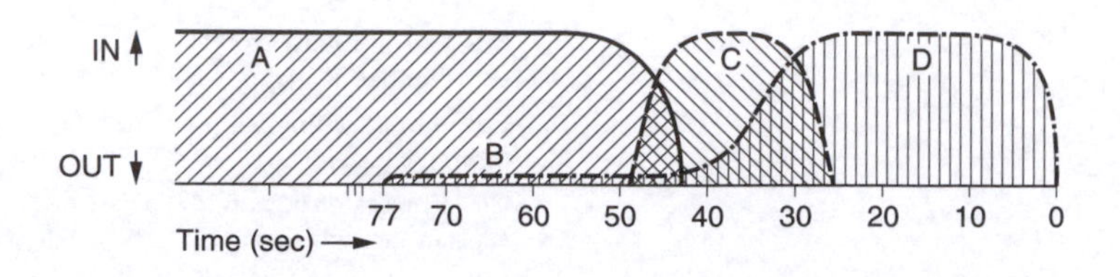

[14.5] **Pre-fading music**
The pre-fade time is the number of seconds from the scheduled end of programme (in this example, 77 seconds). A, Last item of programme. B, Closing music is playing but not faded up. C, Announcer. D, Music is faded up to close programme.

Signature tunes and theme music must be in harmony with, or provide a counterpoint to, the qualities of the programmes they adorn. At the opening they must suggest the mood of what is to follow, and at the end, sum up what has gone before. Another quality often sought in linking music is 'fadeability', with points at which it can be got in and out tidily, all this in a piece of the right duration. At the closing, to reflect the progress of the story, the selection might be a more complex development of the music at the start. For some pieces, this means (conveniently) that the end of the record can be used to end the programme. But it is not so common for the start of a record to provide a good crisp opening signature; more often it is necessary to disregard any introduction and start clean on a main theme.

For the end of live radio shows, closing music is often *pre-faded*: this is a device used to ensure that a slot is filled exactly to the last second. The term 'pre-fade' here means 'not yet faded up'. (An American term for this is *dead-roll.*) In a typical radio pre-fade it was known that the duration from an easily recognizable point to the end of the record was 77 seconds. So the replay was started that same 77 seconds from the scheduled out point. Usually no more than 10–20 seconds was actually needed, and this last remaining part of the record was faded up under the closing words, or a little earlier if a suitable phrase presented itself to the operator, who listened on headphones.

Mixing from speech to music

Some of the ways in which music and speech may be linked together in a record show are outlined below. In this example, linking words cue a pop vocal that has a short instrumental introduction:

- *Straight – or nearly so.* The 'intro' on the record is treated as a music link between introductory words and vocal, of which the level can often be altered to match the speech better. After this link, the record may be faded to a suitable overall level. The first notes of an intro can sometimes be emphasized to give bite to an unusual start.
- *Cutting the musical introduction.* The first word of the vocal follows the cue quickly, as though it were a second voice in a brisk argument. It may be necessary to edge the record in a little.
- *Introduction under speech.* The start of the music is timed and placed at an appropriate point under the specially written cue. The intro is played at a low level, so it does not reduce the intelligibility of the words, but there may be a gap in the cue in which the music is heard on its own for a few seconds: it is

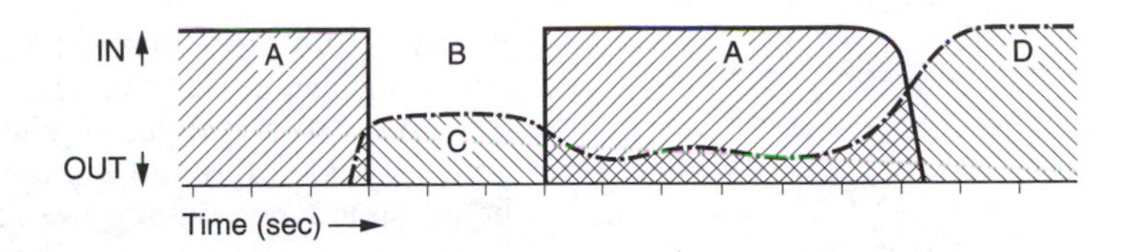

[14.6] **Mixing from speech to music**
A, Speech. B, Pause in speech. C, Music 'intro'. D, Music at full volume. During the pause the music is started and then dipped behind continuing announcement. It is finally lifted on cue to its full volume.

> *established.* The fade-up starts just before the final words, to be at full volume for the start of the vocal, which may follow the cue after a breath pause.

A coarser technique employs compressors arranged so that speech pushes the music level down: an American term for this is *ducking*. With good-quality reception, the effect is unattractive, but it is suited to domestic radio background listening or car radio reception in noisy conditions, and it is operationally simple.

When links are recorded in advance to a play-out system the same type of links can often be programmed in. The final 10 s of the outgoing record is played to headphones for cue purposes, the linking words are recorded with a programmed cross-fade and the process is repeated on the link to the next record. The play-out system stores the links separately and reconstitutes it all, including the fades on transmission.

Joining music to music

The simplest music link is the *segue* (pronounced 'segway' and meaning follow-on), in which the new number follows either after a very short pause or in time with the previous piece. Sometimes an announcement may be spoken over the intro at a convenient point, or over the join.

In some cases it may be better to avoid endings and introductions by *mixing* from one piece to another. Each 2- or 3-minute item on a pop record is, naturally, written and recorded as a complete work in itself, with a beginning, middle and end. In a fast-moving record show only the middle may be wanted, or the intro (or ending) might be too big for use in the show without throwing it out of shape. If the rhythm is strong in both pieces, precise timing is essential. It is easier if the music reaches a point where its form is not too definite; where, for example, the outgoing passage can be faded on a sustained note and the next brought in on a rising arpeggio or glissando. But while all this can be fun to put together, it may not give the audience much added listening pleasure: do not mess records around just for the sake of it.

A segue or mix is possible only if keys are the same or suitably related. It may be that keys that are nominally the same are not close enough in pitch without some adjustment. Expert musical knowledge is not needed, but good relative pitch is essential. Many replay machines have provision for fine adjustment of speed, and so pitch.

In any record programme, too, lend an ear from time to time to the key sequence: a lively variety of keys adds vitality. The question is, how to get to a remote, unrelated key. Within a piece of music it is generally done through a progression of related keys. That may not be possible when linking pre-recorded items, so unless a dissonance is actually wanted, some other means of transition must be found. There is a story of a music-hall conductor who, when faced with the problem of a quick modulation, simply wrote in a cymbal crash and promptly jumped to the new key. The sudden loud noise of indefinite pitch was designed to make the audience forget the key that went before. In radio a marginally more subtle method is used: the announcement. With even a short announcement of 5 seconds or so, difficulties over pitch disappear. The ear does not register a dissonance if two sounds are spaced and attention is distracted by speech.

Adding speech to music

Where speech is *added* to music, as in the case of theme music with announcements or introductory words over, the voice is normally chosen for high intelligibility. The music can therefore often be given full value and dipped just in time for the first words. Similarly, ensuring that only the last word is clear, the music is lifted tight on the tail of speech. Also, in anything longer than a breath pause in speech, the music is peaked up. This is called a 'newsreel' commentary and music mix after the old style of filmed newsreels in which the inconvenience of synchronized speech, or even sound effects, was avoided.

At the end of a programme, an upbeat finish may demand that music, applause and speech all compete: here, similar techniques may again be employed. But in neither case should it be forgotten that, for many of the audience, disengagement is accelerated only by what at home just sounds like a loud and not particularly pleasant noise. An alternative, more relaxed mix has gentle music fades before and after the words.

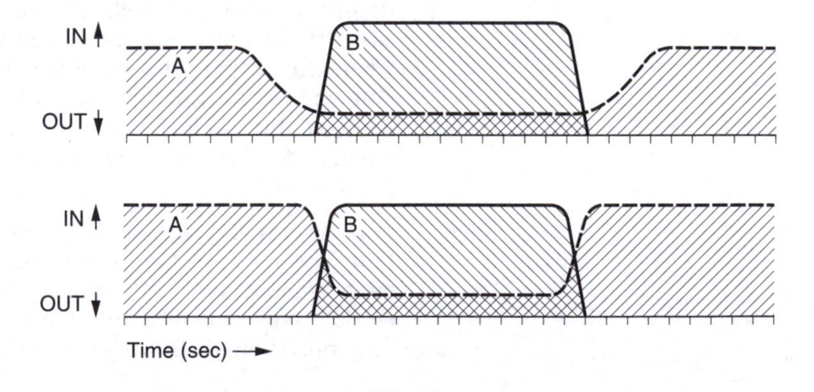

[14.7] **Film music and commentary** *Above*: Normal cueing. Music (A) is gradually adjusted in volume so that commentary or other speech (B) is clearly audible when it starts. After the speech ends, the level is adjusted gently. Alternatively, the music may be written or chosen specially to give this effect. *Below*: 'Newsreel' cueing. Music (A) is held at high volume until a moment before speech (B) starts, and is raised to high level again immediately speech ends. All adjustments are rapid.

Chapter 15

Sound effects

Sound effects can reach the console either from a studio microphone or from a replay machine. There are three main types: spot effects, 'library' recorded effects and actuality recorded effects.

Spot effects – also called *manual effects, hand effects* or *foley* – are those that are created live. In a radio or television studio they are generally produced at the same time as performers speak their lines. In a film sound studio they are created to match action on the screen. For the visual media, they are kept strictly off-camera.

'Foley', originally an American film term for synchronous spot effects, is named after a 1950s practitioner who bridged a gap during which the technique had gone out of fashion in Hollywood. In Britain, where they were never out of use, the long-established name for them was 'spot', so that is what they are called in this book. Spot effects include the noises of doors, telephones, bells, teacups, letters being opened, crashes, bangs, squeaks, footsteps and even that old standby, horses' hooves. They include noises it is simply not worth recording, noises that are so much part of the action that to have them anywhere but alongside it might leave an actor confused and uncertain of timing.

Recorded effects consist principally of those that cannot conveniently be created in the studio: cars, aircraft, birdsong, weather, crowd chatter, and so on. Their field overlaps that of spot effects: for instance, rather than those coconut shells, a recording of real horses' hooves may be preferred – although, curiously enough, some very realistic horses' hoof sounds have actually been recorded with coconut shells. It may seem odd that effects may be created by apparently and in this case ludicrously inappropriate sound sources. But it is important to realize that the purpose of sound effects is not, in general, to recreate actuality, but to *suggest* it. Realism is not always necessary, and at times may even be detrimental to the final result, in that some inessential element of the 'real' sound may distract the listener.

Library effects are those already available on disc (usually CD), hard disk or possibly tape; *actuality* effects are those specially recorded for the occasion (more often for film or video than any other medium).

For any radio production in which effects appear it is probable that both spot and recorded effects will be needed, and a quick skim through a script should be enough to see which category each cue belongs to. The division of labour will often work out about fifty-fifty, with a few cases where the final decision depends on factors such as whether the studio is equipped with the heavier pieces of spot effects furniture or whether a particular recording can be found.

Several people might divide this work: the team leader at the console is responsible for balance; one assistant looks after all recorded effects including on-site editing and another creates effects in the studio – although the spot effects may be pre-recorded (often directly to a digital editing system), then dropped in as required.

The conventional use of effects

Whether for film or television, and even more for radio, effects are rarely used in a strictly realistic way. Even when that is attempted, it is generally a heightened realism that owes a great deal to art.

As an example consider a radio report: an eyewitness account of a 'Viking Raid', re-enacted in a traditional festival. The tape received in the studio could not be used as recorded. The reporter had, rightly, concentrated on getting good speech quality, so that the background noises were too low. There were some good shouts, screams and crowd noise, but following the words 'listen to the clash of their swords', the swords were neither clear nor close enough. Also, there were breaks in the recording, and whenever public address loud-speakers came on to explain the event to onlookers, the atmosphere was destroyed.

To recreate the scene for a radio audience, the recording had first to be edited into a compact, continuous narrative, deleting inessential expressions of enthusiasm – the report conveyed this without the need for words. As reconstructed, the first sound heard was a recorded effect, heavy breakers on the beach (the report referred to gusts of wind up to 40 miles an hour), and these were held behind the studio narrator, who sketched in background information. Toward the end of this the actuality screams were faded in, and the taped report took over. The effect of sea-wash was faded down as the description of the invasion brought the 'Vikings' on to the beach, but an occasional squawk of seagull was dropped into the clamour to keep the sense of location strong. After a first reference to swordplay the words from 'listen to . . . ' were cut, and in their place a closer clash of swords was inserted. The final result was highly realistic: an actual event (with the real shouts and screams) brought into focus by a mix that included a little retouching by adding sounds that could not have been recorded at the time without an array of microphones all over the beach.

That is what might be called a neo-realist technique, well suited to the sound medium in a programme that consists mainly of edited actuality. But for a programme that is created entirely in the studio, we must work within even more artificial – but effective – conventions; we rarely strive after the kind of sound that is found in a real-life recording.

Sound effects must often be held to an unrealistically low level: in real life they may be just too loud. Another reason for this is that masking caused by any background sound is worse in mono than in stereo. The 'cocktail party effect' is the ability of the ear and brain to distinguish a single sound source in a crowded room, rejecting and often almost ceasing to be aware of surrounding noise. When a monophonic link (microphone, recorder or loudspeaker) is introduced into the chain, much of this ability to discriminate is lost. In stereo the sound is clearer: it reduces the need for effects to be held so far back from the rest of the action – although unfortunately, because it may still have to be compatible with mono, it may be impossible to take full advantage of this. When adding stereo effects, listen occasionally to the 'M' signal to check the maximum acceptable level.

Loudness is nothing like so important as the character of a sound, the evocative quality that can make it almost visual. This is true even of effects for which loudness might seem an essential part – for example, a car crash. Analysis of a typical car crash reveals a series of separate components: skid, impact, crunching metal, and broken glass falling to the ground. In a real crash the impact is the loudest part: in fact, it is so loud that if it were peaked exactly to the loudest permissible volume for no appreciable distortion, all the rest would be far too quiet and lacking in dramatic effect. So the whole sound has to be reorganized to ensure that no part is very much louder than any other. Even more important, during this recomposition certain of the characteristics have to be modified to create the right mood for the sound to fit into its context. The comedy actor driving into a brick wall makes a different noise from a payroll bandit doing exactly the same thing. One must be ludicrous, the other vicious and retributive.

The various stages of the crash may be made up as follows:

1. The skid – essential to set up the action. A crash without a skid is almost over before we can tell what is going on. The scream of tyres is a distinctive noise that may not always happen in real life; whether it happens in radio depends on mood. The skid is nemesis, doom closing in on the bandit.
2. A pause. The listener, having been delivered the initial shock, must be given half a second to hold breath and . . . wait for it . . . .
3. Impact. A hard, solid thud – difficult to get just right, but not so important as the other sounds.
4. Crunching metal. In a real crash this can sound like a couple of dustbins banged together, sufficiently ludicrous for the comedian,

but not sinister enough for the bandit. It needs to be purposeful and solid, less like an empty tin can.
5. Glass. Like the skid, this is an identifying sound, and also one that clearly puts an end to the sequence.
6. A period of silence – doom has struck.

No sound effect should be just the noise that would naturally accompany the action. Rather, it must be related to dramatic content and the intended emotional response.

Surrealistic effects

Effects that are intended to be funny are different from ordinary dramatic effects. Action effects may be exaggerated in a subtle way, by slightly emphasizing characteristic qualities and cutting away inessentials, but in the comic effect emphasis of character is carried to an illogical extreme. A whole canon of slapstick, or what in Britain have been called 'cod effects', is created by this method, and in a fast-moving comedy show is deliberately overdone.

Consider an effect in the old Goon Show, where the script simply said 'door opens'. An ordinary door opening may or may not produce a slight rattle as the handle turns, but for comedy this characteristic is seized upon: if it were not for this, it would be hard to find any readily identifiable element in the otherwise quiet sound – a faint click would be meaningless. For a 'cod' door, the volume and duration of this identifying characteristic are exaggerated beyond normal experience – though not so much as to obstruct the flow of dialogue.

Such a method of codding effects works only if the normal spot effect makes use of some slightly unreal element: where the natural sound cannot conveniently be exaggerated, other methods must be found. For example, little can be done with the sound of 'door shuts', which in any case has to be toned down a little to prevent overmodulation. So other methods have to be examined: for example, the creation of completely different and totally unreal sounds. These might include such cod effects as a 'swoosh' exit noise (created, perhaps, by speeding up the change of pitch in a recording of a passing jet plane). Or an object (or person) being thrown high into the air might be accompanied by the glide up and down on a swannee whistle (which has a slide piston to govern its pitch). The villain in a comedy show may be dealt a sharp tap on the Chinese block; or in a children's programme, Jack and Jill may come tumbling down a xylophone arpeggio.

As with dialogue, timing can help to make noises comic. This is typified by the way many a crash ends. An example is Don Quixote being thrown from his horse by a windmill: a whole pile of 'armour' lands on the ground, it is stirred around, followed by a pause. Then one last separate piece of 'armour' drops and rocks gently to rest. To create this effect, a pile of bent metal discs was assembled and threaded at intervals along a piece of string – except for the last single. Any similar pile of junk might be used. The timing of such a sequence is its most vital quality; the essence of its humour could almost be written out like music – indeed, an ear for music is very useful to an effects operator.

Normally a script should be concise, but just occasionally may take off on a flight of fancy: one Goon Show script demanded 'a 16-ton, 1½-horsepower, 6-litre, brassbound electric racing organ fitted with a cardboard warhead'. Here anything but sheer surrealism was doomed to stick on the starting line. The first stage in its creation was to dub together a record of an electric organ playing a phrase of music first at normal speed, then at double, and finally at quadruple speed. Three spot effects enthusiasts dealt with gear changes and backfires, engine noises and squeaks, plus hooters, whistle and tearing canvas – mostly produced by rather unlikely looking implements – and the whole effect was rounded off by a deep, rolling explosion.

Spot effects

A radio drama studio needs a handy supply of spot sound sources. Some may be kept permanently in the studio; others might be collected from an Aladdin's cave of noisy delights – a spot-effects store.

Studio equipment may include a spot door, a variety of floor surfaces, a flight of steps, a water tank and perhaps a car door. It is difficult to find an adequate substitute for the latter, though a reasonable sound can be produced with a rattly music stand and a stout cardboard box: place the stand across the box, and bang the two on the floor together.

A well-filled store (sadly, now rare) can take on the appearance of a classified arrangement of the contents of a well-appointed rubbish tip. The shelves are loaded with stacks of apparent junk, ranging from realistic effects such as telephones, clocks, swords, guns (locked in a safe), teacups and glasses, door chimes and handbells – none of which pretended to be other than they are – through musical sound sources such as gongs, drums, blocks, glocks, tubular bells and the rest of the percussion section – which may appear as themselves or in some representational capacity – and on again to decidedly unobvious effects, such as the cork-and-resin 'creak' or the hinged slapstick 'whip'. In this latter field, ingenuity – or a tortured imagination – pays dividends. One can, for example, spend days flapping about with a tennis racket to which 30 or 40 strands of cloth have been tied (to represent the beating wings of a bird) only to discover later that the same sound could have been obtained with less effort by shaking a parasol open and shut. A classic example is the French Revolution guillotine: slice through a cabbage.

At one time, rain effects were produced by gently rolling lead shot (or dried peas) around in a bass drum, sea-wash by rolling them rather more swooshily, while breakers would be suggested by throwing the shot in the air and catching it in the drum at the peak of the swoosh. This would now be used only for a deliberately stylized effect; for a more realistic sound, a recording would be used. But a bass drum is still a useful item in a spot store: a noise from something on top of it is endowed with a coloured resonance that can overemphasize its character for comic effect. For example, brontosaurus' footsteps can be made by grinding coconut shells into blocks of salt on a bass drum.

A vintage theatrical wind machine has a sheet of heavy canvas, fixed at one end and weighted at the loose end, draped over a rotating, slatted drum. This would now be used only as a deliberately theatrical effect like that in an orchestral work such as Ravel's *Daphnis and Chloë*, where the ability to control pitch is valuable. But realistic wind comes in many qualities, and its tone should suit the dramatic context. A good recorded effects library has gentle breezes, deep-toned wind, high-pitched wind, wind and rain, wind in trees, wind whistling through cracks, and blizzards for the top of Mount Everest or the Antarctic. A wind machine produces just one sound – wind; the same applies to thunder sheets. They are not convincing except for comic effects.

But whether or not such ingenious improvisations have a certain period flavour, there are certain basic qualities of spot effects that are as valid today as they were 50 years ago. With high-quality audio transmission, the acoustic requirements for effects are now more stringent than in the distant past. The need for good timing, however, does not change: it is still the most important quality of a well-performed spot effect – all the more so because it is likely to be a staccato interjection, while continuous background sounds are more often supplied in the form of recordings.

Where feasible, the radio spot effects operator works alongside or behind the performer, their relative positions depending on the stereo image: this helps to match acoustic and perspective, as well as timing. Even large pieces of equipment can be set in the correct perspective. Where this proves impossible – for example, for footsteps that are restricted to an area where a suitable surface is provided, or for water effects that depend on the use of a tank – the effects must be panned to their correct position, and if necessary given stereo spread by using a pair of microphones, or perspective by adding stereo artificial reverberation (AR).

If a separate microphone is necessary it will often be hypercardioid, for best separation. Where the same studio is used as that for action and dialogue, it is usually fairly easy to match the acoustics. Matching that of film has sometimes required more effort: a studio effect can sound too 'clean'. Here AR with a large number of pre-set acoustics may be the solution, and atmosphere of some kind will be added.

Doors

In a studio used for drama, the most obvious item of studio spot-effects equipment is a large square box on castors, generally with a full-size door set in it. The handle is a little loose, so it can be rattled as it opens. Other door furniture might include a lifting latch, bolt, knocker, etc. Also helpful would be a sash window frame (unglazed) fitted in the back of the box. If no spot door is available, do not despair: many existing effects recordings were made with equipment like this, and the principles can still be employed.

The sound of opening has to be exaggerated somewhat: a more natural version may be too quiet to register. The sound of closing may have to be softened a little to avoid overmodulation. Close the door lightly but firmly. The characteristic sound is a quick double-click.

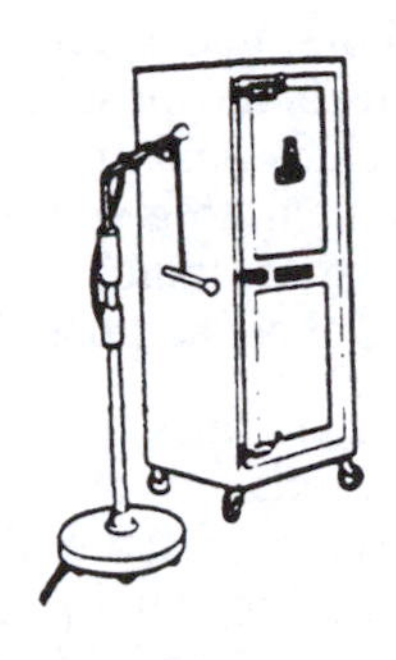

[15.1] **Sound effects door**
This is used in a radio drama studio.

Bad timing will impede the flow of a programme. The sound of a door opening should generally come slightly ahead of the point at which it may be marked in a script; this allows dialogue to carry on without any pause. If a scripted cue follows a completed spoken line of dialogue, superimpose it over the last word, unless there is reason to leave that word cold (e.g. to follow a direction to enter). If the door is to be closed again, allow sufficient time for the actor to 'get through' it. Also, the perspective of actor and door should match, and they should stay matched until the door is shut again, if that is part of the action.

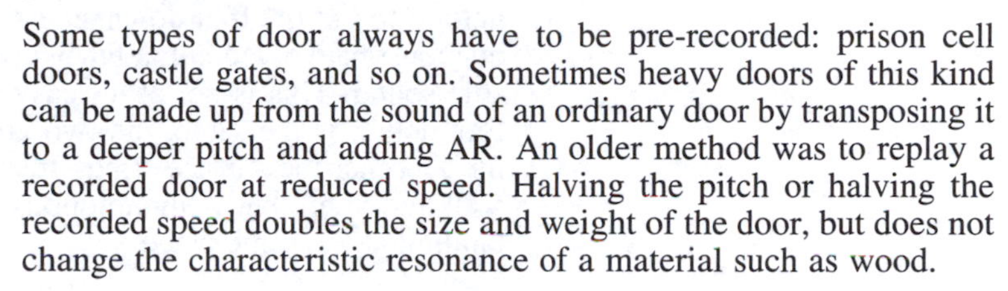

Some types of door always have to be pre-recorded: prison cell doors, castle gates, and so on. Sometimes heavy doors of this kind can be made up from the sound of an ordinary door by transposing it to a deeper pitch and adding AR. An older method was to replay a recorded door at reduced speed. Halving the pitch or halving the recorded speed doubles the size and weight of the door, but does not change the characteristic resonance of a material such as wood.

When knocking, take care to find a place where the door or frame sounds solid enough. Metal knockers, bolts, latches, lock and keys, and so on, should be heavy and old-fashioned; they are best attached to the effects door or, if loose, held firmly against it. The larger area of wood acts as a resonator and gives the sound a truly solid quality. Loose bolts, etc., should be screwed to a small panel of wood.

One last note about doors: when in doubt, cut. Some scriptwriters overdo the mechanics of exits and entrances. An approach on dialogue is often as effective as door-plus-approach. It is better to keep doors, like all other effects, for the occasions when the sound has some real significance in the story. Otherwise, when the time comes that it does, it will not be so easy to establish its importance.

Footsteps

What was a footnote on doors becomes a point of primary importance with footsteps. Are your footsteps really necessary? Go through a script with a 'delete' pencil and make sure that every remaining footstep has direct relevance to the action. In films this effect is often used merely to fill a silence, to reassure the audience that the sound has not broken down at a time when all attention must be concentrated on the picture. There is no direct equivalent to this in the pure sound medium, in which footsteps can all too easily draw the listener's attention in the wrong direction.

Such footsteps as remain may be created as spot effects or played in from a recording: the former is preferable, because the sound effect should tie in exactly with the mood of the moment, while also reflecting character. In the studio the exact pace can readily be set and changes of pace and a natural stop produced exactly on cue.

[15.2] **Gravel 'pit' for footstep effects**
This pit is built into the studio, and concealed below the wooden floor.

A well-equipped radio drama studio or film sound studio should have a variety of surfaces available for footsteps: the most important are wood, stone and gravel. Boards or stone slabs (laid on top of the studio carpet) should be as hefty as is convenient, to give weight to their resonance: the carpet stops them ringing too much. A gravel effect can be created by spreading sand and gravel on the stone slabs.

On these surfaces long sequences of footsteps are difficult. Walking on the spot is not really a satisfactory way to create footsteps, as it takes an expert to make them convincing. Walking along, the foot falls: heel, sole, then heel, sole again. When walking on the spot it is natural to put the foot down sole first, so a special heel-first technique must be worked out: a fairly crisp heel sound, followed by a soft sole sound. This is not at all easy to get just right. Even when the foot action is the same, the two are still readily distinguishable by the fact that the surface 'ring' sounds different on every step when walking along, but is monotonously the same when the foot keeps landing at the same place.

Walking or running up and down stairs are also better done with real steps: there can be a very different quality for each step. But if the effect has to be simulated, the correct footfall is with the sole of the foot, going up, or sole–heel, going down. Again, the result is difficult to get right unless there is some variety of surface.

Attention should be given to footwear. Leather heels and soles are best for men's shoes. High-heeled women's shoes are usually noisy enough without special care in their choice. If only one person is available for footstep effects, remember that it is easier for a woman to simulate a man's footsteps than vice versa. It is not easy to get the right weight of step if shoes are held in the hand. The simplest way of getting two sets of footsteps is to have two people making the sound: except by pre-recording, one person cannot conveniently do the job, especially as the two are not generally in step. But two sets of footsteps will stand for three, and three sound like a small crowd.

Some surfaces are simulated. For footsteps in dry snow, cornflour (in America, corn starch) has been used, but this can be messy. A fine canvas bag of cornflour can be crunched between hands, or a spoon (or two) pressed into a tray of the powder. For a frosty path try salt; for dry leaves or undergrowth, unrolled reel-to-reel tape.

Telephones, bells and buzzers

For electric bells and buzzers it is useful to make up a small battery-operated box that can be used to give a variety of sounds (or make up several boxes, each with a different quality). The mobility of the box allows it to be placed at any appropriate distance, in any suitable acoustic, or against any surface that will act as its sounding board.

For telephones in radio, a standard handset can be adapted, so that the ringing tone may be operated by a press-button. In order to suggest that picking up the hand-piece stops the telephone ringing, try to finish up with half a ring. For example, a common ringing tone in Britain is a repeated double buzz-buzz... buzz-buzz; the ideal interrupted ring goes buzz-buzz... buzz-b... followed quickly by the sound of the handset being lifted. As this last sound is much quieter than the ringing, two things can be done to make it clear. First, the rattle of picking it up should be emphasized a little, and second, the telephone can be brought closer to the microphone at the end of the effect. The first ring is fairly distant (perhaps partly on the dead side of a directional microphone). This is doubly convenient in that it may be interrupting speech, for which purpose it is better not to have a foreground effect. The second ring is closer, as though a camera were tracking in to the phone, to become very close as it is picked up. The words following must also be in close perspective.

The action of replacing the receiver may be balanced at a greater distance. Naturally louder than the pick-up, it is an expected sound that fits neatly in the slight pause at the end of a conversation.

Dialling is more time-consuming than being on the receiving end of a call, simply because most codes are much too long: 10 or 11 digits are commonplace. If there is any way of cutting this sort of hold-up from the script it should usually be adopted, unless dialogue can continue over dialling. Period rotary dials are a menace: codes were shorter, but even so, it is better to dial only a few numbers.

For television, telephones are used so often that it is worth making up 'specials' – a portable kit using real telephones, properly (and permanently) polarized so that they are fully practical, and linked together. This is necessary because good sound separation between the microphones picking up the two halves of the conversation may make it difficult for the parties to hear each other clearly without the telephone. If one of the speakers does not appear in vision it is better to use an acoustically separate booth.

[15.3] **Effects bell**
This home-made device has two press-button controls.

If both speakers do appear in vision at different times their voices are balanced on normal studio microphones, but that of the speaker who is out of vision must be heard through a telephone-quality filter. Manual switching is risky, as the result of the slightest error may be obvious. A better arrangement is therefore for the filter circuit to be linked to camera cut buttons so that the two ends are automatically reversed on a cut. When there are more than two cameras (i.e. more than one at each end) it is necessary to make sure that *any* camera

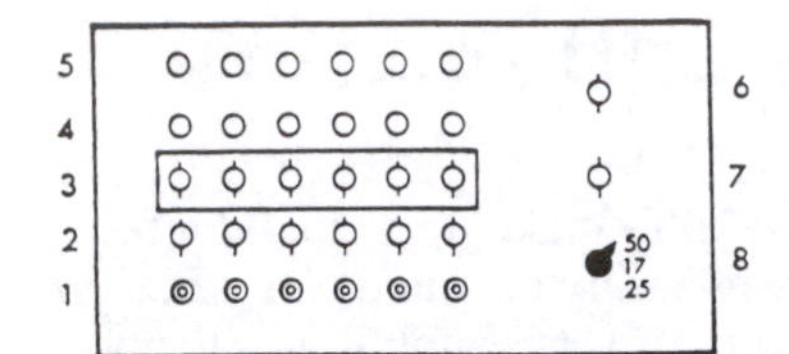

[15.4] **Control unit for telephones used in television plays**
1, Input sockets for up to six lines. 2, Keys to select mode of operation (manual or automatic) for each line. 3, Line keys: down to select, up to cancel. 4, Indicator lamps to show line 'called'. 5, Indicator lamps: 'line calling'. 6, Ringing switch for use by operator: up for automatic ringing; down for manual ringing. 7, 'Auto-dial' switch: when 'on', this permits extensions with dialling facilities to ring each other automatically; in addition, normal dialling tone is heard when a receiver is raised. In all other operations the switch is left 'off'. 8, Ringing frequency selection. A simpler arrangement for interconnecting telephones is often adequate.

that can possibly be used in the sequence is linked in to the switching system.

A kit can be set up with a number of telephones connected through a box that acts as a small telephone exchange that can produce manual or automatic ringing tones at a range of audio frequencies. Lights above each key show the condition, 'called' or 'calling', of the associated line. When the automatic button is pressed the selected telephone rings immediately, beginning with the start of the first buzz, and it goes on automatically until either the operator's finger or the phone is lifted, when the ringing stops in a realistic manner. Using automatic telephones also ensures that the bell is in the same acoustic and perspective as the actor. Take care to use appropriate ringing tones: different countries have their own standard signals.

Most 'far end' dialling, ringing or engaged tone effects can be provided by using a buzzer, etc., on strong filter or by playing a recording.

Personal action effects

There are many sounds that are so individual in their application that it would be almost ludicrous to use recordings: sounds like a newspaper being folded, opening a letter, undoing a parcel, or eating and drinking.

First, paper sounds. Newspaper makes a characteristic noise, but in a spot effect the action may be formalized somewhat, e.g. shake slightly, fold once and smooth the page. Opening a letter should be similarly organized into two or three quick gestures, the last action, smoothing out the paper, being tucked under the words that follow. Work fairly close to the microphone. The paper should be thin but hard surfaced: brown paper can crackle like pistol shots, and airmail paper sizzles like a soda siphon. Keep the actions simple but definite. For undoing a parcel: a cut, perhaps, then two deft movements to smooth away the paper, and a final slower crinkling noise trailing under speech.

Many sounds, such as an orchestral conductor tapping his baton, can be made realistically (using a music stand); similarly for a judge's gavel and block (but soften this just a little to avoid sharp peaks, by binding a layer or two of electrician's tape round the face of the gavel). The surface struck must have the proper resonance: it will not do to hold a small block of wood in the air; it should be on something solid.

Pouring a drink is also a spot effect. The sound of pouring may not in itself be sufficiently definite, and for all but the longest of drinks is over too quickly; so a clink of bottle on glass can help. Even without the clink it is important to use the right thickness of glass. Effervescent health salts do fine for fizzy drinks. A popgun may be a shade less acceptable as a champagne cork than the real thing, but is more predictable. If 'tea' or 'coffee' is to be poured into a cup, the effect can be confirmed by placing a spoon in the saucer afterwards – and again the right thickness of china is essential.

For a scene set at a meal table, the complete spot kit includes knife, fork, spoon and plate, plus cup and saucer. But give the noise a rest from time to time: the right points to choose for action depend on mood and pace of the script. At any point where the tension rises, stop the effect; as the mood relaxes again, the knives and forks resume. It takes only a little of this noise to suggest a lot: listen to a recording of a tea party for six and you will find that it sounds like a banquet for 20.

Gunplay

A rifle shot is one of the simplest sounds you can meet. It is an N-shaped wave like that from a supersonic aircraft: a sharp rise in pressure is followed by a relatively gradual drop to as much below normal pressure. Then there is a second sharp rise in pressure, this time returning it to normal, then nothing – nothing of the original sound, that is. The length of the N (its duration in time) depends on how long the object takes to pass a single point. For a supersonic airliner this is short enough – for a bullet it is almost instantaneous: there is just amplitude and brevity. A rifle shot travels at something like twice the speed of sound, and that from a smaller gun at close to the speed of sound. Whatever the speed, any loudspeaker would have a hard job to match the movement, however simple. It is fortunate that what happens after the double pulse is more complex.

In fact, the very size of the initial pulse is, both in recording or broadcasting, totally unacceptable; it is necessary to discriminate against it. One way is to fire (or simulate) the shot on the dead side of a microphone; another is to take it outside the studio door, or into some small room built off the studio, so its door can be used as a 'fader'. Very often a script calls for loud shouts and noisy action when shots are fired, which makes it easier to deal with the sheer loudness of the effect than if the microphone has to be faded right up for quiet speech. The best balance varies from case to case, so experiment is always necessary, but the shot can be as loud as the equipment can comfortably handle. Check by ear for distortion: the full peak may not show on a meter. Perhaps use a limiter, with rapid recovery, and a second, distant microphone angled to pick up reverberation only.

Experiments are also needed to select the appropriate acoustic: the sound picked up is a picture of the acoustics in their purest form. Shots recorded out of doors are all bang and no die-away, which means that at the level at which they must be replayed they can sound like a firecracker. Always check the acoustic when balancing a studio shot, and watch out for coloration and flutter echoes that may not be noticeable on a more commonplace sound.

[15.5] **Gunshot and other percussive effects**
Above: Slapstick. *Centre*: Wooden board and hard floor surface. *Below*: Gun modified to prevent firing of live rounds.

A third element in the sound is (or may be) a ricochet – the whining noise made by the jagged, flattened, spinning scrap of metal as it flies off some solid object. And whereas the previous sound is just a noise of the purest type, a ricochet is a highly identifiable and characteristic sound. Even if in reality ricochets are less common than they might appear from Westerns, they may be used in practically any outdoor context, i.e. anywhere there is enough space for a ricochet.

Some of the techniques that have been used for gun effects are outlined below:

- A real gun. A modified revolver is better than a starting pistol, which sometimes tends to sound apologetic. Legal regulations must be properly observed and, needless to say, care used in handling the gun – even blanks can inflict burns. Some guns for radio use are modified to allow the gas to escape from a point along the bottom of the barrel, not from the end (so that it is not usable as a real gun). The microphone balance adopted is generally that for a high ratio of indirect to direct sound.
- An electronic gunshot generator or a hard disk variation on this. This is essentially a white noise generator that when triggered produces 'bangs' with a variety of decay characteristics, ranging from staccato to reverberant. There may also be associated circuitry (or a program) for producing a range of ricochets.
- A long, fairly narrow piece of wood, with a rope attached to one end. Put the wood on the floor with a foot placed heavily on one end, then lift the other end off the ground with the rope and let go. The quality of the sound depends more on the floor surface than it does on the type of wood used. A solid floor is best – stone or concrete.
- Cane and chair seat. The sound quality again depends on the surface struck. Padded leather gives a good crisp sound.
- A slapstick. This is a flat stick with a hinged flap to clap against it. Depending on balance and acoustic conditions, it may sound like a gunshot (though not a good one), the crack of a whip, or just two flat pieces of wood being clapped together – literally, slapstick! Narrow strips of thick plywood are as good as anything for this.
- A piece of parchment on a frame, with an elastic band round it – the band can be snapped sharply against the resonant surface. Again, this is more slapstick than even remotely realistic.
- Recordings of real bangs or of any of the above. The apparent size of any explosion can be increased by slowing down a recording – a sharp clap may become a big, rumbling explosion if copied down to an eighth or sixteenth of the original speed. Alternatively, reduce pitch and increase duration independently by digital signal processing.

If the gun itself is used in radio it can be taken well away from the action. For television this may not be possible: particularly if the gun is to be fired in vision, perhaps several times, with realistic flashes and smoke and perhaps with dialogue before, between and after. In this case, again, use two microphones. On the close (dialogue)

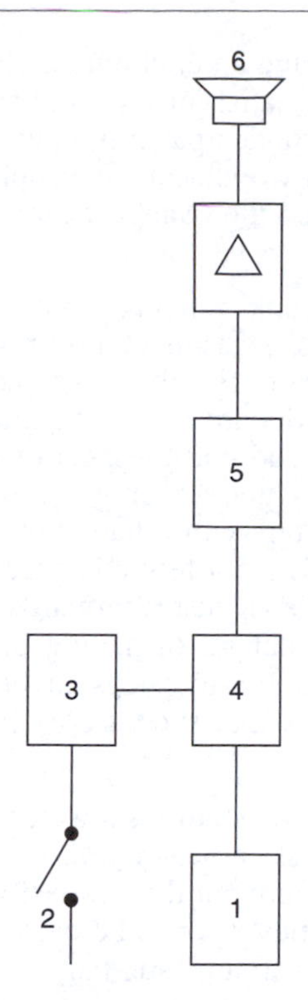

[15.6] **Simple gunshot effects generator: general principles**
1, White noise generator. 2, Push button. 3, Impulser relay. 4, Gate. 5, Low-pass filter. 6, Studio loudspeaker.

microphone set a limiter for 2 dB below full modulation; make the second distant, for a high proportion of indirect sound, perhaps in this relatively dead acoustic with added AR.

For television gunshots out-of-vision or for radio the simplest play-in arrangement is to set up a MIDI keyboard to trigger a variety of gunshot effects. One possibility is to link this directly to a studio loudspeaker that is balanced acoustically. Then, as they can hear the bangs, the actors can react to them directly.

The original sound might again be created by a white noise generator. This passes through a gate (operated by the 'trigger', however this is keyed) and a low-pass filter. The gating circuit allows the white noise passing through to rise sharply from zero to a predetermined peak, automatically followed by an exponential fall that eventually shuts off the white noise once again. With a low-pass filter set for 5 kHz cut-off it sounds much like the close shots of a revolver or automatic; if the filter is set to remove all but the lowest frequencies, it gives distant heavy gunfire – a 'cannon' quality. For a machine-gun effect a device (or program) operates the gate at regular intervals for as long as the key is pressed. In this case the decay is made steeper in order to keep the shots separate and distinct. Ricochets require a separate pulsed output that is rich in harmonics and falls in frequency with time. This might work from the same initial trigger action, but delayed a little.

Creaks, squeaks, swishes, crackles, crashes and splashes

Creaks and squeaks offer rich opportunities for improvisation. Given a little time, the enterprising collector can assemble a vast array of squeaky and creaky junk just by noticing noises as they happen and claiming the object that makes it. Wooden ironing boards, old shoes, metal paper punches, small cane baskets; anything goes – or stays.

Two well-established ways of producing creaks and squeaks are:

- String, powered resin and cloth. The resin is spread in the cloth, which is then pulled along the taut string. If the string is attached to a resonant wooden panel the sound becomes a creaky hinge. Varying the pressure on the string changes the quality of sound.
- A cork, powdered resin and a hard surface, e.g. polished wood, a tile or a saucer. Place some resin on the surface and slowly grind the flat of the cork into it. Again, the squeakiness varies with pressure, speed, type of surface and any resonator that it is in contact with.

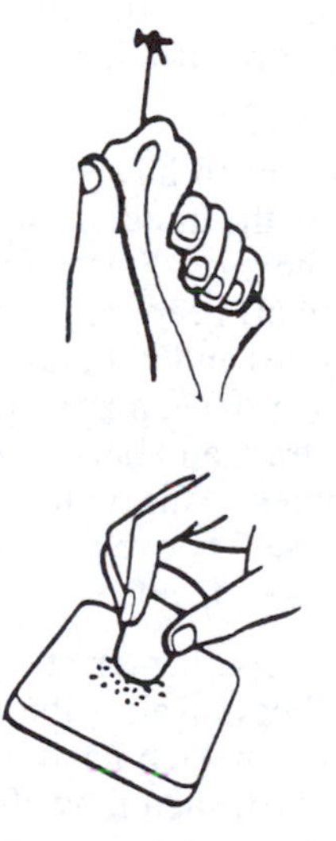

[15.7] **Creaks and squeaks**
Methods using resin and string; or cork, resin and tile.

A bright, crackling, spitting fire may be simulated by gently rolling a ball of crinkled cellophane between the hands.

For the flight of a javelin or arrow, swish a light cane past the microphone. Here realism is not required: in real battles the arrows bounced off stone walls and parapets, making an undramatic clatter. An arrow landing in wood sounds far more satisfying, so convention demands that all the best misses land in wooden panelling and tree-trunks. For this, throw a dart into a piece of wood near a microphone. Proximity makes a sound grow bigger – so the twang of an elastic band may do well enough for the bow.

Crashes are another reason for hoarding junk, ranging from empty cans to pretty well anything that makes a resonant noise when dropped. But real glass crashes demand care. For these, spread out a heavy ground sheet and place a stout cardboard box in the middle. Sprinkle some broken glass in the bottom and place a sheet of glass across the open top of the box. Wearing glasses or goggles and gloves for safety, give this sheet a sharp tap with a hammer. As it shatters it should all fall inside the box. Have a few odd pieces of glass in the free (gloved) hand to sprinkle in just afterwards. Not difficult – but pre-record it, to avoid the danger of having broken glass around on the take (the same goes for footsteps in broken glass). Another reason for pre-recording is that successive crashes are not predictable in quality or volume.

One of the fittings for a well-equipped drama studio is a water tank. For convincing water effects, the tank must be nearly full, to avoid any resonance in the space at the top. For any but the gentlest water effects, protect the microphone – which may have to be close – by hanging a gauze in front of it, or putting it in a plastic bag.

Horses' hooves

Horses' hooves are included here as an example of a specialized spot-effects technique that may seem dated, but is still valid. Coconut shells really can be as good as the real thing, if not better – and are certainly a lot easier to fit with dialogue.

The secret of a good hoof sound lies not so much how to make the shells – take one coconut, saw it neatly across the middle, and eat the contents – as in creating the surface to use them on. Start with a tray about 1 m (3 ft) across and perhaps 10 cm (4 in) deep (e.g. a baker's tray). Fill the bottom with a hard core of stones and put gravel and a mixture of sand and gravel on top. For a soft surface, place a piece of cloth or felt on top of this at one end of the tray, and bury a few half bricks at the other end to represent cobbles. This will give the qualities of surface ordinarily needed for horses' hooves. For horses' hooves in snow, fill a tray with broken blocks of salt.

[15.8] **Horses' hooves**
Traditional method: a pair of half coconut shells.

There are three basic actions to simulate: walk, canter and gallop. The rhythm is different for each. When a horse walks, the rhythm goes clip, clop . . . clip, clop . . . . With a shell in each hand, use one then the other, but with alternate steps from each shell move forward, digging in the 'toe' of the shell, then back, digging in the 'heel'. The sequence is left forward, right back (pause), left back, right forward (pause) . . . . In this way a four-legged rhythm is obtained. For the canter, the same basic backward and forward action is used, but the rhythm is altered by cutting the pauses to give a more even clip, clop, clip, clop, clip, clop, clip, clop. In galloping, the rhythm alters to a rapid clipetty clop . . . clipetty clop . . . 1, 2, 3, 4 . . . 1, 2, 3, 4 . . . . In this case there is hardly time for any forward and backward movements.

For a jingling harness, use a bunch of keys and a length of elastic: hang the keys on the elastic between a little finger and a solid support to one side, so the keys jingle in rhythm with the horse's movement.

A separate spot-effects microphone is needed for hooves, usually a hypercardioid. This should not be too close, and the acoustic should match that used for dialogue.

The recorded sound picture

There are two distinct types of actuality sound recording: some create a *sound picture*; others are selective of a particular *sound element* to the virtual exclusion of all others. These two categories are different not only in content, but are also based on totally different ideas about how sound effects might be used. The sound picture is the complete picture, a story in its own right. But a recorded sound effect, like the spot effect, should be heightened reality: the distilled, essential quality of location or action. It is a simplified, conventional sound, the most significant single element of a complete picture.

A sound picture of a quiet summer countryside may have many elements, with a variety of birdsong, perhaps lowing cattle a field or two away, and in the distance a train whistle. A beach scene may have voices (shouts and laughter), the squawk of seagulls, a distant rush of surf on shingle and, beyond that, the popping exhaust of a motorboat. As pictures in their own right these may be delightfully evocative, but how would they fit into a play?

It is easy to imagine dialogue to match such a scene – but, significantly, the speech would have to fit to the sound effects, not the other way round: in the country picture the cows might play havoc with an idyllic love scene, and the distant train whistle would add a distinctly false note to Dick Turpin's ride to York. Similarly, the beach scene is fine – if we happen to be on the sand and not in the boat. And suppose we want to move about within this field of action? With a sound picture we are severely restricted: the only view we ever have is that seen by the artist from the point where the canvas was set up; we cannot stroll forward and examine one or another subject in greater detail unless this action is chosen for us. But if we ourselves are given paints and canvas we can recompose the scene as we wish. The true sound effect is the pigment, not the painting.

In any programme where recorded effects are to be combined with speech, the final composition will at many points be markedly different from that of a sound picture that has no speech. In the beach scene, the first sound heard might be the surge of surf on shingle, and, as this dies, the hoarse cry of a gull; then, more distantly, a shout and answering laughter. The second wave may not sound so close, and from behind it the distant exhaust note of the motorboat emerges for a few moments. The first words of scripted dialogue are superimposed at a low level and gradually faded up to full. At this stage the composition of the effects might change. Staccato effects, such as gulls and shouting, should be more distant but timed with the speech; brought up highest when the dialogue is at its lowest tension, characters move off-microphone or in gaps in dialogue (other than those for dramatic pauses). The motorboat, too, may go completely or be missing for long stretches, appearing only as a suggestion in a gap. The separate elements can be peaked up more readily than the waves, which cannot be subjected to sudden fluctuations in volume without good reason.

When using effects like this, avoid any that have become radio clichés. On British radio, because the long-running 'Desert Island Discs' starts with seagulls, sea-wash and a record of 'The Sleepy Lagoon', it has been difficult to use seagulls and sea-wash together in that way. Be careful, too, about authenticity in details like the regional location of birds and railways. There are so many experts on these sounds, we should try to get them right – and with birds the time of year as well.

The use of recorded effects

As the aims differ when recording a sound picture or sound effects, so also must the recording techniques. For sound effects, their backgrounds should not be obtrusive, nor their acoustics assertive. Continuous effects should have no sudden peaks or unexpected changes of character.

The best microphone placing for an effects recording may not be the most obvious. For example, a microphone placed inside a closed modern car might not pick up any clearly recognizable car sound, so for a car recording – even for an interior effect – the microphone should be placed to emphasize continuous engine noise. But the mechanical noises associated with gear changes should not be emphasized (except for starting and stopping), as their staccato quality may interfere with the flow of the script.

An effects sequence, like the composite sound picture, is best constructed from several distinct components. Take, for example, a sequence (to be used with dialogue) in which a car stops briefly to pick up a passenger. The elements of this might be as follows:

1. Car, constant running.
2. Car slows to stop.

3. Tick over.
4. Door opens and shuts.
5. Car revs up and departs.

Of course, a recording might be found with all of these with the right timing and in the desired order. But:

- If there is much dialogue before the cue to slow down, the timing is more flexible if the slowdown is on a separate recording. The cue to mix from one to the next should be at a point in the dialogue about 10 seconds or so before the stop, or there will be too big a change in engine quality. As the sound of a car stopping lacks identifiable character, a slight brake squeal or the crunch of tyre on gravel might help to clarify the image. The overall level of effects may be lifted a little as the car comes to a halt.
- The tickover might possibly be part of the same recording as the stop. But the microphone balance for a clear tickover is so different that it is probably better to record it separately. Fade in the tickover just before the car comes to rest.
- Doors opening and slamming must be timed to fit the dialogue. A real car door shutting is loud compared with the door opening and tickover, so it needs separate recording and control.
- Departure could be on the same recording as the tickover, but again timing is more flexible if the two are separate and the departure can be cued precisely.
- Extraneous noises, such as shuffling on seats, footsteps, etc., may be omitted.
- Additional sounds – e.g. another door, a motor horn or the setting of a mechanical taximeter – can be cued in at will.

The best way of putting together sequences like this depends on the replay facilities and sources available. Some constituent effects may originate from hard disk libraries on the Internet, from which items can be downloaded as part of a service for a fee and are then retained on hard disk or another medium such as CD-R. Many users have a more or less limited collection of commercially available CDs (any one of these contains a large variety of effects, each on a separate band) and may buy an additional CD as required. Many make their own recordings, then keep them for possible reuse. The essentials for use are flexibility and speed of operation; it should be necessary to pre-record only the most complex sequences. Hard disk can also be used for shorter sounds, in some cases replacing spot effects, as well as for pre-recorded announcements, jingles and commercials, and for game shows where a limited number of stereotyped musical effects are punched in to signal success, failure or as 'stings' to mark a change of direction.

By far the most flexible method is to transfer each effect to hard disk for cueing via the keys of a MIDI controller – a keyboard for greatest versatility, or (for just a few cues) a picture of the keyboard on a computer screen with a mouse to 'press' the keys. Some can be copied to nearby keys with slightly changed pitch: for example, the same train door slamming twice becomes two different doors (also at different pre-set volumes) as two adjacent keys are hit. A whole sequence can be constructed with overlaps (layered effects) on a sequence of keys with complete flexibility of timing. A performer requires headphones to interact with it. For stereotyped effects and 'canned' applause, too, MIDI provides the perfect can. MIDI is described further in Chapter 16.

A CD or hard disk library of effects can also be used with digital editing (see Chapter 18) or for video soundtracks (Chapter 19).

Recorded effects in stereo

For stereo drama (whether for radio, television or film), it is best to use an effect actually recorded in stereo, steering it to its appropriate position and width. Where no suitable stereo recording is available and it is impracticable to make one specially, it may be possible to adapt existing monophonic sounds.

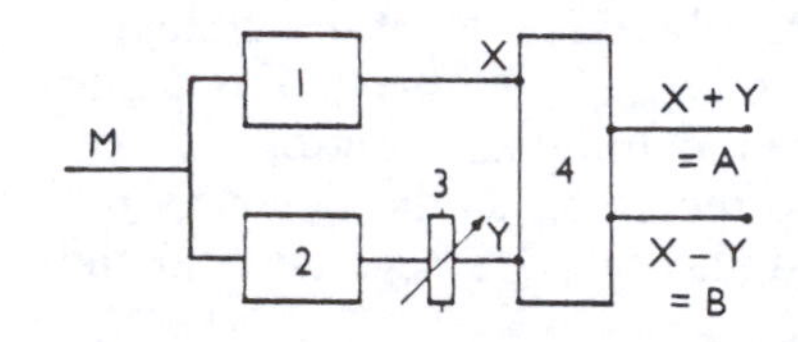

[15.9] **Spreader circuit**
The monophonic input (M) is split equally between paths X and Y. 1 and 2, Delay networks producing differential phase shift. 3, Spread control fader. 4, Sum and difference network producing A and B components from the X and Y signals.

With continuous noises such as rain or crowd effects, a stereo picture can be created by using several recordings played out of sync (or one, with different delay time settings) and steered to different positions within the scene. Rain or crowd scenes could also be fed through a device called a *spreader*, which distributes a mono sound over the whole sound stage like wallpaper.

For a distant, moving sound in the open air (or one that passes quickly), panning a mono source often works well enough. A larger object such as a troop of horses passing nearby would require a spread source to be panned across, with the spread broadened as the troop approaches and narrowed again as it departs. This effect could also be achieved by panning two mono troops across one after the other, while a coach and four might benefit by spread horses from one recording followed closely by wheels from another.

In an example illustrating all of these, a horse and carriage may be heard to pan through a spread crowd, until a shot is heard from a mono point, its reverberation echoing over the whole of the sound stage. Note that in stereo radio drama there is a danger of lingering too long on movement that is meaningless in mono: for the sake of compatibility this should be avoided.

Changing the pitch of an effect

The range of sound compositions available from a given set of effects recordings can be increased if their pitch can be changed – ideally, independently of duration. The digital method is to use a digital signal processor (DSP) that includes a pitch- and time-shift program. The older technique was to change the speed of replay of the recorded effect (shifting pitch and duration together), and this can also be achieved with a MIDI keyboard. Equipment designed primarily to operate at given fixed speeds sometimes has a control that allows continuously variable speeds between, say, half and double the nominal speed.

When the replay speed of an effect is changed (so that pitch and duration are shifted together), the most obvious difference in quality is an apparent change of size: slow down the two-stroke engine of a motor lawnmower and it becomes the lazy chugging of a motorboat, or a small motorboat becomes a heavy canal barge. In the same way, water effects can be increased in scale, just as filming a scaled-down model boat in slow motion may simulate the real thing. Speeding up an effect (raising its pitch and shortening the duration) gives the opposite kind of change – for example, mice talking. For out-of-the-ordinary effects these techniques can sometimes save a lot of work.

Shifts in pitch and duration also change the apparent volume of the room in which an effect was recorded: the characteristic colorations go up or down in pitch with everything else. This means that recordings made out of doors may be more conveniently speed-changed than those with indoor acoustics attached. A good subject for drastic transformation is wind. Record a good deep-toned wind and double its speed to get a high, stormy wind; double again for a biting, screaming Antarctic blizzard.

The recorded effects library

The contents of an effects library fall into a variety of categories.

First, and most important, there are *crowds* – a crowd being anything from about six upward. It may seem odd that the first requirement among recorded effects should be the human voice in quantity, but plainly, the illusion of numbers, the cast of thousands, is often essential if a programme is to be taken out of the studio and set in the world at large. There are many subheadings to 'crowds': chatter, English and foreign, indoor and outdoor; the sounds of apprehension, fear, agitation, surprise or terror; people shouting, whispering, rioting, laughing, booing and applauding. It may even just stand there and listen: the sound of a crowd that is silent except for an occasional rustle, shuffle or stifled cough can at times be as important as the roar of a football crowd. A further subdivision of crowd effects is *footsteps*, in this case the crowd being anything from two pairs of feet upward.

Transport effects include cars, trains, aircraft, boats and all the others (such as horses) doing all the various starting, carrying on and stopping things we expect of them, and recorded both with the microphone stationary and moving with the vehicle, and with interior and exterior acoustics where these are different.

Another important group that has links with both of the above is *atmosphere of location*, or '*atmos*' recordings. Streets and traffic (with and without motor horns), train stations, airports and harbours – all these are needed, as well as the specific identifying noises of London, Paris, New York and other major cities; also the atmosphere of schoolroom, courtroom, museum, shop or concert hall, and the distinctive sounds of a wide range of environments.

The *elements – wind, weather, water* and *fire* – provide another vital field. Then there are the sounds of all branches of human activity: *industrial* sounds, *sport, domestic* effects from the ticking of clocks to washing up, and the sounds of *war*. There are *animals, birds, babies*; and dozens of other classifications of sound, such as *bells, sirens, atmospherics, electrical* sounds, etc.

A good recorded effects library (on whatever medium) should include copies of all these and many more. They may be purchased from a range of commercial sources, and broadcasting organizations, specialist production companies or individual enthusiasts may even add their own. In a large library catalogue it should be easy to find and listen to a dozen different recorded effects that are nominally suitable, to find out which is best and which will not do at all. Typically, it is possible to select two or three specimens of each sound: a likely best bet, plus others sufficiently different in character to act as alternatives in case, on adding sound to dialogue, the first turns out to be wrong in mood, level or timing. A different recording can be readily substituted at any point in a production. But skill in cueing (involving the judgement of timing) is needed for programmes with complex effects.

Acoustic foldback

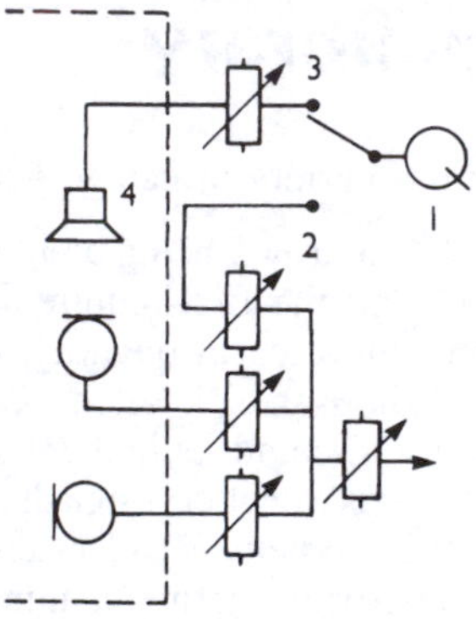

[15.10] **Acoustic foldback**
Reproducers (1), normally fed direct to the mixer via switch contact (2), may be switched (3) to provide an acoustic feed to the studio loudspeaker (4).

Pre-recorded sound effects are sometimes fed through a studio via a loudspeaker and microphone. It is of greater advantage to do this when the original recording is acoustically 'dry' – from the open air, perhaps, or made with a close balance. Passing the sound through the studio adds the same recognizable acoustic qualities that are heard on the voices. In addition, the actors have something to work to, and can pitch their voices against it in a realistic manner. A danger is that the combination of loudspeaker and microphone can introduce coloration.

The volume of the studio loudspeaker amplifier should be set at a level such that most of the range of the console 'send' control is in use. These two controls are used in combination to set the sound level in the studio: experiment to find the optimum settings for the various elements in the chain – if a satisfactory result can be obtained at all. This technique will not work if the effect is already brighter than it should be and the studio is dead.

Effects for television and film

Most of the effects described above work well out-of-vision on television. Recorded effects are likely to be preferred to spot effects, which in turn are more often recorded in advance and played in: this helps to limit the number of open microphones and produces more predictable results when the director has so much to watch and listen to at the same time. However, in-vision sound effects are still produced by some practical props. These need the same care in balancing that they would have in radio, but in more adverse conditions.

Often an effect that can be picked up satisfactorily on a speech microphone sounds much better if given its own closer effects microphone. For example, a discussion about radioactivity using a Geiger counter could have a small electret hidden near the loudspeaker of the counter. For this, a recorded effect could not be used successfully because the rates would not vary realistically; because the people present would not be able to react properly to them; and because a meter may be shown in close-up, working synchronously with the clicks. The effects microphone makes it possible to vary not just the volume but also the quality of sound, so that the clicks could be soft and unobtrusive when they are not central to the discussion, and crisp and strong when they become the centre of interest.

As in radio, recorded effects are used to suggest atmosphere. With crowd effects plus a camera track in over the heads of six extras, the viewer may be convinced it was shot in a crowded bar. Film may establish a wide street scene, then its sound may continue over a studio shot showing a small part of the action within it.

Television introduces an extra dimension to sound effects that radio has never had, the ability to jump easily between different viewpoints – say, between interior and exterior of a car or train. For fast-cut sequences it is convenient to use parallel effects tracks with interior on one and exterior on the other. The operator simply fades quickly between them on each cut or uses a device that automatically links sound and vision cuts. For an effect in stereo, spreaders or stereo reverberation may be added, or a multitrack recording used. In any case, it is useful to pre-record effects sequences for complex scenes, for which little studio rehearsal time may be available.

Footsteps in vision

Where footsteps are to be heard in the studio, the television or film designer should help sound by providing suitable surfaces. Stock rostra are often constructed with a folding frame to support a surface that gives a characteristic hollow wooden sound to footsteps. For an action area a suitable treatment would be to 'felt and clad': this adds a layer of thick felt, with hardboard on top, so that footsteps are audible but no longer with the objectionable drumming sound. If little or no footsteps are to be heard, add a further soft layer of rug or carpet.

For 'outdoor' areas the use of a sufficiently large area of *peat* not only deadens the surface but may also deaden the entire local acoustic. But certain materials commonly found outside cannot be used in television studios. *Sand* cannot be used, as it spreads too easily, gets in technical equipment, and makes an unpleasant noise when a thin layer is ground underfoot. Sawdust is a convenient substitute. For similar technical reasons, salt and certain types of artificial snow cannot be used: again substitutes that do not cause sound problems should be found.

Rocks are rarely used in studios, as they are heavy and inconvenient. Wooden frames covered with wire netting and surfaced with painted hessian have been used, as have fibreglass mouldings or expanded polystyrene carved into suitable shapes. Unfortunately, none of these makes anything like a satisfactory sound, so any climbing, rubbing or bumping must be done with the utmost care by the performers – and then preferably only at places that have been specially arranged not to resonate, perhaps by using peat or sawdust either as a surface layer or packed in sacks. Falling 'rocks' of expanded polystyrene must have recorded sound effects exactly synchronized – which is easier if the action is seen in big close-up, with the blocks falling through frame. Expanded polystyrene can sometimes squeak when trodden or sat on, and particularly when small pieces are ground against a hard studio floor. In such cases, post-synchronization is safer: filming or pre-recording and tracklaying appropriate sound avoids it completely.

Sidewalk *paving slabs* are often simulated by fibreglass mouldings, as these are light and convenient to hold in stock. When laid in the studio they should stand on cloth or hessian to stop any rocking and to damp the resonance. If, in addition, performers wear soft-soled shoes and there is recorded atmosphere, the result should be satisfactory. If clear synchronous footsteps are required, there is no good alternative to real paving stones, despite the inconvenience and weight.

Stone floors can be simulated by the use of a special 12 mm (½ in) simulated stone composition laid on underfelt. But 'stone' steps that are actually made of wood are likely to cause trouble if any sound effect is needed (and even more if not). If a close effects microphone with good separation can be used for the footsteps there is some chance of improvement, as heavy bass-cut can then also be used.

A dramatized series about a prison in a castle had trouble throughout a television season with its simulated cobbles and fibreglass doorsteps around its courtyard. For a second run, the film set was constructed using real cobbles and stone – which, to everyone's surprise, turned out better not only for sound, but also for the programme budget.

Wind, rain and fire in vision

The sound effects for wind, rain and fire create no problem: recordings are used as in radio. But when they are combined with visual effects the device that makes the right picture sometimes makes the wrong noise, which must therefore be suppressed as much as possible, to be replaced by the proper recorded effect.

Wind in vision is generated by machines varying in size from very large propeller fans about 2.5 m (8 or 9 ft) in diameter, through medium sizes, 1 m (3 or 4 ft) across, and which can be moved round on a trolley, down to small machines at 45 cm (18 in), which can be set on a lamp stand and used for localized effects. The bigger sizes are very noisy and push out a large volume of air over much of the studio (if used indoors). Windshields are needed for any microphones within range. The largest machines may be used for the individual shots of a film but are inconvenient in the television studio, where the smallest size is best. When run at low speed, the noise from them can actually sound like wind (at higher speeds blade hum is heard). The visual effect is short-range and the airflow can be controlled to avoid microphones.

Rain, when produced in the television studio, may be far too noisy, so it should be arranged for the water to fall on soft material in a tray with good drainage. Also, if it is in the foreground of the shot the feed pipe may be close to the ideal microphone position, making balance on a boom difficult. Another problem is that it takes time to get going and even longer to stop completely, to the last drip. This may be overcome either by slow mixes to and from effects (including recorded effects) accompanying the transition to and from the rain scene or, more commonly, by discontinuous recording and subsequent editing.

Fire – including coal and log fires – in television studios is usually provided by gas, with any visual flicker effects added by lighting.

Sound equipment in vision

When radios, television sets, tape or video recorders or record players are used as props by actors in television programmes they are rarely practical: obviously, the sound gallery wants to retain control of sound quality (by filtering, if necessary) and to maintain a good mix at all times, and also to ensure perfect cueing in and out. Careful rehearsal is needed for exact synchronization of the action. For any sound cues to the actors, it is fed at low level to a studio loudspeaker.

Pre-recordings using the same studio are sometimes used for out-of-vision sound. This may help to get it in the correct perspective or, more often, is used just to fix one element in an otherwise all-too-flexible medium. Music is often pre-recorded for complicated song and dance numbers – but this may be in a different acoustic and with a close balance, both of which create matching problems.

Television sports and events

Sports and events create sounds that must be captured on television, preferably in stereo. There are two main types:

- Arena events. For these, crowd atmosphere and reaction are taken in stereo and spread over the full available width. Action noise, such as bat on ball, is taken on more highly directional microphones and usually placed in the centre. Sometimes an extra stereo pair is used for action, but this is rarely crucial.

- Processional events, where cameras and microphones are set at a sequence of points along a route. Microphones are usually also set to follow the action, using stereo according to the same conventions as for picture. For television, it is not always necessary for sound to follow action precisely, provided left and right match in picture – it is better to concentrate on dramatic effect. A racing car approaching makes little sound, but its departure is loud. In mono the tail of the previous sound was often allowed to tail over the start of the next (approaching) shot, but in stereo a cleaner cut, and near-silence (apart from stereo atmosphere) on the approach, is effective.

For very precise full-frequency-range pick-up of effects at a distance, with highly effective noise rejection 15° or more off-axis, consider using an adaptive array combination microphone with its digitally processed forward lobe that is only 8° across (switchable to 8° × 30°). This might also be useful for pinpointing small regions with a large crowd or (closer) enhancing the audibility of distant dialogue in noisy surroundings – for both television and film.

Sound effects on film

On film, effects are added in post-production either to maintain a continuous background of sound or to define a particular quality of near silence. In addition, effects of the 'spot' type are made easier by the process of synthesis that is used to create films from many individually directed short scenes, within each of which the sound may be given greater care than in the longer run of the television studio. A live action effect that is recorded synchronously as a scene is filmed can be enhanced in a number of ways:

- it may be changed in volume;
- a copy of the effect may be laid alongside it on another track, so it can be changed in quality and volume to reinforce the original sound;
- the timing may be altered – e.g. when an explosion has been filmed from a distance and the original recording is out of sync because of the time it takes for sound to reach the microphone, the noise can be resynchronized visually;
- the original effect may be replaced or overlaid by a totally different recording, again in exact synchronization with picture.

In the film industry, the provision of footsteps and other spot effects may be a separate stage in post-production, and is the province of spot-effects specialists, often called foley artists. Their skill permits the film director to concentrate on picture and action in locations where there is extraneous noise from traffic, aircraft, production machinery or direction, or where acoustics or floors are inappropriate.

The tracks are recorded to picture and can be moved, treated and pre-mixed as required, before the final re-recording.

Even if an action sound works well on the main microphone, it may be even better if recorded again as wildtrack (i.e. without picture) and given its own special balance. When neither works satisfactorily, some other spot sound simulating the proper noise may be used.

For example, take the case of a bulldozer knocking down a street sign. In the main sync recording the dominant noise is engine and machine noise from the bulldozer: the impact of blade on metal is hardly heard. So a second version of the noise of impact is simulated, perhaps by using a brick to strike the metal, and recorded without any engine noise. This is laid on another track and controlled separately. If the bulldozer knocks down a wall the sound is complicated even more, and the dramatic importance of three sounds may be the inverse of their actual volumes: impact noise may again be the most important; then the falling masonry; then the bulldozer engine and mechanical noise. Several recordings of the principal event could be made with different microphone positions (on either side of the wall), perhaps with additional sounds recorded separately, to be incorporated later.

Commercials and documentary features sometimes make use of punctuating sound effects. For example, an object built up stage by stage by stop-action techniques has a series of parts suddenly appearing in vision, and each time this happens there may be a sharp noise – a recorded 'snap-in', a tapped metal surface, a swish or a click. Cash-register noises accompanying messages about the amazingly low price of a product could be too much: such music may delight the advertiser more than the customer. But there is a vast potential for non-representational, comic or cod effects in commercials.

Chapter 16

The virtual studio

The studio without a studio: welcome to the virtual studio! There is no dedicated acoustic space, just one room where the electrical analogue of sound is created, assembled, manipulated (or processed), mixed and recorded. Real sounds may still be captured by real microphones, but are not at the heart of the operation: they might be recorded during a moment of quiet in the same control space, or they may be imported.

Today, a virtual studio is most likely to be centred on a computer with software that duplicates many of the functions of a real studio – and in its simplest (or purest) form that is all there is. Even the recording process stays within the computer, although if the end-product is to be, say, a CD, the computer turns into a digital audio workstation, and the CD is 'burned' (a term that recalls the record's hot origin under a laser beam) within the same computer. Given the vast and seemingly ever-increasing processing power and memory plus built-in hardware provided inexpensively even in home computers, this is commonplace. The software, including programs on cards that need to be slotted in to the back of the computer, certainly adds to the cost, but the total will still be far less than much of the hardware needed in a 'real' studio.

Virtual studios and their kin, *project studios*, did not start out like this. Their (for many current users) misty origins lie in an era when digital technology was little used. The electrical analogues of sound remained robustly analogue, and the only digits employed were the fingers that turned knobs and operated the switches that literally and laboriously manipulated the processes. One of several early pathways towards the virtual studio was followed mainly by traditionally-skilled European composers who wanted to explore the musical possibilities offered by new technology in order to pioneer their 'modern' music. In Britain the same approach was used to create a combination of music and effects (for radio and television) that was called 'radiophonics'. Since these gave rise to many of the techniques and operations in use today, along with others now less used, but which might still suggest new possibilities, some of this history is described in Chapter 17.

A second and ultimately more powerful influence on the evolution of the virtual studio came directly from what digital technology could offer. Crucial developments along this pathway included the sequencer, which could control a synthesizer or drum machine, and early samplers (initially with very limited capacity) from which the output could be fed to a multitrack recorder, at that time on analogue tape. Even at the stage when such devices were little more than toys in makeshift home studios, it was clear that little more than these basics was needed to make some kind of music. Originally, all of the components with their individual processors were in separate boxes that needed to be connected in some way. This required the development of a commonly accepted 'musical instrument digital interface' – MIDI.

The key developments along this more popular pathway to the virtual studio were nearly all direct consequences of increasing processing power and memory, together with the inventiveness of the programmers using it. Sequencing programs were written for home computers that could trigger sounds from the sampler and synchronizer and sync them to a drum machine's (separately) programmed rhythms. The samplers might also have built-in programs, perhaps to change pitch or duration, and a keyboard could be persuaded to play the often rather irritating results. Effects, too, could be sampled and laid down on spare tracks on a multitrack recorder, prior to mix-down. Dance music produced by such minimal means became immensely popular, making celebrities of, and fortunes for, its most imaginative creators and users.

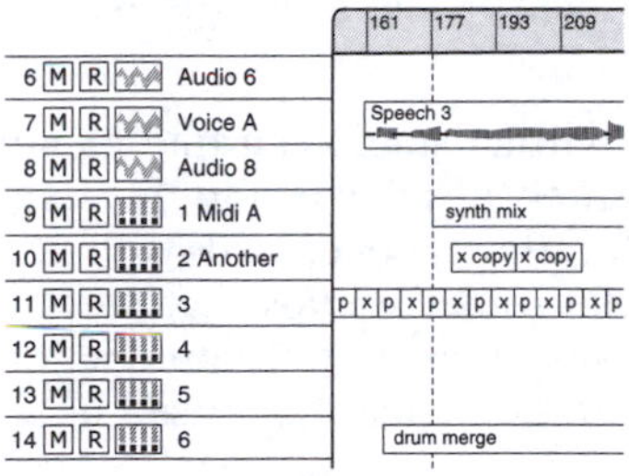

[16.1] **Sequencer timeline screen** The central workspace in a virtual studio is a screen that shows audio and MIDI tracks below a timeline that is likely to be of 'step time' (based on the length of musical notes) rather than clock time (from time code). In this example, audio and MIDI are similar in appearance, except that the audio (an actual recording) can be represented by its waveform, while the MIDI parts (which contain only control data) are shown as rectangular blocks. A smaller transport window, with controls like those of a digital tape recorder, would be inset. The example shown here and below is Logic – one of a small number of software packages that are still described as sequencers, but which have developed into powerful music composition and audio production systems.

In due course, all the separate boxes were gobbled up by the computer, although their functions still exist within its programs. The computer itself became the sequencer, sampler, synthesizer, drum machine and rhythm generator, plus multitrack recorder – and much else. Digital editing emerged almost serendipitously as an extension of sampling, as samples got bigger and the programs to manipulate them grew ever more complex. Almost by the way, these same developments engulfed and swallowed up the previously separate fields of modern classical composition and synthesized music for radio, television and film.

The next major addition to the virtual studio, as to regular studios, has been digital signal processing (DSP). One form in which this may appear is on a '*platform*' within the computer. A manufacturer may offer 'a total production environment' with digital recording, sequencing, editing, mixing, mastering and a variety of signal processors, plus the platform for 'plug-ins' by other software developers. This is much like the way in which studio hardware is centred on a particular console, to which racks full of specialized equipment are literally plugged in.

The visual display screen that comes with a plug-in often represents a set of rack-face controls that an operator might recognize and know how to use. But use them how? It may not be difficult to slide faders and turn knobs with a mouse, but is this the best way to do it? From a state where everything is hidden within the computer with simulated controls on-screen, the seemingly natural next step is a return to at least some of the physical equivalents, using *control surface* hardware that is *reattached* to the computer.

Take this far enough and the ultimate reattachment might include a more sophisticated control surface (we might call this a 'console'), a few microphones, and for the sound sources, an acoustically-dedicated space. In reverse, this is just what happens when a regular studio has a MIDI area within it. In a drama studio, it is convenient to build up complex effects sequences with their timing and pitch played in from a MIDI keyboard, then processed and pre-mixed on its associated control surface. The same keyboard can also provide musical effects.

The more common type of virtual studio is the musical equivalent of a home office. At its heart is a good computer and screen, and often an electronic musical keyboard. One other essential is a good loud-speaker system. But given that much of the background music heard on radio and television, and some on DVD or film, is constructed in this way, the more ambitious studio also has 5.1 surround sound. (One detail in the layout for this is that in such a confined space the subwoofer needs to be positioned with some care. One suggestion on how to find the best spot is: check where in the room the bass is weakest and put it there.)

Readers who want to set up and use a virtual studio can find plenty of advice in the books, magazines and websites devoted to this field and specialized aspects of it. But for those concerned with the operational techniques employed in the sound studios that are the primary subject of this book, this chapter offers a brief introduction to components of the virtual studio that are increasingly used in all studios.

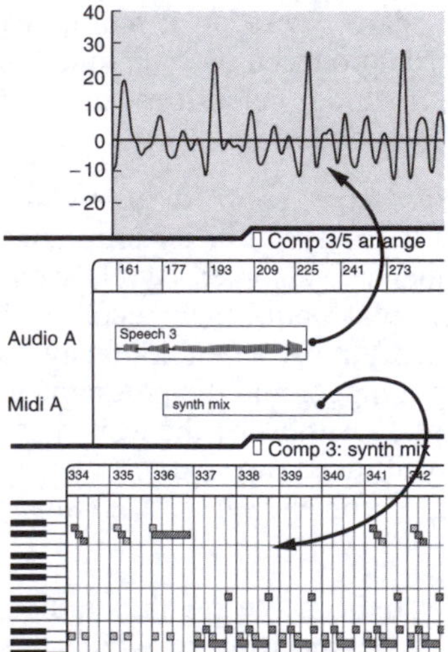

[16.2] **Opening audio or MIDI parts** Clicking on a segment of an audio track opens a window that shows more detail: zoom in to see the waveform in close-up. Similarly, clicking a MIDI part may reveal the keyboard of an instrument and the notes it is instructed to produce. (Other display options include lists or musical notation.)

MIDI plus audio

Originally coined as the name for a way of linking a few things together, 'MIDI' is now understood by many of its users to encompass far more than that – in fact, to include many of the linked objects plus all that is done with them. Tracks associated with digital audio and MIDI events can be shown one above the other on the same screen. They work together and can be played together (subject to a timing correction to adjust for a slight delay in MIDI play-out). 'MIDI + audio' has become a field in its own right, with the usual abundance of literature to consult.

The signature component of a MIDI system is the keyboard derived directly from the synthesizer, and indirectly from the piano, the black and white key layout of which is copied, but extending generally over a more modest five octaves. The processors and controls that go

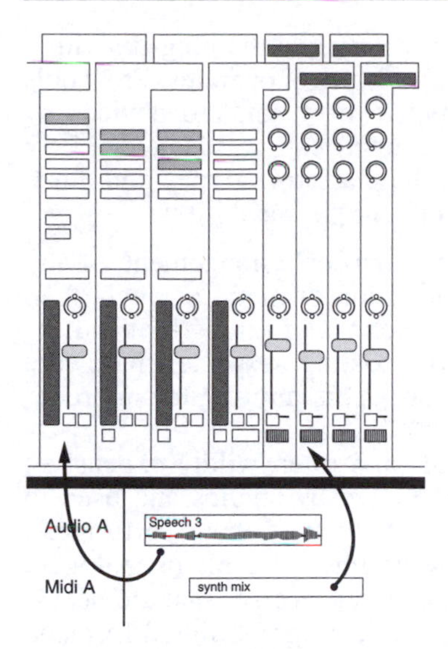

[16.3] **Track mixer window**
Opening a mixer window over the timeline window reveals faders spread out like a mixer console across the screen, with other controls in line above the faders. Each fader is numbered (and may also be named) to match the list of tracks, and each fader operates directly upon its corresponding track. The MIDI strips are simpler than those for audio, which must include EQ, 'sends' to various busses, and insert points for AR, compressors and limiters, and more specialized DSP plug-ins that may be assigned to them. Audio can be either mono or stereo, perhaps with a meter for each leg next to the fader, and it may be possible to mix to 5.1 surround.

with this may fit either in the same box (as in many domestic MIDI systems) or on a separate mini-console. A virtual studio no longer needs a physical keyboard and controls – these can be mimicked on the screen of the computer – but they are more convenient if used often. Also, many users do feel more comfortable, or better rewarded, by having the hands-on keys, faders and switches as well as the screen displays.

With MIDI, keyboards can be used either in a conventional mode with keys used for controlling pitch, or with a series of samples triggered by the keys, or with each key designated separately for some other control function. It follows that either pitch-perfect conventional scales or unconventional progressions can also be constructed and passed through it. An external MIDI keyboard, sequencer, drum machine or synthesizer can be linked to a computer by MIDI cables and a MIDI interface (the MIDI equivalent of a sound card) that translates the stream of MIDI data into a form the computer can work with.

In its early incarnation as a synthesizer, the MIDI keyboard was closely related to the electric organ family, in which musical notes with various qualities were created, with pitch and volume controlled conventionally by the keys. In the software version of this, many of the synthesized notes are replaced by samples created by skilled musicians on real instruments taken either from a sampler program or recorded locally – so the keyboard is used not only for notes from the regular keyboard instruments, but also for strings, woodwind and brass. Full drum-kits can be included. Other programs lay down rhythm, allow a beat or drum taps to be replaced by musical tones or sources, snap off-key voices to perfect pitch and enrich or add depth to their harmonic structure.

Put a chord or random notes in an 'arpeggiator' and it will arrange for them to ascend or descend with metrical precision. Or 'quantize' the notes (or drum taps) to set them in some perfect but expression-free tempo. One way to improve on that is to render it graphically as a full score, on which note order, length, position and musical qualities can be changed at will. Mixing, at best, is fully automated. Editing MIDI + audio is similar to that described in Chapter 18.

MIDI + audio sequencers are now immensely (and ever-increasingly) powerful and versatile packages. The main names are Logic Audio, Digital Performer and Cubase VST (for 'Virtual Studio Technology').

Plug-ins

A 'plug-in' may be completely metaphorical, consisting of a disk that contains the program: this is loaded rather than plugged in to anything, and then runs on the host computer. Or it may be real, a card with a dedicated chip that carries the program: this is literally plugged in to a slot in the computer and runs in parallel with it. There are variations on these themes, but those are the main options.

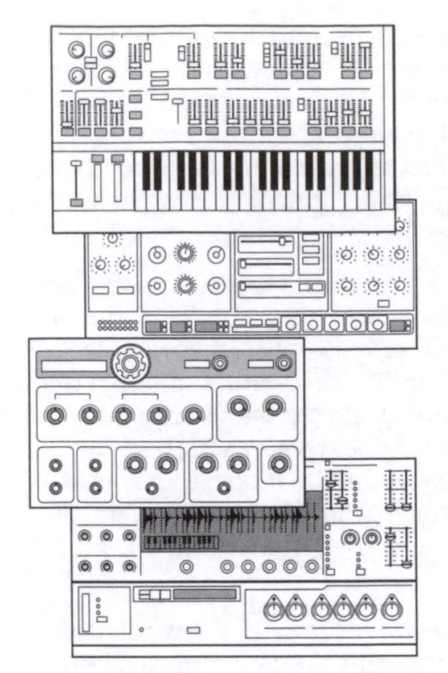

[16.4] **DSP plug-ins**
The sequencer (production system) may act as host to plug-ins, some native (i.e. by the same manufacturer), others from a wide range of sources that use a compatible format, such as DigiDesign's TDM. The function of a plug-in is usually characterized, at least in part, by the graphic style that appears in its window. A 'modern' style emphasizes a function that requires (and has only become possible with) substantial processing power. Some reveal their musical credentials with MIDI-style keyboards.

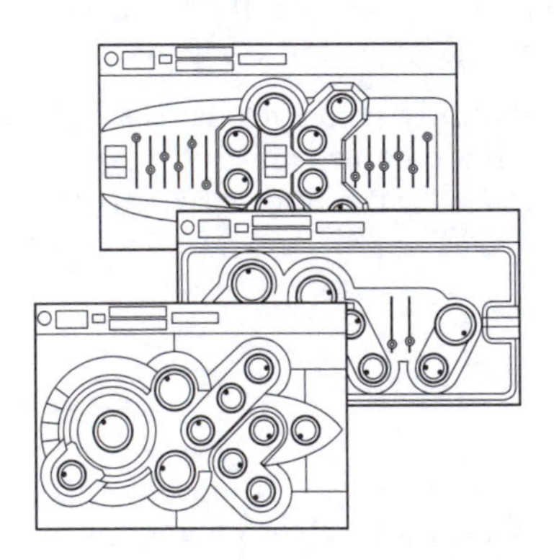

[16.5] **Synthesizers and samplers**
Software-based instruments may be directly available as sources alongside the audio and MIDI input.

The computer, in turn, must be ready to receive the plug-in, with a *platform* for it to link into, and through which it operates. Pro Tools and Logic offer platforms into which over a hundred devices by other manufacturers (as well as their own) can be plugged. Many processes are already available with the platform, so they and third-party designers compete to offer some unique capability:

- to make it possible to assemble any musical arrangement, of any musical quality, with inhuman precision of metre or pitch, then add musical expression and imperfections of timing by hand;
- to emulate the characteristics of existing respected hardware together with all its useful add-ons, plus any digital improvements;
- to emulate historically successful (and often wildly expensive) analogue devices that were once prized by studios and used to create the distinctive sounds associated with famous performers;
- to perform digital functions that were impracticable in analogue;
- to complete all the complex production processes that are necessary in, say, mastering media in a wide range of digital formats.

Some are narrowly defined, for some specialized operation; others come in generous DSP bundles, offering the quality or value that users associate with a particular name. The categories (or functions) include the following:

AR. For some of the earliest DSP hardware, specialist designers created distinctive 'reverb' packages, of which these plug-ins are just the latest versions. The only difference should be in how the controls are presented. One kit, to be different, calls itself 'spatial effects'.

EQ. Filters were some of the earliest analogue devices, with the bonus that the simpler they were, the smoother their operation: a circuit containing just one resistor and one capacitor curved smoothly into a roll-off of 6 dB per octave. Most EQ plug-ins emphasize their 'analogue quality of sound', meaning gentle, smooth transitions unless otherwise stated. Some emulate particular early rack-mounted EQ kits, with photographic screen controls to match, one specifying the simulated use of tube (valve) technology: 'engineers run tracks through this even without EQ, to make them sound better'. Some offer 'sweep' effects. *Parametric equalizers* allow the frequency, bandwidth and gain or loss to be varied continuously, not just at specified settings. Some add *presence* or enhance bass 'without increase in level' and also throw in *de-essing* (reduction of vocal sibilance due to HF peaks).

Compressors, expanders, limiters and gates. Some emulate 'classic' analogue equipment, again with photographically accurate screens. Some even simulate the way in which analogue overload introduced distinctive forms of distortion, so that digital mixes do not completely lose this characteristic ingredient. Others offer digital perfection to fend off the potential disaster of digital clipping. To lift the volume of a track as high as possible, it is necessary to check for ephemeral but potentially damaging peaks, delay the signal, then stop them happening.

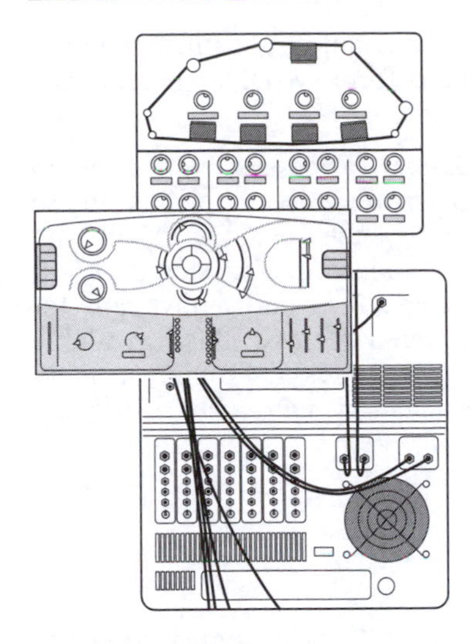

[16.6] **'Retro' plug-ins**
Many DSP plug-ins glory in their 'retro' origins by graphic or even photographic displays based on the control panels of archaic but treasured analogue devices – in some cases even to the extent of showing a simulated tape loop or trailing patch-bay connectors. Despite the period style that recalls the limited capacity of the originals, the digital programs often require prodigious processing power.

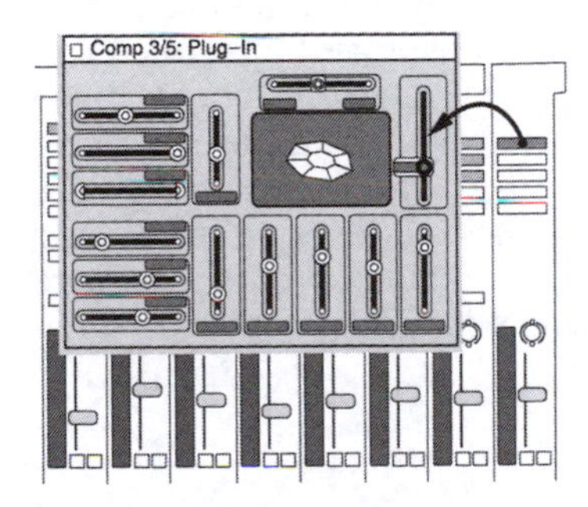

[16.7] **Reverberation insert**
When AR is assigned to an insert tab above a fader, clicking it brings up a window showing controls for the selected program. This one is native to the system shown above, and like most operations within it, can be automated. Other manufacturers' AR software – and many other plug-ins – can also be mounted (using, e.g., TDM, DigiDesign's Time Division Multiplex format) and operated, often also with full automation, from the same group of buttons.

Synthesizers. As ever, these are likely to include oscillators, filters, envelope shapers, chords, and a wide variety of delay effects such as *chorusing, flanging, phasing*, multi-tap and 'ping-pong' repetitions; also 'science fiction' effects on speech such as ring modulator and *Vocoder* (the analogue origins and uses of these are described in Chapter 17). Some plug-ins offer rotating loudspeaker effects or emulate vintage favourites such as the Fairchild or Moog (rhymes with vogue). 'Frame accuracy' may be specified – i.e. no apparently random tiny delays.

Spectral delay. This splits each audio input channel into 1024 separate frequency bands so that delay time and other parameters can be adjusted independently on up to 160 of them at any one time.

Retro quality effects. A number of packages glory in 'down-processing' or the creation of 'lo-fi' effects. These include old guitar amplifier and cabinet effects that are all too easily eliminated by digital technology, together with the deliberate distortion of 'buzz', etc., often added by the player; also variable degrees of tube (valve) 'warmth' to modify drums, vocals, synthesizers, etc. Some add pre-amplifier overload or tape saturation – or the more benign quality of 'vinyl groove'.

Noise reduction. A field in which digital technology excels. This reduces mains hum and its harmonics, air-conditioning, computer whistles, etc. It also restores archive recordings, removing (most of) the clicks, hiss or rumble. These repairs all need to be applied with care: find a setting at which side-effects appear and set back from it.

Pitch and time. Another winner for digital. A number of packages offer variations on pitch correction. Some are designed to 'fix' voices or instruments that are off-key. Also available are pitch transposition and delay. Vocal formants may be shifted, changing apparent gender or even species. Time changes, with or without change of pitch, are usually accompanied by a claim of 'no chipmunk effect', which means that the repeated tiny overlaps are effectively disguised. A related set of effects enriches the timbre of voices by adding harmonics.

Microphone modelling. This converts the frequency characteristics of one named microphone to another. It can also take into account polar response (change of response with angle), distance (for proximity effects) and losses due to windshields. It is based on manufacturers' data on both modern and classic types (i.e. with tube preamplifiers).

Sampling. This is often supplemented by the provision of sample libraries, which may include a wide range of either Western or ethnic musical instruments, or specialize in a particular department such as drum-kits, percussion, saxophones or a range of 'classic' electric guitars. Each pitched instrument is sampled at a succession of points on the musical scale. Because intervening notes are created (generally) by playing them at different speeds, the more original samples there are the better. Another variable is playing styles: some offer sounds created by named players that are copyright-free.

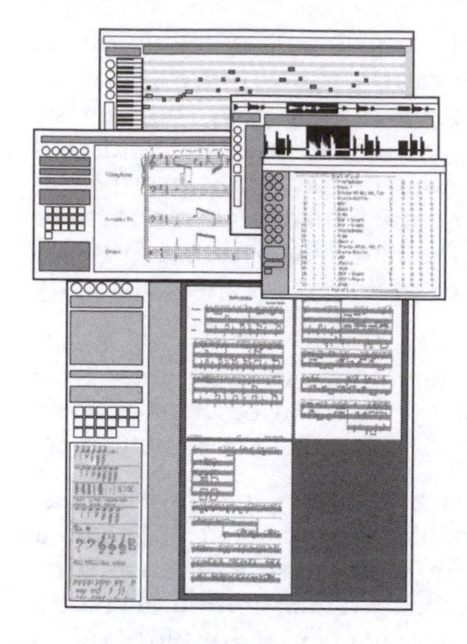

[16.8] **Musical notation**
Traditional notation (available as part of sequencer programs) may be used either to convert a MIDI output display into sheet music or to use it directly as a composition editing tool.

(Taking samples from other sources risks copyright infringement.) Effects libraries are also available in this form. MIDI keyboards may be used with all these.

Also available as plug-ins are technical processes such as *word length reduction, dithering* and *mastering tools*. Typically, these encode 24-bit audio for 16-bit media such as CDs. There are *surround sound tools* and *surround metering* (for stereo, LCR, LCRS, quad, 5.0, 5.1, 6.0, 6.1, 7.0 and 7.1) and a program for downmixing 5.1 to headphones without killing it. Also available are *phase monitoring* and *alignment tools* (to resynchronize or finely control the timing of tracks) and *electrical measurement displays*.

Control surfaces

Adding a small manual mixer, perhaps with eight motorized faders and a master, a selection of function keys, buttons (to select, mute, record or solo on each channel), plus a jog and shuttle (scrub) wheel to wind back and forth against the 'playhead' marker, makes the operation of a virtual studio smoother, faster and more intuitively 'hands on'. Rotary potentiometers, with displays to show selected parameter names and values, may be used to control and automate panning, EQ, dynamics, send levels and other functions on DSP plug-ins. All this may cost little more than some of the software, though each added eight-channel extension costs almost as much again.

For those who want a larger control surface that includes most (or all) of the functions of a multi-layered fully-digital console, complete with automated faders and an extended range of digital processing, this might cost about as much as a well-appointed car. This makes it rather expensive for the hobbyist, about right for much project-studio work, and cheap compared with many broadcast or recording studio consoles. In a virtual studio, where tracks are usually laid down sequentially, the fussiness of layering is less of a disadvantage than it would be in a regular studio, where it is more convenient to be able to reach any channel instantly, without switching layers. And it takes up less space.

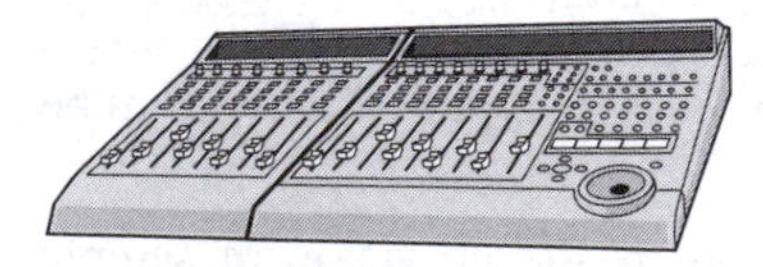

[16.9] **Hardware controller**
A compact unit that features full transport controls, automated touch-sensitive faders (eight channels and a master), rotary potentiometers and displays to control selected (external) DSP functions, and a scrub wheel. This can be extended by adding further eight-channel fader modules.

Digital problems

One of the most important questions about any digital platform is: how stable is it? The next is: what do you do when it falls over? These questions are not limited to the virtual studio, but it is as a good a place as any to ask them.

Computers crash. This is a simple fact of life for anyone who has dealings with them. Computers that are asked to do complicated new things (or sometimes even quite simple things that they were not doing last week) crash in ways and at times that even the most tolerant can find frustrating. Computers used in broadcasting or in other time-limited projects require a history of stability – that is, it must be known from experience that they do not crash very often. They also require redundancy, together with automatic, instant or at least rapid switching so that problems can be bypassed and sorted out later. Project work is (sometimes) more tolerant of such problems than broadcasting is. But tolerance diminishes in the presence of paying clients. These may be vital considerations when setting up any computer-driven system.

Another common problem associated with computers is psychological: the danger of getting locked into a search for bugs (or simply for the correct instructions) that to any observer can seem incredibly time-consuming. It is, of course, essential to be sufficiently familiar with programming that no time is wasted, and to ensure that a crashed system can be reinstated quickly. Consoles with computers should automatically store information at frequent intervals against this possibility, but even so, the recovery procedure requires enough rehearsal to ensure that it does not significantly reduce production time. Handled (and operating) well, computer-driven digital audio is a dream come true; handled badly or failing for perversely obscure reasons, it is a nightmare that makes the apparently simple, understandable faults of analogue audio seem almost forgivable.

Whether in the virtual studio or not, increased digital processing brings with it more and more 'black boxes', items of equipment that most operators cannot repair in the event of failure. Inevitably, operators become the dedicated users of proprietary systems. And new developments in these systems (which appear all the time) require new software, which can all too often lead to more bugs or unfamiliar modes of breakdown. The best time for any user to move in is when a generation of equipment has become reasonably well established, rather than at the cutting edge of the next great advance – unless that advance is vital to the effects required, and the user is happy to help refine the system. Intermittent users of such technology are liable to find it difficult to keep up with what is offered.

In digital technology, rapid turnover is another fact of life. So although individual items are often much cheaper than their analogue forebears, the costs even out. But just occasionally redundant hardware may find a second use. One software item among all those plug-ins converts an outdated G3 Mac computer into a 'high performance standalone non-linear virtual VTR' for use in a digital video production environment.

In general, the shift to digital technology has made possible a greater emphasis on more analytical techniques, towards perfection rather than performance, so project work is likely to be more successful and rewarding when this is the main objective. Fortunately for the users of analogue equipment, the technology time-lag is longer. A microphone, for example, can have a long useful life, and fashions in its use can even bring apparently dead equipment back to life. Or – if it really is dead – fashion may dictate its digital reincarnation.

Chapter 17

Shaping sound

This chapter offers a short digression, to revisit some predecessors of today's virtual studio. The technology was far less powerful, but this did not limit the imagination of its users. Some of what was achieved by slow and complicated analogue procedures is now available as just a small selection from the huge range of facilities on digital signal processors (DSPs), and the names for the effects (often reflecting their analogue origins) are unchanged.

Here we will take a look at how developments in audio technology opened up new ways of shaping sound, and review a few that, although now in decline, might still have uses or could possibly lead in new directions. Long before the digital era, it was clear that anyone with a microphone, recorder and a razor blade (to physically cut and edit tape) could show that the formal organization of any sequence of electronic or natural sounds has some resemblance to music, although how good that was depended on the musical sense of the person making the arrangement.

Some of the methods by which music is now synthesized in the virtual studio were first developed in the earlier field of *electronic music* (at its purest and most austere, in Germany). Similarly, sampled sound began life as *musique concrète* (as the name suggests, in France). These were both used mainly by composers who wanted to experiment with new musical forms. In Britain the BBC, seeking a more practical use for these exotic ways of shaping sound, applied them to the creation of *radiophonic music and effects* for radio and television. Some of the ideas that underlie these (old) new departures are outlined below.

Practically every sound we hear reproduced – however high the quality – is a rearrangement of reality according to particular conventions. For example, monophonic sound is distorted sound in every case except when the effective point source of the loudspeaker reproduces a sound that was originally confined to a similarly small source. Every balance, every carefully contrived mixture of direct and indirect sound, is a deliberate distortion of what we would hear in the same sound, live. This is accepted so easily as the conventional form of the medium that few people are even aware of it. Perhaps distortion would be better defined as 'unwanted changes of quality', rather as noise is sometimes defined as 'unwanted sound'. Sound is continuously being 'shaped', and the listener is always required to accept some convention for its use.

When an audience is presented with some 'new sound', good listening conditions are even more important than for conventional music. A listener with good musical imagination who hears known instruments and established musical forms can judge the quality of composition or performance in spite of bad recording or reproduction. Although they reduce enjoyment, noise and distortion are disregarded. The musical elements are identified by reference to the listener's knowledge and experience: imagination reconstitutes the original musical sound.

Plainly, this does not happen on the first few hearings of music in which the new sound is produced by what, in other situations, might be regarded as forms of distortion. For this, imperfections produced by the equipment are, for the audience, indistinguishable from the process of musical composition: they become part of the music.

A major difference between synthesized or sampled recordings and the conventional music that audiences are more familiar with is to do with performance. The traditional composer commands a large number of independent variables: the choice of each instrument and the way it is to be played; the pitch of each note and its duration; the relationship in time and pitch between successive notes and between different instruments. But however much is specified, there is always room for interpretation by performers. In contrast, an electronic or concrete composition exists only in its final state, as a painting or carving does. Unless it is combined with conventionally performed music or created in real time, for example from a MIDI keyboard, it is not performed: it is simply replayed. In this respect *musique concrète* is like electronic music, from which it otherwise has certain fundamental differences.

Musique concrète was made by sampling the existing timbres of a variety of sound sources, transforming them, then assembling them as a montage. Electronic or synthesized music began, as might be expected, with electronically generated signals such as pure tones, harmonic series or coloured noise (noise contained within some frequency curve). The results are distinctively different.

One characteristic quality of sampled sound is that the pitch relationship between the fundamental and upper partials of each constituent remains unaltered by simple transformation unless any frequency band is filtered out completely. This is both a strength and a restriction: it helps to give unity to a work derived from a single sound or range of sounds, but severely hampers development.

In electronic music, on the other hand, where every sound is created individually, the harmonic structure of each new element of a work

[17.1] **Classification of sound**
The dynamics or envelopes of sounds: early classifications devised by *musique concrète* composers and engineers as a first step towards a system of notation. The terms shown here could represent either the quality of an original sound or the way it might be transformed.

Attack. 1, Shock: bump, clink, etc. 2, Percussive: tapping glasses, etc. 3, Explosive: giving a blasting effect. 4, Progressive: a gradual rise. 5, Level: at full intensity from the start.

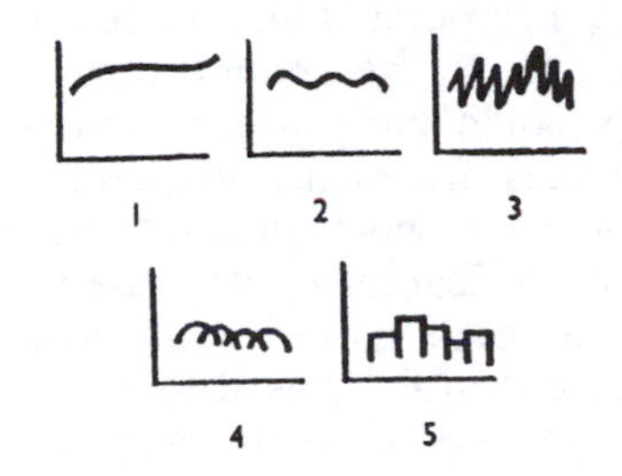

Internal dynamics. 1, Steady: an even quality. 2, Vibratory or puffing. 3, Scraping or tremulous. 4, Pulsing: overlapping repetitions blending together. 5, Clustered subjects placed end to end.

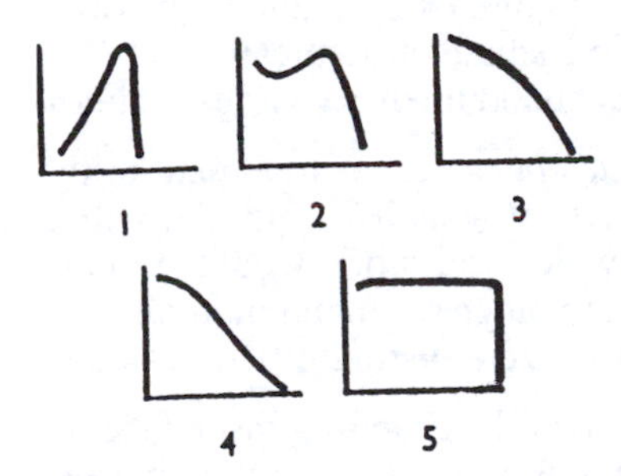

Decay. 1, Reversed shock. 2, Accented. 3, Deadened. 4, Progressive. 5, Flat: cut off sharply.

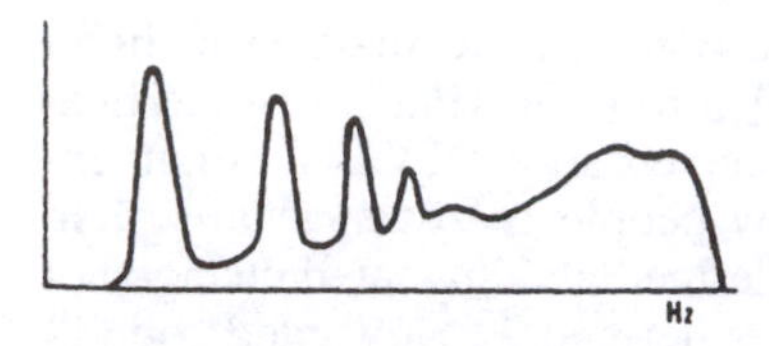

[17.2] **Line and band spectra**
Instantaneous content: this sound is composed of individual tones superimposed on a broad-spectrum band of noise.

is, in principle, completely free. Where *musique concrète* has restrictions, electronic music offers freedom – an essential difference between the two. But together, these techniques and their successors have a proven value in extending the range of sound comment. This developed as a field in which sound effects took on formal musical qualities in structure and arrangement, and musical elements (whether from conventional or electronic sources) complemented them by taking on the qualities of sound effects. A useful attribute for anyone working in this field is a strongly self-critical sense of humour.

Musique concrète

Musique concrète, according to its originators, is (or was) built up in three distinct phases: selection, treatment and montage.

The first characteristic of *musique concrète* lies in its raw material: a sampled sound which, by the choice of beginning and end, becomes complete in itself. In some of the earliest compositions, emotional associations of the original sound were incorporated into the music. There was a later reaction away from this, on the grounds that the lack of an immediate mental association could lend power to sounds. The basic materials might include tin cans, fragments of speech, a cough, canal boats chugging or snatches of Tibetan chant (all these are in a work called *Etude Pathétique*). Musical instruments are not taboo: one piece used a flute that was both played and struck. Differences in balance or performance can also be used to extend the range of materials. All this is very similar to the way that samples integrated into popular music have included news actuality, political statements and fragments of other people's compositions.

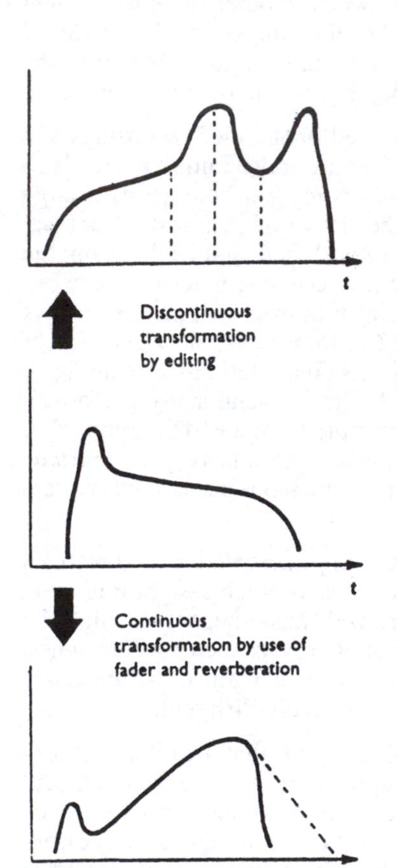

[17.3] **Musique concrète**: **transformation of dynamics**
(Original sound in centre.)

The preliminary *concrète* recording was described analytically (again, by its originators) in terms of a variety of sound qualities:

- instantaneous content – frequency spectrum or timbre (which might contain separate harmonics, bands of noise or a mixture of the two);
- the melodic sequence of successive sound structures;
- its dynamics or envelope (the way sound intensity varies in time).

The second stage in building *musique concrète* was treatment of the component materials to provide a series of what its composers called 'sound subjects', the bricks from which the work would be constructed. To help classify their growing range of 'manipulations', the *Groupe de Recherche de Musique Concrète* distinguished between:

- *Transmutation of the structure of sound* – changing instantaneous sound content to affect melodic and harmonic qualities but not dynamics. This includes transpositions of pitch (whether continuous or not) and filtering (to vary harmonic structure or coloration).
- *Discontinuous transformation* of the constituent parts of a sound by editing – a sound element could be dissected into attack, body and decay (today, divided further into attack, decay, sustain,

release – 'ADSR'), then subjected to reversal, contraction, permutation or substitution. This form of manipulation varies the dynamics of the sound, though at any instant the content is not altered.

- *Continuous transformation*, in which the envelope of a sound is varied without editing, by the use of faders, AR, etc.

In the third phase, the construction of *musique concrète*, the sound subjects were put together piece by piece, like the shots of a film. One method was to edit a single tape by cutting and joining pieces together (montage); another was dubbing and mixing in the way that sound effects sequences might be assembled. Today, they would simply lay tracks in digital memory and mix them down.

In the early days of *musique concrète* the only way to change the pitch of a sound was to change the speed of replay on tape or disc. This led to a more versatile analogue pitch changer in which the signal was recorded on the surface of a rotating drum: with this, pitch and duration could be changed independently – but with a characteristic 'chipmunk' effect from the overlapping segments. Happily this monstrous device has been superseded by a wide range of digital signal processors (see Chapters 12 and 16).

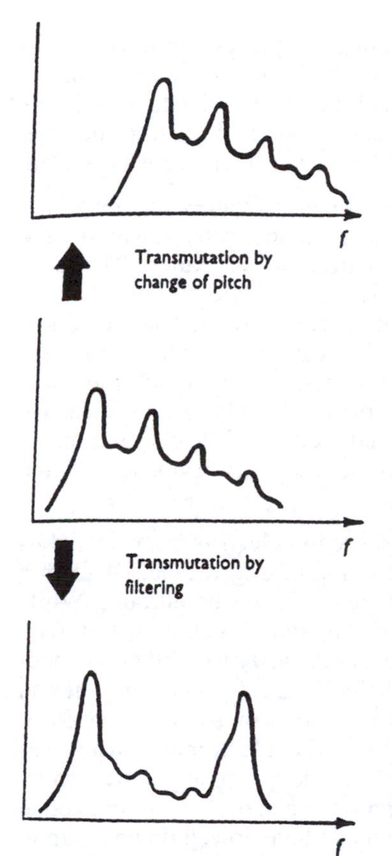

[17.4] **Musique concrète: transmutation of frequency** (Original sound in centre.)

Another analogue device that could change pitch independently of duration was the *ring modulator*. This allows a fixed number of cycles per second to be added to, or subtracted from, all frequencies present (e.g. components f_1, f_2, f_3 become $x+f_1$, $x+f_2$, $x+f_3$, whereas simple alteration of speed multiplies all frequencies by a common factor). Frequencies could also be inverted ($x-f_1$, $x-f_2$, and so on) so that high becomes low and vice versa. With this type of 'transmutation' exotic qualities of consonance and dissonance are obtained. They do not occur naturally (in conventional music) but have been used in some electronic music. Again, this effect is now available on digital signal processors.

Modulators have been used in telephony, so that narrow-frequency bands containing the main characteristics of speech can be stacked one above the other on a broadband line. In their modulated form some speech bands may be transmitted in an inverted form – and it is possible to bring them back down to normal frequencies while still inverted. For example, a band of speech might be inverted so that what was 2500 Hz becomes zero, and vice versa. The most striking result of this is to invert the three main bands of formants, so the speech is translated into a sort of upside-down jabberwocky language – which, curiously, it is actually possible to learn.

Among more complex effects, a filtered human voice can be used to modulate a train siren to make it talk. The effect of an exaggerated machine voice can be obtained by modulating an actor's deliberately mechanical-sounding speech with a tone of 10–20 Hz. In electronic music, two relatively simple sound structures can be combined to create a far more complex form. After using a modulator, it may be necessary to use a filter to get rid of unwanted products.

The *Vocoder* was designed to encode the basic parameters of speech, including its tonal and fricative qualities, so that they could be

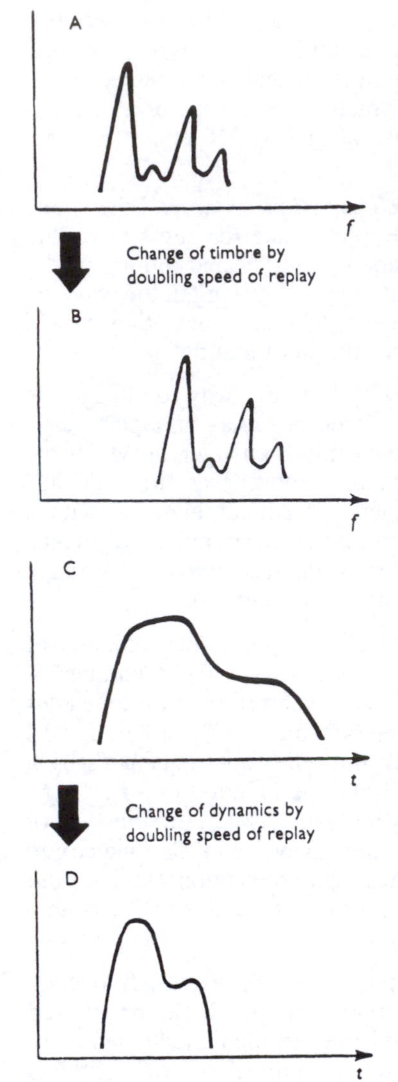

[17.5] **Transposition in pitch**
When a sound is replayed at twice the original speed it changes both the timbre (A and B) and the dynamics (C and D).

imposed on any electronic or other source and give it the characteristics of a voice, including intelligible speech and song. This extended the range of techniques available to films and television programmes that specialize in alien voices and mechanical monsters, which had previously depended heavily on ring modulators and swept filters (the latter usually wobbles a filter up and down in frequency). The Vocoder also offers possibilities for humanizing electronic music by encoding some characteristic 'imperfections' of musical instruments and operating with them on electronic (or other) source material. It could be used in conjunction with changes of pitch or duration, or any other available effect.

When there is picture as well as sound, it is possible to link the two directly, deriving or modulating an audio signal from some parameter taken from the video signal; or alternatively, allowing audio to modulate video or to control the generation or selection of picture elements.

Electronic sources

The creator of *electronic music* has, in principle, complete control of resources, synthesizing every detail of the sound. With no performer in the loop and therefore full responsibility for the finished work, the composer has total freedom – subject only to the limitations of time, patience and equipment. In the early days that patience was needed.

The equipment for this includes sine tone oscillators for producing single frequencies or combinations of tones, plus sawtooth and square-wave generators for richer harmonic structures. The saw-shaped wave contains a full harmonic series, while the square shape includes only odd harmonics, like those of a clarinet. These could be treated by '*subtractive synthesis*', i.e. filtering – which was less time-consuming than an additive process. Precise calibration was essential: if musical instruments are off-pitch, only one performance is affected and the score remains unaltered, but in electronic music the first 'realization' is likely to be the only evidence of the composer's intentions.

White noise is another type of sound used in electronic music and its successors. It could be produced by amplifying the output from a noisy vacuum tube: special tubes were made for the purpose. White noise alone is not particularly useful, but again in combination with filters it could be used to produce *coloured noise*, either of indeterminate pitch or related to a musical scale by placing peaks at particular frequencies and troughs at others, or by using bands of chosen widths. Another effect in early electronic music was achieved by imposing a sinusoidal envelope (or any other waveform) on the noise to create anything from a series of puffs of sound to a tone that could be used musically – then slowed down again to reveal its granular structure.

A characteristic sound in early electronic works was short bursts of tone at various frequencies, sounding like a series of pips, beeps and bloops. If these were assembled from short lengths of tape, variations in attack and decay could be introduced simply by changing the angle at which the tape was cut. A 90° cut produced a click at the beginning or end; other angles softened it. Other kinds of attack and decay were created electronically, producing effects that were described by relating them to those of a musical instrument of similar attack and tone quality.

In parallel with the development of 'pure' electronic music, there had been developments of playable devices – *electronic organs* – with names like Monochord, Melochord and Spinetta, besides the more conventional Compton and Hammond organs. The Trautonium was one of a group of instruments that could be used (and played live) to provide notes of any pitch, not just those corresponding to a keyboard. Organs, as one class, shade gradually into another, the *synthesizers*. Most of the electronics of early synthesizers were 'analogue' also in the special sense that they copied or reconstructed the required waveforms or noises by means that provided a continuous shape, which could also be continuously modified from the start. The next stage of development was to add some digital circuitry – for example, by using a digital control device to program musical sequences of treated sounds, so creating the first *sequencers*.

In its time the Moog was a revolutionary design of analogue synthesizer employing compact multifunction voltage-controlled oscillators and amplifiers with what was seen as a generous ability to modify sound structures in interesting ways. (Now it is found as a digital effect.)

The application of techniques derived from both electronic music and *musique concrète* to broadcasting was pioneered in Britain. At the BBC the product of its Radiophonic Workshop was applied directly to the needs of general radio and television programming. Since it began, in 1958, the BBC's Radiophonic Workshop has updated its equipment many times, in a process of evolution that has tracked the revolution in audio processing and recording until, by the late 1980s, digital and MIDI techniques dominated the field. This expertise has since dispersed, and around the world there are now vast numbers of such studios. Very few are devoted to 'pure' experimental music, a minority taste. But along the way it has been shown that completely new musical forms are possible. One example is outlined below.

New laws for music

Most synthesized music is based on conventional musical scales: these are directly related to human powers of perception and appreciation, which in turn are attuned to harmonic structures that arise naturally. But by starting from pure electronic sources these musical 'laws of nature' can be relaxed – or transgressed. Given how human minds work, whether this is much use depends on the quality of individual offerings.

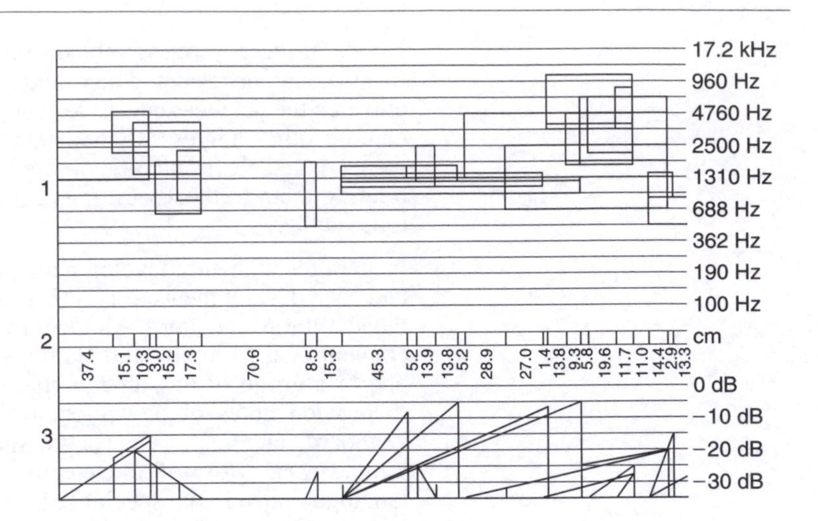

[17.6] **Electronic music**
Score of the type devised by Stockhausen to show the structure of his '*Study II*'. 1, Pitch: this shows the groups of tone mixtures used. Each block represents a group of five tones. 2, Intervals of tape measured in cm. The tape speed is 76 cm/s (30 in/s). 3, Volume: dynamics of the tone mixture.

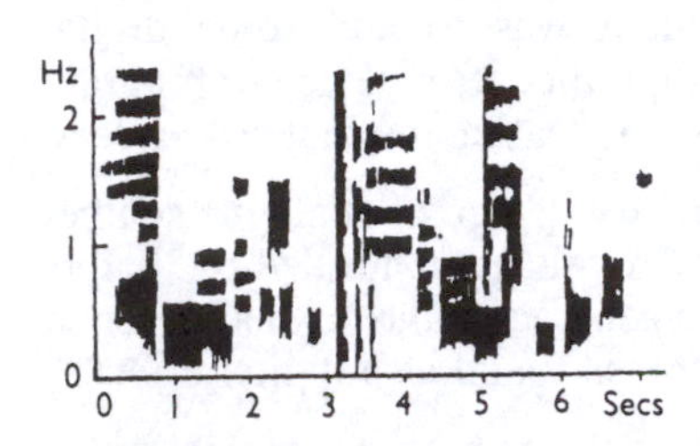

[17.7] **Tone mixtures**
Five typical mixtures used in Stockhausen's '*Study II*': 193 were used in all.

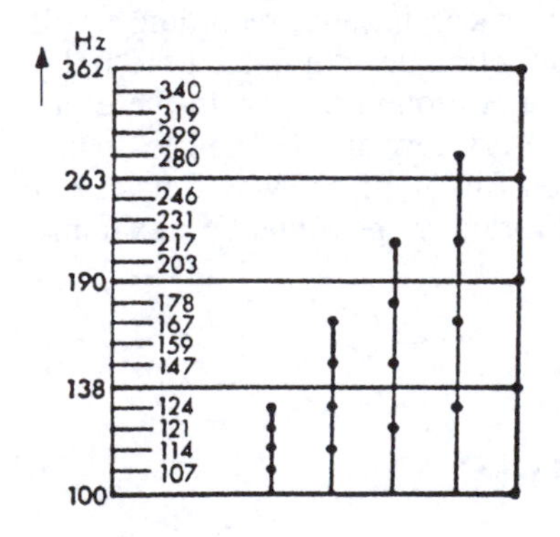

[17.8] **Electronic music and reverberation**
Spectrogram of part of Stockhausen's '*Study II*', showing the softening effect of reverberation. Each of the original groups was replayed through a reverberation chamber (digital AR was not yet available).

All scales used in conventional music have one thing in common: the interval of an octave, in which the frequency of the upper note is exactly twice that of the lower. If the two are sounded together on orchestral instruments the combination sounds pleasant because of the concord between the fundamental of the higher note and the first harmonic of the lower, and also between many other harmonics they have in common. But if one of the notes is shifted off-pitch, these concords are lost; they are present only when the frequency of the two fundamentals can be expressed as a ratio of small whole numbers. Note also that without the harmonics there would be no sense of concord or discord at all, unless the fundamentals were fairly close together. If we start with two pure tones at the same frequency and then increase the pitch of one of them, the first effect we hear is a beat between them. As the difference becomes greater than 15 Hz (two or three times this at high frequencies), the beat is replaced by the sensation we call dissonance. This increases to a maximum and then falls away, until as the frequency ratio approaches 5:6 it is no longer apparent. For pure tones, dissonance occurs only within this narrow range. (There is one proviso: that the tones are not loud. If they are, the ear generates its own harmonics.)

Some experimenters with electronic music, noting all this, have concluded that our conventional concept of scale is just a convenient special case, and that if only we could create harmonic structures that were not based on the octave, completely new scales could be devised.

An early example of electronic music based on a totally new concept of scale has (exceptionally) been published as a complete score, showing everything that had gone into the making of Karlheinz Stockhausen's '*Study II*'. When the work is played to a listener with perfect pitch, or even good relative pitch, it seems immediately that 'something is wrong' – or at least, that something is different. By all conventional musical standards, something is: its arrangement of musical intervals did not exist before the introduction of electronic sound synthesis.

A conventional scale accommodates as many small-whole-number relationships as possible by dividing the octave into 12 equal, or roughly equal, parts. Each note in the chromatic scale is about 6% higher in frequency than the one before, so that the product of a dozen of these ratios is two: the octave. But Stockhausen, for his '*Study II*', devised instead a completely new scale based on an interval of two octaves and a third. This is the interval between a note and its 'normal' fifth harmonic (the interval between any two notes of which frequency ratio is 5). This he divided into 25 equal smaller intervals, so that each successive note on his scale is about 7% higher than its predecessor. In consequence, very few of the intervals based on this scale correspond to anything in the musical experience of the listener.

If music following this scale were played on a conventional instrument, almost any combination of notes attempting either harmony or melody would be dissonant because of the harmonics present. Instruments such as violins or trombones could in principle be adapted to playing it; others, such as trumpets and woodwind instruments, would have to be specially made. But it still would not work, because of the unwanted musical harmonics. The only feasible form of construction for such a work was with completely synthesized sound from electronic sources.

Stockhausen, having devised his novel scale, proceeded to put together new timbres in which the partials would all fit in with each other on the scale. He used just five basic sound groups, each of five tones. The most compact of these contains five successive frequencies ('notes') from the new scale. This consists of a series of dissonant pairs, and its quality could be described as 'astringent'. The next group has members spaced two notes apart, and the internal dissonance has now almost gone; in the other groups, with members spaced at three-, four- and five-note intervals, it has gone completely, and the difference in character between them depends solely on their width of spectrum. But if when groups are played together there are strong adjacent pairs, dissonance reappears. This single example is offered as a stimulus to the reader's imagination. Plainly, the opportunities for shaping sound go far beyond what is conventionally achieved.

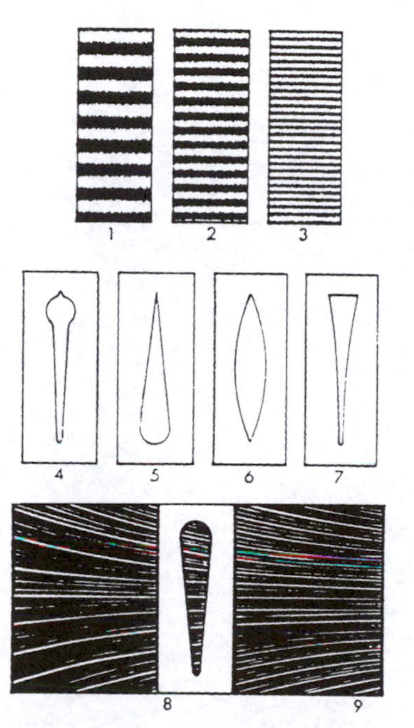

[17.9] **Animated sound on an optical film track**
1–3, Cards from a set giving different pitches. More complex patterns of density give richer tone colours.
4–8, Masks (to be superimposed on tone cards) giving dynamic envelopes of a particular duration. Long durations are obtained by adding together cards that are open at the ends.
9, A card giving a range of pitches of a particular tone colour. A wide range of similar techniques has been employed, using variable density or variable width to control volume.

Animated sound

Finally, in what may be a completely blind alley off this digression, the following is an account of the (literal) use of shape to create new sounds.

For almost as long as film has carried sound, optical tracks have been used as a medium for 'animated' sound. Drawings have been photographed or marks inscribed directly on the soundtrack, either frame by frame or continuously (printed by rotating wheels). Sometimes sound was constructed by filming geometric forms such as rectangles, triangles, ellipses, or lettering, fingerprints, even facial profiles. Pitch was controlled coarsely by changing the size of the shape, and finely by camera distance. Double exposure superimposed one sound on another. Light slits linked to pendulum systems have also been used.

The Canadian animator Norman McLaren made no attempt to imitate natural sounds. Instead, he showed that quite simple shapes could be rich in harmonics, with some producing strident or harsh effects. He sketched a range of 'tone colours' for use with tone contouring masks, superimposed cut-outs that had an attack, mid-note dynamic contour and decay. He shot frame by frame, but sometimes extended an image over several frames to produce a continuous or mandolin quality. For a pause he used black card. The resulting blips, clunks and splinges could sound as though they were squirted in blobs or squeezed like toothpaste – and the difficulty in describing them illustrates his success in getting away from orthodox sound.

Chapter 18

Audio editing

At one time, audio editing almost always meant cutting and rejoining a tape, and many of the terms that are still used refer directly to this physical process. Tape, obviously, is a linear medium: to get to any particular place on it, you must run forward or backward to that point.

Now the same ends are generally achieved by reorganizing and recombining data in computer memory. This begins with recording and laying tracks on hard disk. *Rough editing* is assembling the main body of a programme or sequence in the right order and taking out the longer stretches of unwanted material. *Fine editing* is tightening up this assembled material to provide for the best continuity of action and argument, and also deleting hesitations, errors and irrelevancies that may distract attention and reduce intelligibility. Finally, the edited assembly will usually require *mixing*, in which several (or many) tracks are combined in a final mix.

Hard disk editing is *non-linear*: Any point can be accessed on any track in any order at any time. It is also, generally, *non-destructive*: the original recordings are undisturbed. To edit in this way, you make a series of decisions on what should be done, and the computer demonstrates what would happen if these were carried out. It looks (or sounds) just like the finished product, except that all you have at this stage is a list of instructions for instant non-linear access. This is called an *edit decision list*, or *EDL*. The EDL can be revised again and again without affecting the original; neither is it immediately destructive of earlier versions of the EDL. You can instantly 'undo' several (often many) stages of the edit. If you save an edit from time to time, you can easily return to these even older versions. (And if copies are saved safely somewhere else, you can protect yourself from crashes.)

Tape editing is not the exact opposite of 'non-destructive'. The original would only be destroyed by wiping or recording over deleted material, and although this linear editing process has been used, that is not how it is usually done. Instead, the tape is cut, and deleted material can in principle be joined up on to a second reel for potential reinstatement or reuse. This, of course, is far more laborious than working on disk.

Sometimes a combination of the two processes is used. The main edit is done on hard disk. The result is played out (allowing a final check on the edit) and recorded on analogue tape as backup, prior to broadcast.

For any simple last-minute changes, cutting the tape is fast and convenient. And using tape rather than hard disk for transmission protects the broadcast from crashes.

As computers become more reliable, or (a more likely future) 'broadcast-critical' access to independent backups becomes a safe alternative, the use of tape will diminish further. Its own safety features have helped it survive alongside its digital competitors, but as broadcast play-out systems demonstrate that local computer failures can be bypassed, the additional features that integrated digital systems can offer – including centralized control of every aspect of content from commissioning and scheduling to a final archive resting place, alongside a full account of copyright and history – will make them unbeatable.

Editing techniques will be described below in the order of their historical development. But whatever technique is employed, editing is generally undertaken for one or more of five reasons – which apply to television and film as well as audio for broadcast or recordings:

- *Programme timing* – to adjust overall duration to fit some scheduled space. In radio or television this often overrides all others, whether or not the final stages of editing improves the result – although any original treatment should be written with the intended length in mind.
- *Shaping the programme* – to give it a beginning, a middle and an end, while ensuring that pace and tension are maintained. In actuality speech recordings, any awkward parentheses, repetitions, redundant phrases or even whole paragraphs should go if they obstruct flow.
- *Inserting retakes*. This restores or improves a predetermined structure, especially in music and drama.
- *Cutting fluffs*, etc. Minor faults can lend character to unscripted speech, but too many make it difficult to follow. Mistakes in reading a script do not often sound like natural conversational slips.
- *Convenience of assembly* – to combine location recordings with studio links; to allow material of different types to be rehearsed and recorded separately; or to pre-record awkward sections.

Radio and television techniques that aim to avoid editing presuppose either that the broadcast has a well-rehearsed and perfectly timed script or that those taking part are capable, in an unscripted programme, of presenting their material logically, concisely and coherently, or are under the guidance of a presenter who will ensure that they do. Such self-control under the stress of a recording session is rare except among experienced broadcasters. Even when a recording is made in familiar surroundings, the intrusion of the microphone can produce stilted and uncharacteristic reactions. One purpose of editing is to restore the illusion of reality, generally in a heightened form.

Editorial selection from unscripted material often implies responsibility for avoiding misrepresentation of the character or intentions of participants. On hearing an edited playback, any contributor expects to recognize the essence of the original, within the limits of what was needed for the programme, which should have been made clear in advance. But editing is sometimes used to make content conform more closely to the purposes or preconceptions of those in control of the process. At best, this may be to mould it to some openly declared format. At worst, it subverts the intentions of contributors. Plainly, editing can cause controversy, and may lead to debate on ethics – which may overshadow the process, even as it is happening. But the purpose here is to describe what can be done, and how. Fortunately, in most cases, the purposes are both clear and desirable.

Retakes and wildtracks

Even as the original recordings are made, the first preparations for editing begin. Nearly all recordings benefit from at least a little tidying up, and this is easier if, right from the start, you constantly check what might be needed. What to watch for depends on the type of programme. For example, if a 'fluff' occurs in speech, will it be desirable or even possible to edit, or would a retake be advisable?

On film or during a video recording, the error may be at a point when a picture cutaway is intended or can easily be arranged. In this case a verbal repetition (for example) may be left for the editor to cut, but if this is awkward a retake might be recorded in sound only (*wildtrack*). If a retake in vision is necessary, the director has to decide whether to go again from the start of the scene; whether to change the framing of the picture (often, if possible, to a closer shot or different angle); or whether the same framing can continue after an earlier cutaway. The sound recordist should also (generally without being asked) supply background atmosphere ('*atmos*' or '*buzz track*') at the same level and in sufficient quantity to cover any gaps that may be introduced by editing or where there are more pictures than synchronous sound. This saves time and effort later, at the re-recording ('*dubbing*') session.

In a radio discussion between several people, the director should note whether the various voices have clearly identified themselves, each other, and also what they are going to talk about. It might be sensible to record a round of names and subject pointers in each voice so these can be cut in if necessary. Watch for continuity of mood: if the point a retake is likely to join up to ends with a laugh, try to start the new take with something to laugh about (which can be cut out). Discourage phrases like 'as I've said', which all too often refer back to something that will be cut, and retake if they, too, are not clearly editable.

When editing radio programmes, 'pauses' may be needed. So, here again, record 15 or more seconds of atmosphere. If absolutely necessary, atmosphere for pauses can often be picked up from hesitations between words, but it is better to be safe. Record 'atmos' while the speakers are still there: after they have gone the sound may be slightly different. If much editing is likely, be sparing in the control of levels: over-control bumps the background atmosphere up and down. If the background is heavy or varied, record extra or longer tracks.

At a recording session for classical music it is again useful to record a few seconds of atmosphere following each item or at the beginning of the next in order to have something to join successive pieces. Music retakes should be recorded as soon as possible after the original, so that the performers can remember their intonations at the link point, and also so that the physical conditions of the studio are close to the original. An orchestra quite literally warms up; its tone changes appreciably in the earlier part of each session in a studio.

Tape editing

For many years the standard way to edit mono and then stereo sound was to cut and splice analogue 6.35 mm (commonly known as quarter-inch) tape. The tools required are the following:

- A splicing block with a channel to hold the tape. The channel is lapped slightly so that as tape is smoothed in it is gripped firmly. The block has cutting grooves angled at 45° for mono, and 60° and 90° for stereo. It should be fixed firmly at the near edge of the tape deck.
- A fairly sharp stainless steel razor blade – 'stainless' is non-magnetic. Use single-sided blades (and preferably not too sharp) to avoid cutting fingertips. Store them safely in a holder.
- A soft wax pencil (yellow or white) or felt-tipped pen, for marking edit points on the tape backing.

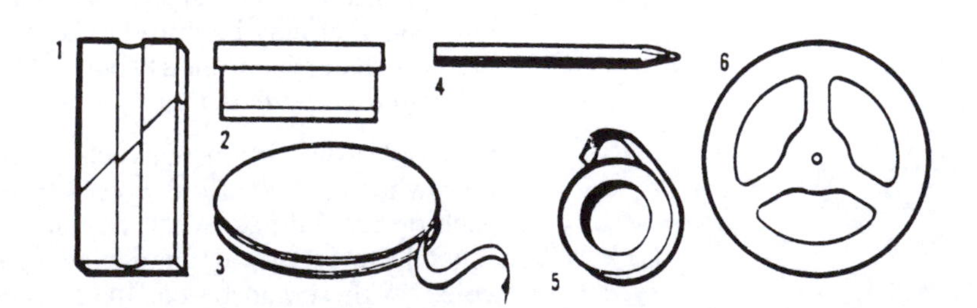

[18.1] **Equipment for cutting and rejoining audiotape**
1, Splicing block. 2, Blade.
3, Leader tape. 4, Wax pencil.
5, Jointing tape. 6, Spare spool.

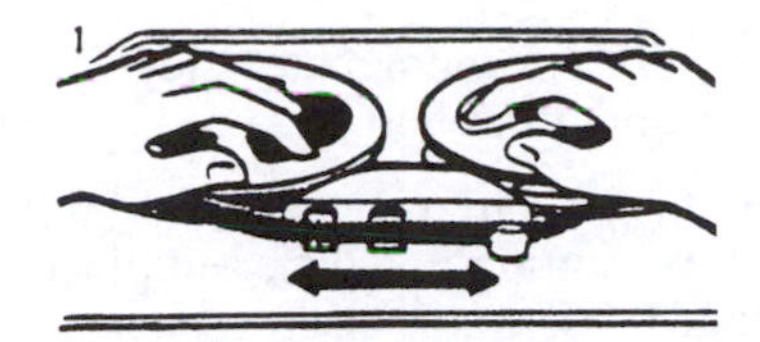

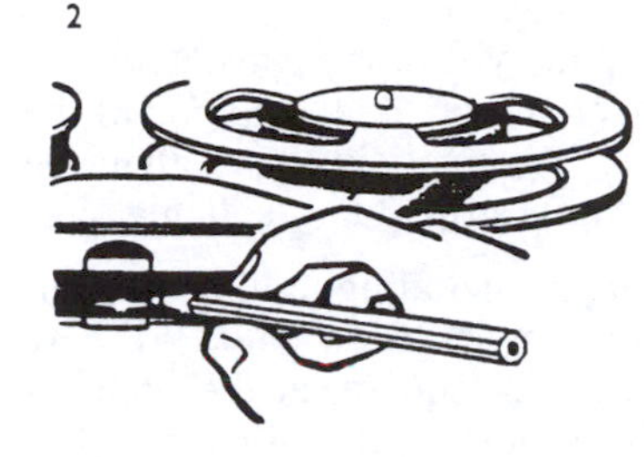

[18.2] **Audiotape editing**
1, Move the tape backward and forward over the heads, by hand, in order to find the exact point. 2, Mark the tape with a soft wax pencil. 3, Cut the tape diagonally using the groove in the block. 4, Butt the two wanted sections together in the block and join with special adhesive tape.

- A roll of splicing tape (non-oozing), a little narrower than recording tape to allow some inaccuracy in use without overlapping the edges. (Pre-cut adhesive 'tabs' in a dispenser are also available.)
- Leader and spacer tapes – white or green for the leader, yellow for spacers and red (or twin coloured, red and white) to mark the tail.

A dusting powder such as French chalk may be used in hot climates, where the adhesive may ooze a little. It is important that tape should not be sticky, or layers will pull as they come off the spool, drag on tape guides and clog the heads.

When preparing to make a cut, first play the tape to the point at which the cut is to be made, and then switch off replay, so the tape can be spooled by hand with the tape in contact with the head. (This head-scrub effect is mimicked by most digital editing systems.) Check the exact point by pulling the tape back and forth over the replay head. Mark the tape on the backing, and always in the same way, e.g. along the lower edge: this ensures that if a scrap of tape is reinstated, it will go in the right way round. A simple edit is made like this:

1. Locate and mark the joint.
2. Switch the recorder to a condition that allows the tape to be lifted from the heads, and place it coated side down in the editing block.
3. Cut. The razor blade should be sharp enough not to drag when drawn lightly along its guide channel. No pressure should be necessary.
4. If a fairly long piece of tape is to be taken out, replace the take-up spool by a 'scrap' spool and wind off to the new 'in' point. Mark this, cut again, and swap back the take-up spools.
5. Place the two butt ends of the tape coated side down in the block, a little to one side of the cutting groove, so they just touch together.
6. Cut off a short length of splicing tape and lay it over the butt, using one edge of the channel to steady and guide it into place.
7. Run a finger along the tape to press the splice firm. If it seems sticky, dip a finger in French chalk and run it along the tape again.

After a splice is made, place the tape back on the heads, and wind back 30 cm (1 ft) or so – usually spooling by hand against the braking effect of clutch plates – then play through the splice, checking that:

- it *is* properly made;
- the edited tape makes sense – you have edited at the point intended;
- the timing is right – that the words after the joint do not follow too quickly or too slowly on those before;
- the speaker does not take two breaths;
- neither the perspective nor volume of voice takes an unnatural jump;
- there is no impossible vocal contortion implied;
- the tone of voice does not change abruptly, e.g. from audible smile to earnest seriousness;
- the background does not abruptly change in quality, nor is any 'effect' or murmured response sliced in two;

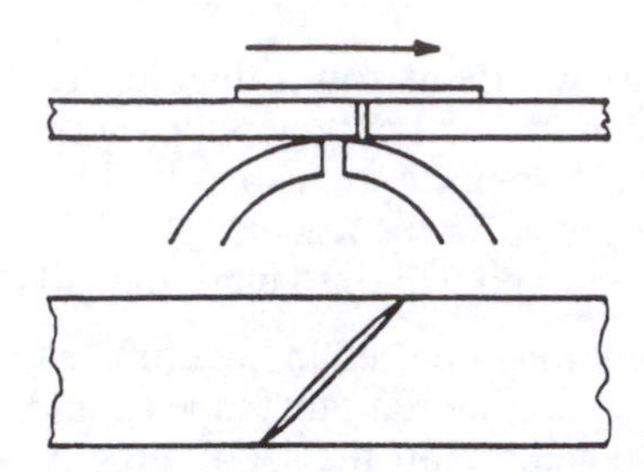

[18.3] **Angled butt joint in audiotape**
As the audiotape (held together by a strip of jointing tape on the outer surface) passes over a recording or replay head, there should be no loss of contact at any time, unless the butt is imperfect. An angled joint minimizes the effect of any slight gap or overlap; it also creates a rapid mix of outgoing and incoming sound. This is suitable for monophonic recordings, but not all stereo (the position of sources in stereo can momentarily 'flicker' to one side).

- (for stereo edits) there is no 'flicker' – a momentary sideways jump that can be introduced by an imperfect butt.

A virtue of this kind of editing is that it is easy to reinstate material if the cut does not work, or to change it by cutting in a different place. You may need to cut down a gap that is too long; insert a pause, a breath or even an 'um' or 'er' to cover an abrupt change of subject or mood; or you may have to restore part of what was cut.

To remake a splice, first mark any new point you propose to cut at, then undo the old join. To do this turn it over and break it; grip one of the points of the butt and pull it away from the splicing tape, then the other. After remaking the cut, back it with a strip of new splicing tape.

It is recommended that monophonic tape (including mono contributions to a stereo mix) should be cut at an angle. If it is cut at 90°, any sound on the tape, even atmosphere, starts 'square' at the point of cut, and any sound cutting in at full volume seems to start with a click. But if a fade in or out exceeds 10 ms in duration there is no click: in fact this is over 20 times more than is actually needed in most circumstances. This duration corresponds to 2 mm of a 19 cm/s (i.e. 0.075 in of 7 in/s) tape and works well for mono. For stereo an angled cut (at 60°) should be used only when editing in a pause or between similar sounds, otherwise a momentary change of position (flicker) will occur. If this seems possible, use a 90° cut (or a mix).

The techniques for rough and fine editing of content were developed using analogue tape, and are equally valid for digital editing.

Rough editing

For this, the simplest and most flexible example is that in which only sound is to be edited, typically for radio or records – but much of what follows applies equally to video editing for television, film or DVD.

When editing begins, it may be tempting to plunge straight in, to work in detail on some accessible sequence – but this might be a mistake. A sequence that is cut so early often ends up with more recuts than any other, as it is weighed and then reconsidered time after time in the context of other sequences. It is generally wiser to begin by putting together material that helps to outline the overall, intended structure. In rough editing, the first task is simply to assemble material in its intended order, at this stage including only those retakes that contain essential highlights. The aim is to get an early glimpse of the ultimate logical or artistic effect, without having to spend much time or effort on the mechanics (or artistry) of smoothly linking it all together.

Remember that that it is easy to *transpose* material. In radio, this can be useful when unrelated subjects appear in an unscripted discussion. Or when editing a recorded 'actuality' event, it may help to explain what is going on if some of the material is taken out of sequence. In any case, factors that govern the order of recording may not hold for the finished programme. For example, it is often best to start with a striking effect that engages the listener's attention and involvement. Then the middle part of a work should, while progressing logically, have light and shade: with variety of pace (in a discussion, of speech lengths and rhythm) and of mood (or emotional development). Ideally, the tension should build to an effective 'curtain': in a discussion, something that contains in a few lines the essence of all that has preceded it, or that will offset and illuminate the conclusion – a 'pay-off' line.

As material is selected it is added to the master recording in programme order, with transpositions and major retakes in their proper places. At this stage use clean, simple cuts: in speech, cut mainly at 'paragraph pauses'. Anything that is not selected in this first pass is retained (and kept readily available) until the work is complete.

Even at this early stage exercise care: try to avoid cutting sound in the middle of a breath, or too close to the first word when there is a heavy background. When joining different sections delivered by the same speaker, check whether voice quality, mood and balance are sufficiently similar: it is all too easy with radio for the listener to get the impression that someone new has entered the conversation and to wonder who that is, instead of listening to what is being said.

If subsequent mixing is envisaged, different components will eventually be laid on separate tracks. For a first, rough edit on quarter-inch tape, sound from each source can (mostly) be laid sequentially on a single track: if necessary, this can be split to two tracks later.

Fine editing

When the rough assembly has been made, give thought (again) to the shape of the programme as a whole. It may now be clear that you have to make severe internal cuts in some sections but less in others; obvious cuts may include overlaps, repetitions or unnecessary elaboration. For many types of programme, ranging from radio discussion to television documentary, you may also want to check the intelligibility and conciseness of individual contributions, and whether the character and personality of participants is adequately represented. Try to keep the theme of what remains tight and unified.

Some other things that may need to be cut are the following (some of these are easier for radio sound than where pictures complicate it):

- Heavy coughs, etc. These hold up action and may cause the audience to lose the thread. But if the speaker has a frog in the throat (and this section cannot be deleted), it does no harm to leave in the cough that clears it. The audience will feel more comfortable too.
- Excessive 'ums' and 'ers' (or any other repetitious mannerisms) should be cut if that improves intelligibility, which it often does if they awkwardly interrupt a sentence. But some actually improve

intelligibility by breaking the flow at a point where it needs to be broken. So be careful: there are people who use these noises to support or modify meaning, or to add emphasis or emotional expression, and while the result may not read well, it sounds right. Others seem to express character in these punctuating sounds. In either of these cases, do not cut. An 'er' moulded into the preceding or following word often cannot be cut anyway.

- Excessive pauses. In real life (or on television or film) we often use pauses to examine and read a speaker's face. On radio, a pause is just a gap, so is a strong candidate for trimming unless it lends real dramatic emphasis to its context. But note that pauses should not be cut to nothing if the effect is of impossible vocal contortions. Try this: record the sentence 'You (pause) are (pause) a (pause) good (pause) editor' and then cut the pauses out completely. The result will probably sound like a case of hiccups. The minimum gap depends on how much the mouth would have to change in shape during the pause. This applies to 'ers' too.
- Superimpositions. Two people talking over each other can be irritating. But edit with care: it may be necessary to leave some overlap, or to take out part of the sentences before and after.
- 'Fluffs' (verbal errors), in cases where the speaker has gone back over the words to correct them. Again, an edit generally improves intelligibility. But take care here too: the first word of the repeat is often overemphasized.

In either discussion or documentary, each different speaker presents a new set of problems and decisions on whether to cut or not. For example, an interview may begin hesitantly and gather speed and interest as it progresses. This is wrong for the start of a new voice on air, when the audience's attention and interest must be captured in the first few words. If you cannot cut this 'warm-up' stage completely, it may be necessary to tighten up the first sentence or two: this can matter more than hesitations at a later stage, when voice and personality have become established.

In certain cases, particularly where the recording is of importance as a document, no editing should be done at all (with the possible exception of cutting down really excessive pauses), and the greatest care should be exercised if any condensing is to be attempted. If in doubt, cut away to another, linking voice.

In any case, do not over-cut. It is useful to have good ideas about where to cut, but it is just as important to know where not to cut.

Finding the exact point

When editing speech internally, there are good and bad places to cut – between or even right in the middle of words. Some suggestions about how to make good edits better are given below.

In a 'sentence' pause, there are two places where you can cut. One is after the last word of the sentence (and any associated reverberation)

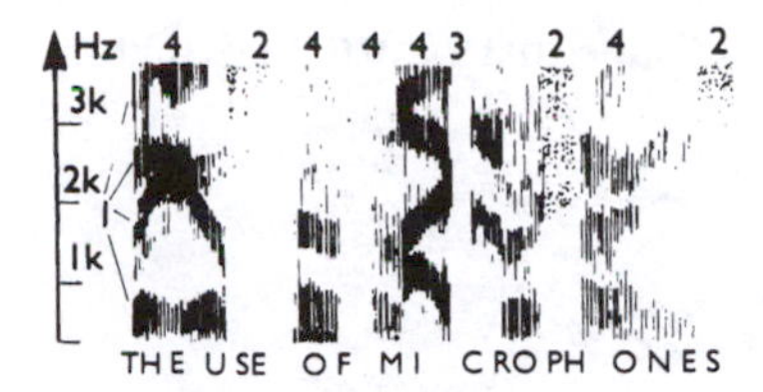

[18.4] **Human speech**
Analysed to show formant ranges. 1, Resonance bands. 2, Unvoiced speech. 3, Vocal stop before hard 'c'. There would be a similar break before a plosive 'p'. 4, Voiced speech. These formants are unrelated to the fundamentals and harmonics, which would be shown only by an analysis into much finer frequency ranges than have been used here. This example is adapted from a voiceprint of the author's voice. Each individual has distinct characteristics.

has finished but before the breath that follows; the other is after the breath and before the next word. In most cases it is best, and safer, to make the cut after the breath: you then retain the full natural pause that the speaker allowed, and can cut as close as you like to the new word, which helps to mask any slight change in atmosphere.

Cutting between the words of a sentence is trickier. In addition, the words themselves sound very different if they are slowed down, so that a suitable point is more difficult to locate. But certain characteristics of speech can soon be recognized at slow speed, e.g. the 's' and 'f' sounds. The explosive consonants 'p' and 'b', as well as the stopped 'k', 't', 'g' and 'd', are easy to pick out (though not always as easy to distinguish from each other) because of the slight break in sound that precedes them. In a spoken sentence the words may be run together to such an extent that the only break may be before such a letter – perhaps in the middle of a word.

Analysis of human speech showing how its frequency content changes as a sentence is spoken makes these transitions very clear. The 'zoomed-in' screen display available on many digital editing programs is not as clear as this, but with a little practice, sounds can be matched to shapes, and tiny gaps can be spotted.

Do not assume that because a letter should be present in a spoken word, it will be there. Whole vowel sounds may prove to be absent. The personal pronoun 'I' is typical of many such sounds. In the sentence 'Well I'm going now', it may be missing completely. If you try to cut off the 'well' you may even find the 'm' was a vocal illusion, no more than a slight distortion of the end of the 'l' sound. Similarly, in 'it's a hot day', spoken quickly, there is probably no complete word 'hot'. Nor is it possible to isolate the word from 'hot tin roof', as the first 't' will be missing and the one that is there will be joined to the following vowel sound. Before attempting fine editing it is worth checking what the beginnings and ends of words sound like at slow speed (and look like on screen), and where words blend into each other and where they do not.

In cutting between words, cut as late in any pause as possible. When searching for the right pause on a recording, do not be misled by the tiny break in sound that often precedes the letter 't' and similar consonants. You can often cut in to a continuous sound, for example where several words have been run together, provided that it can be 'ramped' in over a tiny fraction of a second (the equivalent of what when cutting audio tape would be an 'angled' cut). Do not attempt to cut out of a continuous sound, unless you join straight on to a sound of equal value (one for which the mouth would have to be formed in the same shape).

It is very difficult to insert missing 'a's' and 'the's'; these are nearly always closely tied with the following word, so that unless the speaker is enunciating each word clearly and separately, it is necessary to take this in combination with the sound that follows. The letter 's', however, is somewhat easier to insert or (sometimes) remove.

These principles apply equally to digital and analogue editing. When editing video (or film), the way in which cuts are made has to be related to the frame interval, to ensure that words and picture stay in sync.

Editing sound effects

Editing staccato effects and absolutely regular continuous effects presents no problems. Effects that are continuous but progressively changing in quality can be difficult to edit, as can speech accompanied by heavy, uneven effects or speech in heavily reverberant acoustics.

Rhythmic effects on their own are fairly simple, as they can be edited rather like music (see below). Speech in a reverberant acoustic shares that quality with music, so it, too, can sometimes be edited in the same way as music. With more uneven noises it may be difficult to find two points that are sufficiently similar in quality for a simple cut, so for these it is better to overlap and mix. For example, when cutting a 10-second burst of applause to 5 seconds, a simple jump to a later point would make a blip at the join, but a gentle overlap sounds fine.

Sometimes a jump in the background behind a speech edit can be covered by mixing in some similar effects afterwards. Even so, a change of quality can often show through unless the added effects are substantially heavier, or are peaked on the splice.

There are certain cases where the untidy effect of a succession of sharp cuts in background noise can be tolerated. It may be accepted as a convention that street interviews (*vox pop*) are cut together by simple editing, with breaks in the background appearing as frank evidence of the method of assembly.

Editing music

Reasons for editing music include: to cut out a repeat, or one verse or chorus; to insert a retake; to delete everything up to a certain point so that we hit a particular note 'on the nose'; to condense or extend mood music to match action, and so on. Most of the skill lies in choosing and marking the right part of the right note, and this requires practice.

The important thing in music editing is not that the note after each of the two points marked should be the same, but that the previous one should be – that is, when cutting between two notes it is the quality of the reverberation that must be preserved. Any retake should start before the point at which the cut is to be made: there is nothing so painfully obvious as a music edit where a 'cold' retake has been edited in, resulting in a clip on the reverberation to the previous note. It might be supposed that if the new sound is loud enough the loss may be disguised, but this is not so unless it is cut so tightly as to trim away part of the start of the note. This attack transient is particularly important to the character of a sound. (AR could probably be used to disguise a bad cut, but this is a messy way of working.)

In general, unless the cut is made at the end of a pause that is longer than the reverberation time, the first requirement for a successful music edit is that the volume and timbre preceding both the 'out' and 'in' points on the original tape should be identical.

It may be possible to make an exception to this rule when cutting mood music to fit behind speech and effects, but only if the level of the music can drop low enough to 'lose' the transition. For preference, choose a rest to cut out on and something rather vague to come back to, and if necessary dip the music level at the edit point. However, no such liberties may be taken with foreground music. For example, when linking music is being trimmed to length, any internal cuts must be made with the same care as would be applied to featured music. Speech that has been recorded with music in the background can be almost impossible to edit neatly, perhaps leaving an untidy join in the music to coincide with a slight gap in the speech.

When everything up to a certain point in a piece of music is cut in order to create a new starting-point – e.g. to come in on a theme – the effect of any tail of reverberation under the new first note is often small. Except when this tail is loud, and contrasts strongly in quality with the first wanted sound, it will be accepted by the ear as a part of its timbre. Even then, a quick fade-in may sound better than a cut. To get out of a piece of music at a point before the end where there is no pause, there is no alternative to a fade. When editing on hard disk (or any system where events are controlled by time code), the manual fade that might be used on a live show can be replaced by a programmed fade (a ramp).

When the cut is to be made on tape at a place that is not clearly defined by the music itself, it is often possible to identify and match the cutting points by either timing the distance on the track from a more easily located point, or marking out the rhythmic structure of the music. Of the two methods, the second is the better guide, as the easily located note may not have been played exactly on the beat. On audiotape the lengths are marked out on the back of the tape with a wax pencil, so that the edit point can be calculated and marked for a physical cut; in digital editing it is calculated by time code.

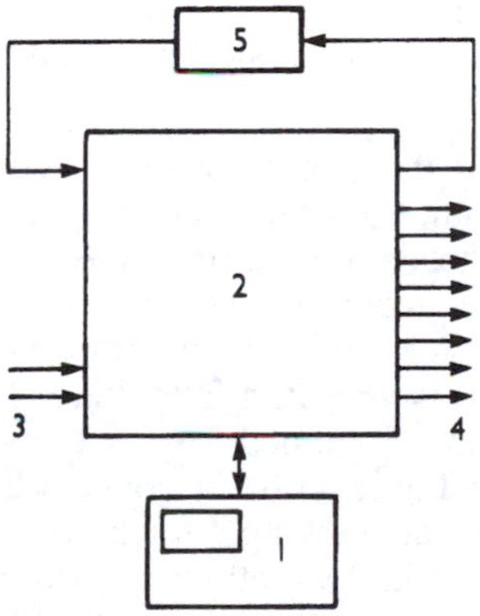

5
2
3
4
1

[18.5] **Early hard disk editing equipment**
AudioFile system layout: 1, Controls ('control surface') and VDU. 2, Equipment rack. 3, Audio inputs. 4, Eight independent audio outputs (allowing the system to operate like a cartridge machine). 5, DAT recorder.

Hard disk editing options

The time code generally used to locate and control audio material was originally designed by the SPMTE (Society of Motion Picture and Television Engineers) to work with picture and sound together. It is usually displayed in an eight-figure format: hours, minutes,

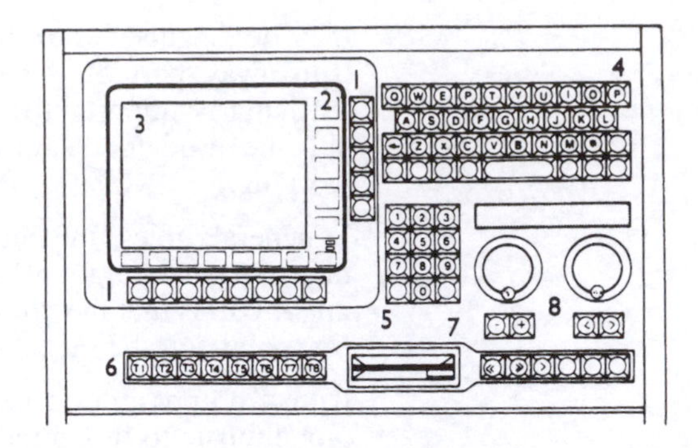

[18.6] **AudioFile editing system: control surface**
1, 'Soft key' function switches.
2, Soft key status displays on VDU.
3, Selected page of information: for example, see below. 4, QWERTY keyboard. 5, Numeric keyboard.
6, Trigger keys. 7, Floppy disk drive.
8, Transport, nudge keys, 'soft wheels' and LED display for locating and shifting material.

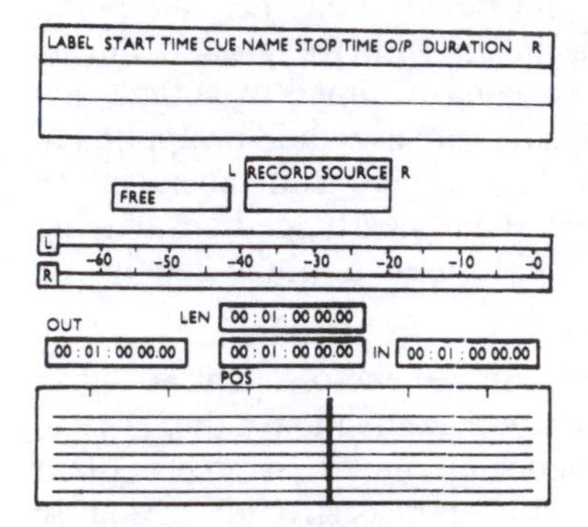

[18.7] **AudioFile control screen**
'Audio record': a typical page shown on VDU. Many have the eight tracks of sound, which 'play' across the screen from right to left, past the vertical line that represents 'now'. There may be many more tracks that are not shown. Positions are based on time codes, but can be measured to greater precision. 'Audio record' is the only page that has a meter display.

seconds and frames (for American video that is 30 frames per second, but in Europe it is 25). Two more subframe figures may be added, based on one-eightieth of a frame (or sometimes one-hundredth). This allows digital audio editing to be flexible to 0.0005 of a second. Time code is described further, in connection with video and film editing, in Chapter 19.

Today, the most common systems for editing on hard disk are those that use the processing power of a host computer. But this is just one of several ways in which it can be organized. Among the earliest were dedicated systems in which the hardware (including all the controls and displays) was built around the software.

In one, the control surface has a keyboard, function keys, two scrub wheels and a floppy disk drive, alongside and below a display screen set in a small panel with more keys and buttons around it. The display has switchable 'pages' that allow the editor to operate in a range of different modes. The sound must first be loaded (digitized). If the sound comes from a source that has no original time code, this must be added, in order to locate and control it. In any such system the precision and repeatability of timings, programmed levels (to a quarter of a decibel), cuts and mixes are far greater than can be expected from manual control.

In this system, the screen shows 16 of the tracks (of up to 24 in memory) in a display that looks like a multitrack tape marching from left to right across screen, past replay heads in a vertical line down the middle. Each track is identified by number, and a suffix, 1 or 2, indicates the left or right track of stereo. A cut in or out is shown by a vertical bar across the track, which travels with it; a fade (here also called a *ramp*) by a diagonal. Sound can be bumped from one track to another instantaneously. Just as easy, but less obvious once it has been carried out, is a shift or *slip* to the side, moving a track (or part of it) backward or forward in time. With minor variations, something similar to this arrangement appears on all multitrack editing displays.

In another, more modestly scaled desktop editor that works with a 16-bit linear (i.e. uncompressed) signal, the hardware looks more like a standalone word processor, but with dedicated control keys and a scrub wheel alongside the keyboard. Its small window displays several tracks or can zoom in to the fine detail of one of a single waveform.

[18.8] **Short/cut desktop hard disk recorder and editor**
Intended as a successor to earlier generations of analogue and digital tape machines, this 44.1 or 48 kHz 16-bit linear (i.e. uncompressed) recorder is designed for rapid editing and replay of near-live news stories and phone calls. The scrub wheel simulates the movement of tape over a replay head. Editing (including volume control, fades and cross-fades) is rapid and precise. It holds 12 hours of stereo, with 10 clips ready for instant playback: an external Zip (or Jaz) drive can be used for archive.

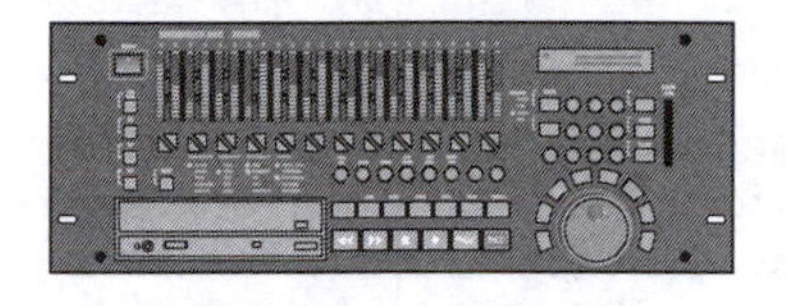

[18.9] **Tascam hard disk 24-track recorder and editor**
Designed to replace both analogue and digital multitrack tape recorders for music recording (24-bit and up to 96 kHz) but with similar transport controls, including a scrub wheel, plus time code, MIDI machine control tools, DVD mastering, and a variety of analogue and digital input and output options. A wide range of music editing operations, including real-time cross-fades, are also controlled from the front panel.

In hardware of this kind, there are *transport* keys (similar to those on a tape machine) to record, play, pause, stop, fast-forward and rewind. There are editing keys to cut, copy, insert, erase, to locate and mark editing points or to repeatedly play a segment as a loop, all of which are non-destructive, i.e. the original recording is preserved. There are also keys to save work, zoom in on the display, or undo previous edits.

This desktop editor is designed as a simple (and speedy) advance on tape editing, but with added features such as cross-fade, fade in or out, gradual rise or fall in level, etc. The internal hard disk holds over 11 hours of stereo. For storage of the edited material, the makers recommend an external Zip disk, or for longer recordings a Jaz disk.

Another hardware-based system is built around higher quality 24-bit, 24-track hard disk recorders, with a separate control surface to display and edit the tracks, which may come from several recorders, stacked to work together.

Several editing applications are designed to operate on a standard office or home computer, controlled either from its mouse and keyboard or more conveniently from a separate control surface that has all the usual dedicated keys, knobs, faders and scrub wheel – although the keyboard is still the fastest and easiest way of typing in titles and other notes. Without the hardware, the keyboard's function and other keys are assigned as 'hot key' transport and edit controls (for which it is convenient to have a mask that fits over the keyboard, so that the function of each key can be marked). But operating faders and rotary knobs by mouse from the screen is inevitably fussier.

It will also often be necessary to have a convenient way of moving a project from one machine to another, from studio to office or home and back, possibly to another studio. One device used for this is a removable hard drive that can be connected by SCSI (from which the 'scuzzy' box takes its name). The hard drive – somewhat exposed when not plugged in – is carried around in its own miniature protective case. Another essential is a means of backing up the original recording and other material in the studio: a DAT, multi-track or some other recorder where it can be kept safely while work on it continues elsewhere.

Digital projects and editing

Digital editing performs many of the same functions as those originally developed for audiotape or separate magnetic sound for film, so the same general principles carry through, with exactly the same stages of rough and fine editing, and the same need to locate and mark editing points in speech and music. For hard disk digital editing, the display on the screen mimics tapes passing replay heads, with the edits clearly visible – but with the added advantages that other information can be superimposed, that a timeline and other

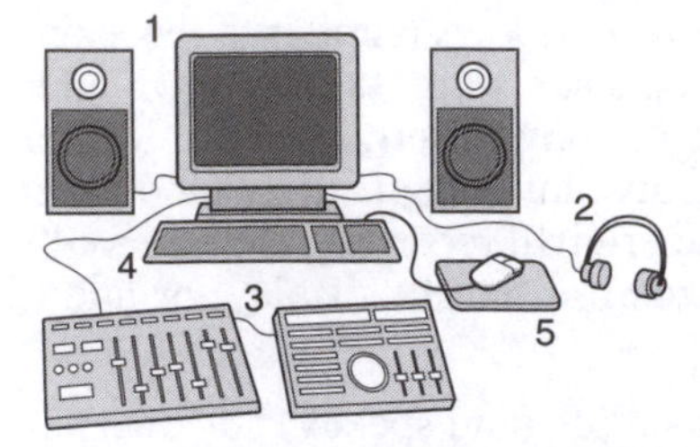

[18.10] **Digital audio edit station**
1, Screen and loudspeakers.
2, Headphones as an alternative to loudspeakers in an active studio.
3, Hardware controllers (optional) add to the cost but are faster, easier and more intuitive to use than a keyboard.
4, Keyboard: more convenient than the hardware for naming. 5, Mouse: used for drag and drop operations. If there is no hardware controller, the keyboard is set up with hot keys. Typically, F1–F12 are used for editing, and the numerals 1–10 in the next row for mixer automation. The arrow keys are used to zoom in or out and move clips up and down, while the keys above the arrows open the clipstore or copy material to it, move the playhead to the start or end of the EDL, or save the whole project. The descriptions here are for SADiE editing.

timings are displayed, that access is far more rapid, and that most of the operations are (usually) non-destructive, instantly playable and can be readily reversed. Most of the controls and procedures are directed towards presenting and working on this pictorial representation of the audio plus any MIDI tracks (slightly different in layout) that may be lined up with them.

The raw materials for editing on hard disk exist within the computer as *components* of a substantial file that for each separate work may be described collectively as the *project*. The project might be a radio programme or series, a record or single element or package, and comprises all the edit decision lists, clipstores and mixes that are created alongside some original audio. It may also include 'metadata', consisting of history, copyrights or even scripts. All this exists as a single named entity within the computer, and can be copied, perhaps on one 'project backup' keystroke, to an external store – still with the original audio. As ever, vital questions are 'How stable is the edit program?' and (when it does crash) 'How easy is it to recover?' and 'Will it let you carry on from where you left off?'

The following describes a typical, well-organized operational procedure: programs differ in detail, but the objectives and order they are tackled in will mostly be similar, and there will usually be a series of things to organize before the interesting part – the actual editing – can begin.

As with any other computer application, a digital editor is started by opening the program, often from an icon that it is convenient to keep on the desktop. Completing any required log-in primes a start-up procedure to open either the current or some other existing project, retrieve a project from backup, or create (and name) a new project.

On creating a new project (which will open up a window for it) the next stages are to record into it and set up an EDL (edit decision list). The main features in the window (or that can be reached from it) are:

- A number of boxes, tracks or 'streams' across the screen (rather like strips of audio tape) that will contain the audio, with (above these) a timeline, and from top to bottom a line (equivalent to a playhead across all the tracks) that represents 'now'.
- In boxes to the left of the streams, a 'playlist' in which they are listed by number (clicking here selects a stream or can be used to combine two adjacent tracks as a stereo pair); there is also a set of small icons that operate on each individual stream, and one of those is a button that will prime it as a target for recording to.
- A large number of icons that change the screen in various ways, edit the playlist, operate audio editing tools, save the work, and so on.
- A mixer window that can be extended or moved around the screen; faders, side by side, are numbered to match the streams.
- A transport control window, with buttons like the controls of a tape machine, plus time code to show where it is up to, and naming tools.

To prepare to record, press a tab marked 'record' near the top of the transport window, then fill in any details required below: one will be

AES31-3 is a public-domain audiofile interchange standard that specifies how random access audio can be described and organized in order to transfer simple audio projects (with an unlimited number of channels) accurately between editing systems. It instantly autoconforms precise audio in and out times, together with linear or shaped fades, plus sources, clip names and remarks. It is also designed to include dynamic level (gain) automation, but not EQ, filter, limiter or compression data, as these can sound different on different platforms. It is used on many (but not all) proprietary systems. A more complex alternative is OMF.

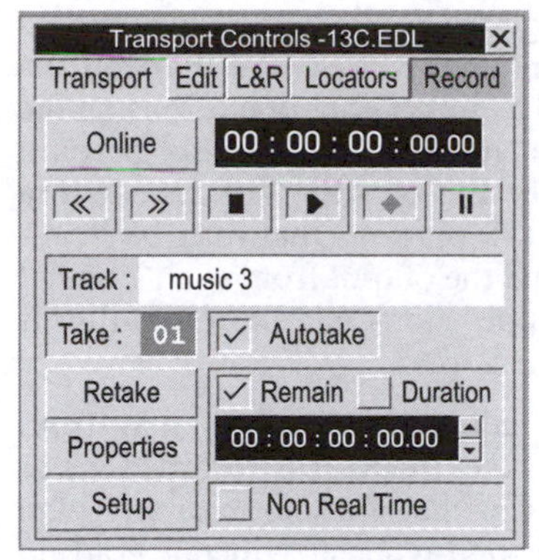

1

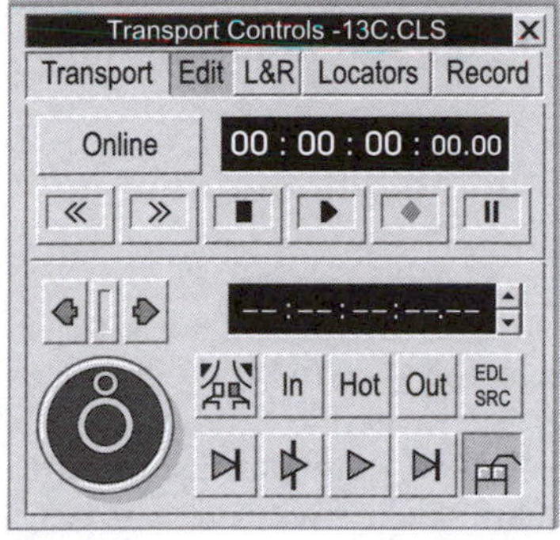

2

[18.11] **SADiE transport controls**
1, By clicking on 'Record', the lower part of the window is configured for information about the track. Release the key after recording is complete. 2, Clicking on 'Edit' configures it to move selected material – e.g. with the scrub wheel (to hear this, click a button near it). To activate the mouse for drag and drop, click on the mouse key in the corner.

a name for the recording. It may also be necessary to bring up a set-up window and choose the type of source: what kind of digital data it is (e.g. AES3) or if it is analogue to be digitized (at some specified sample rate, e.g. 44.1 or 48 kHz). To check that all is ready, try pressing the 'pause' button: a red 'record' button may flash. Then play in some material (or feed sound from the studio) and check that the meter alongside or above the fader shows a signal (which might just reach to the red). If there is a problem at this stage, recheck the routing. If all is well, press the red 'record' button to begin, and when finished the 'stop'. Finally, release the primer button on the playlist.

To begin editing, press the 'edit' tab near the top of the transport control panel. Some systems allow editing to begin even while recording (in 'background') is in progress – also useful for 'bleep-editing' offensive material out of 'near-live' telephoned contributions.

The most useful display for the audio is as a waveform (there may be a waveform icon: click on this first). Clicking on a clip should select it for editing: it should then be possible to move the in or out points while keeping the audio stationary, or move audio relative to a fixed in or out point, or again, move the whole clip over the 'playhead' line. Clicking on a razor blade icon will cut a track at a point marked by the cursor. All edits are provisional: it is easy to recall the original material and use a scrub wheel to perfect the position, or to add back a length that had been taken out. It is usually possible to mark a section, then clip and paste it somewhere else, perhaps in another track, or simply drag and drop it (just as in word processing). It should also be possible to mark and move regions containing several tracks together.

Colours may be used to show the status of each audio clip. Typically, from an original pale blue for mono or green for stereo, when a section between two markers is selected prior to editing it becomes dark red. Bright red shows 'edit in progress', in which state it can be extended forward or back to add more of the original. Dark blue marks a clip that will 'slip' to fill up (or open out) a gap.

In addition to its waveform, each segment between two marks may show the name originally given to it in the top left corner of the clip. In a short segment only a little of this name may be visible, but more will be revealed if the waveform is expanded by 'zooming in' on it. If it helps, the name may be changed to give more information about its source as speech, music or effects. Zooming in vertically at the same time deepens the waveform, making it easier to locate a precise edit point. Zoom in far enough and on a linear (uncompressed) recording the waveform may be seen as a continuous line. (A compressed 'waveform' may appear as a blocky histogram showing the heights of successive samples.) Zooming out may eventually reveal the coarse structure of an entire recording or edit, with quiet and loud passages showing up.

The settings for volume control may also be shown superimposed as a line over the waveform, dropping below it as the sound is faded out. If the start and end of each fade can be separately marked, the

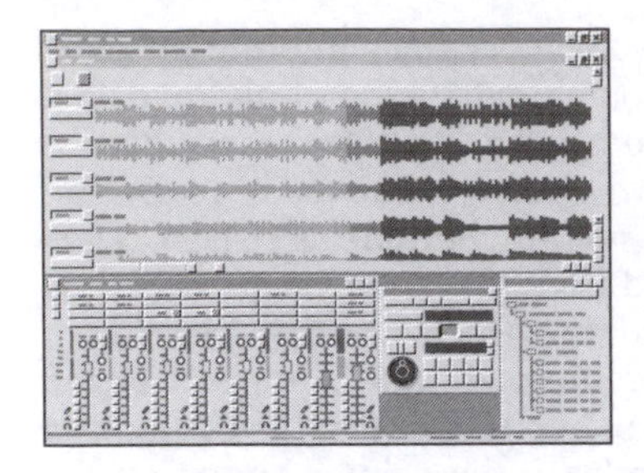

[18.12] **SADiE edit screen**
This shows audio streams set up for editing. Below left is a mixer: each fader (with input and output buttons above them) operates on the track with the same number. Below centre is the transport control window. Below right is the project playlist window. Other windows (such as DSPs) can be opened over this screen.

fades will 'ramp' smoothly. To change the contour of a fade marked like this, it is necessary to change volume over several marks. Cuts are often made a little 'soft', to avoid a click. Other operations on the streams (or channels) are carried out much as they would be on a console, except that controls (e.g. for EQ or AR) will appear in a window on the screen, as do other DSP plug-ins that may be available. There will also be buttons to send the output from each channel to a choice of buses. Many of these operations feel more familiar if an (optional) external control surface is used.

One disadvantage of some digital editing systems is that from analogue sources they must be loaded (i.e. material is recorded in) or played out as analogue in real time. It is therefore necessary to plan for this to be done, if possible, at a time when it will not hold up operations. One use for this time is to check the input or to review an edit. In contrast, a CD-R can be loaded at perhaps 20 times normal speed.

Play-out systems

As an extension of the editing process, some systems can operate as a play-out controller for items either in an active studio or in a digitally automated broadcast system. Edited items are assigned to a playlist, which works rather like a cart stack, except that it is no longer necessary to carry around and load a pile of tapes, cassettes or cartridges. It is also more flexible, in that it is easier to build, edit components, and sign them off as ready for transmission, but still make late changes or switch the order (either in an office environment or at the back of the studio, or at the console). A playlist compiled at the back of an active studio may be 'mirrored' (i.e. shown in another place) at the console, and used to select or play inserts out in order.

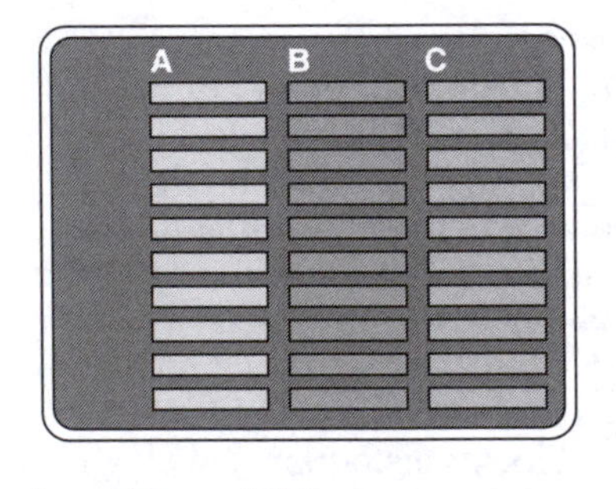

[18.13] **Play-out lists**
For broadcast studio use, this version has three columns. A and B, Alternate ready-to-air segments can be overlapped or cross-faded. C, Station ID, signature music, trailers and other frequently-used items that may be interpolated into the schedule.

Play-out systems work from playlists that automatically shift to the next item as each is completed. It may be arranged that successive items go to alternate channels (e.g. 'A' and 'B') so that an overlap or cross-fade can link them. (If deletion of an unplayed item automatically shifts the channels of later items it is safer to keep both the A and B faders open and control them together as a group. A separate 'C' list of frequently used shorter items might also be played in from a source such as a Zip or MiniDisc cart.

The playlist may be set up as a hierarchical 'tree', with at one level a whole day listed as hourly segments with station identification, etc., at the junctions. The segments may then be opened to reveal their detailed structure with individual and progressive timings, and notes on the content. The items on this list may be audio (such as records, recorded links or completed pre-recorded items), placemarkers for material yet to be selected or completed (with estimated timings), or cues for live links, time-checks, or other notes. An item towards the conclusion of a segment may be of flexible duration, perhaps pre-faded, i.e. set to end, rather than to start, at a given time.

[18.14] **Cart wall**
A touch-sensitive (or mouse-keyed) screen with 49 boxes each holding an item that is ready for instant replay: a two-layer hierarchy of such screens allows rapid access to 2401 (49 × 49) boxes.

Broadcast automation systems have been around a long time, originally using carts of cartridges or cassettes, and then MiniDisc or digital files on, for example, Zip disks. The cart may now be replaced by a cart wall, a touch-sensitive screen with a 7 × 7 array of boxes, to which items are assigned from a database, then replayed instantly as each box in turn is touched. This could be used directly to air, or to assemble playlists ready for broadcast. A valuable adjunct to this technology is that each item can, in principle, be accompanied by metadata that includes all its copyright data and history, which can be continuously updated, then used to 'report' the copyright details and other content automatically after transmission – so that fees due are also calculated (and payment can be set up) without paper. All this assumes that all the different connected devices can be linked by a standard metadata exchange format – or (to add yet another acronym) 'SMEF'.

Broadcast automation in this form is, in principle, capable of managing far wider swathes of broadcasting than just play-out. The seemingly successive processes of commissioning, making, scheduling and transmitting material, then 'reporting' its use and archiving it, plus multimedia offshoots of teletext and pictures and web data, all become elements with a single interlinked process of data management – a big single 'pot' of data with a set of tools attached. There is, of course, one other ingredient that must be added to this pot: far higher 'broadcast-critical' reliability than individual computers can offer.

If even just a playlist is set up as a broadcast schedule, this should be isolated from other uses that could accidentally disturb the order, timing or content of transmissions – 'You wouldn't want your audio out there fighting someone's Friday night Excel spreadsheet'. The system must have proven reliability and redundancy with immediate switchover in the event of failure. It should have procedures in place to protect it from viruses, hacking, deliberate misuse and human error or other ills that may afflict computers. Of several possible arrangements, one has a central server, with a second comparable backup server in another room, but ready to switch in instantly. Another uses multiple similarly paired smaller servers, linked to pairs in other locations, so backup is always available from unaffected parts of the network, and any individual server failure would be on a smaller scale.

Chapter 19

Film and video sound

This chapter offers a compact account of the principles involved in video sound recording and editing, relating this to current television and film industry practice. Note here that as a medium, 'film' now refers more to the distribution system that still uses it, rather than production, in which the use of film is increasingly replaced by video cameras and digital editing. Sound, in either case, is generally recorded and edited digitally. This chapter deals mainly with picture and sound that originates on location (including what is still called a film set); Chapter 20 deals more with sound in the television studio.

Video cameras that are equipped to record picture can always record audio alongside it. For sound this is plainly an economical approach, with the added advantage that there are fewer synchronization problems – at least, not at this stage. It is therefore suitable for television applications such as newsgathering, for which speed, cost and technical simplicity are important. But just how satisfactory that 'simplicity' really is depends on the position of the microphone. If the microphone is on the camera, even though its output can easily be fed directly to a recorder on the camera, this is likely to be unsatisfactory for two reasons: one is that the camera is very rarely the best place for a microphone; the other is that it treats care and attention to the quality of sound recording as an option, not as an essential.

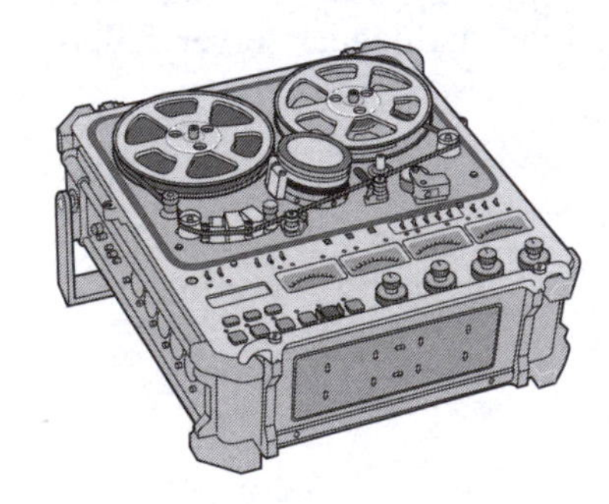

[19.1] **Four-channel digital tape recorder**
A helical scan recorder, like DAT but with a 6.3 mm (1/4 in) digital tape. Records on four channels for 1 or 2 hours, depending on size of spools.

A better alternative, still with the recorder on the camera, is to use a separate microphone, under the control of a recordist who feeds its output by line or radio link to the camera. Again, synchronization is automatic – unless the recordist makes a backup, in which case it should have the same time code on it as that on the camera recording.

For more complex operations, sound recordists prefer to use separate recorders of established quality under their direct control, using the link with the camera primarily for synchronization. Digital audio tape (DAT) has been widely used, especially for television, but in

[19.2] **Removable hard disk cartridge**
A 2.2 GB hard disk holds over 2 hours of high quality (24-bit) stereo.

the film industry, where high definition (HD) video is taking over from traditional film cameras, digital media with four or sometimes more tracks are increasingly employed (exceptionally, 24 tracks), and in these cases any sound fed back to the camera recorder is for guide and viewing purposes only. Other options include two channels on a removable 2.2 GB hard disk cartridge. And an interesting extra facility with one DVD recorder is hard disk backup with up to 10 seconds of sound recorded from before the 'start' button is pressed.

Time code

A time code, as applied to video, is an eight- or ten-character number indicating hours, minutes, seconds, frames and subframes. It can be linked to a 24-hour clock, and so reset automatically to zero at midnight. Alternatively it can be started from an arbitrary positive number (such as 01 00 00 00, which means 'one hour exactly'), after which the numbers will continue to rise in real time as the machine runs but will hold while it is switched off, then continue in sequence when it is run again: this therefore measures total recording (and so replay) time. If the tape starts at a number sufficiently far above zero, it can park at a point before the nominal start and be ready to run up without ever encountering negative numbers.

As an alternative to elapsed time, a recording can be marked with 'time of day', which is safest and clearest when several recordings are made, and it also means that sequences separately recorded over a long period can be identified more easily when consulting a log.

[19.3] **BITC**
Burned-in time code offers eight numbers for time and eight as 'user bits' (sometimes available for a second time display). The position on screen is arbitrary. On professional equipment the data is kept separate and combined, on demand, only on the screen display. The combined signal can be fed to a domestic-quality recorder, where it will then be permanently superimposed, and can then be viewed away from studio or cutting room.

The first form of time code was *linear* or *longitudinal* (LTC). This is recorded continuously (originally on an audio cue track, where it made a high-pitched buzz if replayed as audio). It can be 'read' when a tape is played or spooled, but this breaks down at low speeds. On videotape, time code provides a method of individually marking every frame and for this LTC is generally supplemented by *vertical interval time code* (VITC), which is recorded intermittently in the video signal just above the top of the picture; VITC still replays at low speeds or even when the tape is on 'pause'. Either can be used to generate characters that feed number displays on the equipment itself or superimposed on the picture as *burned-in time code* (BITC). A video system with time code has separate feeds, with and without BITC. If BITC is re-recorded on a domestic VCR, it permanently overlays part of the picture.

Different time codes are used for film (24 fps – frames per second), European television (25 fps) and American television (30 fps or an offset 'drop frame' version of it that runs at 29.976 fps – often written as '29.97' or '29.98').

The digital numbers that contain the time code have some spare capacity, called *user bits*, that can be used to assign tape numbers.

Time code for audio clips
Because audio clips can be cut anywhere, their time codes must contain more information than those for video or film. AES31-3 uses the sample-accurate Timecode Character Format (TCF), which may appear as HH:MM:SS:FF/1234. Over a 24-hour clock, this shows hours, minutes, seconds and frames. It allows for the various video or film formats, including 'dropframe', in which a second is divided into 1001 parts. It is also used, typically with 25 notional 'frames' per second, for sound-only editing. The 'slash' divider signifies that in this example a 48k sample rate is being used and the last four figures count all the remaining samples beyond the last full frame, in order to define a cutting point absolutely. Audio Decision Lists (ADLs) using this format are used to transfer data rapidly and accurately between a wide variety of proprietary editing systems.

It may be arranged that at the press of a button the real-time flow of numbers pauses, so that the time of a particular event can be logged. When the flow resumes it jumps on, to the continuing real-time display. In this way logged material can be readily relocated. More importantly, however, it also means that computer-controlled operations can be initiated and related to any given frame on a tape. For tape replay any required time code can be keyed in and the control system set to search for that address. Once found, the tape automatically sets itself back and parks, ready to run up and display the required material.

Digital video editing systems again lead to the generation of time-coded edit decision lists (EDLs), which are used to control sound and picture recordings both for the mix and in a final conforming process.

Time code can also be used for time-cued play-out in automated broadcasting systems. It also synchronizes video with multitrack tape or digitally recorded sound files in order to locate separate soundtracks (e.g. the commentaries made for different users of a common picture of some event or sporting fixture), and television with stereo sound that is broadcast simultaneously by radio. In addition, it provides an accurate measure of duration.

Picture and sound standards

The sound recordist may be faced with any of a wide range of cameras and frame rates. Originally, film was used by television (mainly 16 mm) as well as by the film industry (with 35 mm as standard). In television, film is now rarely used (except where its 'special' picture quality is sought): it was replaced first by analogue videotape, and this has given way to digital formats. The principal format is now what is called standard definition (SD) digital, and this is occasionally backed up by a more compact, lightweight DV CAM format. For film industry purposes, 35 mm cameras are no longer manufactured (but many directors still want to use them, despite the high cost of film stock); increasingly the high definition (HD) video format is replacing it. So far, this leads to three main picture definition standards (there are others, which are mostly variations on them and which will be disregarded here – as will the heated debates on the virtues of film compared to video).

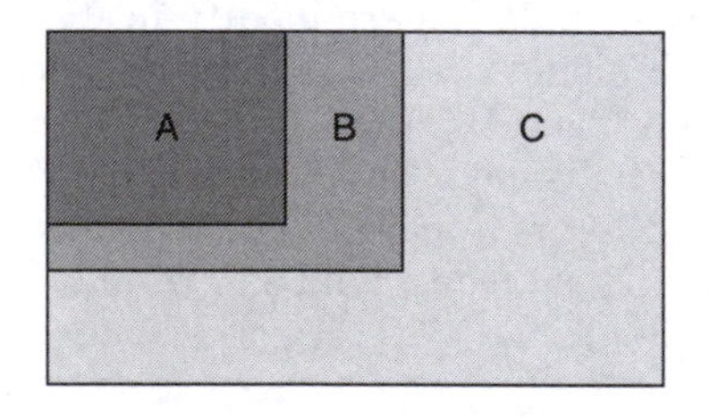

[19.4] **Digital video picture areas compared**
A, Standard video: 720 pixels wide, 678 lines deep (covers PAL and NTSC formats).
B, DTV format: 1280 × 720.
C, HDTV common image format (CIF): 1920 × 1080.

Unfortunately, other developments have led to a proliferation of standards even for the camera types already in use. An important question, which the sound department is unlikely to have much input on, is choice of frame rate. The simplest case is for television in countries where the electricity supply frequency is 50 Hz and 25 fps has been adopted as the standard for television. Each frame is scanned twice, once for the odd-numbered lines, then for the even numbers and the two 'interlaced' fields make up what the eye sees as a single complete frame. This is now called '50i' ('i' for

[19.5] **Frame rates**

Interlaced (television format):	
50i	50 fields/s (PAL)
60i	60 fields/s
59.94i	NTSC offset
Progressive scan (like film):	
24P	24 fps standard
23.98P	24 fps offset
25P	Euro TV standard
30P	American, 60 Hz
29.97P	Offset for NTSC

Choices available on a high definition camera.

interlaced) – and the leading picture and sound format using it is PAL. Low-budget features may also be shot at this frame speed, so their transfer to television is easy (although adjustments must be made if they achieve cinema distribution). As it happens, DAT recorders were also originally designed to operate at the European 25 fps.

In America and other countries that have the 60 Hz mains frequency television, this becomes 60 interlaced ('60i') fields per second, which makes 30 fps. The NTSC 'off-set' variation on this actually runs at 29.97 fps and the interlace is '59.95i'.

The film industry had already fixed on 24 frames per second, and for the HD video version of this now prefers a picture that is scanned just once, from top to bottom: this is 24P (the 'P' is for 'progressive scan'). HD cameras are now available to operate at a wide range of scan and frame-rate combinations.

There is an established process for converting between the 24 and 30 fps formats, and a slight difference in playing speed converts 30 to 29.7: the sound drops in pitch by a virtually insignificant 0.1%. Running 24P at 25 fps or vice versa changes both action and the rate of delivery of dialogue by 4.1%. Television audiences experience this regularly but audiences do not seem to mind, even though a rise in pitch by about two-thirds of a semitone shifts male voice formants a little closer to the female range, and music is lifted almost into a different key. This effect can be corrected by a digital signal processor set to control pitch and duration independently.

There have been several, successive analogue videotape recording standards. At one time television pictures and sound were recorded on bulky and very noisy 2-inch tape machines (the Imperial measure of width has always been used to describe this format). Special transfer machines are now required for replay of any material that has to be retrieved from 2-inch tape. After this came 1-inch, also retained now only for archive material that has not yet been re-recorded, and ¾-inch, now used mainly for domestic or viewing purposes.

For professional digital recordings ½-inch (and, on some cameras, narrower) metal tape is used. Because most errors on digital recordings are automatically identified and corrected, dropout has less effect on them, and uncompressed data suffers no significant loss of quality over several generations of copying. The number of channels of sound has increased from one, then two, to four (or even eight), to include a main stereo pair – most users preferring M and S – in addition to any tracks for a microphone on the camera (or elsewhere), as well as the essential time code.

The analogue Betacam SP camera, once widely used for documentary and news, has four audio channels, but only two are linear (i.e. along the length of the track and independent of picture), while two are coded into, and so combined with, the picture recordings. Stereo recordings are generally taken on linear tracks (1 and 2) and copied to the coded tracks 3 and 4, but other formats are possible and should be indicated in the documentation. Later systems are more versatile if all four (or more) audio tracks are independent of picture: there are fewer problems when the same tapes are used for editing. All professional video recording systems also have recorded time code.

More compact formats include plug-in hard disk cartridge.

Sound is synchronized with picture either by a physical link (a sync-lead) or by radio link. Another system that has been used for film is a pair of crystals, matched to vibrate at the same frequency, that run recorder and camera at the same speed.

In the rare cases where television material is still shot on film (for its 'special' picture quality), it is almost always immediately transferred to video for digital post-production, after which it stays in the television video format, ready for transmission. Many of the basic principles for editing picture and sound together were originally developed for use with film, but have now been adapted to the video editing techniques that are used even when film is the original source.

Digital editing systems are almost universal, with videotape (copy) editing still being used only in the simplest cases. At one time, a professional standard of digital editing could be achieved only on specialized equipment, but for all but the most complex work this is now challenged by software that can be run on a domestic computer, although the more memory the computer has the better.

This chapter ends with a brief account of flatbed film editing, which is still occasionally used. The techniques described apply equally to 16 and 35 mm, but where slight differences exist the method for 16 mm is described. Film (as well as film sound) is most conveniently edited using guillotine tape joiners and viewed on equipment that allows continuous movement of picture, stabilizing the image by the use of rotating prisms. This allows good-quality sound to be played and heard in sync without the loud machine noise of old intermittent viewers.

But this account must begin at the place where the combination of picture and sound first appears. This is often far from a studio (in which sound could, in principle, be controlled with relative ease), at some location where the first requirement all too often has to be for good pictures, while the achievement of even acceptable sound may have to begin with an exercise in damage control.

Choice of location

When a location is chosen for video or film shooting, it is essential to check its sound qualities and potential for intrusive noise. At worst, bad location sound can make it necessary to replace dialogue and effects completely: one branch of the film industry specializes in this. But plainly it is simpler to record and use the original location sound if this is possible. For documentaries this is almost always preferred, not least for reasons of cost.

Location surveys should establish that aircraft are not likely to interrupt sound recordings. Direct observation is a good start, but in addition a glance at an air traffic control map will show what type of disturbance there might be – passenger aircraft along main airways and in terminal control zones, and military or light aircraft elsewhere. The survey should also ensure that road traffic noise is not obtrusive (or, if traffic is accepted as a visual element, that it is not excessive), and that machinery noise from factories, building sites or road works will not create difficulties, and so on. If the investment in good sound is high, a visit to the site at the same time of day in the appropriate part of the week is a useful precaution.

In many cases where people unconnected with the production are (in the course of their own normal activities) making noises that mar a performance, a courteous request to stop often produces results. If so, courtesy also demands that the noisemaker should be told when shooting is complete or when there are long gaps. It is, of course, unreasonable to expect people to stop earning their living in order that you may earn yours, so in such cases (or where it is the only way of gaining silence) payment may have to be offered.

For interiors, acoustics as well as potential noise should be surveyed. On this will depend the choice of microphones, type of balance, and remedial measures against deficiencies in the acoustic environment, followed by debate (and often compromise) with people responsible for camera, lights and design and, where necessary, recommendations to the director on how sound coverage could be improved.

Film and video sound recording

On location a video or film unit includes a director, a cameraman, a sound recordist and an electrician (for anything more than minimal lighting), plus as many assistants in each of these departments as is necessary for efficient operation. There are few occasions when a sound recordist actually needs an assistant for simple television location work, except as a mobile microphone stand for certain types of dramatic subject. For more complex set-ups, including many of those for feature film shoots, a larger sound team may be required to help with rigging microphones, cables and radio links, including the link between sound control and camera.

Although radio links are now generally reliable, cable links (send and return) between the sound equipment and the camera are, in principle, safer. One potentially awkward decision is whether to lay cables early, in which case on a big shoot others may lay their own cables and equipment over the top, making changes difficult, or whether to leave it late in order to secure easier access, but at the risk of running short of time if there are unexpected demands on sound.

A feature film may have a whole 'video village' generating electronic and mechanical noise (e.g. from cooling fans on equipment), as well as operational noise: plainly, this will not be a good place for sound, but the sound department must collaborate with it.

Camera noise has long been a problem for sound – from the clatter of the film camera onward. Noise from the camera has been reduced by draping it with blankets (which is obviously an encumbrance for the operator) or by wrapping it in a sound-absorbent 'barney'. The eventual near-solution came with better design: a self-blimped camera. It might be assumed that the switch to video cameras would remove the problem entirely. Larger cameras (of which some of the worst culprits have been in the first generation of HD cameras) may require cooling fans. These can create a high-pitched squeal, perhaps at about 12 kHz, well within normal hearing. If muted by draping it with a coat, this may reduce it briefly, but then as the camera heats up again, a second fan cuts in, shifting the problem to post-production – where a notch filter will be required (but getting rid of the noise is still not easy). Again, the only good solution is better design in later-generation cameras: some already do not have fans. But HD monitors, too, may have noisy fans.

A number of things will (or should) have been settled in advance; these include the standards for the distribution of digital sound and the camera frame rate, which will determine the time code generated in the camera, and how that is distributed (possibly by radio link). The frame rate for the edit and the rate for showing the finished work may be different again. With good planning, well-matched equipment will keep problems to a minimum, but failing that can normally be sorted out later: some time code gymnastics may be necessary anyway. There are potential pitfalls: if picture is transferred to NTSC for viewing, a delay line may be required to resynchronize sound. On large feature film productions there may be a specialist in charge of the entire chain: this certainly helps.

Film and video recording routine

Consider first the case where a film director has control over the action before the camera.

If there is any rehearsal, this allows the recordist to obtain a microphone balance and set the recording level. In a full rehearsal the action, camera movement and lighting are close to those to be adopted on the actual take, so the recordist has a final chance to see how close the microphone (or microphones) can be placed without appearing in vision or casting a shadow, and can also judge the ambient and camera noise levels in realistic conditions.

When the rehearsal is complete and performers are in place, a film camera is made ready for identification of the shot, and the lights (if used) are switched to full power. The recordist listens for and warns the director about any rise in noise level from traffic or aircraft, or unwanted background chatter. The director has checked with each department – often very informally – then (in the conventional start to a scene shot on film) the camera assistant holds a clapperboard up to the camera, and the recordist or sound assistant points a microphone to it.

On the director's instruction 'turn over', the sound recording tape (or other medium) is run: note that on digital audio tape a few seconds run-up is required before the first usable sound in order to lay down time code. Otherwise (for analogue sound) tape and film are run together, and the recordist and camera operator check as their equipment runs up to speed and the recordist checks for sync. With 'speed' confirmed, the camera assistant is told to 'mark it'. Note that although time code should be clear enough, a traditional clapperboard is still preferred by many filmmakers, as it shows unambiguous shot and take numbers and sync sound confirmation. It is always used when the picture is on film.

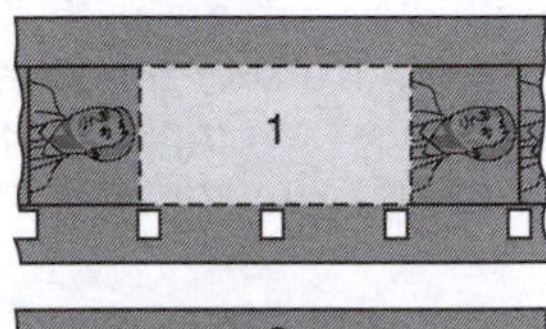

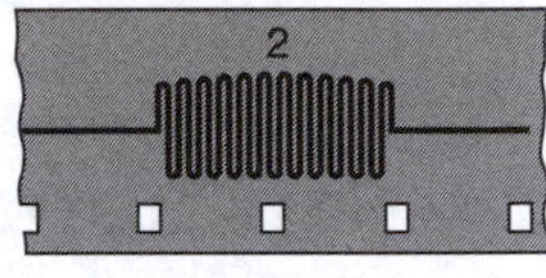

[19.6] **Synchronization of separately recorded sound**
A, Clapperboard showing scene and take number (and other information). B, Simple 'blip sync'. 1, A light flashes to blank picture. 2, Matching blip, recorded on sound. This lacks the precision of identification given by a clapperboard, but offers greater operational flexibility, and less film stock is used.

Marking a shot (*boarding* or *slating* it) consists of stating clearly the shot and take numbers, e.g. 'one-three-seven take one', while holding the board up to the camera with the bar open. The board is then clapped, the assistant and the camera operator both taking care that the action of the board closing stays clearly in vision, and the recordist noting that the clap is identifiable and distinct. If any fault is observed, a second board can be called for and the camera assistant puts the board back in shot, saying something like: 'one-three-seven take one, sync to second clap', then clears the shot quickly but quietly, while the cameraman checks the framing and (if necessary) focus, and the recordist rechecks the microphone position and gives a final thought to noise problems – as the director may also do at this time.

The director checks that the performers are absolutely ready, waits for a signal that the camera is running at its proper speed, notes that there are no new shadows or unwanted sound, then calls 'action'. At the end of the take the director calls 'cut': the camera stops running first, and as soon as it has, the recordist stops too – or, if there is some continuing noise, may signal for everyone to hold positions while extra sound is recorded for a fade or mix.

Mute (US: *MOS*) shots, i.e. picture filmed without sound, are marked with an open board (perhaps with fingers blocking any clap).

Wildtracks are recordings made without a picture. Identify them by relating them to a particular shot number, e.g. 'Wildtrack, extra background of car interior with gear change for shot one-two-eight' or 'Wildtrack commentary for one-three-two onwards'. Alternatively, use letters – e.g. 'Wildtrack B: atmosphere quiet schoolroom'.

Atmosphere or *'buzz' tracks*: the recordist should make a separate recording of the characteristic continuing sound of each different location – this may be called 'atmos' or 'room tone' – unless specifically told it will not be wanted, and may also wish to re-record important effects that can be improved by a balance in the absence of the camera, together with any additional noises the director wants.

In the case where the director does not have complete control over the action, it may still be predictable to the degree that a front board is possible, or failing that, an end-board, which is inserted and identified with the words 'board on end'. With film, the recordist should call for the camera to keep running if the director calls 'cut' before the shot has been slated. The director may instruct the camera operator to take cues from the action: in this case the recordist follows their lead. If a film synchronizing mark is missed, it is often simplest to run the shot again and board it properly, rather than leave the editor to find it.

Where a scene to be filmed would be disturbed by the usual procedure for marking shots, it can be simplified to the recordist speaking the shot number and tapping a microphone in vision. For this, keep the numbering simple, with no 'take two'. A handclap, pencil tap on a table or any similar action can also be used for rapid synchronization. When informal action is covered by a video camera with time code, to be followed by digital editing, conventional shot-numbering may be omitted but will generally be replaced by a comprehensive, time-coded log, made (or checked) by using viewing tapes, if possible before editing begins.

Film and video documentation

The director's assistant, the camera assistant and the recordist each have their own documentation to complete. The production team can work with domestic-quality viewing copies that have BITC (time code in vision); these can also be used to make time-coded transcripts of dialogue. The director's shot list should include the following:

- Videocassette number or film roll.
- For video, detailed time codes. For film, the board and take number of each shot (perhaps adding some indication such as a suffix 's' or 'm' to the shot number for sync or mute; also 'eb' for end-board).
- Shot description – opening frame (on start of action); action within the shot (including camera action; pans, tracks or zooms, timed if possible); continuity points; and faults, errors or sound problems.
- The overall duration of significant action.
- For film, the duration or footage used for each take (optional) and 'P' ('print'), 'OK' or a tick, or 'NG' for a shot not worth printing. (When 16 mm film is printed as low-cost cutting copy it is usually printed anyway, but the 'P', 'OK' or 'NG' serves as an editing guide.)
- Wildtrack details (in the shot list between the sync takes at the places where the sound itself is to be found).

The sound recordist's report should include as many technical details as may be useful later, including camera speed, sound format and standards adopted. Analogue systems (if still used) may have been controlled by time code or FM pulse, while two-channel DAT has a single worldwide standard. Each shot for which sound is recorded is listed, followed by a list of takes. 'Good' takes are marked: one way is to put rings round those numbers only. Wild-tracks are included on the report in the order in which they were recorded, and notes on quality may be added (see also notes on documentation in Chapter 20).

In addition, the recordist identifies each new roll of (separately recorded) tape over the microphone, with a standard list of details that might include the recordist's name, name of production, any job number or costing code, date, location, recorder type and number, and (in case of doubt) camera speed. This may be followed by a 10-second burst of reference tone at an agreed (or stated) standard volume such as 'minus 8 dB' (i.e. 8 dB below nominal maximum).

Unless otherwise agreed, the film sound recordist is responsible for getting tapes back for transfer within the shortest convenient time: the simplest arrangement is for them to accompany videotapes (or film, which if possible is sent for overnight processing) and then forwarded with the rushes for transfer while the picture is viewed with any guide audio from the camera (or while film rushes are viewed without sound). This assumes a fast, routine operation, but even in a slower schedule, film would still be viewed for technical faults as soon as possible, though sound might not be synchronized with it until later.

The first, preparatory stage of editing is to identify all raw materials clearly, and then to review and 'log' them. '*Prepping*' includes checking through content in enough detail to know (with the help of timed notes) where any major component is to be found. The log may include timed 'in' and 'out' cues, as well as highlight points and notes on participants and content. Ease of editing will often depend on how quickly obscure material can be located – often things such as a single spoken word with a particular intonation that would not usually be included in a log.

If time and budget allow, a full transcript of unscripted speech can save a lot of time. The transcription typist should be told in advance who the voices are (and how to identify them), be provided with dates, times, places, subject matter and unfamiliar spellings; and also advised on how much to include from off-microphone questions, director's instructions, noises off, background atmosphere, interruptions, and errors and idiosyncrasies of delivery. Transcripts should be supplied on disk as well as on paper. Users often have their own preferred layouts and may want to be able to select material from it, to assemble a script of work in progress or the final edited product.

A transcript can be marked up by eye, rapidly skimming through it prior to logging, then checked out in more detail against recordings. Individual words and phrases, too, can often be found by eye. Word processor 'find' searches can be used for this, but are fallible unless a transcript has been checked for errors that might not show up on common spell-check programs.

Videotape editing

A variety of techniques has been used for editing picture and sound recorded on videotape. One major division is between *off-line* and *on-line* techniques. Where editing is done off-line the on-line process that follows it is essentially a computer-controlled copy-editing system. Editing is further divided into *linear* and *non-linear* techniques. The term linear means that everything downstream of a cut has to be recopied, as in any copy-editing system. In non-linear editing a change can be made in the middle of a section without affecting material on either side. It is non-destructive. Film editing is a non-linear process.

On-line linear editing may begin in a two-machine suite, progressing to three or more replay machines at a final (and more expensive) on-line picture editing session. But in any linear editing system flexibility is severely limited, either to few editing stages or in complexity. Film, in contrast, becomes easier to recut in the later stages, so the process of refinement is not halted by diminishing returns for the effort and time expended. Film editing was an early example of non-linear editing.

Given its greater flexibility, film editing was unlikely to be fully superseded by videotape editing. However, digital editing systems are non-linear, just as film is, but much faster, which makes it far easier to try out different ideas for both picture and sound, and also speeds consultation between editor and director. As a result, digital off-line video editing is increasingly used in the film industry as well as in television. If (in many cases) it now takes as long as film editing once did, this is because it allows more opportunities to work on finer detail and complex effects. As with so many advances in digital technology, its main benefits have not been in cost or time reduction, but in permitting more ambitious and sophisticated results, while maintaining technical quality.

Digital video editing

Professional digital video editing requires a substantial investment in dedicated hardware and software – carrying the associated bugs and glitches that inevitably seem to emerge in such a rapidly developing field in which frequent upgrades are offered. Initially, the most severe limitation was in memory for picture, but in off-line systems about 10 GB of memory offers adequate (reduced) picture quality for editing most television productions of an hour or so if material is loaded selectively, while 60 GB gives much greater flexibility, and access to a server increases this further.

As computer processing power and memory (with recordable data tape stream as backup) have dramatically increased, off-line non-linear editing can also, if required, be replaced by on-line non-linear. This has extended the duration, from initially just component items, up to complete productions for which no picture compression is necessary. It also means that relatively simple editing (to a standard that just a few years earlier would have been regarded as highly sophisticated) can be achieved even on good domestic equipment by using editing software that costs very little, compared to the top professional systems.

The director's time-coded shot list is used first to make a preliminary broad selection from the material available. From the shot list (and transcripts, if available) the director can search out the preferred scenes, takes, segments of interview or other selections. The in and out time codes are entered into a program that can be used to copy from the original tapes directly into the memory of the digital editing system without the intervention of the editor other than to feed in the tapes on demand. As the selections are loaded, this is also a good opportunity for the director and editor to view the material together and discuss their objectives. If there has been no time for preparation, the time codes may be entered manually as the rushes are reviewed.

Audio tracks are loaded in a digital format. If the sound is sampled at 44.1 kHz (as used on CD) or 48 kHz (the standard for DAT) it will remain at broadcast quality, and can be treated as on-line, i.e. the product of the edit can be used in the final version of the sound. Lower sampling rates are acceptable for editing where sound as well as picture will be recopied (conformed) later to match the edit decision list (EDL). On-line-quality sound can be cut alongside either edit-quality or on-line-quality picture.

Picture and sound are loaded together, if this is how the material was shot, or separately in the case of mute picture or wild sound. It is helpful if the system also records and holds a few seconds more sound at either end of the selected picture in and out points: this will allow for mixes or 'soft' cuts without having to take extra care, except in cases where there is unwanted sound very close to a cut that will become a mix. During replay, a marker travels along time-line view-strips that look (for those with experience of film editing) much like film and audio tracks laid out across the screen one above the other. These strips are also used to show cuts and fades. A simple display might have just the picture and one (mono) or two (stereo A and B) tracks, but with provision for many more. In principle it is possible to have an unlimited number of tracks – although with a visual limit to the number that can be seen on the screen at any one time.

To make an edit, a shot (or some other length of picture or sound) is marked, then some chosen operation carried out. In this way, complex edits can be undertaken, and if they do not work little time is lost, as almost any change can be reversed by an 'undo' command, and the choice of system will dictate how many stages it will be possible to go back in this way. (But if a long sequence of operations is envisaged, the edits up to the starting point should be saved, so

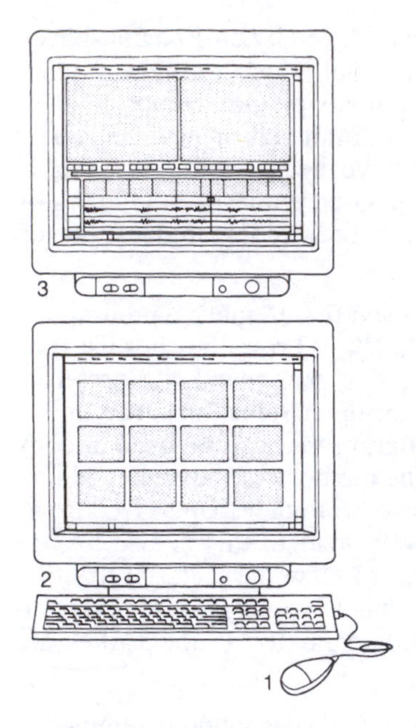

[19.7] **Avid hard disk editing**
1, Keyboard and mouse to control a range of display windows on two screens. 2, Picture monitor. 3, Edit monitor, here operating in 'trim' mode, showing pictures before and after a cut, with a bin of clips and the picture and sound timeline below.

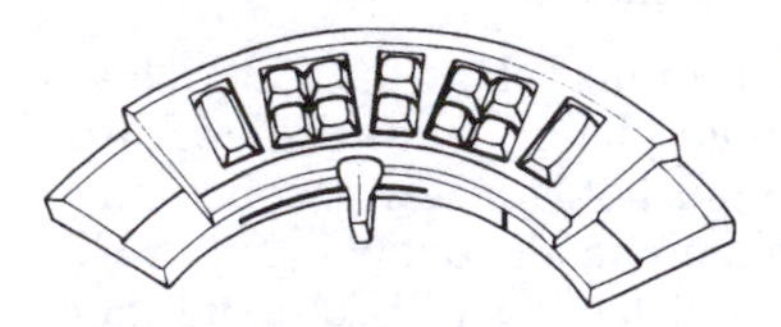

[19.8] **Avid 'manual user interface'**
This is one of a range of control options.

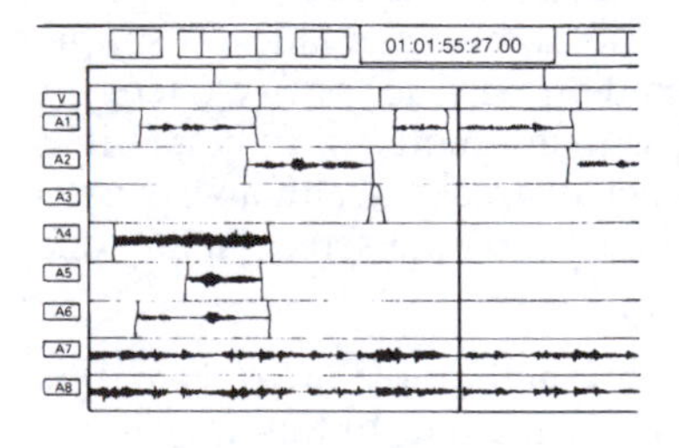

[19.9] **Avid sound editing and tracklaying displays**
Picture and tracks are lined up in sync, as for film. The audio waveforms help the editor to find cutting points to within a small fraction of a frame.

they can be reloaded, if necessary.) This, together with the usual computer drag and drop and cut and paste facilities, makes digital editing both easier and much faster than the conventional film editing it is otherwise very similar to.

One of the few disadvantages of digital over film editing is that computer programs can crash, so it is essential to back up regularly to ensure that a recent edit and perhaps also some earlier versions are always available. Another is that the increased speed reduces thinking time. A documentary, for example, may require commentary, and if this is written in the editing suite it may slow down other work, so it can sometimes be helpful to divide the editing with a gap of, say, a week or more in the middle – although this may be impracticable for scheduling reasons.

Editing sound to picture is more complicated than sound on its own (as described in Chapter 18) in that, as the picture is opened out or condensed, all the tracks of sound must have a corresponding length (an integral number of frames) inserted or deleted. Shots or sequences can be inserted, pushing the existing sound or picture apart, or replaced, overlaying it. Keeping it all in sync requires care.

Digital editing systems must operate in a range of modes, which include initial loading (which for film or analogue sources includes digitizing), then editing and, after that, tracklaying. It is best to do most of the editing with just one track or a stereo pair, at first mostly with parallel cuts, then at a later stage putting in simple overlays. When the picture edit is virtually complete, tracklaying begins. Even then it is still possible to recut the pictures, although more laboriously and with an increased possibility of confusion over sync.

One leading system (Avid) has two screens, one showing pictures where the effects of small changes can be seen, the other offering information (timeline, access menus, bins of clips, etc.) and some of the control icons for editing. As for the audio editing described in Chapter 18, the sound waveform is shown. By using variable-speed replay the sound can be finely matched to the waveform, and a cutting point selected. As usual, for a clean cut, avoid cutting where the waveform is dense: a gap is better. Again, there is a 'zoom' that condenses or expands the timeline. Zooming in makes detail of the waveform clearer, so that a finer editing point can be located. Zooming out reduces detail but puts more (eventually all) of the production on screen, which is useful for finding and moving to a different place.

An alternative system (Lightworks) employs a single, large screen to display pictures, tracks and a wide range of filing systems, each with a separate window that stands out like a three-dimensional block seen in perspective, with control icons that may be ranged along the top and down one side. There is also a keyboard with function buttons and an external VDU that displays data. It is useful to add an external monitor for the output picture and a conventional PPM.

The main screen window areas include one or more 'viewers' for displaying selected individual pictures, 'racks', or filing cabinets of

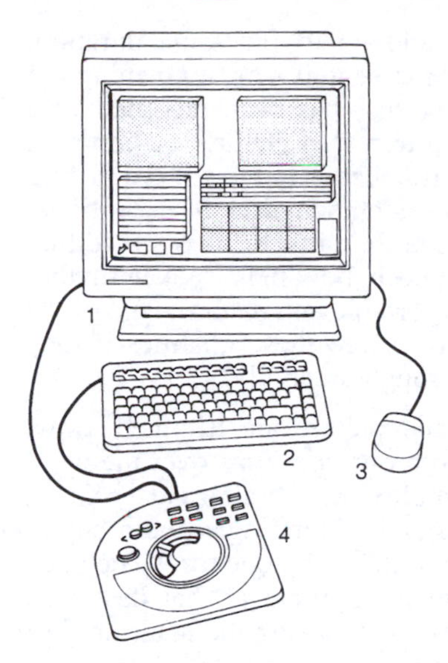

[19.10] **Lightworks editing system**
1, Editing screen, with layout chosen by the editor, here including two 'viewers' (windows displaying pictures running in sync); a 'rack' containing the material and information on selected groups of shots; a 'gallery', here of eight shots; and a view-strip with a segment of the timeline of picture and two audio tracks. 2, Keyboard. 3, Mouse. 4, Flatbed variable-speed control. At low speed the sound 'scrub' is heard at proportionally reduced pitch, as with film or audiotape, helping the editor to locate sound cutting points precisely.

time-coded and annotated lists of shots and 'galleries' of scenes (banks of tiny initial frames with their designated titles below them). For sound, the most important window is the strip-view with a scale of seconds along it. In its simplest form the strip corresponds to 10 seconds or so of picture with the sound below it and, as usual, it is possible to zoom out for a broader view or in for detail. Cuts are indicated by changes of colour and cues by displacement of the strip.

From the start, Lightworks mimicked film transport by including a large flatbed-style lever control to run picture and sound at variable speed back or forward in sync, in order to be able to stop precisely at the point required for a perfect sound cut, even though this might not be on a junction between two frames. Gaps, sound effects or changes in background or voice quality can be quickly located by listening for them, as with film. This kind of facility has been widely adopted.

Plainly, as sound and picture editing progresses, it is vital to maintain synchronization of all tracks downstream of the point on an individual track or stereo pair where changes are being made – which means that an editor must also be skilled in recovering sync if by mischance this is lost. Further, any technical problems encountered when learning such new techniques can be frustrating and also expensive in studio time, if this is locked into the use of the digital equipment.

For sound with picture, as for sound alone, the editing process – the stages and objectives at each stage – are much the same. Except for the simplest edits, it begins with a rough assembly, followed by more detailed editing of individual sequences.

Rough and fine cutting

The editor usually works with the picture and as few sound tracks as possible until the picture cut is complete. The first stage of editing is to make a rough cut, almost invariably at this stage cutting sound and picture parallel, making 'straight' or 'editorial' cuts. Where picture and its related sound have been shot (or recorded) separately, and so may differ in length, matching is sometimes just a matter of making the longer one equal to the shorter with a parallel cut. Where picture is longer than the sound intended to cover it, the picture can usually be allowed to run its proper length, with a gap in the sound that will later be filled out with matching atmosphere or effects.

Cuts may be made by removing corresponding lengths of sound and picture. A short cut in picture can often be accommodated by making

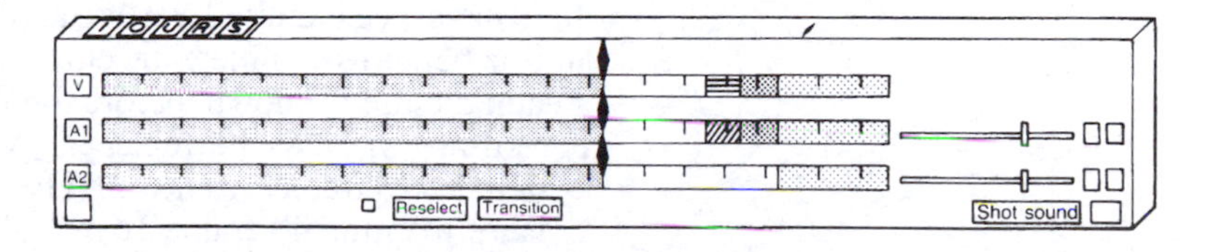

[19.11] **Lightworks view-strip**
V, Segments of picture, each represented by a different block of colour. A1 and A2, The associated sound, here as two tracks with, *right*, faders that can be clicked and dragged to new settings. The upper and lower parts of the fader 'knob' can be moved separately to set volumes before and after a fade, the ends of which are marked on the associated strip.

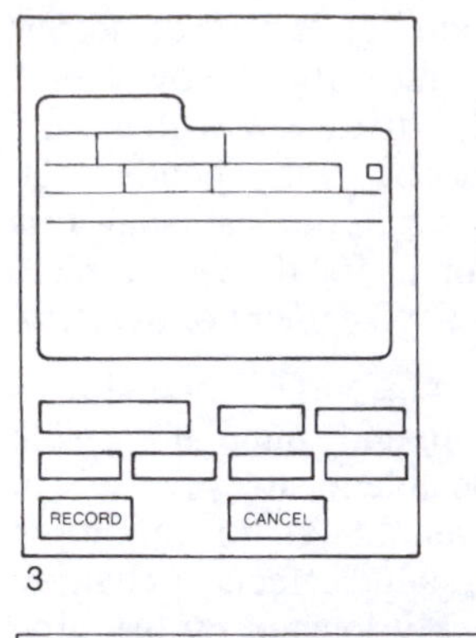

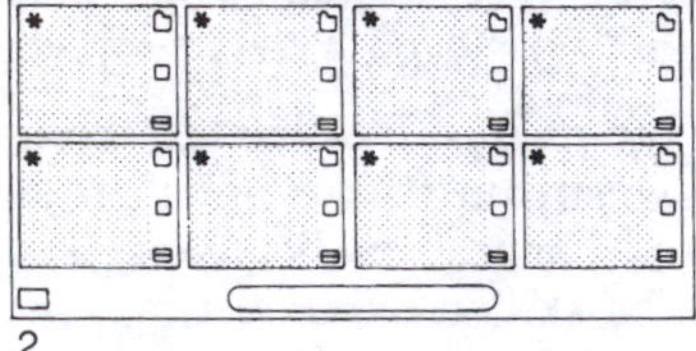

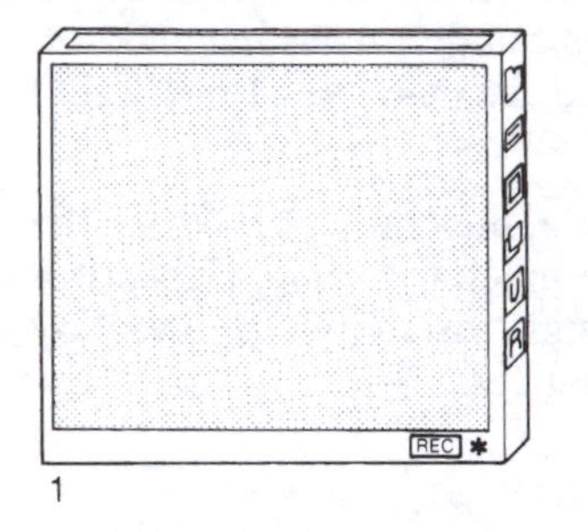

[19.12] **Lightworks display elements**
1, Viewer: a simulated 3-D box with the picture on the front and control icons and data displays on two sides and plinth. 2, A gallery, in this case of eight selected shots, with the names or numbers that have been allocated to them and icons that allow the editor to identify, select and use them. 3, Part of a recording panel, with controls and a file card.

several shorter cuts in sound that add up to the same duration, perhaps by taking out hesitations or inessential words. Opening out sound, and in particular commentary that has been recorded wild (i.e. not to replay of picture), in order to match picture is a common and often desirable device, because it is better to have more picture than words. Lengthening picture to match sound is an exceptional measure – and requires a better reason than just to solve an editing problem. In all cases, the timing of speech, both in its spoken rhythm and in its cueing, should remain comfortable and natural. One of the objectives of the fine cut will be to restore these qualities if they have been lost or diminished in the rough assembly.

At the fine cut, as further sound and picture are trimmed away, sound overlays (where the sound from one shot runs over the next) may be introduced. Before any complex non-parallel cut is made, the editor ensures that sync can easily be found again both before and after the region of the cut. Much of the technique for the detailed editing of sound on video and film is identical to that described earlier for audio tape: the procedures for choosing the place are just the same, but there are a few additional points to note:

- Words, syllables or pauses can often be lifted from rejected takes; even if it results momentarily in a slightly awkward mouth movement, this may still be preferable to inappropriate words.
- A change of sound that happens with a cut is much more natural and acceptable than an unexplained change at another point.
- Short gaps in background atmosphere created by lengthening picture (e.g. with a cutaway that needs to be seen for longer than the master action it replaces) can sometimes be filled with closely matching or looped sound. This is better than trying to find similar effects from library sources, as it takes a lot to obliterate a drop-out of atmosphere, particularly where there is a continuous steady noise such as machinery hum or the whine of jet aircraft.

In professional work, there is often a stage at which the work is reviewed by someone with executive responsibility for the result (and the power to require changes). It may be sensible to leave a final stage of fine cutting until all major editorial changes have been made. These may also affect overall timing.

Shot list, commentary and music

A timed shot list of the edit (often read as progressive time code from the first frame) may be required for a variety of reasons: for documentaries these include commentary writing. Narration will often have been drafted earlier and revised as editing progressed, but it is better to complete the edit, so that the exact length and value of each picture is known, before the final version of the words is set. (For film, time may be replaced by 'footage', where each foot is taken to be 16 frames, as it would be on 35 mm film: these are a finer measure than the 40-frame 16 mm alternative.)

The shot list can conveniently be combined with a script of the dialogue as cut and notes of other important sound. This can be laid out to look like the standard television script with picture on the left of the page and the corresponding sound on the right. Times should be noted not only at cuts in picture, but also at important points within the action of a shot (e.g. when some feature first appears in vision) and also at the start and end of dialogue, strong effects or dominant music.

Commentary can then be written accurately to picture: for a fast-speaking narrator allow a maximum of about three words per second (on film, two per 35-mm foot). For slower speech or more expression, these figures must be reduced. It is far better to write too little than too much; in any case, strong pictures are often better without words.

The audience's primary need from a commentary is that it should interpret the picture: this is a more effective use of the medium than when the pictures are cut like wallpaper to cover the words ('wallpaper' is a term of derision). The worst possible case is when pictures and words do not agree, either because the commentator wants to talk about something else or to go on about something that is no longer in the picture. In these cases an even moderately interesting picture demolishes the commentary entirely: the words are not heard.

There is a particular danger when a presenter records part of a commentary in vision, either in the studio or on location, and another part later in the confined space of a commentary booth. Even if the voice or microphone quality can be matched by equalization, there may be an abrupt change of acoustic. Short links should, if possible, be planned in advance and recorded at the location; for safety, a variety of versions of the words might be recorded.

At best, a succession of different acoustics on the same voice, for no apparent reason, is mildly distracting. At worst it is destructive of artistic unity and also confusing: the audience may be unsure whether a new sound is the same voice or some new and unexplained character. On the other hand, well-chosen words, written to fit and complement the pictures, may add far more than is lost by the mismatch in sound. If so, one compromise is to allow a gap in time, a music link or strong sound effect to intervene: the change will then be less noticeable.

If it is obvious from the picture that the voice-over was not filmed on location, the change in quality may be helpful: the commentator finishes speaking in vision, turns and walks into the action, and the dubbed voice takes over: here a clear convention has been adopted, and the change of acoustic confirms it.

Specially composed film, television and video music is often written at a late stage. The composer requires a copy of edited picture and sound that is as complete as can be arranged, a cue list with time codes or footages and also a script (for documentary, a shot list with narration that is as advanced as possible). However, after the music has been composed and recorded, the editor will often find that the interaction of picture, words and music suggests new possibilities: it is wise to allow editing time to allow for this. Final changes in the picture may in turn require changes (further cuts or extensions) to the music, so because it is often written and recorded to a tight deadline, it is worth making sure that there will be sufficient flexibility.

Music from existing sources should be selected earlier in the editing, so there is more time for the pictures to be accommodated to it.

Tracklaying

Tracklaying is the first stage of preparation for dubbing (making the final sound mix). Here again the standard techniques were pioneered with and are still used for film, and in digital editing are functionally almost identical. In all media, when the fine cut is complete, the constituent parts of the cut sound are separated on to several tracks: this is called 'splitting the tracks' (on film, a physical process).

For as long as the sound continues at the same level and quality, whether of the main or background sound, it is kept on the same track, but at points where these change, the sound is switched to another track alongside it. If, by simple adjustments of level and equalizer settings, it is possible for the sounds to be matched, a parallel shift from one track to the next may be all that is necessary. But if there is a more drastic change that is not immediately explained by an equally strong change of picture, then extra sound is required for a cross-fade. These fades are generally made quickly, but it is better to have too much than too little, or the sound may be heard cutting in or out at low level. The chopped-off sound appears unnatural if it is not at a cut.

For film, extensions had to be found from a trim bin. On digital editing systems they should be easier to find: in some, overlaps reappear automatically as the sound is split onto neighbouring tracks.

During the early stages of editing, extra sound that had to be synchronized with action may not yet have been cut in, perhaps because there was enough other synchronous material for viewing purposes. At the tracklaying stage extra tracks are created for them, and also for any commentary that has been pre-recorded.

Music may also be laid: where it is dominant it will already have been arranged for the picture cuts to be related to it, or alternatively for specially composed music to match the picture. The precise manner in which picture and music should be co-ordinated is a complex study, but a strong rhythm suggests that some, but not too many, of the cuts are related to the beat; similarly, strong notes may cue a cut.

The editor accurately lays as many of the sound effects as possible; this includes both library effects, which should be selected and transferred for the purpose, and wildtracks that were specially shot.

Some continuing sounds may be formed as loops. The name reflects the way this was once done: by making up lengths of recorded track with the ends joined back to the beginning. Now this is replaced by a digital 'loop', using material selected and repeated from a sampler or disc. Loops with strong distinguishing features must be long, so that the regular appearance of that feature does not become apparent. If it is likely to, it could perhaps be trimmed out, though this would shorten the loop even more. In joining the ends of loops (i.e. by selecting the in and out points of a sample) take care to find points where volume and quality are matched, or where overlaps can be mixed.

The documentation required to go to a dub varies both with media and with the complexity of the operations that will follow. With film (or in any case where the tracks do not appear on screen) the editor or an assistant prepares a chart showing the main points of the action, and parallel to this the sound on each track, or source, showing the cuts in and out of sound with symbolic indications of fades and mixes and clearly marked time codes or footages, together with notes on loops and tape and disc or hard disk effects. If the tracks do appear on a screen, this is (in principle) simpler – although aids (such as cues and colour codes) may be added, to make them easier to read. A marked-up script is helpful – or essential, where commentary is to be added.

Conforming sound and picture

When an off-line or digital (or videotape) edit is complete, the picture must be conformed (matched and copied) from the originals: this is equivalent to the negative-cut and showprinting of film.

When the off-line picture has been accompanied by on-line quality of sound, this must be transferred back in sync with the picture. In the more general case, the sound as well as picture will be conformed, which means that it must be separately recopied from the original recordings. Tracks which have been laid off-line as a provisional set of editing decisions are used to create a corresponding set of full-quality audio tracks in a format that will serve as sources to be used in the mix. For this, there are two main options. The older method is to copy the originals to multitrack tape; the alternative is to use hard disk. An edit decision list (EDL) for picture and audio is used to control the conforming process automatically.

To do this, the EDL finds and displays the first time code, including the user bits that identify the source: in effect, it 'asks for' a recording. When this has been inserted, it will then cue and transfer all the material on it, leaving gaps for material that will be inserted from other recordings; then it will request a second source and so on. Picture is generally conformed to a single, continuous recording, and dissolves are programmed or put in manually later. In contrast, sound is transferred with the tracks alternated in a checkerboard A- and B-roll arrangement, so that they can be separately mixed and controlled.

Where television to be transmitted with stereo sound has been edited using equipment on which the editor, for simplicity and to minimize the danger of losing sync, has been working in mono, a clearer idea of what is available in stereo or surround sound may emerge only at this stage. Any stereo in the MS format is converted to A and B. Note that for the benefit of television viewers who do not have stereo decoders, television sound is still also transmitted in a form in which the A and B signals have been recombined to form a 'compatible' mono signal. Where encoded (and so compressed) surround sound is transmitted, in order for it to be heard, this will require decoding at the receiver; without that, normal stereo will still be available. Stereo and surround should not be made so extreme that mono listeners cannot hear all the information they need to make sense of and enjoy the production.

Complex sound preparation is now often undertaken using a digital audio editing program, either a separate stage as described in Chapter 18 or (on a digital console) as an extended first stage of the process of mixing. Even if the previous tracklaying has been comprehensive, preparation for the mix may be a time-consuming process that requires equipment to replay audio from a range of different formats. From being one of the more neglected components of video editing, 'prep' has developed to the stage where it may occupy a room in its own right, in effect a small studio. Depending on the facilities available, some tracklaying may be left until this stage; some of the sounds may be processed here, automatic dialogue replacement (ADR) and spot effects ('foley') laid and edited, and some sequences pre-mixed. In such a case, the preparation may be undertaken by a specialist sound editor who carries this to a stage where the final mix is much simplified – except that the theatre also serves as a meeting place where production can hear what is offered and either accept or change it on the spot.

Post-synchronization

Post-synchronization or automatic dialogue replacement (ADR; not necessarily very 'automated' by current standards) may be needed in order to re-record voices that were inadequately covered or overlaid by inappropriate background noises, or because a director has given priority to a costly setting or unrepeatable event. It may employ an improved version of the original speaker's voice or substitute a more appropriate accent, vocal quality or performance. It may replace one language by another. ADR is a specialized job that must be done well if it is to be effective.

Good post-synchronization has to be done phrase by phrase using words that give reasonable approximations to the original mouth movements. A loop of film or its digital equivalent is repeated over until the best match is obtained, then the next phrase is taken with the next loop. Any final trimming or tracklaying to match lip movements perfectly can generally be done afterwards. Near-perfect results can sometimes be achieved by digital editing.

Post-synchronization of spot effects ('foley') is – for a few talented individuals – a well-paid job in the film industry. These experts are capable of seeing through a scene or sequence a couple of times, then collecting together a pile of odds and ends which they manipulate in time to movements both in and out of picture, in a convincing simulation of the casual sounds of live action. In a low-budget production this may be sufficient, but in the more general case, the sounds will be refined by further editing, tracklaying and treatment.

A complementary technique (at a much earlier stage) is shooting picture to playback of pre-recorded sound – typically for musical production numbers (miming), where the quality and integrity of the sound are of primary importance. It also permits live action (such as dance) to be combined with song in a way that would be difficult if not impossible in real life. In miming, however, the illusion can be shattered by a sudden change of acoustic between speech and song, emphasizing and making awkward a transition that should be smooth.

Mixing sound for picture

To mix sound for picture, the theatre (dubbing theatre) has equipment to display picture and run many synchronized audio tracks, and then to record it. There is often also a studio, if only for a single commentary voice, and provision for playing CDs, tapes, digital audio files, etc., together with all the additional facilities on the control console that can make operations simpler and more flexible.

Since digital processes have entered the dubbing theatre they have allowed video and film dubbing to converge and become independent of both the original recording medium and its final format (film,

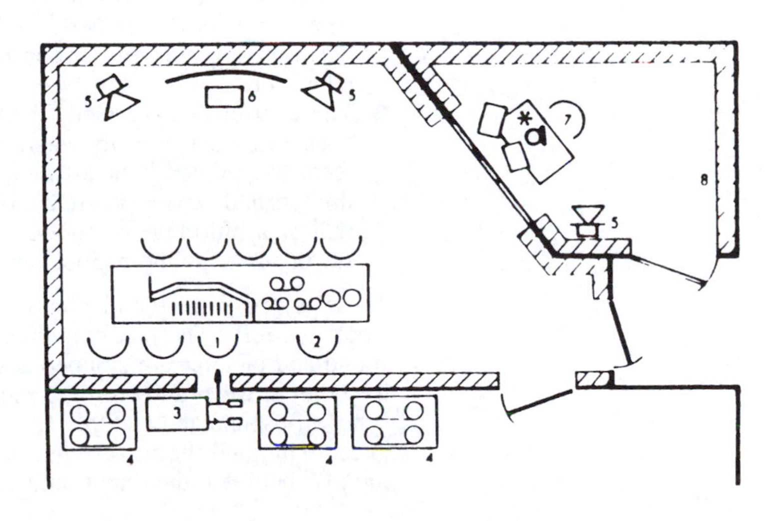

[19.13] **Film re-recording studio (dubbing theatre)**
1, Mixer at control console. 2, Assistant, with tape, cassette decks, turntables, etc. 3, Film projector and footage counter (here with video cameras to relay picture and footage as required). 4, Magnetic film sound reproducers and recorder. 5, Loudspeakers. 6, Screen and footage monitor. 7, Narrator with desk, microphone, cue light and monitors for picture and footage. 8, Acoustic treatment.

DVD, videotape or television transmission). These are some of the components that may be required:

- Picture display – in digital systems, a simple monitor, plus another for any narrator or commentator if the main screen is out of sight.
- A counter showing timings or time codes in minutes and seconds should be displayed close to or superimposed on picture, perhaps with repeaters on the console and in the studio. A narrator or musical director works to these numbers together with other cues.
- Replay and recording equipment. This may be under local control while being loaded, but then switched to remote operation so that synchronous replay of all sources and recording machines is controlled from the console.
- Ancillary replay equipment. There must be the normal provision for replay of various recorded formats, for original recordings, music and effects. There should also be provision for the replay of sampled sounds as loops.
- Commentary booth, usually with a dead acoustic. This ensures that narration does not carry an ambience that is alien to the picture it accompanies; for example, by overlaying a live, indoor acoustic on to an exterior scene. A dry acoustic also seems to help the voice to cut through background effects or music.
- Cue keys and lights (for a narrator to be cued from the console). The usual system is 'green for go'; the cue light should be close to the reader's sight line. The narrator will often require a headphone feed of music, dialogue and effects, with or without voice feedback (whichever is preferred). Before rehearsal begins, check that the sound level fed to the studio headphones is comfortable. There is also talkback between control console and studio.
- Provision for the production of spot effects ('foley') with a range of floor surfaces, perhaps in the same acoustically separate booth.
- Monitoring equipment – high quality, and usually set loud enough for imperfections in the original material or of the mix to be heard easily. Also, as usual, there should be provision for the sound to be switched to domestic quality loudspeakers to check what most of the audience will hear – particularly important when mixing dialogue or commentary with music or effects for television.
- The control console (see Chapter 9). This has the usual fader channels, each with its standard facilities for equalization and feeds to graphic filters, artificial reverberation, etc. Compressors, limiters and expanders are available, and also a noise gate. There will be a full range of controls for stereo and generally also 5.1 surround, including panpots and spreaders.

There is provision for discontinuous recording, sometimes called 'rock and roll'. The mix continues until errors are made, when the sound and pictures are stopped and run backward until well before the point at which re-recording must restart. A counter runs back in sync with sound and picture. Then the whole system is run forward once again, and the mixer checks levels by switching the monitor rapidly between the input and the previously recorded signals,

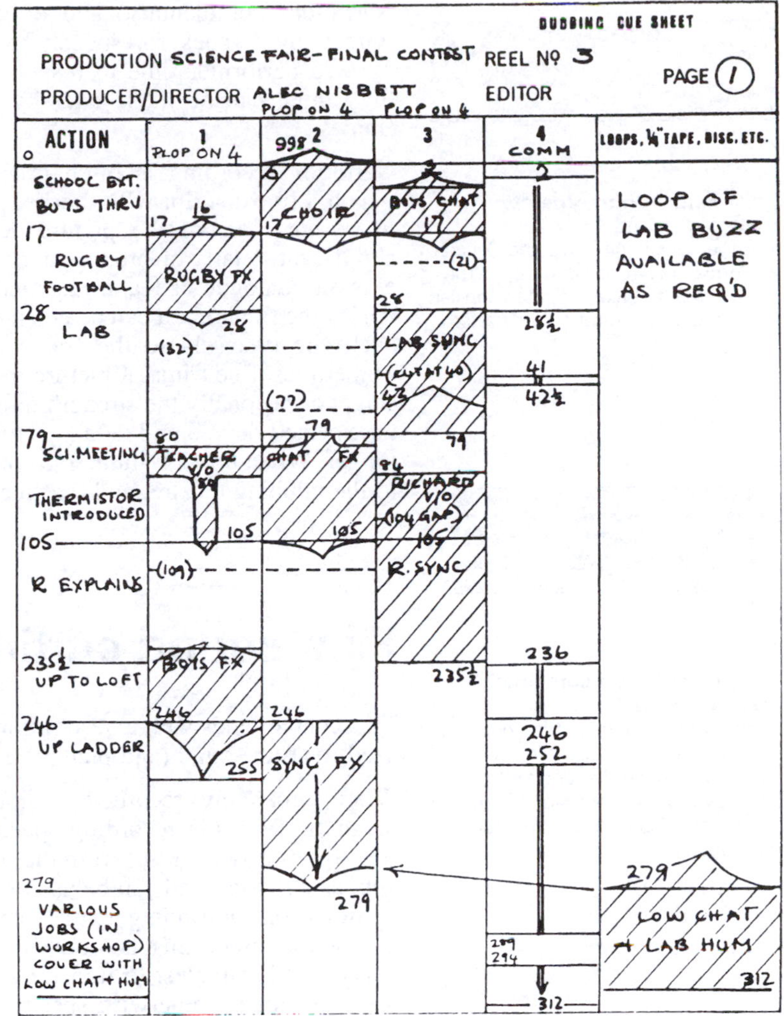

[19.14] **Film dubbing cue sheet** The original dub on which this was based has provision for more sound rolls. Here 35 mm film footages are given (i.e. 16 frames = 1 ft) but timings may also be used. The action column does not list every picture cut, just sequences within which sound should remain roughly the same. In some cases sound has been laid to several seconds beyond a picture change, but will be faded out quickly on the cut (this is represented by the shallow V). As when dubbing in any format, fine points are discussed or decided by consultation on the spot. For example, here it was not made clear on the chart, but the choir was taken behind the boys' chatter between 3 and 17 and was faded quickly after 17, although it was originally laid as far as 32. The rugby football sound (at 16) preceded the cut at 17 and, in effect, killed the sound on tracks 2 and 3. 'V/O' means 'voice-over'. At 80–84 the teacher's voice was to be heard over background chatter and effects, and from 84–105; this in turn was taken to low level while Richard's voice was heard (out of vision) talking about what is happening at this point in the story. 'Hum' and 'buzz' are terms used to indicate low-level background sound. The commentary voice had been recorded in advance and laid (on track 4).

adjusting faders to match them, then switches back to record, and from that point the new mix replaces the old. The switch should not be instantaneous, of course, or a click might be introduced.

Documentaries are often mixed in stages, with a so-called music and effects (M and E) mix being made first (in fact, this would also include any sync speech, and there might be no music). The final stage is to add the commentary; but the M and E track should still be retained against the possibility of changing the commentary later, or making versions in other languages, or for use when extracts are taken for other productions. In fact, on some desks, a mix at any stage can be held in memory and subsequently read out automatically if needed (for example) to fulfil contracted specifications for 'deliverables'.

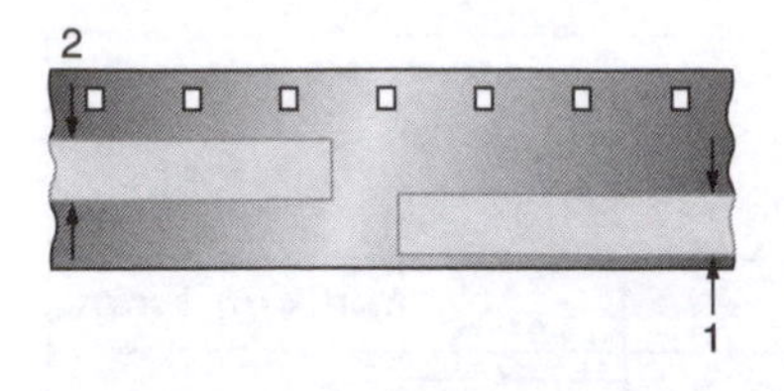

[19.15] **Film: 16 mm separate magnetic sound**
1, Edge track used in US and Canada. 2, Centre track used in Europe. The full width of the film is coated and can therefore be used for either system.

Narration (commentary) and effects can be added at this stage. 'Green light' cues for speech have to be given a little early to allow a performer time to respond; if desired, fine adjustments to the timing can be made later by shifting the recording by a few frames.

Narration recording is often treated as a separate operation, for which a timed script with marked cues is provided. This script can leave out most of the picture details, and show only the most significant details of other sounds, such as the in and out cues of existing dialogue or featured sound effects. If narration is recorded before the final mix it can be taken back to the cutting room to adjust each cue precisely to the frame at which it has the best effect. Sometimes, fine trims of picture may be removed or restored at this stage: occasionally the strength of the word enhances a picture to an unexpected degree and demands that it remain on the screen a little longer. When this fine-tuning is complete, the tracks are all returned to the dubbing theatre, and the final mix is completed.

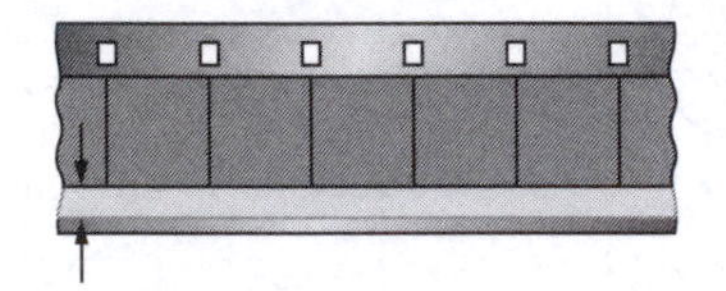

[19.16] **Film: 16 mm combined magnetic sound**
The narrow magnetic stripe occupies the space formerly used for optical sound. Picture and sound are displaced to allow for the two sets of equipment required for synchronous reproduction.

Film sound editing

This chapter ends with a description of some of the older techniques used for film sound (and picture) editing. This begins with transfer.

When separately recorded audiotape is transferred to separate magnetic film, the recording speed is controlled by time code (or sometimes a sync pulse) from the original. The original recording is kept as a master, and can be used for retransfer of worn or damaged sections, or for finding extra sound for transfer to digital sound file or for direct, unsynchronized replay to fill gaps at a dubbing session. It is simplest to transfer stereo – MS or AB – in the form received, unless otherwise specified, but MS will be converted to

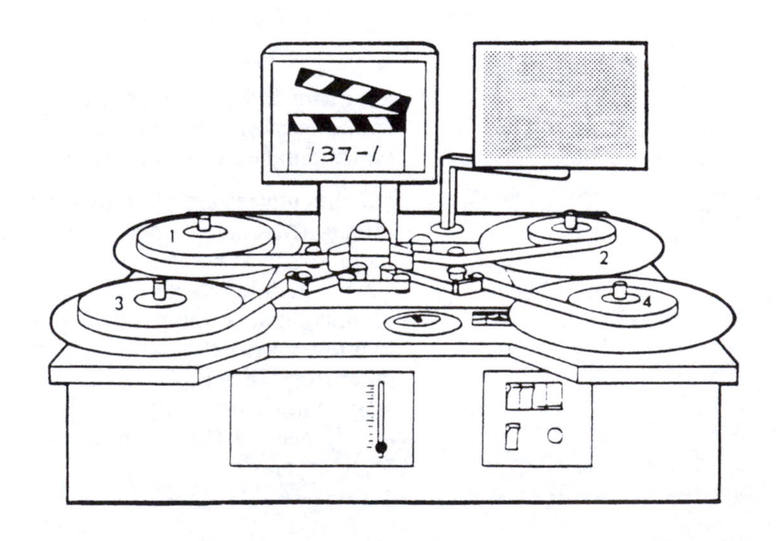

[19.17] **Flatbed film viewer**
Once very widely used for film editing, this is a simple example of the equipment for viewing 16 mm film at normal speeds and with moderate picture and sound quality. 1 and 2, Picture feed and take-up plates. 3 and 4, Sound feed and take-up plates. Lacing is simple, with drive spindles on either side of the central sound and picture heads. It is not necessary to have rolls of film as shown here: the machine requires only enough head or tail of film to be held by the drive rollers. Typically, the film can be run at normal and 'fast' speeds, and either forward or backward. For editing, an extra sound path is needed, so most flatbeds have six plates. Further variants allow for stereo sound and other features.

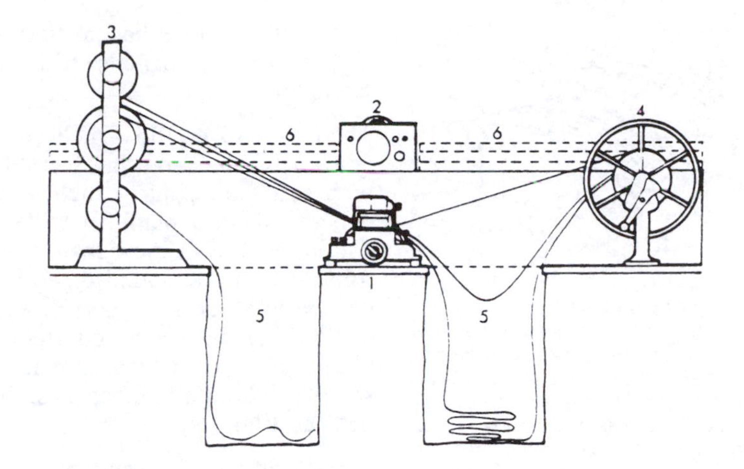

[19.18] **Editing bench**
1, Picture synchronizer.
2, Amplifier and small loudspeaker.
3, Horse supporting rolls of film and spacer. 4, Winding spindle with 'split' reels fitted. 5, Bins lined with soft fabric, to hold excess film that has not been wound up. 6, Racks for small rolls of film awaiting selection of material.

AB when sound is being prepared for the dubbing (re-recording) session.

In the cutting room the film editor receives:

- film positive, 'rush prints';
- the original sound tapes;
- sound transfers on the same gauge as the original film;
- documentation on the film and sound, and perhaps also laboratory and viewing reports on the rushes;
- a shot list;
- a script, or sometimes a transcript of the sound;
- alternatively or additionally, a cutting order, which may include shot descriptions from the shot list and dialogue from the transcript.

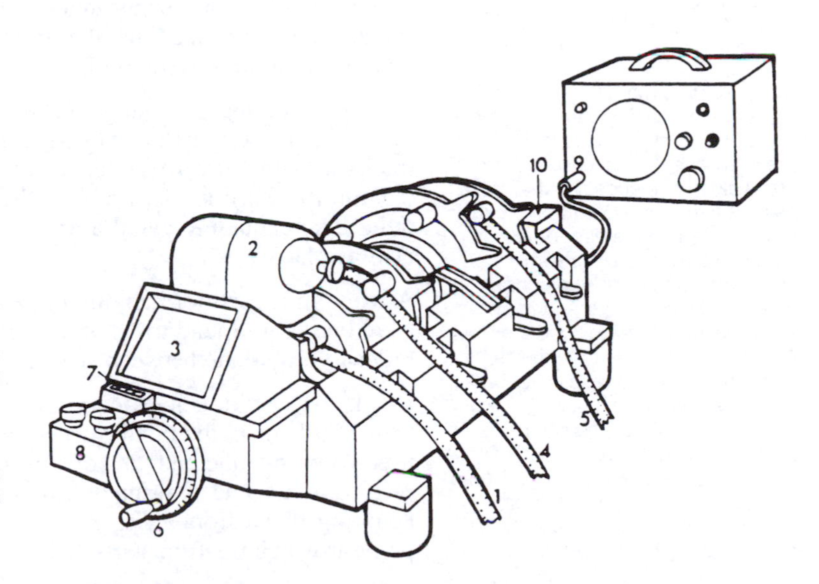

[19.19] **Synchronizer**
'Pic-sync' with simple loudspeaker.
1, Picture track. 2, Lamp projecting picture via rotating prisms to the screen (3). 4 and 5, Sound paths over replay heads on centre line: each path passes over a wheel with (for 16 mm film) 40 sprockets, a '16-mm foot'.
6, The wheels are on an axle that is turned by hand, but some models also have a motor and clutch plate, so edits can be checked at normal speed, and a disc indicates the frame within each foot. 7, Counter showing feet or time from when it was set to zero. 8, Simple low-level sound mixer. 9, Loudspeaker input.
10, Power transformer. Synchronizers may have two picture heads or more sound paths and a more sophisticated sound system.

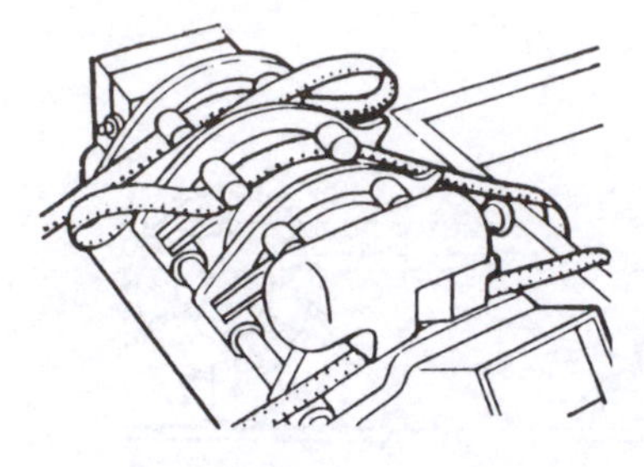

[19.20] **Looping back**
A mechanical technique for matching sound to picture: these are in sync on both sides of the 'pic-sync', but at the picture join there is an excess of sound, which has been taken up in a loop between two of the sound paths. In this way, picture and sound can be rolled back or forward together until suitable points for cutting the sound are found. Similarly, excess picture may be removed to match sound.

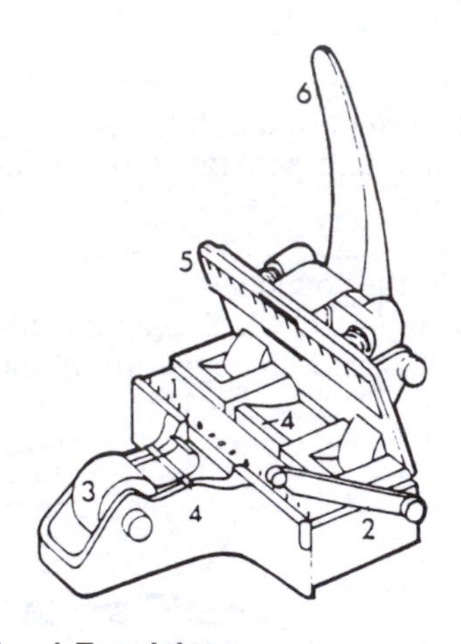

[19.21] **Tape joiner**
1, Picture or magnetic track is placed on the lower plate and located precisely over the sprockets that stand along one edge. 2, After knife has cut film, it is moved to the centre of the plate and the film to be joined with it butted to it, again with sprockets holding the two ends precisely in position. 3, A strip of adhesive tape is drawn across the join. 4, Tape also adheres to two bars, one on each side of the base plate. 5, The upper head is then brought down on to the film. 6, When the spring-loaded arm is pressed down, two guillotines cut the tape on either side of the base plate, and a punch perforates the adhesive tape at the sprocket holes near the centre of the plate.

The cutting room has a flatbed viewing and editing machine, tape joiner, wax pencils, trim bin, leader film, spacer film and a waste bin.

There is also a film synchronizer (with its own small amplifier and loudspeaker) on an assembly bench. This has two bins, lined with soft fabric, into which picture and film sound can fall, and between them a platform wide enough to hold the synchronizer. On either side of the bench, outside the bins are supports for the rolls of film and sound to allow them to spool across the bench. On the left side is a 'horse', several uprights with holes, so rolls of film can be supported freely on spindles passed through the plastic cores on which film is wound. On the right is a take-up spindle to hold a number of split spools, with a winding handle to take up film.

The synchronizer has sprocketed wheels, rigidly mounted on an axle so they are mechanically synchronized. Each wheel is 1 ft in circumference, so that for each revolution 1 ft (16 frames of 35 mm or 40 frames of 16 mm) of film passes through the synchronizer. A counter shows feet of film. Even for 16 mm film this is often calibrated in '35-mm feet', so that 16 frames equals one digit on the counter. There is also a disc that numbers each frame on the circumference of the wheels: on 16 mm equipment these run 1–40. Picture is carried on the front film path and sometimes on the second as well. The other paths have sound replay heads. The synchronizers can be turned either by hand or by a motor that is engaged by means of a small clutch so that picture and sound run at their normal speed – forward or backward.

Film and sound are cut by stainless steel (and therefore non-magnetic) blades on a tape joiner (guillotine splicer), which then holds the ends in position while tape is stuck across the join. The tape recommended for joining film is similar in appearance to transparent plastic domestic tape but is polyester, very thin and strong, with a non-oozing adhesive. Tape joins can be undone, to remake edits or reconstitute the original without loss of frames. Wax pencils are used to write on film 'cel'.

The standard angle of cut provided on some joiners used for magnetic film sound is 10°. A mono recording on separate magnetic film track is 5 mm (0.2 in) wide: a cut therefore fades in over about 0.5 mm (0.02 in); for 16 mm film the duration of the fade is 2.5 ms, quite long enough to avoid a noticeable click. A right-angled cut is also available.

A trim bin is used for hanging up sections of picture and sound that have been cut. Sound trims may be needed later to overlap sound in order to cross-fade between scenes.

Blank spacer film is inserted in the sound assembly where no sound is available or in the picture where a shot is missing. Spacer is also used to protect the ends of rolls of assembled film. Leader film for picture and sound is printed with start marks and 35 mm footages counting down from 12 to 3, after which the print is black to the point at which the film starts (which is where zero would be on this

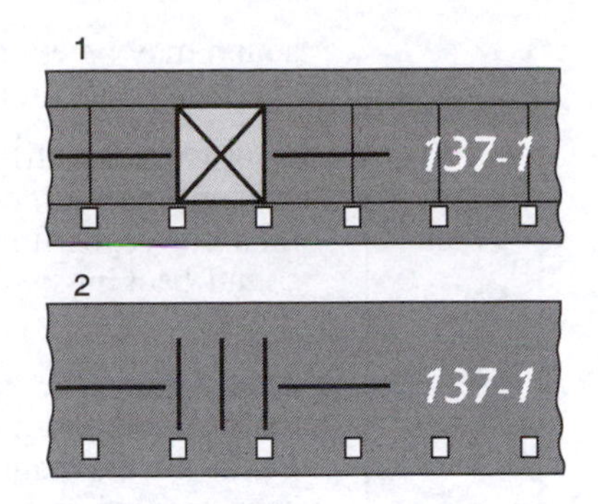

[19.22] **Synchronization marks**
1, Marks on picture. 2, Marks on separate magnetic sound. In the conventional clapperboard system these are marked at the points where picture and sound close. When electronic synchronization is used, they mark the flash on the picture and the corresponding buzz on sound.

[19.23] **Identification marks on film trims**
A, Picture trim: 1, Edge number (and bar-code on extreme edge above it) on picture only, printed through from negative and used by neg-cutters working with a cutting copy to identify the original.
B, Magnetic sound trim: 2, Code number ('rubber number') printed on edge of both sound and picture of synchronized rushes. Numbers repeat every 16 frames, so sound and picture can easily be resynchronized. 3, Shot number, written on trims by hand for rapid identification (not essential if trims are rubber-numbered). 4, Stereo sound, cut square to avoid 'wiggle'.

count). Sometimes a frame of tone (a 'sync plop') at 1000 Hz (for 16 mm) or 2000 Hz (35 mm) is cut in on the sound leader a few frames before the film starts. As it passes through prior to being switched, the presence of the tone confirms that sound is live on the replay itself.

The principles of cutting separate magnetic film sound are mostly similar to those used for audiotape, though a few extra problems (and possibilities) emerge from the combination of sound with picture.

Film picture available to the editor includes:

- film shot with synchronized sound – lip-sync sound, synchronized effects or music, and material where picture and sound are related, but for which there is no need to maintain perfect sync;
- 'mute' (silent) film, for which no useful sync sound could be recorded;
- library film for which suitable sound may or may not exist.

Film sound available to the editor includes:

- sync sound;
- wildtrack commentary or dialogue that will generally be played only with pictures that do not show the speaker's mouth;
- wildtrack effects that will be synchronized to action on mute film (e.g. a gunshot) or sync film for which the effects recorded at the time were inadequate (e.g. too distant);
- Location atmosphere tracks;
- Library sound effects, atmosphere or music.

The film editor (or an assistant) synchronizes the rushes or, if this has already been done, checks sync. To synchronize film, run the picture and find the frame where the clapperboard has just closed. With a wax pencil, mark this frame with a boxed 'X'. On the roll of transferred sound, mark the corresponding frame with three bars, 'III', on the backing with a thick felt marker pen (wax can stick to the magnetic coating of the next wind). Write the shot and take number ahead of these marks on both picture and sound. For continuous replay the spacing is then adjusted so that this picture and sound follow in the same synchronization as that of the preceding shot on the rolls of rushes. Intervening mute film shots and wildtrack

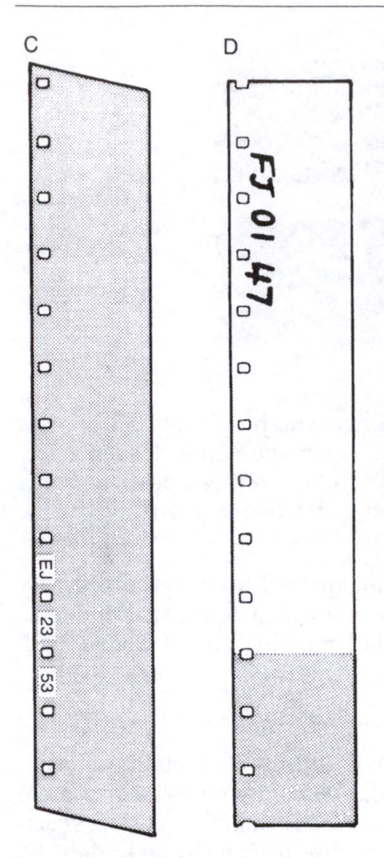

[19.24] **Magnetic film trims**
C, Mono sound (cut at an angle) identified by its 'rubber number'. D, Stereo sound trim with no rubber number printed on this short segment: the spacing attached to it shows where the nearest number (written in by hand) would be.

sound may be cut out and wound on separate rolls: all good takes of sync material can then be seen in the order in which they were shot.

In order to identify trims of sound and picture at all later stages of editing, the paired rolls are sent to be run through a machine that inks corresponding serial numbers on the edges of both film and sound backings, so that they fall between sprocket holes every 16 frames.

At this stage the director and editor may view the film together, so that the content, intention and cutting style can be discussed. Meanwhile, the assistant logs the material and breaks down rolls that have been viewed into single shots or related groups of shots and rolls them up with picture and sound, if any, together. These are kept in large film cans: each can contains a substantial number of small rolls, each marked with its shot number, and the cans are labelled on the edge with the shot numbers they contain. The assistant looks after this rough filing system from which material can be taken, restored and retrieved as necessary.

Up to here much of the work can be done on a synchronizer. Most of the subsequent rough and fine editing is carried out on the flatbed machine, with the synchronizer used mainly for identifying additional shots and their sound. From this stage on, the general principles and techniques parallel those described for digital editing.

Chapter 20

Planning and routine

Before any sound or television studio session, some planning is likely to be necessary – if only to ensure that all technical facilities that might be needed are available and ready for use. If a script is provided in advance, note and roughly mark up the parts that directly affect you. For television there may be a planning meeting to co-ordinate what is expected from all the technical and design departments. Later, in the studio, there are routines for line-up, cueing, timing, and completing documentation, all of which are essential to its efficient operation.

These preparations and procedures apply not only to radio or television, but often also (with minor changes in terminology) to record company recording sessions and to the film and video sound dubs (re-recording sessions) described in Chapter 19.

Planning: radio or recordings

For radio or other sound recordings, when there is a range of studios available the request for a booking is generally accompanied by a brief description of its use, e.g. the number of speakers or musicians and type of balance required, such as 'discussion, 6 people' or 'novelty sextet with vocalist'.

Except for the simplest programmes, the production office prepares a running order or script that (except for the most topical material) is sent to the programme operations staff. Many productions follow an established routine, so little discussion is necessary before meeting in the studio; however, if there is any doubt, check in advance. For example, before recording an unfamiliar musical group, the balancer should find out what instruments are expected and what sort of balance the producer wants to hear; if, for example, it were required to sound like an existing record the balancer would listen to it. Is the music to be featured, or for background use? Should it be brilliant and arresting or soft and undisturbing? Will it be heard with voices over or with announcements at certain points within it? Will there be applause?

The recording arrangements might also be discussed: its probable duration, whether it will be stop–start or continuous take, the medium and number of copies; and arrangements for and time of delivery.

Any unorthodox or experimental layout would be discussed with the musical director in sufficient time to allow for the provision of chairs, music stands, acoustic screens and other furniture, and studio instruments such as a piano. The balancer also sets up microphones (or checks their layout) and arranges their output on the console, working out and plugging up (patching) a suitable grouping of channels, and checks that equalizers, compressors and limiters are ready for use; that reverberation and other devices are switched to their appropriate points; that talkback is working; and that foldback or public address levels are set.

Television planning

For a complex television production the team might include an operator to replay records and recordings and boom operators (if booms are in use). Second in command and the most versatile in duties is the sound assistant, who often leads the sound department's forces on the studio floor, supervising the rig for musical productions, distributing radio microphones for a game show, operating the main boom on drama, or sitting alongside the supervisor to handle the audience microphone pre-mix, public address and foldback for a comedy show.

Representing the facilities that the team can offer, the sound supervisor attends the planning meeting (several if necessary) called by production to co-ordinate design and operational plans. For even a simple programme, this may involve the director, designer, lighting supervisor and technical manager (in charge of technical services other than lighting and sound). More complex productions may also require the presence of the floor manager, chief cameraman, costume and make-up supervisors, and others who deal with house and studio services such as audience management, the assessment of fire risks, allocation of dressing rooms, and any special catering due to breaks at unusual times or the arrival of distinguished guests. The sound supervisor will then ensure that all non-standard equipment and facilities are booked and available, and that microphones and other equipment are set ready for studio rehearsal on schedule.

Planning meetings also serve a secondary purpose: as a forum where ideas can be proposed and discussed, then adopted or abandoned. The director should listen and arbitrate rapidly between conflicting aims: these must be resolved in compromise and co-operation. Sometimes the director modifies the original plan, to make things easier for all. But if, in a complex situation, the director does not do this, it is often the sound supervisor who is left with the trickiest problems to solve: many television directors think picture comes first and sound second. Sound supervisors who make light work of awkward problems are therefore valued members of any production team. Equally, those who accept unsatisfactory compromises and then explain after the programme that poor sound was inevitable will not be thanked.

At the planning meeting the sound supervisor starts to build (or re-establish) a working relationship with other members of the team that will carry on throughout the term of the production.

The television director

The sound supervisor meets various kinds of director (and producer). Those who outline their needs concisely and precisely and then stick to them are easy to work with. They will budget for adequate staff levels, or ensure that time is allowed for any overlap of function; conversely, they are unlikely to waste others' efforts or demand extra staff and resources that will not be needed. Other directors may have few of these obvious virtues, but still be worthy of all the help the sound supervisor can offer. These may want to maintain flexibility up to a late stage in preparation so they can then take advantage of elements of the production that turn out better than might have been expected, or cut losses on ideas that do not shape up. In these circumstances the sound supervisor has to plan by intelligent anticipation, expecting some waste, and accepting that the means can be justified by results.

Inexperienced directors may need more help from the sound supervisor, but might be persuaded toward simpler solutions so that things are less likely to go wrong. Then, when they do, the problem is relatively easy for everyone (including the director) to cope with.

For some productions the sound supervisor sees a script before the planning meeting, and will already have questions for the director – who will probably be ready with answers without being asked. The director may start with an outline of content and its intended treatment, and will show a plan (perhaps even a scale model) of the studio layout. The sound supervisor should pay particular attention to descriptions of visual style: a 'strong' style with fast cutting, mixtures of close and wide shots, unusual angles and shifting focus could require different sound coverage from a simple and more logical progression. At this stage the director may welcome advice on sound coverage. This might involve pre-recordings or extra expense for equipment or staff, so the director must have the opportunity to decide whether the effect being sought is worthwhile.

Before the planning meeting, the studio sound supervisor may also have been consulted on matching between material recorded on location and the studio. Mismatches of background sound or voice quality might be avoided by an early choice of appropriate microphones or technique. For example, a close balance might be possible on location, but should be avoided if it gives a bad sound cut with a related studio shot. Or specially recorded sound effects might be requested.

A running order for rehearsal and recording is also agreed at the planning meeting: here the director needs to hear whether (for example) a recording break will save equipment and staff, and so also save money. On the other hand, to bring a television studio to a halt for 10 minutes for the lack of one person could be an expensive economy. Note, however, that a sound supervisor's desire to provide perfect results without regard to cost may conflict with the director's responsibility to make the best use of limited resources.

Before they reach the studio, performers may have a period of rehearsal, towards the end of which the sound supervisor should attend and watch if time and budget permit. At this stage, scenery is indicated mainly by floor markings, but much of the action will already be set. This includes speed of movement and relative positions, and the quality of the voices and direction of projection. It should be clear whether any already-planned sound coverage is likely to work in practice and also how it may be affected by evolving plans for lighting and camera movement. Any substantial consequent changes in sound coverage should be discussed with the director and specialist departments, and entered on the studio plan. It should also be possible to visualize how equipment such as property telephones or acoustic foldback will be used in the studio, and get a feel for the timing of speech and action relative to music and effects. Sometimes recorded sound is taken to the rehearsal and tried out with the performers.

Rehearsal-quality sound equipment used outside the studio, for example to work up sequences mimed to sound replay, may be collected from the sound department and operated by an assistant floor manager – but if a sound assistant rehearses its use and timing with the performers, this may save studio time later.

The television designer

The television designer produces a scale plan of the proposed studio layout, to which symbols for booms, slung and stand microphones, and loudspeakers will be added by the director after their positions (boom tracking lines, etc.) have been discussed with the sound supervisor.

Rostra, ceilings, doors, archways and alcoves are all accurately marked on the plan, as are features such as simulated tree-trunks or columns that may obstruct a boom arm, or areas of floor covered by materials that make it impossible for sound equipment to follow. Steps and upper working levels are also shown. At this stage the plan is not final: a ceiling may be redesigned using an acoustically

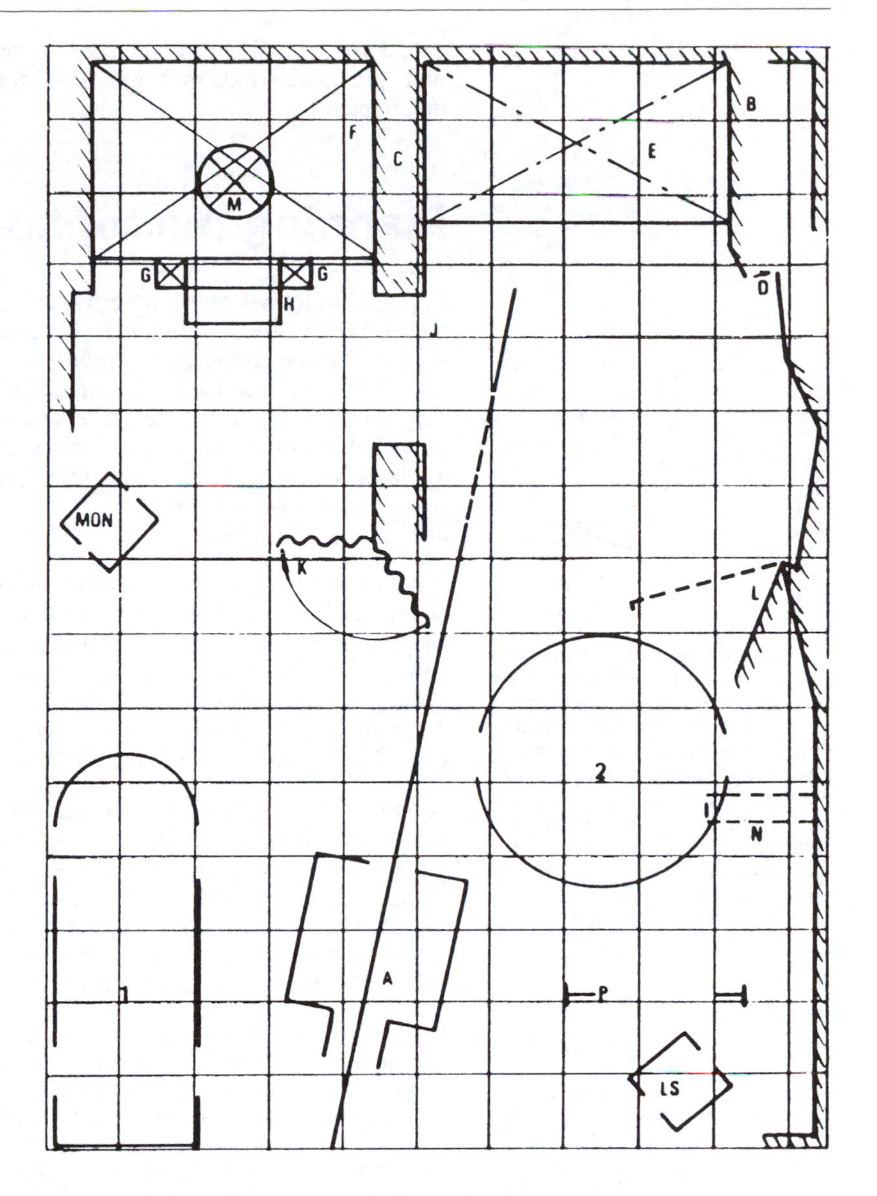

[20.1] **Studio plan**
Corner of set in a television studio. 1, Camera One: a small motorized tracking camera (other cranes require much more space). 2, Camera Two: a pedestal camera, the circle represents the working area it requires. A, Boom, capable of further extension (to reach beyond the back wall of this set if required). M, Slung microphone. MON, Monitor. LS, Loudspeaker. P, Lighting barrel overhead (one of many that will be shown all over the plan). Details of set: B, Single-clad flats. C, Double-clad flats. D, Door. E, Ceiling over (broken lines). F, Rostrum (height will be marked). G, Columns. H, Steps. J, Arch. K, Drapes on gallows arm. L, Swinger (a flat that can be pivoted in and out of position). N, Bracket.

transparent material so that a microphone can 'see' down from above; a tree may be reduced in height so that a boom can swing over it; or a potted plant may be strategically repositioned so that it can bear improbable fruit in the shape of a concealed microphone.

Certain types of set reflect and therefore reinforce sound: this is particularly true of flats made of wood. These may help musicians or other performers, but can cause trouble if arranged in a curve that will focus sound (whether this be close dialogue or distant noise). Floor and other surfaces used in television are often painted or made up to look other than they really are. Consequently, footsteps may be audible when they should not be heard at all, or for which the natural sound is inappropriate to picture. Here again the designer can help sound by the suitable choice and treatment of surfaces. Together with practical wind, rain and fire effects, this has been described in Chapter 15.

Planning microphone coverage

In planning television sound coverage, at one time the first question was often: can it be covered by a boom? But as studio-based dramatic productions have given way to location recordings that are shot more like films, the need for booms is reduced, and the skills employed in their use get less exercise – a pity, because booms still offer the most versatile coverage in a studio – as they also do on a film set. Static or 'personal' microphones (often as radio microphones) are, of course, widely used, and for many productions may also appear in vision.

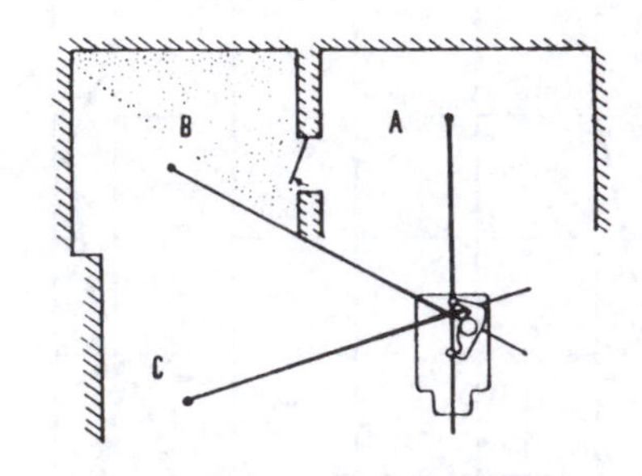

[20.2] **Boom operator in difficulties**
The boom is positioned for use in area A. An operator who is asked to swing to area B is blind to all action that is obscured by the wall, and so cannot judge the angle and distance of the microphone. If the action continues to area C without movement of the platform, only the boom and a toehold will support the operator. Alternative provision should be made for coverage of B and C before the position for A is decided: this must be planned, as cameras and lighting are also involved.

Watch out for minor changes in a script that have a big effect on coverage. For example, it may seem reasonable to plan for speeches by performers who are separated from each other to be covered by a single boom because the script appears to allow time for it to be swung. But suppose that at a late stage it becomes apparent that, dramatically, the timing is wrong and must be cut? Plainly this kind of problem should be anticipated.

A boom may require no operator (if it is simply placed in relation to a performer who does not move), one operator (for normal operation) or two (if it must be tracked). The sound supervisor must ensure that an operator can always follow the action, and is also responsible for the operator's safety: as the boom swings the operator must not run out of platform to stand on – with some types this can happen all too easily. Positions and tracking lines must be agreed with cameras.

It is also essential that the angle of the boom is agreed with lighting, so that shadows are not thrown on to performers or visible parts of the set. To avoid these, it is necessary to understand a little about film and television lighting (much of what follows applies to both).

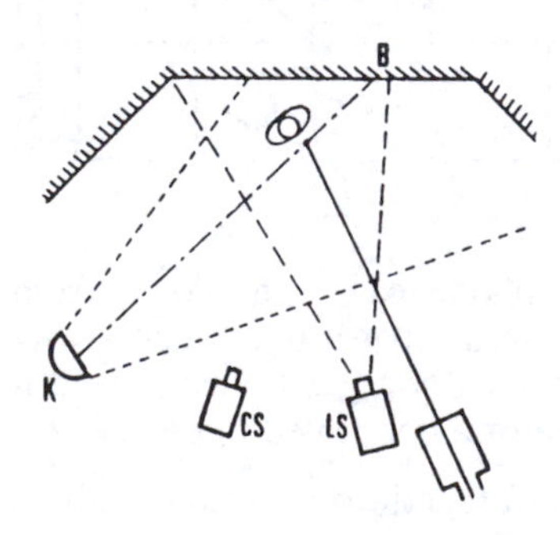

[20.3] **Lighting, cameras and microphone: problems**
K, Key light. CS and LS, Cameras taking close and long shots. The performer is too close to the backing and the microphone has followed in an attempt to get good sound: a shadow is visible at B in the long shot.

Performers are generally lit by two frontal lights. One, the *key*, is the main light: it is the strongest, and casts sharp-edged shadows. The second frontal light is the *fill*, and ideally this should be from a source that has a large area (e.g. an array of relatively weak lamps) so that it does not cast a hard-edged shadow. In addition there is often rim lighting from the rear to outline the figure and separate it from the scenery (which is lit separately). For portrait close-ups, the camera and the key light are not widely separated in their angle to the subject (but will be far enough apart to give satisfactory modelling to a face). If the boom also comes in from this side, it might cast a shadow on the subject's face. So the first rule is that *the boom should come in from the side opposite to the key light* (perhaps at 90–120° to it).

Provided that the same camera is not used for wide-angle shots, there should be no danger of the shadow being seen against the

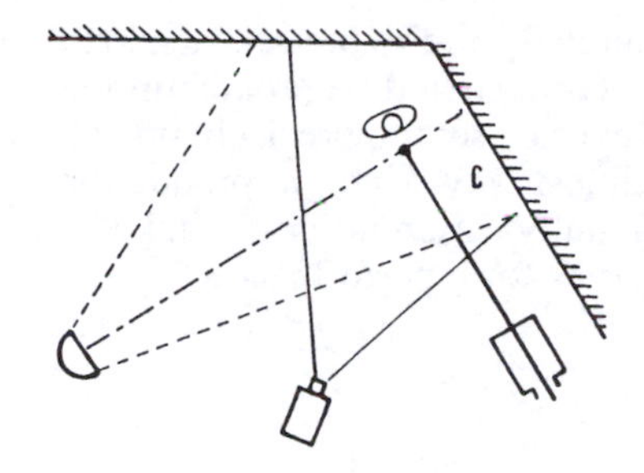

[20.4] **Boom shadow**
C, Boom close to a long wall. There is a danger of a shadow along the wall on wide shots. If the boom is very close, even soft 'fill' lighting will cast a shadow.

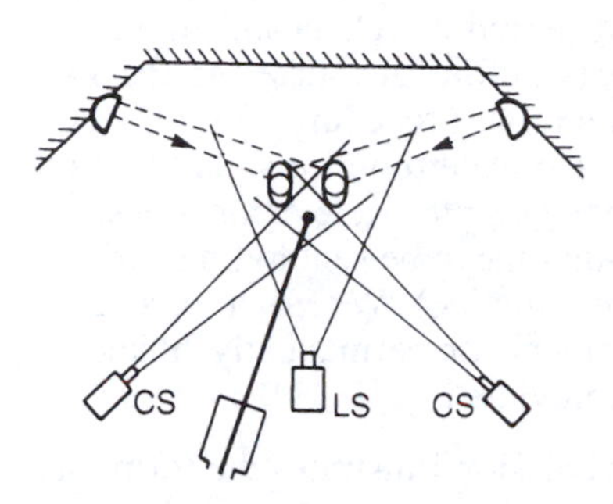

[20.5] **Rear key lighting**
The two people facing each other are both satisfactorily lit, and there is no danger of microphone shadows as these are thrown to the front of the set.

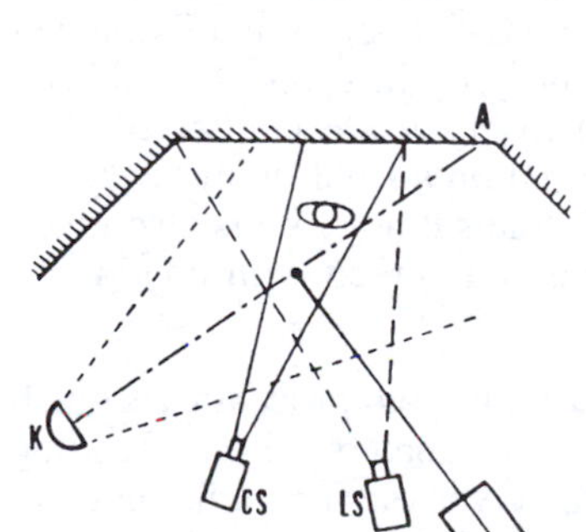

[20.6] **Lighting cameras and microphone: the ideal situation**
K, Key light. Cameras: CS takes the close shot, LS the long shot. The close-shot camera has near-frontal key lighting. The microphone shadow is thrown on to the backing at point A, where it appears in neither picture. Few cases are as simple as this.

backing. This leads to the second rule: *long shots should, if possible, be taken from an angle well away from the key light*. Then, if the boom comes in from roughly the same angle as the long-shot camera, the shadow is more likely to be thrown out to the side of the shot, not into it.

In television, lighting often has to serve several purposes in order that a scene can be shot from a variety of angles. A light used as a key for one performer may be the rim for another facing in the opposite direction. As scenes become more complex in their camera coverage, what would be easy in shot-by-shot filming may become tricky in television, and require compromise. A third rule (which generally makes things easier) is that *plenty of separation should be allowed between the main action and the backing*, on to which a boom shadow could easily fall. This is, in any case, desirable for good back lighting.

As a corollary to this, note that the boom arm should not be taken close to a wall or any other vertical feature of the set that appears in vision. If the arm is close enough, even soft fill lighting begins to cause shadows. Slung microphones may also cause shadows, especially if they hang close to a wall. Sometimes there is no convenient alternative to having a microphone shadow (perhaps from a fill light) in vision: it may be the only way to get close enough for good sound. If so, arrange for the shadow to fall on a point of fussy detail rather than an open, plain surface, and keep it still.

In the planning stage, if action seems to be getting too close to the backing, the lighting and sound supervisors are likely to join together in an appeal for more room to work. The director might then change the action or timing of the words, move action props away from the wall, or arrange with the designer for the working area to be enlarged.

Preparations for rehearsal

During the first run-through of a complex programme the sound supervisor (or 'studio manager' in BBC radio) glances ahead down each page of script, quickly estimating what will be needed, to ensure that the right microphone channels are open and to get the feel of overall and relative levels, and such operations as fades and mixes. At this first attempt the results are rarely exactly right, but each operator quickly notes anything out of the ordinary in some personal shorthand: it might be the level of a recorded effect or the setting for the first words that are to be heard in a fade-in. In radio, extra words may have to be written for a fade, or an adjustment made to a performer's distance from a microphone: such questions are referred to the producer. In television, the sound department is more inclined to try to solve problems with the minimum of reference to the director. There, the difficulties can be more severe, often raising the question of how to get adequate sound coverage without a microphone getting in the picture or throwing shadows.

Difficult sequences or unusual techniques require pre-rehearsal and perhaps pre-recording. Awkward material should be broken up into tractable segments. Nothing should ever be set too precisely prior to a recording or transmission: in almost any kind of production, emphases and levels may be substantially different on the 'take' as participants let loose their final reserves for a performance.

Line-up

In any studio session, time must be allocated for *line-up*: this is the check that, for a standard signal at the start of the chain, all meters read the same, showing that nothing is lost or added on the way. If extra equipment is brought into the system, this too must be lined up.

What this means in practice is that when a separate recording session or broadcast transmission follows a period of rehearsal, time must be scheduled for further line-up tests before the 'take' or the live show. Television line-up (involving similar procedures for the video signal) lasts longer, so there will be adequate time for sound line-up within that. (However, in the more complex television consoles provision can also be made for line-up tone to be sent to line without disturbing normal sound rehearsal circuits.) Where a recording machine is part of the studio equipment, or permanently linked to it, the line-up procedure may be abbreviated.

The procedure for a BBC radio transmission illustrates the form that such routines may take. At the end of rehearsal, and a few minutes before the scheduled air time, continuity calls the studio manager on the *control line* (an ordinary telephone circuit). Details of the recording are checked, and the equipment is lined up using zero-level 1 kHz tone, which is sent from the studio. This reads '4' on a mono or 'M' signal PPM or 3 dB below this on the stereo A and B channel meters. The circuit is checked by identifying it, using the talkback key. Cue words for the transmission to begin are confirmed. The previous, on-air programme is monitored on a small loudspeaker in the console and fed to any presenter in the studio, who is also given confirmation of the cue and, if known and agreed with continuity, the expected start time.

To make a recording in another studio the line-up procedure is much the same, but may include sending '*level*': someone in the sending studio is asked to read a few lines. This gives the far studio an extra check on quality. For an interview there will be a check that voices both ways are reaching everyone who should hear them.

If a recording is to be made at the far end, after a final check that everyone is ready, the studio manager might say over studio talk-back (to be heard at both ends), 'We'll be going ahead in 10 seconds from . . . now', then switches on a red light to warn people outside the studio that a recording or transmission is in progress (this also changes the state of the console so that talkback and monitoring circuits are re-routed), fades up the studio (if necessary), flashes a green cue light – and the recording is under way. For an interview the arrangements may be less formal, simply ensuring and confirming that the recorder is running, switching on the red light and giving a verbal go-ahead.

For a transmission, once the programme is on air, the studio manager switches back to studio output. Since at every stage between studio and transmitter everyone is listening to their own output, the source of any fault can be rapidly located. The studio is switched back to network output just before the end of the transmission.

In television line-up, the video and sound lines are tested (again the BBC uses zero-level tone for sound line-up). At the start of a continuous ('as-live') recording a video clock giving production details is shown for 30 seconds or so prior to the start of the required material. (This is used for transmission line-up, and may be added or updated at any editing stage.) On an open microphone the floor manager calls '30 seconds... 20 seconds... 10–9–8–7–6–5–4–3...' and then both sound and picture are faded out for the last 3 seconds. As an additional sound check, 10 seconds of tone (lined up to zero level at the studio console) are sent to line between the counts of 20 and 10, and this, too, is recorded as a final reference level.

At the end of a recording comes the reckoning: the duration checked and possible retakes discussed. If necessary, further cues are given and the extra material is added to the recording, to be edited in later. Apart from documentation, the recording session is now complete.

Production timing

Timing may mean either of two things: one is the artistic relationship in time between the various components of a production; the other is the tyranny of the stopwatch over broadcast durations. In broadcasting, for networking or where commercials must fit into the schedules with split-second precision, each second must be calculated. An American director once described himself as 'a stopwatch with an ulcer'.

Today, the stopwatch is probably no longer in that director's hands: it is likely to be digital, possibly as time code and perhaps in a control unit. For the purely mechanical business of getting programmes, short items, music, commercials, time checks and station identification announcements on the air, many stations are now largely automated, leaving the staff formerly occupied by this work free for more useful work, such as (according to the manufacturers of one broadcast automation system) selling more air time to advertisers, or giving news stories on-the-spot coverage. Or perhaps the director still has both the ulcer *and* the stopwatch, because the segments from which productions are built must still be created, and often to a strict overall duration: automation can – so far – only build transmission schedules from material that already exists.

This preoccupation with the value of time may also remind us to tighten up our material, to make sure that every part counts. Although 'time' and 'timing' are two separate things – and often are at odds with each other – they are closely interconnected. Differences between rehearsal and transmission timings may explain a fault of pace. The relative lengths of different scenes or segments of a production may throw light on faults in overall shape. In short, the obvious and overwhelming reason for counting seconds is not the only reason.

When timing a radio item, it is usual to set the clock going on the first word and not before it – not, that is, when the cue is given – and it is stopped after the last word is complete. In this way, if timings are added together in sequence, any pause between items is counted only once. Television timings normally begin with the first usable picture.

There are various ways of marking rehearsal and recording timings on a script. It can be done at minute or half-minute intervals, at the bottoms of pages, paragraph by paragraph, or by scenes or segments or at significant points within them. Timings can also go on a running order, cue sheet or recording form (along with rough editing notes). Marking times on a script each minute makes calculation easier, but marking the time at natural junctions probably has more actual value.

If a production is being timed and an error makes it is necessary to break off, first stop the clock and mark the time in the script at the point it was stopped, even if this is after the mistake. Then as that same point is passed, restart the clock. This way, even with a substantial delay, the true duration is probably close to that noted.

Where a production consists of separate segments, as in a magazine programme, two (or more) sets of timings may be required, including one for the whole, and another for the individual items. In a live broadcast, as recorded inserts are played in, the director's assistant keeps everyone informed over talkback how many more minutes or seconds there are (useful if moves or microphone changes have to be made in the studio) and may count down the last seconds.

Another way timings may be used in radio is to give a performer the last minute (say) of air time. A stopwatch or clock is started and placed where it can be seen by the speaker, who then knows, as it ticks up to the minute, exactly when to stop.

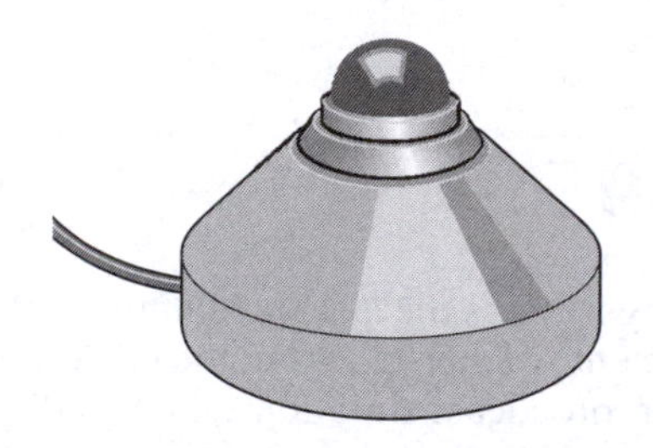

[20.7] **Cue lights**
Top: The table cue light with its massive, stable base can easily be moved to ensure that it will be in the eye-line of the speaker.
Bottom: Wall-mounted cue light.

Cueing methods

For cueing, there are two principal systems of communication between director and performer. These employ either cue lights (good in radio) or hand signals (used universally in television).

Many studios are equipped with green cue lights, operated by switches at, or sometimes by foot, under the console. A separate cue light is provided for each studio area (or table), placed where the performer can see it without turning. One green flick means 'go ahead' and is held on long enough for the least observant speaker to see it, but not so long as to hold up the person who thought you said 'when it goes off'. An alternative convention may be adopted for fast-moving news and actuality. For this the presenter may be given a steady green 'coming up' cue 10 or 20 seconds (as agreed) before the word cue is expected, and goes ahead when the cue light is switched off.

Some hand signals are also widely used: the equivalent hand signal for 'go ahead' is to raise the arm for 'get ready' and then drop the arm, pointing at the performer.

For radio talk, cue lights are also used to help regulate timing and pace. A series of quick flicks means 'hurry up', and longer steadies mean 'slow down': A mnemonic for this is 'quicker-flicker, slow-glow'. The equivalent hand cues are: hand and forefinger rotating briskly in a small circle for 'speed up'; while a hand, palm down, gently patting the air means 'slow down'. These are both used even when cue lights are available. In television the director calls out cues by production talkback to the floor manager, who signals them to the performer.

In discussions the moderator may wish to receive these signals and others for precise information about timing, e.g. two steady lights at 2 minutes before the end, one at 1 minute to go, and flicks at 30 seconds. To give the signals by hand the director must stand where clearly visible, and then indicate 'minutes to go' by holding up the appropriate number of fingers, and 'half' by crossing the forefinger of one hand with the forefinger of the other.

Other hand signals sometimes used are: pulling the hands apart repeatedly for 'stretch' (but have care, as it might also be taken for 'move back from the microphone'); moving the hands together may be used to indicate 'get closer' – either to the microphone or to another speaker; waving arms across each other with palms down, in a horizontal plane, can be used for 'stop'; and the side of the hand drawn across the throat predictably means 'cut'.

Crowd noise in a television studio may be waved out, but in radio a series of quick flicks may be used to indicate 'the studio is faded out: everybody stop', and can be followed quite soon by the single flick for the start of the next scene or sequence.

There is one special case in radio: when the speaker is blind. The director, or an assistant, can stand behind the speaker's chair to give cues by laying a hand on the blind person's shoulder or arm. Complex cues can be given in this way if arranged beforehand.

Cueing applause

The management of audience applause is important to many radio and television productions.

It is possible that a studio audience will wait politely for a speaker to stop, and then respond appropriately, but rather late. If they are uncertain whether to interrupt or not, applause may be delayed or may seem or actually be half-hearted. The audience may even completely fail to applaud (particularly if the performer is inexperienced in this branch of stagecraft). For any of these, the effect is lack of warmth and failure to generate the feeling of an event with unity of location and purpose. As a result, there is nobody that the home audience can identify with, to feel vicariously that they, too, are taking part.

At the other extreme, the studio applause may be excessive to the extent of breaking the flow of a performance. If applause is injected one sentence before the best point, then the true climax may not be applauded at all, and might as well be dropped, or it may get reduced applause, so it becomes an anticlimax. If the reaction holds up the action it may alienate the home audience. Excessive applause may be inappropriate in its context, or simply unsatisfactory for sound – for example, when young people stamp their feet on the wooden or metal rostra that support the seating in many television studios.

In some quarters, the overt management of applause is viewed with caution. It is, however, necessary where the presentation becomes clumsy without it; it is necessary in the case of small audiences of, say 100 or less, which may lack enough of the people who help to start a crowd reaction; and it may be useful where 'atmosphere' requires that applause continues unnaturally long, such as at the end of a show, behind closing music and titles or announcements, until sound (and, where appropriate, vision) finally fades out. The management of applause is right and proper where it leads the local audience to react in places where reaction would in any case be natural, where such applause helps to interpret the material to which it is a reaction, and where it helps by pacing a production and allowing the home audience time to absorb and assess what has gone before.

The behaviour of the audience may be partisan, either in the sense of being composed of admirers of one performer, of a participating team, or of a political party: for any of these, it is reasonable to show the outward trappings of an existing genuine relationship. But for all of them, applause should be handled with responsibility towards both truth and intelligence.

Appropriate signals (in television from a floor or stage manager) might be as follows:

For 'start applauding', the stage manager stands in clear view of the audience, raises hands high and starts clapping (not too close to a live microphone). A folded script in the hand may help to draw attention.

For 'continue applauding', the same action may continue, or be replaced by waving the paper in a small circle, still well above head height (this adds variety to the stimulus and allows the other hand to be used for cueing). Both hands high, palms inwards, will also keep it going. But turning the palms outwards will bring the applause down, and waving the hands downwards at the same time, or perhaps more vigorously across the body, should bring it to a stop.

Other signs may be used: the only requirement is that the audience understands and follows. Avoid the ludicrous 'applause' placards that are sometimes seen in old films. It is better to lead an audience than to direct it. For natural, warm applause it is often best if there is a slight overlap between speech or music and the applause. This implies careful timing and clear understanding by the sound supervisor about whether the last words must be intelligible or whether it is permissible for them to be partly or completely drowned. It must also be clear whether the first words of a new cue starting over applause require moderate or high intelligibility: ideally, for 'warmth', it should be the former, with redundant information in the words spoken until the applause has almost completely died away.

The 'warm-up' is an essential part of any production with an audience. Often the sound department plays records of a suitable style until the warm-up proper, which may include information about the production and the audience's vital role in it, and, for a comedy show, 'warm-up' jokes.

Documentation

After a recording is complete, various details must be written up for later reference, either on paper or (within an organization that has a system for attaching data to audio and passing it on from one device to another) as digital metadata that always stays with it. Some of these details are technical responsibilities and some are dealt with by production. Typically, they may include:

- Production title plus any reference or costing number.
- Date and place of recording.
- Name and address of production company or owner of material.
- Any broadcast transmission information, with contact details for whoever is responsible at the time of broadcast.
- Name (if necessary, with contact details) of recordist.
- Technical notes. Recording reports will include:
  - level of line-up tone;
  - format: mono, stereo, surround, Dolby 5.1, etc.;
  - roll, reel or disc number if more than one (e.g. 'reel 2 of 5');
  - recording and other equipment used, including identification numbers where possible: this helps trace faults back to source;
  - any noise reduction process used;
  - whether CD is normalized or not.
- List of contents with 'in' and 'out' cues and durations, plus details and timings of any technical imperfections.
- Names (and, where necessary, addresses) of participants.
- Music details – record, composer, writer, performer, copyright, etc.
- Notes on material that is not original.
- Timed script or cue sheet, with editing notes and stage reached.

- Any other information required for costing.
- Provision for adding details of edits and later history.

From this list, note particularly the need for precise timings (of the whole work and of each segment), clearly identifiable 'in' and 'out' cues, and level of line up (reference) tone. These are easily forgotten.

In addition to this, write summary notes on the box (e.g. on the back cover) and separately on the label of any recording. The main things here are reference number, reel number, technical information, duration and date.

On the spine of a box put the title and identification and reel numbers.

When no further editing of a work is needed, this must be absolutely clear. A completed work (ready for transmission or distribution, etc.) must be signed off by the person responsible.

Radio and television stations have to 'report' output as it is broadcast: this is needed in order to pass information to copyright owners and collection societies that the organization has an agreement with, and to calculate fees that may be due. In America a log is also required by the FCC: this can be used, for example, to check the total duration of commercials per hour or the proportion of sustaining (i.e. unsponsored, public service) material in a schedule. Since compiling this information benefits no one at the station directly, it is not surprising that even its storage space is grudged. However, stations with digital archiving or broadcast automation systems may be able to log details automatically from the attached metadata: this takes very little storage space. Better, a full archive of digital audio recordings of original material plus associated data is a valuable, instantly reusable resource that makes modest demands on space, and a log can be taken from it on demand.

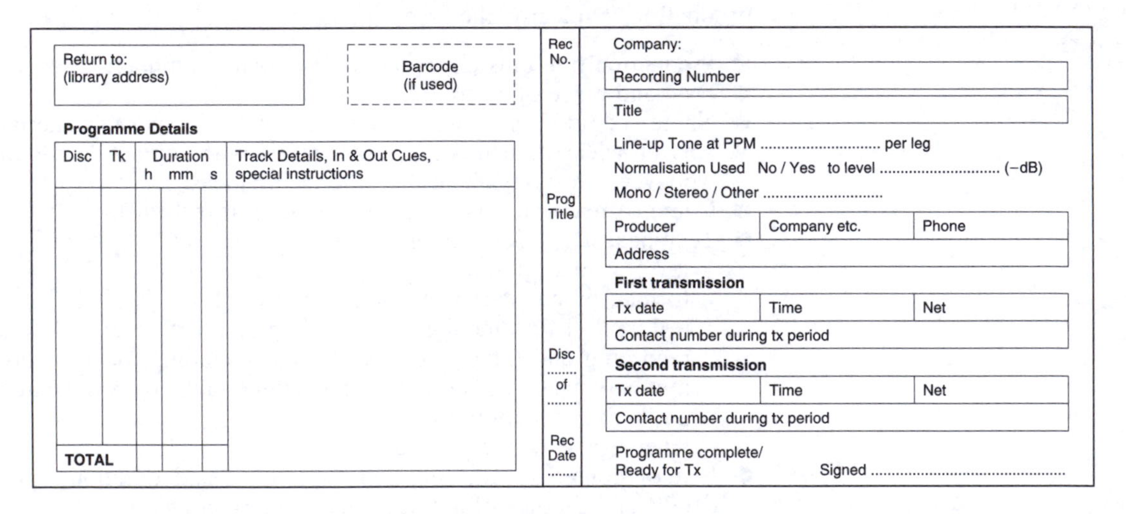

Return to:
(library address)

Barcode
(if used)

Programme Details

Disc	Tk	Duration h mm s	Track Details, In & Out Cues, special instructions
TOTAL			

Rec No.

Prog Title

Disc .... of ....

Rec Date ....

Company:

Recording Number

Title

Line-up Tone at PPM .... per leg

Normalisation Used No / Yes to level .... (–dB)

Mono / Stereo / Other ....

Producer	Company etc.	Phone
Address		

First transmission

Tx date	Time	Net
Contact number during tx period		

Second transmission

Tx date	Time	Net
Contact number during tx period		

Programme complete/
Ready for Tx Signed ....

[20.8] **Recording report**
A CD recording report with details for front (right), spine and back of case. A further sheet shows 'Programme History' with boxes for the original recording session and mix/edit sessions, each with date, studio/channel and comments on recording quality and equipment used. A final box shows final mastering details: date, studio/channel, write speed, CD burner type and burner ID number.

Chapter 21

Communication in sound

The main subject of this book is audio techniques. But – just how important *is* technique? Some of what follows is my personal view, but it seems clear that we do need to see technique in perspective.

It is essential that the mechanics of sound production should be seen in their true relationship to the full process of creation; that they should be given neither too much deference nor too little. Plainly I think technique is important, otherwise this book would not have been written. But technique is a means to an end, not an end in itself: and its major object must be to concentrate the audience's attention on the content, and to do so in a particular way, with particular emphases. In such a supporting role technique is indispensable.

When combined with picture, in television or film, sound should always support, occasionally counterpoint, and rarely dominate; its most effective role is to extend, strengthen and more closely define the information that comes first through vision.

It is relatively easy for a skilled professional to make technique serve the programme; work is rarely done just to display the operational ingenuity of those involved. But for the newcomer, much of that work must necessarily also serve as exercises in handling sound. To novices, technique is a vital need, perhaps even their most vital need – because they may have no lack of ideas, but little or no knowledge of the means of expressing them. They must learn the techniques, not merely of construction, but also self-criticism and correction. Once the techniques are mastered, they must learn to apply them, to make them serve the needs of each successive production. The medium itself is *not* the message.

What makes a good production? The answer is of importance not only to the writer or director, but to everyone in a studio. It concerns them all in two ways: firstly, so that they can see their own contribution in perspective, and to know in what ways this must be subordinated to the whole; secondly, as their creation's first audience.

Over the years, people in daily contact with a medium become sensitive to virtues or defects that even the director may be unsure of, and this front-line audience can be of immense value in cross-checking the director's calculation or intuition.

The things that matter most in any programme are the ideas within it (its content) and the way they are presented to the audience (its form). Some subjects are better for sound than others, because they make contact more easily through the listener's imagination. Others are more suitable for television or film, demanding the explicitness or possibilities for visual imagery that those media give.

Many subjects can be treated adequately in any medium – adequately but differently, because the medium dictates certain possibilities and limits others; or it presents alternatives, where one path presents only difficulties, while another luxuriates in opportunities for attractive or artistic results. To this extent at least the medium does condition and reflect itself upon the message.

But so far as technique is concerned, there is one major condition that, in conventional (if not in modern experimental) terms, must be fulfilled: whatever the choice of subject, it must be presented in such a way as to create interest and maintain it throughout the production.

Sound and the imagination

The greatest advantage of the pure sound medium lies in its direct appeal to the imagination. A child once said of radio that, compared with television, 'the scenery is better'. This may not be a conventional view, but it is true that in radio the scenery is built in the mind of the listener. It is magic scenery, which can at times seem so solid that you feel you could bang your head against it, but in a moment might dissolve into an insubstantial atmospheric abstraction. It is scenery populated not just by people but also existing as a place where ghosts, ideas, emotions or impressions can be conjured up as easily as the camera can show reality.

Well, perhaps it is not quite as easy as that, because sound does require a greater contribution from the listener. To compel this involvement and also to deserve it, the material coming from the loudspeaker must maintain a high level of interest throughout. If it drops for one moment, the listener's imagination is turned off like a lamp, and all real communication is immediately lost. The magic scenery is built and painted by the artists and craftsmen in the studio, but it is illuminated by the imagination of the listener.

The ways in which imagination can be stimulated are many. To some extent they depend on form, which depends in turn on techniques. Apart from this, what is done, rather than the way it is done, is a matter for individual judgement or intuition.

In an analysis of the qualities of any programme, we should perhaps take 'interestingness' as the starting-point, because the one factor that all successful productions have in common is the ability to engage and retain the interest of the audience. For a start, it is clear that interest depends on previous experience. What we know, we may like; what we do not know is more likely to leave us relatively indifferent – this is true even of most people who regard themselves as being 'interested in anything', for very few people indeed are interested in things that cannot be explained in terms of what is already understood.

The reason for this is simple enough: any reference to something within a listener's experience can call up an image in the mind. When we listen to someone talking, we may soon forget such superficial qualities as accent and voice timbre (unless we are forcibly reminded of them at any stage), nor do we listen closely to the actual words; instead, we go directly to the meaning. The words are forgotten almost the moment they are spoken; the meaning we may retain. This meaning, together with its associations in the listener's mind, forms an image that may be concrete or abstract, or a mixture of the two. And if it is possible to present a subject in terms of a whole series of such images, a considerable amount of interest can be aroused.

The audience

From the start, a radio or television director must think about not only what is to be said, but also who to: who is the intended audience? This is not merely a matter of their intelligence, but also of what sort of background they have; what they are likely to have learned at school and in later life; what sort of places they know and people they might have met or seen in the media; what sort of emotions they are likely to have experienced. It is also related, most powerfully, to the things individuals within that audience might like to do, consciously or not – ranging from taking up a life of meditation and prayer to winning a major prize for saying the Eiffel Tower is in Paris.

In practice many programmes develop a distinctive personality to accord with characteristics of the audiences at which they are aimed, and may be further modified in style according to the time of day when they are transmitted. But, all in all, the dominant factor in programme appeal is the previous background and knowledge of the listener; and failure to take proper account of this is one of the most serious mistakes any director can make – if the aim is communication. We are not really concerned here with works of art where the artist is *not* interested in communication, and this is not to decry them. But if they are works of importance to anybody but the artist, then teachers and critics will inevitably busy themselves with explaining and relating until, in the end, communication between artist and public is re-established, just as though it had always been intended.

In one investigation into the effect of educational broadcasts, the main issue was intelligibility, but sidelights were thrown on 'interestingness'. It was noted that while too many main 'teaching points' (say six or more in a 15-minute programme) were bad for the programme, the presence of what sometimes seemed to be an unduly large number of subsidiary points did not seem to result in confusion; and also that whereas long, complex sentences with vocabulary that was difficult for the target audience, and also lots of prepositions (used as an index of complexity of construction), both did reduce intelligibility a little, the presence of numerous adverbs and adjectives did not.

It seems reasonable to conclude that an apparent excess of facts, figures and descriptive terms is no disadvantage, because although they may not be assimilated, they help to maintain interest during the necessary gaps between the essential 'teaching' points. Apart from any other function they may serve, these are the things that provide 'colour'; they are part of that vital series of images that the listener's imagination needs. Remember here that what is true for educational programmes (the immediate subject of this study) is also likely to be true for a wider range of material. On the larger canvas we can take 'adverbs and adjectives' to include any descriptive, or even simply amusing, piece of illustration in sound or picture.

In that survey, it was also found that concrete subjects were preferred to the abstract; and, when ideas were explained, concrete examples – references to people, incidents and actions – were preferred to 'clever' metaphors. This, too, must be a more general requirement. For apart from the genuine absent-minded professors of this world – the people for whom the world of the mind is more real than the world of reality – there can surely be few people for whom abstractions (except those dealing with the emotions) are more telling than are concrete references. Indeed, one danger was noted: that well-developed illustrations can sometimes be so effective as to attract attention away from the main point.

If there's no picture – where is everybody?

Some problems in communication apply to one medium more than another. Before we turn to the more general case, let us consider one or two that apply particularly to the pure sound medium.

We have noted that an advantage of sound lies in its direct appeal to the imagination, and that the really successful programme has one quality above all else: the ability to conjure up an unbroken succession of powerful images in the mind of the listener. These are not just imagined visual images, there to replace the concrete visual evidence that the eye is denied: they are much more. Nevertheless, we must now recognize that some of the limitations and disadvantages of sound do stem from the confusion caused by its lack of exact pictorial detail.

In ordinary life we take this detail for granted. When we take part in a conversation with half a dozen other people we are never in any doubt as to who is speaking at any time; our eyesight combines with our faculty for judging direction aurally, and no thought is necessary. Indeed, we would have no difficulty in coping with almost any number of people; the ears localize and the eyes fix on the speaker immediately.

Without vision, the number of voices that we can conveniently sort out is much reduced. In the case of stereo sound, half a dozen people, spread out and not moving around much, should be fairly easy to place; but with more people (or more movement) the picture becomes confusing. When this happens, imagination fails: the voices no longer emerge as identities. Then, instead of generating a mental picture of a group of individuals, we register the sound as a quantity of depersonalized speech coming from various points in space.

In mono sound the picture is even more restricted: we have room for only three or four central participants in any scene at any one time, and even then the voices and characters must be very clearly differentiated. Additional voices may come and go, provided it does not matter if they do so in a fairly disembodied manner. The butler who comes in to say 'My Lord, the carriage awaits' does not count as an extra character to confuse the ear; he is little more than a sound effect. Apart from such peripherals, the basic group must be small.

It must also be remembered that from the moment people on radio fall silent, the audience begins to 'lose' them. Leave them out of a conversation for long enough and they become invisible, so that when one of them speaks again that sudden re-entry can come as a surprise; it seems to leap in from nowhere.

From this, we already have one marked restriction on subject and treatment. Beware: any feature that involves many-voiced scenes is in danger of sounding vague. From a discussion programme involving half a dozen speakers the listener will probably identify only about two or three clearly. The extreme case is an open debate: this will be merely a succession of voices, though it will help if each speaker in turn is named and labelled. For clarity this must generally be done even when a voice has been heard before.

Often, of course, it does not matter to the listener that voices are not identified individually; but the director must know what is intended – and also what has been achieved. Otherwise the audience may be living in a place of shadows, while the artist thinks he or she is communicating a world of imagined substance.

The next difficulty with sound is that of establishing location. This must be fixed as early as possible, and then reaffirmed by acoustics, effects and references in the text – anything to prevent the scene slipping back into a vague grey limbo and becoming once again no more than a set of voices from a box.

On the other hand, it often helps if the listener is not unnecessarily burdened by pictorial detail. During a television talk, the eye is captured by the picture but may wander within it. It may shift to the clock on the shelf behind the speaker; the shape of a lampshade or the pattern on wallpaper may distract it. Without all that – given sound alone – the speaker becomes part of the atmosphere of the listener's own private space.

These weaknesses and strengths must be seen before the treatment of a subject is chosen; it is important to allow for the frequent reiteration of locale in the scene of a play, or the formal disregard of surroundings in a straight talk.

Plainly, it is impossible to establish one picture, and one only, in the mind's eye of every single listener. So it is wise to avoid subjects that require accurate pictorial detail, or treatments where the impact of a thing that is seen has to be explained in dialogue: in fact, any stories that depend heavily on accurate visual information. As an example of this, imagine the first meeting of boy and girl in a crowded room: eyes meet, fall, and meet again. It needs a Shakespeare to express verbally the magic of such a moment, and even in *Romeo and Juliet* the visual element adds immeasurably to the impact. Suppress it, and the shape of the whole scene is altered.

This example also illustrates another point: that direct emotional effects are best expressed visually. The close-up of a human face, and especially the eyes, can touch an audience directly. In terms of pure sound, using only the voice, parallels are not easy to find.

The emotional link between screen and viewer can be strong. Vision is the dominant faculty, and it must be either fully engaged or completely suppressed if each individual in the audience is to identify with a performer. Through film or television, the link is simple, direct and potentially compulsive. In sound only, the link has to be established indirectly but no less firmly through the imagination. The results may be as effective, but they are far more difficult to achieve.

The time factor

As a medium, sound has much in common with print. Both printed and spoken word can be presented as straight argument (or narrative) or as dialogue. There are comparable forms of illustration: the sound effect and the graphic display. And further, because of the restriction to a single channel of communication, neither medium can engulf the senses without first making a strong appeal to the imagination. The analogy is so strong that it is worth considering where the differences lie: the more so as there is inevitably a tendency to assume that what can be done in print can equally be done in sound.

The great difference lies in the dimension of time. The printed page leaves the time element in the hands of the reader. You can read at any speed you like: you can read a novel in a few hours, or take months over a textbook; you can read one paragraph quickly and linger over the next. You can also stop at any time and go back, when the meaning of something is not clear at first reading. Again, in the process of reading, the eye is able to run on ahead in order to see how a sentence is shaped, and so take in its meaning as a whole.

For the radio (or television) audience the time relationship between successive elements of a story is fixed. A recording may be stopped and reviewed, but this disrupts the flow more than for print. A listener, having decided *when* to play a record, has little control over any other aspect of time: for technical reasons, the effect of speed on pitch, it cannot be played faster or slower than was originally intended.

To avoid ambiguity and aid understanding (if that is the intention), ideas must be presented in clear logical sequence. It is not just the order of the arguments that must be clear; the style in which they are presented must itself be lucid and lively. If possible – that is, if the subject matter does not militate against it, and sometimes even if it does – speech should be grammatical, as well as vivid in expression. Realistic dialogue, with its broken sentences, repetitions, fractured logic and other inconsistencies, is often regarded as essential. But if it is used carelessly it can also confuse the audience: when this happens the listener will hear not meanings but only strings of words, and particularly any imperfections in the way they are spoken. The most frequent complaint about this sort of dialogue is 'inaudibility', even when the sound would be adequate in other circumstances. This is not to say that it should never be done, just that it should be done only for good reason, with the degree of communication calculated.

At the other extreme, in the 'radio talk' by a single speaker with a prepared script, good delivery is still important. For some listeners, an awkward speaker who is obviously reading, and doing so from material written in a flat literary style that makes no concession to colloquial speech, will seriously diminish enjoyment. The listener might reasonably ask, 'Why bother to present this in sound at all, when it is so obviously designed for print?' But this may be a minority view, because quality of delivery has very little effect on intelligibility, and does not worry most listeners any more than would any other relatively unimportant fault of technique. For example, many of the film stars of the 1930s now seem curiously stilted when we see revivals of their films, but this never prevented huge audiences from adoring them. Only if a programme is thrown completely out of balance, for example by speech that is too fast to follow, is faulty or 'unrealistic' delivery found upsetting.

Because both radio and television are essentially one-shot media, it is reasonable to demand that the basic message of a piece should be understood in a single hearing. By all means include deeper layers of meaning for the benefit of the people you hope will be shouting for an encore – or for whom your main points may be old hat. But a sensible assumption is that almost no one will hear your creation twice.

If you find yourself telling a baffled audience, 'You'll probably see what I'm getting at if we hear it through again', then you may take it that you are working in the wrong medium. Only if you recognize this will it be at all likely that anyone will *want* to hear you twice. Of course, there are many important exceptions; and it is always tempting to hope that one's own work is one of them. This is almost invariably a major error: so aim to make a major impact at the first hearing.

Understanding and remembering

Control over the rate at which ideas are presented is completely in the hands of the programme maker, who must be particularly careful not to use this power in such a way as to impair intelligibility. The temptation to disregard this can arise in two ways. One is the quest for 'pace' – but as we shall see later, pace is not achieved by over-condensing vital information.

The second factor that may influence a director is more subtle: it is the wish to present a subject in a tidy, well-shaped manner – in itself an unexceptionable aim. But, being (in most cases) literate both by nature and education, an author all too often falls into the trap of assuming that what looks well shaped in the form of a script will become so in its finished form, once such minor details as logical presentation and colloquial speech values have been attended to.

Unfortunately, this may not be the case – because it also depends on the rate at which new ideas are presented. To take an extreme case: on the printed page, a concise, elegant argument can be presented neatly, tautly, as finely shaped as a mathematical theorem – and the eye will dwell on it until it is understood. In a broadcast, such a construction is out of the question.

In order to ensure at least a moderate degree of intelligibility in this constantly progressing medium, the argument must sometimes be slowed down to what in literary terms might seem like an impossibly slack pace. The good magazine programme, play or documentary may appear in script form to have a much looser construction than a well-written article or short story: there may perhaps be an apparent excess of illustration and padding. This is necessary because it seems that the brain will accept and retain material that is new to it only if that material is presented at no more than a certain speed. Even if a listener is intellectually equipped to follow a concise, complex line of argument, there is also a need for time – time to think, which may be more time than it takes to present the bare bones of the argument – if it is to sink in.

There is a clear analogy with the teaching process. But learning – in the sense of the *permanent* retention of fact and argument – may not be the immediate result. *Repetition* plays an important part in learning: if something that is clearly understood in the first place is repeated after a few days, and yet again after a few weeks, there is a fair chance of it being retained. Nevertheless, radio or television may help in the learning process, for although very few individual items will be given a second hearing by the audience, their content may be related to what they have read somewhere or heard in a previous programme, or know already. In such a case the content of the programme stands a much better chance of being understood and making some impact.

Plainly, learning is very much a matter of chance, or perseverance on the part of the audience. But the immediate effectiveness of any particular presentation is, or should be, more predictable: the programme-maker must calculate each effect, allowing sufficient time for the audience to take each point as it is made. Certainly, this will affect the amount that remains in the mind after a programme is over, or the following day – but that is only half the story. If there is a logical argument or a narrative line, a single failure may render the whole programme incomprehensible, or substantially change its meaning. In the action of a play, for example, an understanding of points made early on is often essential to being able to follow developments at a later stage. Fail to establish a single important plot point and the whole play fails.

This is a matter of vital importance to any major broadcaster, so it is not surprising that experiments have been carried out to determine the intelligibility of programmes and the rate at which fact and argument can be absorbed.

One test was made with news broadcasts to determine how many items and how much detail within them could be recalled by an attentive listener immediately after the end of the transmission. It turned out that only a small proportion of either was retained. This can easily be checked. Listen to and at the same time record 10 minutes of news. Then, before playing it back, try to write down all the items of news in as much detail as possible (the exact words are not important; the gist will do). Playback will confirm just how much was missed.

Another way of trying this experiment is in a group, as a sort of party game. One person writes down the subject matter and timings of the items; the others just listen to the whole sequence, and then try to reconstruct it by discussion but without help from the running order. It may be instructive to record the discussion as well as the original news, and then play them both back. Such experiments are called aural assimilation tests. Marks may be awarded for major and minor points that are recalled, and subtracted for errors. Many tests of this sort have been carried out, and the marks for untutored audiences are generally low. The ear is not efficient at taking in ideas.

The intelligibility of educational and current affairs material was also studied. The aim of a series of 5-minute single-subject broadcasts was that most listeners should understand them without difficulty. This series reached a substantial audience: it was respected, enjoyed and thought to be interesting and informative. The enquiry was thorough: each talk was divided into main and subsidiary points and marks for understanding them were allocated according to their importance. The result of this study was that individual talks scored between 11% and 44%. The average was 28%.

It was quite clear that, simple as these talks already were, they could with benefit be simplified still further. If this could be done without giving an appearance of talking down – always to be avoided – the intelligibility could be increased. Direct faults in presentation were found in some cases: such things as assuming too much background knowledge; the use of unfamiliar words, jargon, flowery analogies or abstractions; poor logical construction or unnecessarily complex sentences. One of the major conclusions was that the presentation of the more important points should be better organized. Major points should be emphasized and repeated in such a way as to make it quite clear which, out of all that is said, is the important part to grasp. And a summary at the end would improve understanding further.

This particular survey was carried out for a specific programme and target audience, but would be valid for almost any type of programme (not just serious ones) and for any audience, provided account is taken of possible differences in background knowledge, literacy and intelligence.

The communication of ideas

Experience, together with the above experiments, leads to a working rule: that *not less than 3 or 4 minutes of programme time should be devoted to any point of importance.*

This is a generalization, but if it errs, it is on the side of allowing too little time: these are likely to be *minimum* times for points of average complexity. Of course, for a simple point it may be possible to get away with spending less than 3 minutes, while a complex point may need much longer to establish fully. In the latter case it may be better to break the argument into stages.

In a 7-minute item, for example, the 'working rule' suggests that it is possible to make two major points. The rest of the 7 minutes must be spent in supporting and linking those points. If a further major point is introduced in that same time, then it is likely that one of the three points will be lost. To take an extreme case, if the 7-minute piece is made to consist of seven important points, most will be lost, and the listener may be left with a frustrating feeling of having missed something. Alternatively, all that 'sticks' may be just a single thought that happens to appeal, plus the idea that this was part of a pattern that included several other points that on first hearing seemed plausible. In such a highly condensed case there can be no exact or predictable control of what the listener picks out. It may be something totally unsuitable from the point of view of the programme-makers.

One can point to Sir Winston Churchill as a broadcaster who triumphantly mastered the technique: a study of recordings of his Second World War speeches shows how he took each point in turn and attacked, analysed and presented it in a variety of ways before moving on to the next. Despite a difficult voice and an almost complete refusal to make any concession to radio (oratory is one of the 'don'ts' of the medium) or even replacement by the voice of an actor, the results remain masterpieces.

It seems that all broadcasters who want to get something over to their audience must follow the methods of a schoolteacher – though not too overtly, because unlike the teacher, they have no captive audience. A good teacher, at whatever level, aims to set out a subject in just this way. However, in the classroom the pace of delivery is subject to immediate correction by the pupils. In contrast, when speaking to an invisible audience, there is a feeling of pressure to 'get on with it', because everyone is behaving as though air time is valuable (which it is) and because there is often more to say than time to say it.

This rule about good layout is more obviously valid for programmes that seek to educate and inform, but it is also true for most of those whose outward purpose is solely to entertain, and especially those with a story to tell.

Pace

The term *pace* has already crept into this discussion, and it is worth thinking about what that word really means.

To maintain interest, a story or an argument must be kept moving at a fairly brisk rate. The secret of pace is *to allot just enough time to each point for it to be adequately understood* – and then move on. Pace is impossible without intelligibility. It is perhaps odd that it should depend in part on not going too fast. But there it is: pace depends on precise shaping of the programme in time, and it provides the basic foundation on which the finer details must be built.

When we talk of lack of pace, we are often thinking of another, secondary, defect which frequently afflicts the work of those without access to skilled critical analysis: a superficial slackness that follows from a poor sense of timing. But it is the other more subtle fault, lack of organization in the underlying shape, that is more likely to mar professionally produced shows that otherwise seem slick enough.

To sum up: in the construction of almost any successful programme, we must have a series of basic units that each occupies a few minutes. Each unit should contain one major point and sufficient supporting detail to help establish it as clearly as possible. There may also be material that retains the attention of the listener in between the major points that are to be made, but which may not be of direct importance to them, and which must not actively draw the listener's attention away from them. Quite apart from anything else, this sort of construction helps to provide necessary light and shade within a piece – that is, some variation in the pressure of attack. And that again is absolutely essential to a feeling of pace in the final product.

All this affects the choice of form for the presentation of a particular subject. An argument that is by its nature concise may have to be filled out by extra material: illustration perhaps, repetition in a different way, anything. It may turn out that an extra point that seems to be creeping in may have to be cut, because it is interesting enough to distract the listener from the main line that is being followed, but cannot be allowed time for its own full development.

Certain types of construction, whether in radio, television or film, simply do not work: they contain too much argument in too little time. The cool, swift logic of a mathematical theorem, or a solid, indigestible, encyclopaedic mass of essential detail are both death to clear, intelligible construction in sound.

Of course, there are exceptions to all such rules. As a television producer, director or writer who enjoys dealing with complex themes, I never construct by rote, but instead work outward from the story, the meaning, the material itself. Often the results may be criticized as overturning some of the tenets of communication, and testing others to the limit, which at least avoids deadening uniformity or predictability. But I believe that the programmes that seem to have succeeded despite breaking all the rules are in reality those that stick most closely, if subtly, to the ideal form.

Glossary

As well as short definitions, some indication of the context in which terms may be used is given. Unless otherwise indicated, the term is usually in international use, although local variants may exist. Where usage in the United Kingdom (*UK*) and America (*US*) differs markedly, both forms are given. No attempt is made here or elsewhere in this book to enter into a study of basic electrical theory, or to describe equipment or circuits in technical detail.

'A' signal. In stereo, the signal to be fed to the left-hand loudspeaker of a two-speaker system.

AB. Stereo system in which left and right components are generated and/or processed and transmitted separately, as distinct from MS (middle-and-side).

Absorption coefficient. The fraction of sound absorbed on reflection at any surface. It therefore takes values between 0 and 1, and unless otherwise stated, is for a frequency of 512 Hz at normal incidence. *Soft absorbers* depend for their action on such things as the friction of air particles in the interstices of the material. This friction increases with particle velocity, so the absorption is greatest where particle velocity is greatest, i.e. at a quarter wavelength from the reflecting surface. Soft absorbers (glass wool, fibre board, etc.) in layers a few centimetres in depth are poor absorbers at low frequencies and are fair to good at middle and high frequencies. If an outer surface is hard (e.g. painted) and unperforated, the high-frequency absorption will deteriorate. *Membrane and vibrating panel absorbers* are, in effect, boxes with panels and cavities that can be tuned to resonate at particular frequencies or broad frequency bands. They readily remove sound energy from the air at their resonant frequencies and this is then mopped up within the absorber by various forms of damping. Such absorbers may be efficient at low and middle frequencies. *Helmholtz (cavity) resonators* respond to narrow bands and if damped internally remove them selectively.

Accent microphone (*US*). See **Spotting**.

Acoustic reproduction of effects, etc. (from recordings). A feed of pre-recorded sound to a loudspeaker in the studio instead of directly to the mixer. The sound picked up by the studio microphone depends on the characteristics of the loudspeaker and microphone, and also the studio acoustics and balance. It may be noticeably different from the original sound even if a close balance is used (often the loudspeaker is balanced at a separate microphone). Acoustic reproduction techniques may be used either to modify the sound or to provide cues when no other method (headphones, cue lights or hand signals) is possible. In the latter case, volume should be kept as low as practicable, otherwise there may be a change of quality due to acoustic coloration. See **Foldback**.

Acoustics. The study of the behaviour of sound. The acoustics of an enclosed space depend on its size and shape, and the number and position of absorbers and reflectors. The *apparent acoustics* of a studio depend on the effect of the acoustics on the indirect sound reaching the microphone, and also on the ratio of direct to indirect sound. Thus, apparent acoustics depend on the microphone position, distance and angle from the sound source (i.e. the balance), and also on the level of sound reproduction. Besides this, the actual acoustics of the studio may be modified close to the microphone by the placing of screens, etc.

Actuality. Recording of the actual sound of an event, as distinct from a report, interview or dramatized reconstruction.

A/D. Analogue-to-digital signal conversion.

ADR. Automated dialogue replacement, also called 'looping' or 'post-sync'. A short loop or long digitally recorded sample of the original picture is replayed repeatedly so that new dialogue can be rehearsed in order to attempt to match the original lip movements and duration. When an acceptable version has been recorded, fine adjustments can be made at a later, tracklaying stage before the final remix.

ADT. Automatic double tracking, doubling. This adds a copy of the original signal with a delay of about 4–20 ms, which creates the effect of doubling the apparent number of musicians – for example, the strings, woodwind, horns or (in popular music) backing vocalists. Acoustic doubling or tripling can also be achieved by recording or copying extra, slightly displaced tracks on subsequent passes.

AES. Audio Engineering Society. Along with the EBU (European Broadcasting Union), sets standards (such as AES3) for two-channel digital audio transmission and for the embedded word clock that accompanies it. (AES/EBU word clock can also be carried between items of equipment on a separate cable.)

AIFF. An Audio Interchange File Format (suffix: '.aif') used for recording uncompressed audio on Mac computers. See also **WAV**.

Ambient noise. In any location, from quiet studio or living room to busy street, there is a background of sound. In a particular location the ear automatically adjusts to and accepts this ambient noise in all except the loudest (or quietest) cases. In a monophonic pick-up, noise at a natural level generally sounds excessive. One purpose of microphone balance is to discriminate against such noise, where necessary. Unexplained noises should be avoided. See also **Atmosphere**.

Ambisonics. A sound system that (unlike most other surround sound systems or stereo) is fully three-dimensional. To define the direction of any sound source requires a minimum of four microphones, typically in a tetrahedral array. See also **Soundfield microphone**.

Amplifier. A device for increasing the strength of a signal by means of a varying control voltage. At each stage the input is used to control the flow of current in a circuit that is carrying much more power. Precautions (feedback) may be necessary to ensure that the degree of amplification does not vary with frequency. In successive stages the magnification of the signal (i.e. ratio of output to input) may be much the same, so that the rise in signal strength is approximately exponential. Relative levels may therefore conveniently be measured in decibels (q.v.).

Amplitude modulation (AM). A method whereby the information in an audio signal is carried on the much higher frequency of a radio wave. The envelope of the amplitude of the radio wave in successive cycles is equivalent to the waveform of the initial sound. Historically, AM was the method first used for radio, and such transmissions still crowd the short, medium and long wave bands. This crowding restricts the upper limit of the frequencies that can effectively be broadcast without the transmitter bandwidth overlapping an adjacent channel. The random background of radio noise that a receiver is bound to pick up at all frequencies also appears as minor fluctuations in the amplitude of the carrier wave, and cannot be distinguished electrically from the audio signal. For an improved quality transmission, AM has largely given way to FM (frequency modulation) or digitally coded systems such as NICAM, etc.

Analogue. Any system in which a continuous electrical waveform mimics that of the original sound. Microphones and loudspeakers are analogue transducers.

Artificial reverberation (AR or 'reverb'). Simulation of the natural die-away of sound that occurs in a room or any other enclosed space (e.g. in a cave, down a well, etc.). Multimicrophone techniques for music need AR, as the studios used are normally made much less reverberant than those used for natural balance. The technique is also used to permit a different treatment to be applied to each microphone output.

Atmosphere. The background sound at any location. This can enhance authenticity and give listeners a sense of participation in an event. Even so, it may be advisable to discriminate against background sound in the master recording, and record atmosphere separately for subsequent mixing at an appropriate level (which may not be held constant). When good monitoring conditions are available at the time of recording (or broadcast), separate microphones may be set up and the mixture again judged by ear. *Studio atmosphere* is the ambient noise in a studio, which may at times be obtrusive even in a soundproofed room, particularly if omnidirectional microphones are used for other than close balances. Clothing or paper rustle, the flow of air for ventilation or even heavy breathing may create problems for the balancer. See also **Ambient noise**.

Attenuation. Fixed or variable losses (usually in an electrical signal). As with amplification it is convenient to measure this in decibels (q.v.). See **Fade**.

Axis (of microphone or loudspeaker). Line through centre of diaphragm and at a right angle to it. It is usually an axis of structural symmetry in at least one plane, and thus an axis about which the polar response is symmetrical. Also, it is generally the line of maximum high-frequency response.

Azimuth. The angle that the gap of a recording or reproducing head makes with the line along which the tape moves. For audiotape, this should be exactly 90°. Misalignment may be due to incorrect setting of the head, or to the tape transport system not being parallel to the deck, and may be corrected by adjustment to either. Adjustment of a reproducing head may be made using a standard recording of high-frequency tone (i.e. one recorded with a correctly adjusted head). Maximum output, which may be judged by ear or by meter, occurs when alignment is correct. A separate recording head may then be adjusted by recording tone with it and checking for maximum output from the reproducing head. Azimuth misalignment results in reproduction that is apparently lacking in high frequencies.

'B' signal. In stereo, the signal to be fed to the right-hand loudspeaker of a two-speaker system.

Backing track. A pre-recorded accompaniment to a singer (etc.) who listens to a replay on headphones while contributing to a performance. The two will be mixed to give the final recording. The performer's (and other musicians') time is saved, and adequate separation is achieved without the use of separate studios. *Back tracking* is the technique in which a preliminary recording by the same artist is used as the accompaniment.

Backup recording. A second recording, made elsewhere at the same time as the master, to safeguard against faults in the recording process or medium or subsequent loss of data. A *backup copy* (or simply *backup*) is a subsequent safety record. The variety of available media seems endless, but for archive purposes DVD and CD are among the best. Digital audio workstation (DAW) projects can be backed up on removable hard drive. Some people still like to keep analogue tape copies.

Baffle. (a) A rigid screen around a loudspeaker diaphragm, extending the acoustic path from front to rear in order to reduce the flow of air around the edge of the speaker at low frequencies, an effect that causes serious loss of bass. (b) A small acoustic screen that causes a local variation in the acoustic field close to the microphone diaphragm. A baffle of hardboard, card, etc., when clipped to a microphone produces a 6 dB peak at a frequency about 1/300th of its linear dimension in metres (1/1000th in feet). In some microphones, gauze baffles are used as an integral part of the design to enhance the response.

Baffled omni pair. See **Dummy head stereo**.

Balance. (a) The relative placing of microphones and sound sources in a given acoustic, designed to pick up an adequate signal, discriminate against noise, and provide a satisfactory ratio of direct to indirect sound. This is the responsibility of a *balance engineer, sound supervisor* or (in BBC radio) a *studio manager*. (b) The distribution of a sound between A and B stereo channels.

Balance test. One or several trial balances, judged by direct comparison.

Band. Separately recorded section of a vinyl disc, of which there may be several on each side of a record. By extension, the term may also mean an individual section of a tape recording bounded by spacers. Locations may be indicated on a script in short form, e.g. 'S2B3' means 'side 2, band 3'.

Bass. Lower end of the musical scale. In acoustics it is generally taken as the range (below 200 Hz, say) in which difficulties (principally in the production or reproduction of sound) are caused by the large wavelengths involved. Loudspeakers that depend on the movement of air by cones or plates (e.g. moving-coil or condenser types) become inefficient in their *bass response* at wavelengths greater than their dimensions. This is due to too small a piston attempting to drive what, at these frequencies, is too pliant a mass of air. Certain systems of mounting (or cabinets) extend the bass response by means of a resonance in the octave below the frequency at which cut-off begins, but this results in a sharper cut-off when it does occur (e.g. in bass reflex and column resonance types). *Bass lift* may be introduced, but this requires a more powerful amplifier (most power is concentrated in the bass of music anyway) and a loudspeaker system capable of greater excursions without distortion. An efficient way of producing bass is by a large acoustic exponential horn with a low rate of flare. This ensures that coupling between the moving elements and the air outside is good, up to wavelengths approaching the diameter of the mouth of the horn.

Bass management. A system using crossover networks at about 120 Hz or below that splits off low-frequency sound in stereo or surround channels and routes it to a subwoofer with a separate amplifier. The crossover frequency chosen depends on the practicable lower limit for the main speakers. The subwoofer also receives any separate LFE (low-frequency effects) track (usually with video or film).

Bass tip-up (*UK*) or **proximity effect** (*US*). Increase in bass response of directional microphones when they are situated in a sound field that shows an appreciable loss of intensity in the distance travelled from front to back of microphone. The operation of the microphone depends on the pressure gradient and is exaggerated at low frequencies. At middle frequencies the effect is masked by the phase shift that occurs in this path distance in any case: at high frequencies the mode of operation is by pressure and no longer by pressure gradient.

Beat. If two tones within about 15 Hz of each other are played together, the combined signal is heard to pulsate, or beat, at the difference frequency. Two tones may be synchronized by adjusting the frequency of one until the beat slows and finally disappears.

Bias. In tape recording, a carrier frequency with which the audio signal is combined in order to reduce distortion. Its importance is greatest in recording low frequencies. It is generally of the order of 50–100 kHz and is produced by an oscillator associated with the tape-recorder amplifier. It mostly vanishes from the tape soon after recording, by a process of self-demagnetization. However, a little of the bias signal can still be heard if recorded tape is pulled slowly over a reproducing head. The bias oscillator also supplies the erase head.

Bidirectional microphone. One that is live on the front and back faces, but dead at the sides and above and below. The polar response is a figure-of-eight.

Board, clapperboard, slate. On film, a means of marking the start of each shot with scene and take numbers. It is held open for mute scenes (those shot without sound); for sync takes, the scene and take number are identified on sound and the board is then clapped in vision so that sound and picture can subsequently be matched.

Boom. A telescopic arm (often attached to a floor mounting) from which a microphone is slung.

Boomy. Subjective description of a sound quality with resonances in the low frequencies, or a broad band of bass lift. Expressions with similar shades of meaning are 'tubby' or, simply, 'bassy'.

Boundary microphone. One placed very close to a hard reflecting surface, so that interference effects are limited to very high frequencies. See also **Mouse**; **Pressure Zone Microphone**.

Bright surface. A surface that reflects sound strongly, particularly at high frequencies.

Broadcast chain. The sequence, typically of studio, continuity suite, transmitter, receiver, together with any intermediate control or switching points through which a signal passes.

Bus. A line to which a number of separate channels are fed and combined: this may go on to an amplifier, recorder, effects processor, group or master fader, remote location, or broadcast or other output.

Butt changeover. A changeover in mid-replay from one tape, disc or other recording to another, the second being started as the out cue on the first is about to come up. The crossover from one reproduction to the next should not be noticeable.

Cancellation. Partial or complete opposition in phase, so that the sum of two signals approaches or reaches zero.

Cans. Headphones.

Capacitance. The ability of an electrical component, or components, to store static charge. Charges present in a conductor attract opposite charges to nearby but not electrically connected conductors; changes of charge induce corresponding changes of charge in nearby conductors. A signal may therefore cross between components where there is no direct path. Capacitors are used in electronics for coupling, smoothing and tuning purposes. In lines, capacitance is a mechanism whereby the signal is gradually lost (capacitance between the wires or to earth). Capacitance varies with distance between plates, or other components: this variation is used in condenser microphones and loudspeakers. *Stray capacitance* is a means whereby audio signals or mains hum is transferred unintentionally from one component to another nearby.

Capacitor microphone. See **Condenser microphone**.

Capstan (*US*: **puck**). Drive spindle of tape deck: the tape is driven between the capstan and an idler pulley that presses against it when the machine is switched to record or replay. The diameter of the spindle and the motor speed determine the speed of the tape. If the rotation of the spindle is eccentric a rapid fluctuation in tape speed causes a flutter in the signal.

Capsule. (a) Removable pick-up head. (b) The diaphragm assembly of a condenser microphone.

Cardioid microphone. Microphone with a heart-shaped polar diagram, which may be achieved by adding omnidirectional and figure-of-eight responses together, taking into account the phase reversal at the back of the latter. In condenser microphones (q.v.) the combination occurs in the principle of construction. See also **Pressure gradient microphone**.

Cart. A play-out device or system that allows rapid access to a range of recorded items for broadcast in any order. Many formats have been used, from tape cartridges or cassettes to MiniDiscs or Zip disk for short items or CD or DVD for longer segments. They are also used in broadcast automation systems. Hard drive play-out systems and some digital editing systems can also do this directly.

Cart wall. A screen showing an array of boxes, each labelled with an item that is ready for immediate replay by random access from hard disk.

Cassette, cartridge. A fully encapsulated tape system that does not need to be laced up. Cassettes (or cartridges) have been used for endless tape loops, in this case often employing standard-width tape. To avoid confusion it is probably best to use the term cartridge for the endless loop system, and cassette for the encapsulated double-spool system.

CD. Compact disc. A digital medium, recorded ('burned') and read optically by laser, that has largely replaced vinyl discs as a medium for recorded music. The 12 cm (4¼ in) disc holds 650 MB of data, plays out a signal at 44.1 kHz (over twice the highest recorded audio frequency, as is necessary for digital sampling), and carries up to 74 minutes of audio. It is also used to carry data for such applications as computer programs or small libraries of information (the text of many individual books would fit on an older 1.44 MB floppy disk; by comparison, the CD would accommodate a small library of books).

CD-R, CD-RW. CDs that are recordable (once) or rewritable (several times). Term also used for the PC, domestic or higher quality equipment to record/rewrite CDs.

Ceramic microphone or pick-up. This is similar in principle to a crystal type, but with a barium titanate element that is less sensitive to temperature and humidity.

Channel. (a) Complete set of (professional) recording equipment. (b) Recording room (or part of recording room that can operate independently to record a programme). (c) The series of controls within an individual signal pathway in a control console: see **Microphone channel**.

Chorusing. Combining an original track with a slightly de-tuned and delayed copy in order to add strength and harmonic complexity.

Clean feed (*US*: **mixed-minus**). A cue feeding back to a programme source that includes all but the contribution from that source.

Clean sound. Actuality sound of an event, without superimposed commentary.

Clock. See **AES**; **Video clock**; **Time code**.

Coating (emulsion). On a tape or magnetic disk, a layer of finely divided magnetic material that carries the magnetically recorded signal.

Cocktail party effect. The means whereby a listener can concentrate on a single conversation in a crowd. It depends largely on the spatial spread of sound, a quality that is lost in monophonic recording, so that if a 'cocktail party' type of background sound is recorded at a realistic level in mono it can be unpleasantly obtrusive and impede the intelligibility of foreground speech.

Codec. A compression system that removes information that is physically present in the original sound but may be masked or otherwise redundant when it reaches the ears. This makes data storage and transmission over the Web (e.g. by MP3, q.v.) or ISDN lines more efficient. 'Codec' is commonly used to describe not only the original, but also similar systems that can be used to store or transmit data (such as music or video) with the same equipment, but which may not be so good. In general (but not always), the more kilobits per second (kbps), the higher the quality, up to 160 kbps, which is difficult to distinguish from CD quality. If data is repeatedly encoded and decoded, perhaps only three or four times, the masking effect soon disappears, and the reduced quality becomes obvious – and the higher the initial quality, the worse this is; for example repeated encoding from 48 kHz does more damage than from 32 kHz.

Coincident pair (*US*: **X/Y technique, intensity stereo**). In stereo, a pair of directional microphones or a double microphone with two directional elements mounted close to each other (usually one above the other) and at an angle to pick up A and B signals. This technique avoids the phase interference effects produced by a spaced pair.

Coloration. Distortion of frequency response by resonance peaks. Marked acoustic coloration in a studio may be due to the coincidence of dimensional resonances, to wall-panel resonances, or to frequency-selective excessive absorption of sound. Some mild acoustic coloration may, however, be beneficial. Coloration is often exhibited by microphones and loudspeakers, and in these is often unwanted.

Comb filter. The effect of adding identical signals with a small time difference, so that there is cancellation, taking the form of 'notches' throughout the audio-frequency range. These are distributed in arithmetic, not logarithmic (i.e. not musical) progression.

Compact disc. See **CD**.

Compatibility. In stereo, the requirement that a recording or broadcast can also be heard satisfactorily in mono.

Compliance. The ratio of displacement to applied force. It is the inverse of the (mechanical) stiffness of the system. Compliance is the acoustical and mechanical equivalent of capacitance and is a measure of the performance of microphone, loudspeaker or record-replay components.

Compression. Control of levels in order to ensure that (a) recorded or broadcast signals are suitably placed between the noise and distortion levels of the medium, and (b) the relationship between maximum and minimum volumes that is acceptable to the listener who in the home may not wish to hear the greatest volume of sound that would be acceptable in a concert hall, and who may be listening against background noise. *Manual compression* seeks by various means to maintain as much of the original dynamics as is artistically desirable and possible, and varies in degree with the expected listening conditions. The purpose of *automatic compression*, in short-wave transmissions, is to bring peak volumes consistently to full modulation of the carrier. It is also used for more efficient transmission by radio microphones, and in pop music to limit the dynamic range of the voice or a particular instrument. Above a selected onset volume, say 8 dB below full modulation, compression is introduced in a ratio that may be selected: e.g. 2:1, 3:1, 4:1 or 5:1. *Digital compression* of video or audio signals may be used in off-line editing to speed processing and increase the capacity of hard disk storage. The reversible compression of a digital audio signal increases the capacity of audio data files or ISDN lines. With a ratio of 4:1 (or with well-encoded masking, 10:1) the effects should not be perceptible.

Computer mix-down. See **Mix**; also **Digital audio techniques**; **Pulse code modulation**.

Condenser microphone (electrostatic or capacitor microphone). In this, the signal is generated by the variations in capacitance between two charged plates, one of which is rigid and the other flexible, acting as a diaphragm in contact with the air. If the space between the plates is enclosed, the air vibrations can affect one side of the diaphragm only and the microphone is pressure operated (and substantially omnidirectional). If the back-plate is perforated the diaphragm is operated by a modification of the pressure gradient principle. It can be arranged that the distance that sound has to travel through the back-plate (via an acoustic labyrinth) is the same as the effective path difference from back to front of the plate: in this case the pressure gradient drops substantially to zero for signals approaching from the rear of the microphone, and a good cardioid response is obtained. If two such cardioids are placed back to back, the characteristics depend on the sense of the potential between the centre plate and the two diaphragms. The doublet can then become bidirectional, cardioid or omnidirectional by varying the size and sense of the potential on one of the diaphragms. A condenser microphone has to be fitted with a head amplifier (a single stage is sufficient) in order to convert the signal into a form suitable for transmission by line. For both condenser microphone and loudspeaker a polarizing voltage has to be supplied unless the polarization is permanent, as in an electret (q.v.).

Cone. In a loudspeaker, a piston of stiff felted paper or possibly of plastics. It should be light and rigid. Paper cones are often corrugated to reduce any tendency to 'break up' radially and produce subharmonic oscillations. Elliptical cones give a greater spread of high frequencies out along the minor axis: they should therefore be mounted with the major axis vertical.

Console. A desk (or 'panel' – *BBC*) that includes a mixer and associated controls.

Contact microphone. Transducer attached to a solid surface.

Continuity. Linking between programmes, including opening and closing announcements (when these are not provided in the studio), station identification, trailers and other announcements.

Continuity suite (sometimes shortened to 'Con'). A centre through which programmes are routed or where they are reproduced to build a particular service ready for feeding to a transmitter or to provide source material for other programme services.

Control. The adjustment of programme levels (in the form of an electrical signal) to make them suitable for feeding to a recorder or transmitter; where necessary, this includes compression. Mixing involves the separate control of a number of individual sources. At a *control console* or *control desk* (a mixer) there are faders (individual, group and main control), with their associated sound treatment, feed, cue and monitoring systems, and a variety of communications equipment. A *control line* between one location and another is a telephone circuit on which programme details may be discussed, and is so called to distinguish it from a broad-band (i.e. high-quality) audio 'music line' along which programme is fed. A control line may, of course, be narrow-band, but for outside (remote) broadcasts it is safer to use lines of equal quality so that the two are interchangeable.

Control cubicle (*BBC radio*). The soundproof room occupied by production and operational staff and equipped with control console and monitoring equipment.

Control room. In BBC radio, a switching centre; otherwise, the sound mixing room.

Copyright. The law in relation to the ownership of creative works, which is initially vested in the author, composer or artist. In Britain this extends to 50 years after the author's death (expiring at the end of the calendar year). For private study, research, criticism or review, 'fair dealing' provisions generally permit extracts to be used without payment. Otherwise payment must always be made, the details being arranged by negotiation with the author or his agent. Sound recordings, films and broadcasts are also protected by copyright and are not subject to the 'fair dealing' exceptions. Thus, a commercial recording such as a CD may have several copyrights, including those of the composer and of the company that made the recording, and these two may extend from different dates. Public performance of music and records, etc., is easy to arrange, and fees are small (usually paid through a Performing Rights Society), but recording or re-recording except of music recorded specially for the purpose (mood music) is permitted only with elaborate safeguards, and even so permission may be difficult to obtain. Copying of records and films for private purposes is not permitted. In Britain, the BBC and its contributors' unions do permit the recording of schools and further education programmes by schools and colleges, provided that the recordings are used only for instructional purposes and provided that they are destroyed at the end of specified periods. There is no copyright in events, e.g. an impromptu speech or interview without a script or casual sound effects (though recordings of them are). Dramatic or musical performers are also protected: their permission must be obtained before a record is made of their performances other than for private or domestic purposes. Copyright laws are easily evaded and owners are justifiably concerned at the losses they undoubtedly sustain. Copyright laws vary in detil from country to country: the above notes relate mainly to British law.

Cottage loaf characteristic. See **Hypercardioid**.

Crab. Move camera or sound boom sideways relative to the performing area.

Crossed microphones. See **Coincident pair**.

Cross-fade. A gradual mix from one sound source or group of sources to another. During this, both faders (or groups) are open at the same time. The rate of fade can be varied manually if desired for artistic effect.

Crossover. The frequency at which a signal is split in order to feed separate parts of a loudspeaker. *Crossover network*: the filter that accomplishes this.

Crosstalk. In stereo, breakthrough between channels. The separation (in dB) between wanted and unwanted sound is checked by feeding tone at zero level through one channel and measuring its level in the other. Outputs of −37 dB at 1 kHz rising to −30 dB at 50 Hz and 10 kHz are reasonable. Crosstalk also means breakthrough (or 'induction') of signal between any other pair of lines, e.g. on a telephone circuit.

Crystal microphone or pick-up. This generates a signal by means of a crystal bimorph – two plates cut from different planes of a crystal such as Rochelle salt and held together in the form of a sandwich. Twisting the bimorph produces a voltage (termed a *piezoelectric* voltage) between foil plates on the top and bottom surfaces of the sandwich. In a record player pick-up the frequency response is a fair match to the characteristic needed to reproduce records, so the output level after equalization is improved and smaller amplifiers are needed. The assembly can be very light in weight and cheap to produce. Crystals were once used in inexpensive microphones but have given way to electrets.

Cue. Signal to start. This may take the form of a cue light or a hand cue, or may be related to a prearranged point in a script. Headphones or, exceptionally, a loudspeaker may also be used for cueing.

Cue material. Introductory matter supplied with a recording for the scriptwriter to fashion into a cue for announcer or narrator to read.

Cue programme. A feed provided for cueing purposes.

Cycle. One complete excursion of an air particle when vibrating in a sound, or the corresponding signal in electrical or any other form; one full period of a sine wave.

Cycles per second (cps). Now written Hz. See **Frequency**; **Hertz**.

D/A. Digital-to-analogue signal conversion.

DAT. Digital audio tape. The standard two-channel tape is 4 mm wide and for four channels 6.3 mm (¼ in) tape is used. It is a high-quality rival to compact disc, but with a variety of sampling rates. DAT is very widely used in broadcasting and for other professional uses, where a sampling at 48 kHz is commonly used. It has had little success in the domestic market, for which pre-recorded tapes with a sampling rate of 44.1 kHz can be coded to reduce illegal copying. To permit a low tape speed, it employs helical scan and a rotating head that records the signal in separate bursts, which are recombined and smoothed on replay. In its principal mode of operation the ultra-compact tapes can record up to 2 hours. The low, normal speed is combined with a search facility nearly 300 times faster, operating on subcodes that identify the tape, carry timings and mark points within it.

DAW. Digital audio workstation. A system that uses a hard disk for recording, mixing, editing or creating digital masters.

dB. Decibel (q.v.).

Dead acoustic. An audio environment in which a substantial loss is introduced at every reflection. For studio work this is the nearest approximation to an outdoor acoustic (in which little sound is reflected).

Dead-roll (*US*). See **Pre-fade**.

Dead room (anechoic chamber). A space enclosed by very thick soft absorbers (typically about a metre deep), used for testing the frequency response of microphones and loudspeakers. It is unsuited to use in a studio because of its claustrophobic effect.

Dead side (of microphone). The angles within which the response of a microphone is low compared with its on-axis response. In figure-of-eight microphones this is taken to be about 50–130° from the axis, in cardioids the entire rear face. In the dead region there may be rapid changes in response with only slight movements of position, and the frequency response is more erratic than on the live side(s). In a studio these effects may be masked by the pick-up of reflected sound on the live sides.

Decay characteristic (of studio). The curve that indicates how sound intensity falls after a sustained note is cut off. Plotted in decibels against time, decay should (roughly) follow a single straight line, and not be broken-backed, but is generally slightly erratic.

Decibel (dB). A measure of relative intensity, power or voltage. Sound intensity is the power flowing through unit area and is calculated relative to a reference level of 2×10^{-5} Pa (Pascals, q.v.). The threshold of hearing at 1000 Hz is at about this level for many young people. However, the aural sensation of loudness is not directly proportional to intensity or power, but to the logarithm of intensity or power. It is therefore convenient to measure differences in loudness in units that follow a logarithmic scale, i.e. decibels. Differences of intensity in decibels are calculated as $10 \log_{10} (I_2/I_1)$. In a circuit the power is proportional to the square of the voltage, so that for a given impedance a power ratio may be expressed in decibels as $10 \log (V_2^2/V_1^2)$, or $20 \log (V_2/V_1)$. Note that the voltage scale dB(V) is used for microphones, and that this is not the same as the acoustic intensity scale dB(A). In acoustics the decibel is also convenient in that it can be regarded (very roughly) as the minimum audible difference in sound level. A change in speech or music level of 2 dB or less is not ordinarily noticeable to anyone who is not listening for it. Loudness levels have sometimes been measured in phons (q.v.). For dB(A), dB(N) and PNdB, see **Noise**.

De-gauss or de-flux. Demagnetize (e.g. a tape recording head, or scissors or blades used for cutting tape). Wipe or erase (tape) in bulk.

Delay line. Device that electronically delays a signal – for example, in analogue format, by (switched) multiples of 7.5 ms. The term is also sometimes used (by analogy) for digital delay using RAM.

DI. (a) Direct injection (box) used in popular music balance: on the output from an electronic instrument, this intercepts the high-level, unbalanced signal that goes directly to its amplifier and speaker cabinet, sending a low-level, balanced and electrically isolated signal to a console. (b) Directivity index: in microphone design, this is a measure of forward response.

Diaphragm. The part of a microphone upon which the pressure of a sound wave acts. In size it should be small enough not to suffer substantial high-frequency phase cancellation for sound striking the diaphragm at an angle, and big enough to present a sufficiently large catchment area to the pressure of the wave. In microphones working on some principles (e.g. condenser microphones) the diaphragm mass may be small, so there are minimal inertial effects and therefore a very good transient response.

Difference tone. A perceived frequency 'heard' when two tones are played together: e.g. if 1000 and 1100 Hz are played, 100 Hz may also be perceived.

Diffraction of sound. A sound wave re-forms after it has passed an object that is small compared to its wavelength, almost as though the object were invisible to it. When sound passes through a small hole or slit, this radiates sound as though it were a new source. Where an obstacle is large compared to the wavelength of sound, so that its edge casts a shadow, the sound waves that pass by also bend into the shadow a little. These effects are similar to the way light diffracts in similar circumstances.

Diffusion of sound. The degree to which sound waves are broken up by uneven surfaces, and by absorbers that are scattered in all parts of a studio rather than clustered together. A high degree of diffusion is desirable in a sound studio.

Digital audio techniques. In these, an audio signal that was initially obtained as an electrical *analogue* of the sound is converted into binary data, in which numerical state it may be operated upon, fed to random access memory, etc. The analogue signal is sampled at a rate which must be well over twice the highest frequency required, e.g. at 32 kbps for 15 kHz. Other sampling rates are 48 kHz for professional DAT, 44.1 kHz for CD, 15 kHz for client presentations of digital video edits, and 38 kHz for applications demanding a signal that is flat and distortion-free to 16 kHz; 96 kHz is favoured by purists who claim to perceive an audible improvement in acoustic quality. The signal must be converted back to analogue form before it is broadcast or fed through components designed to operate in that form.

DIN. Deutsche Industrie Norm. German industrial standard, applied to film speeds, tape equalization characteristics and items of equipment such as plugs and sockets.

Directivity pattern. Polar diagram (q.v.).

Dissonance. The sensation produced by two tones that are about a semitone or full tone apart. As the tones get closer together they beat together and finally become consonant; as they get further apart the dissonant sensation disappears, and for pure tones does not return with increasing separation. However, dissonances and consonances also arise from the presence of harmonics. The sensation of dissonance is sometimes described as unpleasant, but 'astringent' would perhaps be a better antithesis to the sweetness of consonance.

Distortion. Unwanted changes of sound quality, in the frequency response, or by the generation of unwanted products. *Harmonic distortion* may be caused by flattening the peaks or troughs of an analogue waveform. The human voice and nearly all musical instruments possess a harmonic sound structure that tends to mask this distortion unless it is severe: 1% harmonic distortion is not usually noticeable. *Intermodulation distortion* is more serious because it includes sum and product frequencies that are generally not harmonically related to those already present.

Downmixing. A process that converts surround 5.1 to stereo. In some domestic equipment, left and right signals are combined and the centre track divided equally and added to them, while the LFE (subwoofer) signal, if included in the process, is also divided equally between them. The result is often unsatisfactory, and balances generally create a separate stereo mix, to be used by formats that allow it. More sympathetic downmix programs (even for conventional stereo headphones) are available.

Dropout. Loss of signal often due to a localized fault in tape coating or other media.

Dry acoustic. Lacking in reverberation. By way of mixed metaphor, the opposite is 'warm' or 'bright'.

DSP. Digital signal processing. This is performed by digital devices that offer a wide (and apparently ever-increasing) range of effects. Some (for which digital processing is particularly well adapted) are delay-related: these include EQ, doubling, flanging or pitch (or duration) changing, etc. Multiple-function DSP may replace a number of more limited devices. Also often called 'plug-ins'.

Dubbing (*UK*). Copying, to arrange for material to be available in convenient form, e.g. transferring music from disc to tape. Also, copying together and combining effects or music with pre-recorded speech. In film, the mixing of all sounds to a single final track (or the two or more required for stereo or surround sound).

Dummy head stereo. One of a group of stereo pick-up techniques that employ microphones separated in roughly the same manner as human ears. Although not among the most widely accepted approaches to stereo, and so not recommended in this book, it does have strong support by some users. It is perhaps best when the output is fed to headphones – although this has two disadvantages: the image is centred within the head rather than in front of it, and any head movement will swing the whole sound stage with it.

DVD. The V might be supposed to stand for 'video', because that is how many of its users see it: *DVD-Video* is designed to carry a complete feature film picture together with surround sound. But in fact it is more 'V for versatile' than that. Digital Versatile Discs fit into drives that can also carry CDs, but hold far more data: up to 17 GB in its most complex layout, which has eight times the density of information per layer, two layers (at different depths) per side, and by using both sides, it carries about 25 times more data than a CD. As *DVD-ROM*, this is enough for a substantial text library of well over 10 000 books. In addition to video (either taken from commercial films or prepared directly for release in this form), it is also good for high-quality stereo and/or 5.1 surround sound with data compression (slightly reduced quality). Like CD-RW, DVD-R (recordable DVD) is standard on good domestic computers.

DVD-Audio. This offers a variety of (uncompressed) options, some with far higher fidelity than standard CD.

Dynamic microphone. One that responds directly to the air pressure on its diaphragm on one side only. Most moving-coil microphones work in this way, but other principles can also be adapted to do so.

Dynamic range. The range of volumes in an audio signal. It may be measured as the range of peak values (i.e. the difference between the highest PPM readings at passages of maximum and minimum volume) or, alternatively, the range of average volume. It may refer either to the range of the original sound, or to what remains after compression.

Dynamics. The way in which volume of sound varies internally within a musical work (or in a speech or speech-and-music presentation). It may refer to the variation of levels within the work as a whole, or from note to note, or in the envelope of a single note.

Earphone, earpiece. A single headphone (q.v.) or earfitting of the hearing-aid type.

Earth loop (*US*: **ground loop**). A circuit created when a signal path is linked to earth from more than one point. If a mains cable passes through the loop, it will induce mains hum. This is severe if one of the earth connections is before a preamplifier, as this will not only boost the hum but also generate harmonics.

EBU. European Broadcasting Union. Along with the Audio Engineering Society of America, the EBU sets standards for digital audio transmission. EBU time code is related to the European transmission standard of 25 fps.

Echo. Discrete repetition of a sound, produced by a single reflected sound wave, or a combination of such waves in which the return is coincident in time and at least 20 ms after the original sound. Delay of, typically, 35 ms or more may be used to create the effect of an echo. Colloquially, 'echo' is used inaccurately to mean the sum of such reflections, particularly in artificial reverberation.

Echo chamber. A room for producing or simulating the natural reverberation of an enclosed space. Such chambers are rarely of sufficient size, and have mostly been superseded by electronic artificial reverberation.

Echo plate. Better known by its proper name, reverberation plate. Like the echo chamber, largely superseded by electronic devices.

Edit decision list (EDL). A list of time codes at which edits are programmed to take place. It is generally generated from computer memory and stored on disk, but may also be noted and written manually, perhaps by reading numbers superimposed on picture.

Editing block. A metal plate with a channel the width of magnetic tape running along the centre, with lapped edges to grip the tape, and a cutting groove to guide a (razor) blade across it.

Effects. Simulated incidental sounds occurring (a) in the location portrayed (usually *recorded effects*) or (b) as a result of action (usually *foley* or *spot effects* – those created on the spot). A heightened realism is generally preferred to the random quality of naturally occurring sounds, but effects have sometimes been taken a stage further and formalized into musical patterns such as *musique concrète*. *Comedy* or '*cod*' effects are those in which some element, a characteristic natural quality of the sound or a formal element such as its attack or rhythm, is exaggerated for comic effect. In this category are simple *musical effects*, in which a sound is mimicked by a musical instrument or group.

Eigentone. The fundamental frequency of any resonance associated with a dimension of an enclosed space. Its wavelength is twice that dimension. It can form effectively only between parallel reflective surfaces.

Electret. Electrostatic microphone capsule in which the diaphragm or base-plate has a permanent charge and therefore requires no externally applied polarizing voltage.

Electronic music. A work constructed from electronic source materials by arranging them in a formal pattern (which may be beyond the range of conventional instruments or musicians). See also **MIDI**.

Electrostatic loudspeaker. An application of electrostatic principles to the movement of air in bulk. This is done by means of a large charged diaphragm suspended between two perforated plates. As the alternating signal is applied to the outer plates, the diaphragm vibrates; and so long as the excursions are not too large the transfer of power is linear (the use of two plates rather than a single back-plate helps considerably in this respect). The problem of handling sufficient power and transferring it to the air at low frequencies is not fully solved in any such speaker of convenient size (though baffles reduce the immediate losses). Electrostatic loudspeakers should not be placed parallel to walls, or standing waves will form behind them. With careful positioning, however, the quality of sound achieved is very clear. As with a number of other transducer principles, this can also be applied in reverse, as a microphone.

Electrostatic microphone. See **Condenser microphone**.

Enclosure. Commonly, a loudspeaker cabinet. Its most important function is to improve bass response. It may do this by acting as a baffle and increasing the distance rear-emitted sound has to travel before it can combine with and cancel or reinforce forward sound (thus lowering the frequency at which cancellation occurs). It may be a ported or bass-reflex cabinet, from which a low-frequency resonance is emitted more or less in phase with the forward output from the cone. In another design the box entirely encloses the rear of the loudspeaker: in this 'infinite baffle' the rear-emitted sound is supposed to be absorbed entirely within the box.

End-fire. Orientation of the diaphragm within a long microphone such that the axis of directional pick-up is along the main axis of the casing. An end-fire microphone is therefore 'pointed' toward the sound source, an arrangement easily understood by performers.

Envelope. The manner in which the intensity of a sound varies with time. Graphical representation of the envelope (or dynamics) of a single note may show separate distinctive features in its attack, internal dynamics and decay. The term may also refer to the envelope of frequency content, or of an imposed frequency characteristic, as in a formant.

Equalization (colloquially, EQ). The use of a filter network to compensate (a) for any distortion of the frequency response introduced by a transducer or other component (e.g. a landline), or (b) for a recording or transmission characteristic employed to ensure an efficient, low-noise use of the medium (as in disc recording or FM transmission). It is also used to produce changes in frequency response that are subjectively evaluated as improvements or are otherwise artistically desirable.

Erasure. The removal of recorded signals from a medium so that it is ready to reuse. On tape (for example) this is done automatically on most recording machines as the tape passes the erase head, which lies between the feed spool and recording head. This head is similar in structure to the other heads, except that the gap is wider. The pure bias signal fed to it magnetizes the tape first in one direction and then the other until it approaches the end of the gap, where the magnitude of the oscillations dies away, and the tape is left demagnetized. With a *bulk eraser*, the tape as a whole is given a similar 'washing' treatment, often being physically removed from the region of alternating current before that is switched off.

Establish. Establishing a sound effect, etc., is allowing it sufficient time (and volume, which may be greater than that subsequently used) for it to register in the listener's mind. Establishing a location may be helped by the above technique, but equally may be a matter of supplying sufficient 'pointers' in scripted speech.

Exponential. An exponential curve is one that follows the progress of a particular form of natural, unrestrained growth or decay. If, in exponential growth, something has doubled in time t, it doubles again in further time t, is eight times its original size at time $3t$, and so on. Exponential growth inevitably overhauls all other forms of regular growth (unless, as happens in nature, some regulating factor intervenes). Exponential decay halves in time t, halves again in further time t, and so on: this must eventually become the slowest regular form of decay. A *logarithmic scale* (e.g. on a graph) is one in which the scale is gradually reduced according to the same principles. The spaces between 1, 2, 4, 8, 16, 32, etc. are all equal. This form of representation reduces an exponential growth to an apparently linear growth. The ear judges both changes of volume and changes of pitch by ratio; thus, a logarithmic scale is more relevant than any other to perceived sound. The doubling of sound frequency is equivalent to an interval of one octave. A 10-fold increase in the intensity of a sound is equivalent to an interval of 1 bel (10 decibels).

Extinction frequency. Frequency at which there is complete loss of signal when the dimension of a component matches signal wavelength.

Fade. Gradual reduction or increase in the signal. This is accomplished by means of a *fader* (i.e. a potentiometer – or 'pot' for short). A *logarithmic fader* is one in which the ratio of gain or loss is the same for equal movements of the fader. Such a fader may be marked off in a linear scale of decibels. Faders used in audio work are logarithmic over their main working range.

Feedback. The addition of some fraction of the output to the original; this in turn may contribute to a further addition, and so on. If the feedback is mixed in at a higher level than that of the original signal, a howlround occurs; if at a level that is sufficiently lower, it gradually dies away. In amplifiers a signal may be fed back to an earlier point: *negative feedback* (in which the signal fed back reduces output) is used to control distortion introduced by the amplifier. See also **Spin**.

Field pattern (*US*). Polar response. See **Polar characteristic**.

Figure-of-eight. See **Bidirectional microphone**.

Film leader. Section of film used at the head of picture or magnetic track showing a start mark, then 'footage' marks at 16-frame intervals between 12 and 3, after which the leader is black until the first frame of action. The same intervals (representing feet on 35 mm film) are used for all film gauges.

Filter. When dealing with analogue signals, a network of resistors and condensers (inductances could, in principle, also be used) that allows some frequencies to pass and attenuates others. The simplest form of filter (one resistor and one condenser) rolls off at 6 dB/octave above or below a certain frequency. There is an elbow in the curve at the turnover point: the nominal cut-off frequency is that at which the loss is 3 dB. For many audio purposes this gentle form of filter is quite as satisfactory as the sharper cut-off that can be obtained with a more complex network. The terms *bass-* and *top-cut filter* and *stop-* and *pass-band filter* are self-explanatory. An *octave* filter is one in which the signal is divided into octaves whose levels may be controlled separately. Digital filters are more flexible, but there are advantages in mimicking the smooth analogue curves.

Flanging. Effect produced by playing two recordings almost in sync but with one running slightly slower than the other, so that phase cancellation sweeps through the audio-frequency range. Originally produced by manual drag on the flange of one of the spools, but now usually simulated electronically, and generally included as a facility in a digital signal processor.

Flat. (a) On a vinyl record replay stylus, a surface of wear that appears on the two sides of the tip after some period of use. (b) On the rubber tyre of an idler wheel, a flat is an indentation that may form if the idler is left parked in contact with drive spindle, etc. It causes a momentary flutter in recording or replay.

Flutter. Rapid fluctuation in pitch having a warble frequency of, say, 8 Hz or more due to a fault in equipment, such as an eccentric drive spindle on an analogue tape drive.

Foldback. A feed of selected sources to a studio loudspeaker or headphones to be heard by performers. It may be fed from outside sources, or from tape or disc. If it is taken from a microphone in a distant part of the studio, it may help to overcome problems of audibility or (in music) time-lag. See also **Acoustic reproduction of effects**; **Public address**.

Foley artist. US film industry term for a post-production specialist in matching footsteps and other sound effects. Tracks are laid to the replay of edited picture, and may be further edited, adjusted or treated before the final mix. See **Spot effect**.

Footage. Length of film expressed in feet: either in terms of its own gauge or the equivalent for 35 mm film (on which metric and Imperial measures seem to be mixed indiscriminately). One 35 mm foot is 16 frames. One foot of 16 mm film is 40 frames long, but a 16 mm film leader is marked at 16-frame intervals, i.e. in 35 mm feet.

Formant. A characteristic resonance region: a musical instrument may have one or more such regions, fixed by the geometry of the instrument. The human voice has resonance regions associated with the nose and sinuses, which are fixed, and the mouth and throat cavities, which are capable of more or less variation in size and shape, permitting the formation of the vowel sounds and voiced consonants. The range of the formant regions is not directly related to the pitch of the sound on which they act. If the fundamental is well below or low in the formant range, the quality of sound produced is rich, as harmonics are clustered close together within the formant; if the fundamental is relatively high, there are less harmonics in the range and the quality is thinner.

Frequency. The number of complete excursions an air particle makes in one second (formerly described as cycles per second, c/s or cps, now as hertz, Hz). The only kind of sound that consists of a single frequency is a pure (i.e. sinusoidal) tone. This corresponds to the motion of an air particle that swings backward and forward in smooth regular cycles (simple harmonic motion, s.h.m.). The velocity of sound in air is about 340 m/s (1120 ft/s), depending on temperature. Frequency (f) and wavelength (λ) are related through the velocity of sound (c) by the formula $c = f\lambda$, so any frequency can be represented alternatively as a wavelength. A complex sound can be analysed in terms of its frequency components, and plotting volume against frequency gives the *frequency spectrum*. A sound that is composed of individual frequencies (fundamental and harmonics or partials, or a combination of pure tones) has a *line spectrum*. Bands of noise have a *band spectrum*.

Frequency correction. The change required in the frequency characteristics of a signal to restore it to its original form. The term is also used ambiguously to indicate the application of desired deliberate distortion of the response (often by increasing some range of the upper middle frequencies). A *linear response* in any process is one in which the frequency distribution is the same at the beginning and end. See also **Equalization**.

Frequency modulation (FM). A method whereby the information in an audio signal is carried on the much higher frequency of a radio wave. The frequency of the audio signal is represented by the rate of change of carrier frequency;

audio volume is represented by amplitude of frequency swing. The maximum deviation permitted for FM transmission is set at 75 kHz. Overmodulation does not necessarily cause immediate severe distortion, as with amplitude modulation (AM). Noise, which appears as fluctuations of carrier amplitude, is discriminated against, though it does produce phase-change effects that cannot be eradicated. Pre-emphasis of top (i.e. prior to transmission), with a corresponding de-emphasis at the receiver, helps to reduce noise further. Unless two carriers on the same wavelength have almost the same strength, the stronger 'captures' the area: there is only a small marginal territory between service areas.

Frequency response. Variation in gain or loss with frequency.

Fundamental. The note associated with the simplest form of vibration of an instrument, usually the first and lowest member of a harmonic series.

Fuzz box. Used in the pick-up and reproduction circuitry of some 'electric' instruments (electric guitar, piano, etc.), this intentionally heavily overloads the system, thereby filling in the texture of the sound with a dense array of distortion products. A buzzing effect is achieved.

FX (indication in script). Effects.

Gain. Amplification: ratio of output voltage to input voltage. It is most conveniently calculated in decibels.

Gate. A switching circuit that passes or cuts off a signal in response to some external control, e.g. to whether or not an applied voltage is above or below a given threshold.

Graphic filter. Filter in which the signal is divided into narrow bands, perhaps of a third of an octave, each controlled by a slide fader. Seen side by side on the face of the instrument, the fader knobs form a curve approximating to the response of the filter.

Grid. A framework below the roof of a theatre or television studio from which lighting and microphone leads may be suspended. A television studio normally has some of its microphone points in the grid.

Groove. Track on a record that carries a mono audio signal in the form of a lateral displacement, and stereo in a combination of two 45° displacements. *Coarse groove* was used for 78 rpm recordings, with a pitch of approximately 100–150 grooves per inch (*US*: lines per inch). *Fine groove, microgroove, minigroove* was used for 33⅓ and 45 rpm recordings: the pitch, about 250–330 grooves per inch, varies with modulation.

Ground loop (*US*). See **Earth loop**.

Group fader (*US*: **submaster fader**). A fader though which the combined output of several individual faders is fed.

Guide track. A separate track on a recorder that is replayed to artists to assist synchronization, but is not itself included in the finished performance.

Gun microphone. Moving-coil or condenser microphone fitted with an interference tube leading out along the axis. Sound enters the tube through a series of ports. That approaching off-axis reaches the diaphragm by a range of paths of different lengths, and mostly cancels out. The microphone is therefore highly directional except for sound that is of wavelengths substantially greater than the length of the gun.

Haas effect. When a sound is heard from two loudspeakers in different directions and there is a time delay on one path, all the sound seems to come from the other loudspeaker unless the delayed sound is increased in volume to compensate. The effect grows rapidly to a maximum as the delay increases to 4 or 5 ms (equivalent to a difference in path length of about 1.5 m, or 5 ft). At the maximum, an increase in volume of 10 dB in the delayed sound recentres the image, and a greater volume swings it to the louder source.

Harmonics. A series of frequencies that are all multiples of a particular fundamental frequency. They are produced by the resonances of air in a tube (e.g. woodwind, brass, organ) or of a vibrating string, etc.

Head. Transducer that converts electrical energy into magnetic or mechanical energy or vice versa. Thus we have tape recording and reproducing heads and disc cutter and pick-up heads. The electromagnet used for erasing tape is also called a head.

Headphones. A pair of electro-acoustic transducers (e.g. moving coil) held to the ears by a headband. Alternatively, devices similar to hearing aids may be used in one ear (e.g. for listening to talkback instructions). Earphones are efficient and produce the best quality when the channel from transducer to ear is completely closed, as with moulded plastic ear-fittings. It is best to have a volume control in circuit. When headphones are used mainly for communication purposes, intelligibility is more important than high quality.

Headroom. The safety margin (in decibels) allowed, primarily when recording an audio signal. If digital signals exceed the maximum coding level, this causes an unacceptable loss of data, so for unknown material headroom must be greater than for analogue recordings, where exceeding the maximum level causes harmonic distortion (best avoided, but not so great a disaster). When CDs are mastered the material is known, so levels can be raised (normalized) to effectively eliminate the headroom.

Hearing. Essentially, the ear acts like an instrument for measuring frequencies and volumes of sound logarithmically (so they appear linear in octave scales and decibels). The subjective aspect of frequency is pitch: judgement of pitch is not entirely independent of volume. The subjective aspect of sound intensity is loudness. Below about 1000 Hz the ear is progressively less sensitive to low volumes of sound: above 3000 Hz the hearing may also become less sensitive, but not necessarily following an even curve. The upper limit of hearing for young ears may be 16 kHz or higher, but with increasing age this is gradually reduced. It seems reasonable, therefore, to suppose that no audio system needs to exceed a range of 20–16 000 Hz, implying that the signal may be cut off at this upper frequency (though some dispute this). The middle ear contains a group of bones that match the mechanical impedances of the eardrum, a diaphragm that is in contact with the outside air, and a second diaphragm, which transmits sound to the liquid of the inner ear. The physiological construction of the inner ear is known, but the mechanism of aural perception is complex.

Hertz (Hz). The measure of frequency (q.v.) formerly expressed as cycles per second, a cycle being one complete excursion of simple harmonic motion (s.h.m.). It is more convenient for mathematical analysis to consider frequency as an entity rather than as a derived function.

Hiss. High-frequency noise.

Howlround or **howlback**. Closed circuit (either wholly electrical or partly acoustic) in which the amplification exceeds the losses in the circuit. In a typical case, if the loudspeaker is turned up high in a monitoring cubicle and the microphone in the studio is faded up high for quiet speech, then if the acoustic treatment is inefficient at any frequency (or if both connecting doors are opened) a howl may build up at the frequency for which the gain over the whole circuit is greatest.

Hum. Low-frequency noise at the mains frequency and its harmonics.

Hybrid transformer. A transformer with two secondary windings, so arranged that there is little crosstalk directly between the secondaries.

Hypercardioid. Cottage-loaf-shaped polar response of microphone, intermediate between figure-of-eight and true cardioid.

Idler. On a tape deck, the idler presses the tape against the capstan when the drive is switched on, but does not itself transmit rotation.

Impedance. A combination of d.c. resistance with inductance and capacitance that responds to an alternating current as a resistance does to a direct current. An inductive impedance increases with increasing frequency; a capacitative impedance decreases with frequency. Either type introduces change of phase. See **Matching**.

Indirect sound (in microphone balance). Sound that is reflected one or more times before reaching the microphone.

Inductance. The resistance of (in particular) a coil of wire to rapidly fluctuating alternating current. The field built up by the current resists any change in the rate of flow of the current. This form of resistance increases with frequency.

Insert point. A point in a signal path where there are two jacks, wired so that when an external device is plugged in, the signal is diverted through it – or where a pair of switches performs the same function.

Intensity of sound. The sound energy crossing a square metre. Relative sound intensities, energies or pressures may all conveniently be measured in decibels. See **Decibel**; **Wave**.

Intermodulation distortion. See **Distortion**.

ISDN. Integrated Digital Switched Network. ISDN telephone lines offer broadband Internet connection (for rapid downloading) and can also be used to transmit natural-sounding speech – although when an ISDN line is used in broadcasting as an outside source, its quality depends on the headset and encoder used for transmission: sometimes this may be little better than an ordinary telephone circuit. See also **Codec**; **MPEG**.

Joint. A place on a tape where two physically separate butt ends have been joined together.

Jointing tape. A specially prepared, non-oozy adhesive tape, slightly narrower than magnetic tape, used to back two pieces of tape that have been butted together.

Jukebox. In radio, a play-out device or system that can be set up for instant replay of station ID, signature tunes and jingles, commercials, trailers, and standard announcements.

Jump cut. Originally a cut in replaying a disc, made by lifting the stylus and replacing it in a later groove (at which the cue has been checked in advance). There is a momentary loss of atmosphere while the replay is faded out. In film, a cut between scenes or two separate parts of the same scene in such a way that continuity of action is deliberately lost.

kc/s. Kilocycles per second. Now written kHz. See also **Frequency**; **Hertz**.

Lavalier. Personal microphone originally worn suspended around the neck, like pendant jewellery; now a microphone in any similar position.

Lazy arm. Simple form of boom consisting of an upright and a balanced cross-member from which a microphone may be slung.

Leader. White uncoated tape that may be cut on to the start of a spool of recorded audiotape and on which may be written brief details of the contents of the sound recording. Yellow spacer is also used. See also **Film leader**.

Level. Volume of electrical signal as picked up by microphone(s) and passed through preamplifiers and mixer faders. Typically, this volume may be calculated in decibels relative to a *reference level* of 1 milliwatt in 600 ohms (also called zero level). *Zero level* in this convention corresponds to 40% modulation at the transmitter: 100% modulation is 8 dB above this. To *take level* is to make a test for suitable gain settings, in order to control the signal to a suitable volume for feeding to transmitter or recorder. This test may sometimes be combined with a balance test. *Voice level* is the acoustic volume produced by a voice in the studio.

LFE. Low-frequency effects, generally below 120 Hz, on the sixth track in 5.1 surround sound: this goes directly to a subwoofer, where bass fed from the loudspeaker system's bass management system (q.v.) is added.

Limiter. An automatic control that stops volume exceeding a predetermined level, for some artistic purpose or to prevent overmodulation. Typically, it uses feedback: the signal is monitored by the limiter and if it exceeds the set volume it creates a corresponding increase in feedback, which reduces the signal. If this operates on a marginally delayed signal, even the start of the higher volume is controlled. After a while, the operation of a recovery device allows the gain to return to normal.

Line. A send and return path for an electric signal. In its simplest form a line consists of a pair of wires. A *landline* is a line (with equalizing amplifiers at regular intervals) for carrying a signal across country (cf. **Radio link**; **Broadcast chain**).

Line microphone. Gun microphone (q.v.).

Line up. Arrange that the programme signal passes through all components at the most suitable level. In a broadcast chain this is normally at or about zero level (see under **Level**).

Line-up tone. Pure tone, typically at 1000 Hz, fed through all stages of a chain. It starts at zero level, and once set should read the same on a meter at any stage. A subsequent drop or jump in the level of the tone between successive points in the chain may be due to a fault in the line or in other equipment.

Linear audio. Digital audio in which the data is uncompressed. Ideally, professional equipment, processing and storage media (particularly for archive storage) should be linear (and also have a frequency range of 20 Hz–20 kHz). Unfortunately, many distribution media are designed to lower standards.

Lip-ribbon microphone. A noise-cancelling microphone placed close to the mouth, the exact distance being determined by a guard resting against or close to the upper lip. The microphone's directional properties discriminate in favour of the voice; close working also discriminates against unwanted sound, and low-frequency noise is reduced still further by the compensation necessary when working very close to a directional microphone.

Live angle. Angle within which reasonable sensitivity is obtained. See **Cardioid microphone**; **Bidirectional microphone**.

Live side (of microphone). The face of the microphone that must be presented to a sound source for greatest sensitivity.

Logarithmic scale. See **Exponential**.

Loop. Originally a continuous band of tape made by joining the ends of a length of tape together; the same effect achieved by a sampling device within a CD or MiniDisc replay machine or a DSP program. A loop may be used (a) to provide a repeated sound structure or rhythm; (b) to add an 'atmosphere' track where this is regular in quality; (c) using tape, to create a short delay.

Looping. Repeatedly copying a short segment of a recording. See also **ADR**.

Loudness. Subjective aspect of sound intensity. See **Meter**; **Decibel**.

M signal. The combined A + B stereo signal. Corresponds to the signal from a single 'main' microphone. See also **S signal** and **AB**.

M and E. A pre-mix, nominally of music and effects. In practice, in documentary it is a mix of all but commentary or narration.

Magneto-optical recording (MO). This uses a combination of laser and magnetic field to record an audio signal on a disc within a spring-loaded cartridge; a laser is used to replay it. Useful (among other applications) for backup recording. Discs hold up to 2.6 GB (or a possible maximum of 5.2 GB) of data.

Main gain control (*US*: **grand master** or **overall master control**). Final fader to which the combined outputs of all group (*US*: submaster) or individual faders are fed.

Matching. Arranging that the impedance presented by a load is equal to the internal impedance of the generator. Unless this is done there will be some loss of power, and the greater the mismatch the greater the loss. Often, when there is adequate gain in hand, some degree of mismatch is not critical, but the input impedance should be lower than that of the following circuit: 'low into high will go'. However, in the case of microphones and similar generators, the signal is low to start with and any immediate loss will result in a poorer signal-to-noise ratio. Matching is done by means of a small transformer. Where a microphone impedance is strongly capacitative (e.g. in a condenser microphone), its output voltage controls current in an external circuit.

MD. See **MiniDisc**.

MDM. The modular digital multitrack recording system is a way-station on the relentless advance of digital memory and processing power. At a tiny fraction of the cost of the previous generation of tape-based digital multitracks, the eight-track MDM modules (each using an inexpensive domestic-style videotape) can be stacked to make available 24, 32 or more (up to 128) recording tracks.

Meter. Device for measuring voltage, current, etc. In audio, several types of meter are used for measuring programme volume. The *VU (volume unit) meter* is used on much American equipment. Over its main working range it is linear in 'percentage modulation'. It is not, therefore, linear in decibels, and this means that for all but a very narrow range of adjustments of level the needle is either showing small deflections or flickering over the entire scale. The VU meter is not very satisfactory for high-quality work where a comparison or check on levels (which are not always close to the nominal 100% modulation) is required. In Britain the *peak programme meter* (PPM) is used: this measures only the highest peaks, so is not a good indicator of average volume. The PPM is linear in decibels over its main working range.

Micron (μ). A thousandth of a millimetre.

Microphone. An electro-acoustic transducer. A microphone converts the power of a sound wave into electrical energy, responding to changes in either the air pressure or the pressure gradient (q.v.). Principal response patterns are omnidirectional, figure-of-eight, cardioid; principal types are moving-coil (dynamic), condensers (electrostatic) and ribbon (q.v.). *Microphone sensitivity* is measured in dB relative to 1 volt/Pa. See **Pascal**.

Microphone channel. The preamplifier, equalization circuit, fader, etc. in a mixer that are collectively available for each microphone. The channel may also include facilities for 'pre-hear', and feeds for echo, foldback and public address.

MIDI. A 'musical instrument digital interface' is data communications hardware that is used to link electronic musical instruments by means of a digital signal with each other and with computers. A special performance-related language is used. However, the term has come to symbolize far more than this: as a major step in the evolution of the virtual studio, MIDI technology is seen as the leading alternative to natural acoustics, conventional analogue sound sources and digitized audio signals. There is now an abundance of books on the subject, though some are rather technical and heavy with acronyms. For those who like hands-on musical sources, the original, apparently indispensable MIDI device will continue to survive its apotheosis, to be found – typically – linking a synthesizer keyboard to a computer. For others, the keyboard will appear only as a simulation on the screen, often with each key linked to data streams that are shown on the same timeline as audio tracks, and the MIDI hardware has disappeared.

MIDI cable. A shielded, twisted pair of wires, up to a maximum of 15 m (50 ft) in length. On the male five-pin DIN plug at each end, only three of the five pins are connected.

Midlift. Deliberate introduction of a peak of 'presence' (q.v.) in the frequency response in the upper-middle frequency range (somewhere between 1 and 8 kHz). A group of midlift controls may be calibrated according to the frequency of the peak and degree of lift.

Mil. A thousandth of an inch, about 25 microns.

MiniDisc (MD). Another way-station in the rapid advance of recording systems: its sub-compact size has allowed it some success in domestic (and portable) use, while in radio studios, a MiniDisc cart is a significant advance on its clunky predecessor carts of cartridges. It has what is described as near-CD quality, but in order to allow reasonable recording time (74 minutes for two tracks, down to 18½ minutes for eight) data is compressed 5:1. In the latest versions of this process, the effects of compression are almost impossible to hear – but the recording will still not survive many generations of copying. It is difficult to see how this can compete with any uncompressed recording system in a world where memory is growing so fast. MiniDisc has also been used as an inexpensive means of recording in the field, but if quality rather than cost is the criterion, a better miniature medium is digital audiotape (DAT).

Mix. Combine the signals from microphones, signal processors, replay devices and other sources. To *mix down* or *reduce* is to combine separately recorded tracks in a multimicrophone balance. *Computer mix-down* is an aid to mixing multitrack recordings. A trial mix is fed to the memory and then successive changes are made to improve it. This ensures that each individual stage is operationally simple: complexity does not grow, and so does not become a limiting factor.

Mix-minus, mixed-minus (*US*). Clean feed (q.v.).

MO. See **Magneto-optical recording**.

Modulation. Superimposition (of sound wave) on a carrier, which may be a high-frequency signal (e.g. by amplitude modulation, AM, or frequency modulation, FM, q.v.) or (on a record) a smooth spiral groove. When the carrier or groove has no added signal, it is described as unmodulated; 100% modulation is a designated maximum amplitude for any recording or transmission carrier system.

Monaural sound. Alternative term for monophonic sound, suggested by the analogy that listening with one microphone is like listening with one ear – which is not completely accurate. See **Mono**.

Monitoring. (a) Checking sound quality, operational techniques, programme content, etc., by listening to the material as it leaves the studio (or at subsequent points in the chain, or by a separate feed, or by checking from a radio receiver). A *monitoring loudspeaker* in the studio is usually as good as the best that listeners might be using (but remember that adverse listening conditions might radically alter the listeners' appreciation of sound quality, and also that they may be listening at a level lower than that at which the monitoring loudspeaker is set). (b) Listening to other broadcasts for information.

Mono (monophonic sound). Sound heard via a single channel. This is defined by the form of the recording or transmission, not by the number of loudspeakers. Several microphones may be used and their outputs mixed, and several loudspeakers may be used and their frequency content varied, but this is still mono unless there is more than one channel of transmission. Multiplex radio transmissions and the single groove of a stereo disc each contain more than one channel. In mono the only spatial movement that can be simulated is forward and backward. There is no sideways spread; in particular, there is no spatial spread of reverberation or of ambient noise. The balance of sound in mono is therefore not natural but reformulated in terms of a special convention. In television, 'mono' also means monochromatic, i.e. black and white pictures. See also **Cocktail party effect**.

Mouse. A microphone laid flat on the floor, inside a foamed plastic shield that is flat on the underside. As the capsule is very close to a hard reflecting surface, interference effects are limited to very high frequencies. See **Boundary microphone**; **Pressure Zone Microphone**.

Moving-coil microphone, loudspeaker or pick-up. These all use a small coil that can move within the field of a permanent magnet. In the microphone or pick-up the movement generates a current in the coil; in the loudspeaker the current causes movement that is transmitted to a cone (q.v.) that drives the air.

MPEG (pron. 'em-peg'). The Moving Picture Experts Group, established by ISO and IEC, has developed a range of digital coding standards for audio and video. MPEG-3 deals with data compression standards for audio. See **Codec**; **ISDN**; **MP3**.

MP3. ISO-MPEG Audio Layer 3: a compressed audio format which reduces data required for transmission over the Web by factors of 10:1 or more, without seriously degrading music quality. It allows rapid downloading of files with the suffix '.mp3'. On some links, this also allows music to be heard in real time. Other popular Web formats include the somewhat higher quality Windows Media Audio (WMA) and RealAudio, which is sometimes used for music previews. Audio on the Web, a rapidly developing field, is largely outside the scope of this book.

MS. Middle-and-side (*US*: mid-side): a stereo system in which the components are generated and/or processed and transmitted as 'M' and 'S' signals.

Multiplex stereo. Radio transmission carrying A − B stereo information on a subcarrier above the A + B signal, or other combined signals of this type.

Music line. Full audio-frequency circuit for carrying programme (including speech) as distinct from a telephone line, which may occupy only a narrow band.

Musique concrète (*French*). A work in musical form constructed from natural sounds that are recorded and then treated in various ways.

Mute film (*UK*). Scenes shot without sound (*US*: MOS).

Network operation. A broadcasting system involving local stations and transmitters that may join together or separate at will.

Newtons per square metre (N/m^2). A unit of sound pressure. See **Pascal**.

NICAM. Near-Instantaneous Companded Audio Multiplex: a complex compressed digital transmission system. NICAM 728 is a version of this adopted for stereophonic television broadcasting in the UK. With a sampling rate of 32 kbps it is limited to a bandwidth of 15 kHz, but is otherwise comparable in quality to CD reproduction. There is no perceptible lag behind picture, but enough to cause severe comb-filter effects if decoded NICAM is mixed with synchronous stereo.

Noise. This is generally defined as unwanted sound. It includes such things as unwanted acoustic background sounds, electrical hiss (e.g. caused by the random flow of electrons in a transistor, or the random placing and finite size of oxide particles in tape coating), or rumble, hum or electromagnetic noise (background noise picked up on a radio receiver). In all electronic components and recording or transmission media the signal must compete with some degree of background noise, and it is vital in radio and recording work to preserve an adequate signal-to-noise ratio at every stage. *White noise* contains all frequencies in equal proportion. *Coloured noise* (by analogy with coloured light) has an uneven frequency response. Acoustic noise levels are often measured in dB(A) (or SLA); 1 dB(A) = 40 dB relative to 2×10^{-5} Pa at 1 kHz, and at other frequencies is weighted (electrically) to be of equal loudness. The PNdB (perceived noise decibel) scale employs a weighting function that can be used to place aircraft and other noises in order of noisiness. The analysis of sounds in PNdB can be complex; a similar analysis uses an electrical network giving adequate results in units that are designated dB(N).

Noise reduction systems. When applied to analogue tape recording (especially multitrack), these add some 10 dB to the signal-to-noise ratio, of which part is used to hold down recording levels that might cause print-through. The signal is encoded by splitting it into frequency bands that are separately compressed. Digitally encoded audio signals are intrinsically less susceptible to noise and distortion.

Noise rejection. The difference (in dB at a given angle off-axis) between on-axis wanted sound and off-axis unwanted noise. Noise rejection can be achieved by the use of directional microphones and combinations of close working and compensation for bass tip-up. An adaptive array microphone also uses digital signal processing to juggle the response of inner and outer components of the array. See also **Lip-ribbon microphone**.

Normalization. A process, used in CD mastering, in which the highest instantaneous level is determined and contained, so that the transfer never exceeds the maximum coding level but the need for headroom is eliminated.

Normalled. Pairs of jacks, usually one above the other in a rack, that are connected to consecutive points along a continuous signal path. The path is broken when a plug is inserted in either jack, and the pair can be used to plug in an external device. In a half-normalled pair, the path can again be diverted by using both jacks, but headphones (etc.) can be plugged in to the upper jack to monitor the signal without interrupting the circuit. If, instead, the jacks are connected 'in parallel', they can both be used for headphones without breaking the circuit.

Obstacle effect. Obstacles tend to reflect or absorb only those sounds with a shorter wavelength than their own dimensions; at greater wavelengths the object appears to be transparent to the sound. For example, a screen 27 in (67 cm) wide is effectively transparent to frequencies below 500 Hz, but will absorb or reflect higher frequencies; 15 kHz is equivalent to a wavelength of just over 2 cm (about 0.9 in) – a microphone containing parts of about this size is subject to effects that begin to operate when this frequency is approached.

Off-microphone. On a dead (i.e. insensitive) side of a microphone, or at much more than the normal working distance on a live side.

OMFi. The Open Media Framework interchange facilitates the transfer of audio data between different manufacturers' hardware.

Omnidirectional microphone. One that is equally sensitive in all directions. In practice, there is a tendency for this quality to degenerate at high frequencies, with the top response progressively reduced for sounds further away from the front axis. Microphones with small capsules maintain their omnidirectional characteristics better.

Optical sound. A film recording system that is replayed by scanning a track of variable width (or, sometimes, density) by means of a lamp, slit and photocell.

ORTF stereo. Named after the French broadcasting organization that developed it, this uses a pair of microphones 17 cm apart (at roughly the separation of human ears, but without the baffle effect of the head). While there is no doubt that acceptable stereo balances can be obtained in this way, the acoustically cleaner and simpler coincident pair technique is preferred in this book.

Oscillator. A device that produces an alternating signal, usually of a particular frequency (or harmonic series). An audio oscillator produces a pure sine tone at any frequency in the audio range. *Square wave* and *sawtooth generators* produce waveforms of roughly those shapes (these correspond to harmonic series). Other oscillators (e.g. tape bias and radio-frequency oscillators) provide signals at higher frequencies.

Outside source (*UK*) or **remote** (*US*). A source of programme material originating outside the studio to which it is fed, and appearing on an individual fader or stereo pair on the mixer panel just like any local source (except that no preamplifier is needed, as it is fed in at about zero level).

Overdub. Added recording (usually on a separate track) made while listening to replay of tracks already recorded.

Overlap changeover. A type of changeover from one recording to the next in a series that may be used when the two have, say, half a minute or so of material in common. Before changing over, the first recording is heard on the studio loudspeaker, and the second pre-faded on headphones, or on a loudspeaker of different quality. The overlap is used to adjust synchronization before cross-fading from one to the other.

Overmodulation. Exceeding the maximum recommended amplitude for recording or transmission. With analogue audio distortion may result, and in certain cases damage to equipment or to the recording itself (limiters are used where this could occur). On digital recordings about 10 dB of 'headroom' is allowed so that peaks are unlikely to reach the maximum coding level.

Overtone. A partial in a complex tone, so called because such tones are normally higher than the fundamental.

Pad. Attenuator of fixed loss.

Panning or **steering**. Splitting the output from a monophonic microphone between stereo A and B channels. *Panpot*: a potentiometer (fader) arranged to do this. For surround, an extra panpot (or a joystick) may be used.

Parabolic reflector. A light rigid structure (of aluminium or other material) that reflects sound to a focus at which a microphone is placed. The assembly is very strongly directional at frequencies for which the corresponding wavelengths are less than the aperture of the reflector. Unlike gun microphone systems, (a) there is acoustic amplification as the sound is concentrated and (b) it is not degraded by interference from reflecting surfaces.

Parity. Whether the number of ones in a binary number is odd or even. This is designated by a 1 or 0, which is attached to the end of the number as an error check. If the parity of each row and column of a matrix of digital numbers is recorded, any single bit (a 0 or 1) that is missing from a row or column can be detected and restored.

Partial. One of a group of frequencies, not necessarily harmonically related to the fundamental, appearing in a complex tone. (Bells, xylophone blocks and many other percussion instruments produce partials that are not harmonically related.)

Pascal. International unit of sound pressure, equal to 1 newton per square metre or (in older units) 10 dynes/cm^2. A common reference level, 2×10^{-5} Pa approximates to the threshold of hearing at 1 kHz. Microphone sensitivity is calculated in dB relative to 1 V/Pa: this gives values 20 dB higher than those using the older reference level, 1 V/dyne/cm^2. Characteristically, microphones have a sensitivity of the order of −50 dB relative to 1 V/Pa or −70 dB relative to 1 V/dyne/cm^2.

Patch (*US*). Cross-plug.

PCM. Pulse code modulation (q.v.).

Peak. A short period of high volume.

Peak programme meter (PPM). A device for measuring the peak values of audio volume. In Europe it is the main volume control aid. See **Meter**.

Perspective. Distance effects, which may be simulated by varying the levels and the proportions of direct and indirect sound. In a studio, this can be done by a performer moving on- and off-microphone. In dead acoustics the ratio of direct to indirect sound cannot be varied, as indirect sound must be kept to a minimum. In this case, try 'throwing' the voice, to simulate raising it to talk or shout from a distance. In television, sound and picture perspectives are sometimes matched by the adjustment of microphone distance, or by placing a fixed microphone sufficiently close to a camera so that as a performer moves toward the camera, the sound also appears closer.

PFL. Pre-fade listen. See **Pre-hear**.

Phase. The stage that a particle in vibration has reached in its cycle. Particles are *in phase* when they are at the same stage in the cycle at the same time.

Phase shift. The displacement of a waveform in time. If a pure sine tone waveform is displaced by one complete wavelength this is described as a phase shift of 360°. If it is displaced by half a wavelength (i.e. through 180°) it has peaks where there were troughs and vice versa. If two equal signals 180° out of phase are added together, they cancel completely; if they are at any other angle that is greater than 90°, partial cancellation occurs. Some electronic components introduce phase shift into a signal, and the shift is of the same angle for all frequencies. This means that the displacement of component frequencies in the wave is different (depending on their wavelength) and distortion of the waveform results. This will not normally be of importance, because the ear cannot detect changes in phase relationship between components of a steady note. Interference between two complex signals, similar in content but different in phase, results in the loss of all frequencies for which the two signals are 180° out of phase. Cancellation may occur when two recordings are played slightly out of sync (this is used creatively in popular music), or when a signal has to be sent by two different paths (this can make the transmission of stereo signals by landline difficult when lines with broad transmission characteristics are not available). It occurs in radio reception when a sky wave interferes with a direct signal, etc. In a studio, cancellation may occur when one of a pair of bidirectional microphones has its back to a sound source. Care must always be taken when there is any chance of pick-up on two such microphones at the same time. It is possible to check this by asking someone to speak at an equal distance from both, fading both up to equal volume, and mixing. If the sound appears thin and spiky, or direct sound is completely lost, one of the microphones should be reversed. (Reverberation and random sound generally are not affected by the cancellation.) Pairs of loudspeakers may also be out of phase. If a monophonic signal is fed to an in-phase pair the sound appears to come from behind them: if they are out of phase the sound appears to be thrown forward to a point between speakers and listener. Correct phasing of speakers is vital to true stereo reproduction.

Phon. A somewhat archaic unit of perceived loudness. Phons are the same as decibels at 1000 Hz, and at other frequencies are related to this scale by contours of equal loudness.

Pick-up. The electromechanical transducer of a gramophone. The movement of a stylus in the record groove gives rise to an electrical signal.

Pilot tone (sync pulse). A signal related to camera speed that may be fed to a tape (separate from that on which picture is recorded) on which the associated sound is recorded. When the sound is subsequently transferred for editing, the pulse controls the speed of re-recording. For analogue sound this employs a tape format originally developed to use the full width of the tape for mono, but with the pilot tone superimposed along the centre. It has two narrow centre tracks each of 0.25 mm (0.01 in) with a similar gap between them. Pilot tones on the two tracks are recorded out of phase, so that on full-track replay of the recorded sound the two cancel and are not heard. For stereo, if the outer tracks carry the A and B signals, full-track replay will produce an A + B, i.e. mono, signal.

Pilot tone system (broadcasting). Stereo broadcasting system in which the A − B signal is transmitted on a subcarrier, and a pilot tone is also transmitted to control the phasing of the two signals.

Pitch. The subjective aspect of frequency (in combination with intensity) that determines its position in the musical scale. The pitch of a group of harmonics is judged as being that of the fundamental (even if this is not present in the series). Dependence of pitch on intensity is greatest at low frequencies. Increase in loudness may depress the (perceived) pitch of a pure tone 10% or more at very low frequencies, although when the tone is a member of a harmonic series the change is less apparent. Pitch does not depend on intensity at frequencies between 1 and 5 kHz, and at higher frequencies increase in loudness produces a slight increase in pitch.

Pitch changer. A digital device that samples a succession of segments of a sound from random access memory and changes their pitch independently of the duration of the recombined sound. It can generally also be used to change its duration without change of pitch, or both, independently, at the same time.

Plug-in. A computer program that is used on a 'virtual studio' computer (or a digital console) designed to act like a device that would be plugged into an analogue console to treat the sound or add effects. The computer must have a program or '*platform*' that accepts such plug-ins. A control panel appears on screen that sometimes mimics the appearance and operation of the controls on a physical unit. It involves digital signal processing (DSP).

Polar characteristic, polar diagram (*US*: **field pattern**). The response of a microphone, loudspeaker, etc., showing its sensitivity (or volume of sound) in relation to direction. Separate curves are shown for different frequencies, and separate diagrams are necessary for different planes. Such diagrams can also be used to indicate the qualities of reflecting screens or the radiation pattern of transmitter aerials, etc.

Popping. Breakup of the signal from a microphone caused by blowing the diaphragm beyond its normal working range. This may occur when explosive consonants such as 'p' and 'b' ('plosives') are directed straight at the diaphragm at close range.

Portamento. Used in musical performance, a slide in frequency to reach a note, or from one note to another. Also provided in some synthesizers. Measured in octaves per second.

Pot (potmeter, potentiometer). Fader (q.v.).

Pot-cut. Dropping a short segment of unwanted material out of a programme without stopping the replay, by quickly fading out and fading in again. It results in a momentary loss of atmosphere. Editing is preferable if time is available.

Power (of sound source). The total energy given out by a source (as distinct from intensity, which is energy crossing unit area). Power is the rate of doing work, and in an electrical circuit equals voltage times current.

PPM. Peak programme meter. See **Meter**.

Practical. A prop (property) in film or television that works partly or completely in its normal way, e.g. a telephone that rings (and on which the ring is interrupted when the handset is lifted), or an office intercom that transmits a voice.

Preamplifier. Amplifier in circuit between a source and the source fader.

Pre- and post-echo. An 'echo' of a particular programme signal that may appear before or after the signal. On tape it may occur one turn of tape before or after the normal signal, and is caused by printing; on discs it may appear one groove before and after as molecular tensions in the groove walls relax. See also **Print-through**.

Pre-emphasis. Increasing the relative volume of part of the frequency response (usually HF) in order to make the best use of some segment of a recording or transmission system (e.g. VHF radio). This is compensated by matched de-emphasis after the link or component in question.

Pre-fade (*UK*) or **dead-roll** (*US*). Playing closing music from a predetermined time in order to fit exactly the remaining programme time, and fading up at an appropriate point during the closing words or action.

Pre-hear (Prefade-Listen, PFL). A facility for listening to a source either on headphones or on a loudspeaker of a quality characteristically different from the main monitoring loudspeaker. The source can thereby be checked before fading it up to mix it in.

Pre-mix. A mix made of several but not all components of a final mix, e.g. a music, effects and dialogue mix for a documentary (the commentary being added subsequently). It is sometimes used for complicated effects sequences or dialogue with numerous cuts and quality matching.

Preparation, 'prep'. In broadcast production, the stage at which material is reviewed, logged, checked for cues, and provisionally discarded or selected to go in a cutting order. In studio operations, the stage at which microphones are positioned and assigned to channels on the console, balance confirmed, levels set, cues checked, and so on: anything short of a full and formal rehearsal or run-through. In film and video sound editing, it is the stage at which tracks are converted to the format required for sound mixing. In hard disk editing, this becomes an integral (and potentially major) stage of the sound editing.

Pre-recording. Recording made prior to the main recording or transmission, and replayed into it.

Presence. A quality described as the bringing forward of a voice or instrument (or the entire composite sound) in such a way as to give the impression that it is actually in the room with the listener. This is achieved by boosting part of the 1–8 kHz frequency range. In fact, emphasis of a single component (or several, at different frequencies) in this band gives greater clarity and separation, although at the expense of roundness of tone. Presence applied to the entire sound (e.g. by a distorted loudspeaker response) achieves stridency but little else.

Pressing. To service a niche market, 'vinyl' records are still made, as they always were, by stamping a plastic material such as polyvinyl chloride (PVC) in association with other components; PVC-based pressings have low surface noise (provided they have not been played with a heavy pick-up) and are not breakable, but are easily damaged by scratching or heat. Formerly, shellac was used for pressing records and, having much greater elasticity, was suitable for record materials when only very heavy (low-compliance) pick-up heads were available. The shellac was combined with a large proportion of inexpensive filler (e.g. slate dust) which made them hard, served to grind the stylus to the shape of the groove, and also contributed the characteristic surface noise of 78s. The recordings are analogue – which ensures that, as replayed (and unlike CDs, MiniDiscs, MP3, etc.), they are uncompressed. Various forms of noise and distortion may be added, but (in principle) nothing has been taken away. Paradoxically, using the powerful digital programs designed to clean up archive discs must risk taking away more than just the defects.

Pressure gradient microphone. One with a diaphragm open to the air on both front and back, and which therefore responds to the difference in pressure at successive points on the sound wave (separated by the path difference from front to back). The microphone is 'dead' to sound approaching from directions such that the wavefront reaches front and back of the diaphragm at the same time. A pressure gradient microphone measures the rate of change of sound pressure, i.e. air particle velocity. This is not quite the same as measuring intensity (as the ear does) but differs from it essentially only in phase, and as the ear is not sensitive to differences in phase, this does not matter. Pressure gradient operation begins to degenerate to pressure operation for wavelengths approaching the dimensions of the microphone. This may help to maintain the response of the microphone at frequencies such that cancellation would occur due to the wavelength being equal to the effective path difference from front to back of the diaphragm. See also **Figure-of-eight**; **Cardioid microphone**.

Pressure microphone. One with a diaphragm open to the pressure of the sound wave on one side and enclosed on the other. If the diaphragm and casing is sufficiently small, the microphone is omnidirectional (q.v.).

Pressure Zone Microphone (PZM). Proprietary name for a range of boundary microphones that are attached or very close to a hard reflecting surface, so that interference effects are virtually eliminated.

Print-through. The re-recording of a signal from one layer to another in a spool of tape. The recorded signal produces a field that magnetizes tape separated only by the thickness of the backing. (Other factors affecting printing are temperature, time and physical shock.)

Programme (*UK*). (a) Self-contained and complete broadcast production. (b) The electrical signal corresponding to programme material, as it is fed through electronic equipment.

Pro Tools. A proprietary set of computer programs that simulate many of the operational systems available on sophisticated consoles. The simulation includes visual displays that often look like the physical controls they replace. Also acts as the host platform to further programs that may be added as 'plug-ins'.

Proximity effect (*US*). See **Bass tip-up**.

Public address (PA). A loudspeaker system installed for the benefit of an audience. The output of selected microphones is fed at suitable levels to directional loudspeakers.

Pulse code modulation (PCM). An efficient means of using available bandwidth. The coded signal is less susceptible to noise and distortion. On a multitrack recording it can be used for noise reduction (q.v.). A coded signal can travel considerable distances, and television sound may be interleaved with the picture signal: the signal is reconstructed at its destination and broadcast in the normal way.

Punching in. Performing an editing operation by switching in and out manually in real time, rather than by identifying the end points and marking them up by computer. It allows the exercise of skill together with direct aural judgement of the result, and also retains the more satisfying feel of rehearsal and performance that can be lost in many digital operations.

Quad, quadraphony. A system that provided four-channel sound surrounding the listener. Two speakers were placed to the front (left and right) and two to the rear. The rear pair could be used to enhance the apparent acoustics of the listening room, or all loudspeakers might carry direct sound as well as the acoustic enhancement. Quad was commercially unsuccessful, although it lasted long enough for a substantial number of recordings to reach the market. Techniques originally developed for quad underlie those now used in 5.1 surround sound (q.v.). The four channels of LCRS surround, now used with domestic viewing equipment, are more closely related to 5.1 than to its quadraphonic ancestor.

Radio link. A radio transmission focused into a narrow beam and directed toward a directional receiver placed in line of sight. Used in place of a landline.

Radio microphone. Microphone and small transmitter sending a signal that can be picked up at a distance of up to perhaps several hundred metres (provided that the two are not screened from each other).

Radio transmission. A system for distributing audio information by modulating or encoding it on to a carrier at a particular frequency that is then amplified to a high power and broadcast by means of an aerial. Transmitter and receiver aerials act in a similar way to the two windings of a transformer; they are coupled together by the field between them. The field at the receiver is extremely small: it diminishes not only as the square of the distance, but also by the action of other 'receivers' along the path. In passing over granite, for example, a great deal of a medium-wave signal is mopped up. For VHF, which does not penetrate so deep into the ground, the nature of the terrain does not matter – but high ground casts a 'shadow'. This highly inefficient transformer action is improved if the receiving aerial is so constructed as to resonate at the desired frequencies of reception, and it will discriminate against other signals if it is made directional. It is important that the aerials should be parallel (both vertical or both horizontal). Systems of carrier modulation include amplitude modulation and frequency modulation (q.v.). Digital broadcasting systems are less susceptible to noise.

RAM. Random access memory: an electronic component storing digital information that can be retrieved at any time and in any order.

Ramp. Fade, in programmed (hard disk) editing.

Recording. An inscription of an audio signal in permanent form, usually on magnetic tape or disc or hard disk. The term *record* implies commercial disc.

Reduce. See **Mix**.

Reinforcement (in sound balance). The strengthening of direct sound reaching a microphone by the addition of indirect sound. This adds body without adding appreciably to volume. See **Resonance**; **Reverberation**.

Remote (*US*). Outside broadcast or outside source (q.v.).

Resistance. The ratio of EMF (electromotive force) to current produced in a circuit; the ratio of voltage drop to current flowing in a circuit element.

Resonance. A natural periodicity or the reinforcement associated with it. The frequencies (including harmonics) produced in many musical instruments (e.g. vibrating strings or columns of air) are determined by resonance.

Response. Sensitivity of microphone, etc. See **Frequency response**; **Polar characteristic**.

Reverberation. The sum of many reflections of sound in an enclosed space. This modifies the quality of a sound and gives it an apparent prolongation after the source stops radiating. *Reverberation time* is the time taken for sound to die away to a millionth of its original intensity (i.e. through 60 dB).

Reverberation plate. A metal plate, held under stress and fitted with transducers, one to introduce sound vibrations, and others to detect them elsewhere on the plate. By this means the decay of sound in an enclosed space is simulated. This technique has been superseded by digital reverberation programs (in which 'plate' may appear as a special case).

Ribbon microphone (or **loudspeaker**). The ribbon is a narrow strip of corrugated aluminium alloy foil suspended in a strong magnetic field provided by a powerful horseshoe magnet with pole pieces extending along the length of the ribbon. It is made to vibrate by the difference in air pressure between front and back of the ribbon. If both surfaces are open to the sound wave, the resulting motion is in phase with the velocity, and not with the amplitude of the sound wave. In a ribbon loudspeaker the same principle is used in reverse, but is not suitable for handling considerable power, and so has been used only in a tweeter, with a flared (exponential) horn to improve coupling. See also **Figure-of-eight**; **Pressure gradient microphone**.

Ring. Undamped resonance.

Rock and roll. A facility for discontinuous film sound re-recording (dubbing) in which the recording is stopped after an error is made, the picture and tracks run back in sync and a new recording begun from an earlier point. An essential characteristic is that the join must not be perceptible.

Rumble. Low-frequency mechanical vibration picked up by an audio system.

S signal. The stereo difference signal A – B. The S does not stand for stereo (of which it is only part) but for 'side' response, such as may be obtained by a side-fire bidirectional microphone – used sometimes in combination with a 'main' microphone to produce stereo.

SACD. Super Audio Compact Disc. An advanced CD format with two layers on one side of the disc. One layer can be read by a standard CD player for good CD quality. The other layer is encoded by Direct Stream Digital (DSD), which offers a 120 dB dynamic range and 0–100 kHz frequency range: it carries two-channel linear (i.e. uncompressed) stereo and six-channel surround sound.

Sampling. Recording a short audio segment and storing it for subsequent processing and playback (often several times) as part of another work. Originally done on analogue media (usually tape), sampling was more widely used when digital media (and digital signal processing) came into use. Some CD and MiniDisc players can also pick out samples or create loops by repeating them. As digital memory increased, techniques originally developed for working with samples were adapted for digital editing and, conversely, digital editing equipment can now be used to collect samples.

Satellite. In this context, a stage in the transmission of audio or television signals from one part of the globe to another by two line-of-sight paths. Using a synchronous satellite (i.e. one for which the orbital speed is the same as the Earth's rotational speed, so that the satellite remains stationary with respect to a point on the surface of the Earth) there will be a delay of nearly a quarter of a second, so that there is a noticeable gap when 'near' is followed by 'far' (but not between far and near).

Scale. Division of the audio-frequency spectrum by musical intervals (i.e. frequency ratios). An octave has the ratio 1:2, a fifth 2:3, a fourth 3:4, and so on: common musical intervals are derived from series of ratios of small whole numbers. The *chromatic* or *12-tone scale* is a division of the octave into 12 equal intervals (semitones in the equal tempered scale). On this scale most of the small-whole-number intervals correspond (though not exactly) to an integral number of semitones, which may, of course, be measured from or to any point. Certain scales omit five of the 12 notes of the octave, leaving seven (plus the octave) into which most of the small-whole-number ratios still fit, but in these it is no longer possible to start arbitrarily from any point. Instead, the interval must always be measured from a *key* note, or a note simply related to it. Music that for the most part observes the more restricted scales is called *tonal*, while that which ranges freely over the chromatic scale is called *atonal*. In *12-tone music* an attempt is made to give each of the 12 notes equal prominence.

Screen (*US*: **flat, gobo**). A free-standing sound-absorbent or reflecting panel that may be used to vary acoustics locally, or to cut off some of the direct sound travelling from one point to another in a studio. It may be moved about the studio at will. An object (such as a script) is said to *screen the microphone* if it lies in the path of sound coming directly from a source.

Screening (electrical). Protection from stray fields. It may take the form of an earthed mesh wire surrounding a conductor carrying a low-level signal. Valuable tapes may be screened in a metal box when sent by air.

Script rack. An angled rack on which a script may be placed in order to encourage a speaker to look up at an angle that is closer to the microphone. It should be transparent to sound.

Scrub wheel (Search wheel). The audio application of the jog and shuttle wheel allows a sound waveform to pass a play marker (often described as 'the playhead') in the same way that tape can be 'scrubbed' manually over a physical play-out head, in order to locate a precise point on the track, or to shuttle (at random speed, but still with the track audible) to find another.

SCSI (pron. 'scuzzy'). Small Computer Systems Interface for high-speed data transfer, connecting through a double row of pins to the back of a computer or recording device. A removable hard-drive in a tray that uses this kind of connector is also commonly called a SCSI.

Segue (pron. 'segway'). Musical term meaning follow on.

Sensitivity. The ratio of response to stimulus, usually measured in decibels relative to some given reference level. See **Pascal**.

Separation. Ensuring that the sound from one source does not spill excessively on to the microphone for another. Signals from neighbouring microphones must not interfere with each other, as this can cause phase cancellation. One way to control this is by the '3:1 rule', which is described for pairs of omnidirectional microphones and for sources of similar volume. It states that for each unit of distance between the two microphones, there should be no more than a third of that distance between each microphone and its intended sound source: this ensures a separation of at least 10 dB. With directional microphones, the rule can be relaxed a little, but with caution. Using flats also helps to improve separation.

Sequencer. Originally, a digital program for ordering data from electronic sources in order to create a musical work. Now it may be a powerful software system for integrating digital audio and MIDI, also providing related musical processes (and even notation).

Shelf. At high or low frequency, a pre-set higher or lower level that lies beyond a rise or fall from the main volume level.

SI. International system of measurements based on the metric system, used by scientists and engineers. See, e.g., **Pascal**.

Sibilance. The production of strongly emphasized 'f', 's', 'sh' and 'ch' sounds in speech. These may in turn be accentuated by microphones having peaks in their high-frequency response. One method of control is to use a frequency-selective compressor (a 'de-esser').

Side-fire. Orientation of diaphragm within a microphone such that the axis of directional pick-up is at a right angle to its length. The forward axis is usually marked.

Signal. The fluctuating electrical current or digital data stream that carries audio information (programme).

Signal-to-noise ratio. The difference in decibels between signal and noise levels. A standard reference level is 1 Pa (94 dB at 1 kHz).

Sine tone (pure tone). A sound (or electrical signal) containing one frequency, and one alone. This corresponds to an air particle executing simple harmonic motion, and its graphical representation (waveform) is that of s.h.m. (i.e. a sine wave). In principle, all sounds can be analysed into component sine tones. In practice, this is only possible for steady sounds, or for short segments of irregular sounds: noise contains an unlimited number of such tones. Sine tones are useful for studying frequency response and (as 'tone') for lining up equipment.

Slew. Frequency shift, portamento (q.v.).

Slung microphone. One that hangs by wires or by its own cable from a ceiling or grid fitting (or from a boom or lazy arm).

SMPTE. Society of Motion Picture Engineers. Defined the American 30-frame time code standard, but 'SMPTE' may also be used to refer to time code at any frame rate. See **Time code**.

Snake. A cable carrying a number of microphone output channels.

SND. Sound Designer II. A file format (suffix: '.snd') for sound on hard disk, developed originally for Pro Tools and digital editing. See also **WAV**.

Solid-state device. A circuit element such as a transistor (q.v.) or a complete microcircuit combining a number of circuit elements and their connections. It is characterized by small size, low power consumption and high reliability.

Sound. A series of compressions and rarefactions travelling through air or another medium, caused by some body or bodies (*sound sources*) in vibration. At any place it is completely defined by the movement of a single particle of air (or other material). The movement may be very complex, but there are no physically separate components due to different sound sources. Such contributions may be isolated by mathematical analysis, and the brain also acts like a mathematical computer that perceives the different components as separate entities. See also **Hearing**; **Decibel**; **Level**; **Frequency**; **Wave**; **Pitch**; **Phase**; **Formant**; **Sine tone**; **Velocity of sound**.

Sound card. A circuit, plugged in to a computer, that converts between analogue audio and digital data (A/D and D/A conversion).

Sound effects. See **Effects**.

Soundfield microphone. Proprietary name for microphone with a tetrahedral array of diaphragms, the output of which can be combined to create any desired (primary) polar response in three dimensions.

Spaced pair. Two separated microphones used to pick up stereo. Phase distortion effects occur, but many balancers regard these as tolerable, or even as adding to the richness of the stereo sound.

Spacer. Uncoated tape, which may be yellow or other colours (other than white or red, which are used mainly for leaders and trailers), cut into a spool of tape to indicate the end of one segment and the start of another.

Spatialization. Phase-change effects that distort a stereo image to make it appear wider than the distance between the two loudspeakers. It should be used as a studio technique with caution, as it can confuse the stereo image, and if replayed in mono, cancellation effects may appear. If included on domestic equipment it is used at the discretion of the user.

Sphere microphone. See **Dummy head stereo**.

Spill (*US*: **leakage**). Sound reaching a microphone other than that intended, thereby reducing separation.

Spin. The combination of original and delayed signals with multiple repetitions. Decay (or continuation of the effect) is controlled by a fader in the feedback circuit. Eventually the quality changes to express the frequency characteristics of the feedback circuit itself.

Spin-start. Quick-start technique for disc, requiring no special equipment, in which the record is set up with the stylus in the groove and the motor switched off. On cue, the motor is switched on and the turntable boosted to speed by hand. This boost is not necessary for turntables with rim drive.

Spool. Reel for carrying tape (or film). *Cine spools* for tape are similar in design to 8 mm film spools. Some spools have a larger hub size, so that the tension does not vary so much as it would with a large-diameter cine spool. Another design consists of a hub and a one-side-only backing plate.

Spot effect (*US*: **foley artist**, q.v.). A sound effect created in the studio. It may be taken on a separate microphone or on the same microphone as the main action.

Spotting. Reinforcement of a particular element in a stereo balance using a monophonic microphone (*US*: accent microphone). The balance is usually very close, to avoid 'tunnel' reverberation effects and other problems.

Standing wave. See **Wave**.

Steering. See **Panning**.

Stereo (stereophonic sound). A form of reproduction in which the apparent sources of sound are spread out. The word 'stereo' implies 'solid', but in fact the normal range of sound in two-channel stereo is along a line joining the loudspeakers, to which an apparent second dimension is added by perspective effects. Stereo may be used to simulate the spread of direct sound of an orchestra or a theatre stage. Whereas each individual source may be fairly well defined in position, the reverberation associated with it is spread over the whole range permitted by the position of the loudspeakers. But this range is small in comparison with what happens in a concert hall, where reverberation reaches the listener from all around. *Stereophonic sound in the cinema* often uses surround sound (generally with 5.1 multitrack, multi-loudspeaker systems). This ensures that members of the audience who are not ideally seated for two-channel stereo still hear sound that is roughly coincident with the image, and also from the sides of the auditorium. A simplified domestic version of this has four channels, LCR and S. *For a computer*, an add-on speaker system may be in the intermediate 2.1 layout (left and right channels plus subwoofer).

Sting. Musical punctuation pointing dramatic or comic mood.

Stops. A term referring back to the stud positions on a stud fader, and the markings associated with them, and so, by extension, to the corresponding arbitrary divisions of a continuous fader. Thus, to lift programme level 'a stop' is to increase it by moving the fader (potentiometer) about 1.5–2 dB.

Stray field. Unwanted a.c. field that may generate a signal in some part of the equipment where it should not be. The use of balanced wiring discriminates against this, and so also, where necessary, does screening (in which an earthed conductor surrounds parts that might be affected).

Studio (sound studio). Any room or hall that is primarily used for microphone work. Its most important properties lie in its size and its acoustics – the way in which sound is diffused and absorbed, and the reverberation time (q.v.). A *studio manager* in BBC radio is the balancer in operational charge of the broadcast or recording. (At the BBC the studio manager is not an engineer, which is a separate function.)

Stylus. The needle of a gramophone pick-up. Materials most used are diamond and sapphire: diamond lasts about 20 times as long. Wear depends on playing weight (and record material) and is most rapid in the earlier hours of playing.

Subharmonic. A partial of frequency $\frac{1}{2}f$, $\frac{1}{3}f$, $\frac{1}{4}f$, etc., that lies below the fundamental f. The subharmonic $\frac{1}{2}f$ can be generated in the cone of a moving-coil loudspeaker.

Subwoofer. A loudspeaker delivering frequencies so low that their wavelengths are comparable with the size of a small auditorium. Its position should not be obvious: in a large auditorium it will generally be at the front, but in a smaller room it can be anywhere convenient. This frees other, more directional loudspeakers from extremes of power (and overextended excursions of the diaphragm) that might interfere with their response at mid-frequencies. The frequency below which a subwoofer radiates sound may typically be set at about 120 Hz, but can be 90 or even 60 Hz.

Surround sound. One of several systems that feed loudspeakers set on all sides of the audience. The most common system is 5.1 surround sound, which has left, centre and right loudspeakers at the front (where a cinema screen might be), and left and right surround speakers at the sides, towards the back. The '0.1' component (so called because of its restricted frequency range) feeds a subwoofer (q.v.). In a wide auditorium the rear signals can be matrixed to go to an extra loudspeaker at the back, in a 6.1 system, or to two more in 7.1. A simpler domestic version is LCRS, in which there is only one signal for side or rear speakers. All of these systems operate essentially in a two-dimensional plane; another dimension is added in Ambisonic systems, for which a tetrahedral (or tetragonal?) arrangement of four microphones such as the Soundfield (q.v.) provides a fully three-dimensional source.

Sync pulse. See **Pilot tone**. *House sync* is a mains frequency tone that is fed to all equipment linked to a common server system in order to ensure that the frames defined by time code all start together.

Sync take. Film shot with sound, using a synchronous recording system.

Talkback. Communication circuit from control console to studio, used mainly for the direction of performers. *Reverse talkback* is a secondary circuit for communication from studio to console. Both forms of talkback are independent of live programme sound circuits, although in unfavourable circumstances breakthrough (or crosstalk) can occur.

Tape. Recording medium consisting of a magnetic coating on a plastic backing. Thinner (long play) analogue recording tape is more susceptible to printing (q.v.). *Videotape* is a similar medium. Metric equivalents: the standard width of ¼ in = 6.35 mm, ½ in = 12.7 mm approx., and so on. Audiocassettes and DAT (digital audio tape) use a narrower tape, 3.8 mm wide.

Tape joiner. Film joiner using a guillotine blade and polyester tape. The film is not overlapped, so no frames are lost.

Tape speeds. For analogue tape recording, these were originally based on an early standard of 30 in/s (inches per second, Imperial measure being part of the usual description). Improvements in tape, heads and other equipment led to reductions in speed to 15 and 7½ in/s (used for music and speech) and 3¾ and less (used domestically). Metric equivalents: 15 in/s = 28 cm/s; 7½ in/s = 19 cm/s; 3¾ in/s = 9.5 cm/s, and so on.

Telephone adaptor. A pick-up coil that may be placed in the field of the line transformer of a telephone or in some similar position where a signal may be generated.

Telephone quality. A frequency band of about 300–3000 Hz. Simulated telephone quality should be judged subjectively according to programme needs. ISDN allows a telephone line to carry broadcast-quality speech.

Tent. A group of screens, arranged to trap (and usually to absorb) sound in the region of the microphone.

Timbre. Tone quality. The distribution of frequencies and intensities in a sound.

Time code. A recorded time signal that can be electronically read and displayed. It may be the real time at which the recording was made or, alternatively, tape time (duration) from an arbitrary starting point. It is used for automatic control and synchronization, and also for timing duration, automatic logging, etc. Normally eight figures are shown, as HH:MM:SS:FF, where HH is hours on a 24-hour clock, followed by minutes, seconds and television frames. The American Society of Motion Picture and Television Engineers (SMPTE) originally specified 30 frames; the EBU adapted this to the European standard of 25 frames: the term 'SMPTE time code', confusingly, may refer to either; 'EBU time code' is clear. A further subdivision is into one-eightieth of a frame (though some equipment calculates this as one-hundredth).

Time constant. For a capacitor of C farads, charging or discharging through a resistance of R ohms (as in a PPM, limiter, etc.), the time constant, $t = CR$ seconds. (Similarly, for an inductance of H henries and resistance, R ohms, $t = H/R$ seconds.)

Tone. Imprecise term for sound considered in terms of pitch (or frequency content). A pure tone is sound of a particular frequency. See **Sine tone**.

Tone control. Preamplifier control for adjusting the frequency content of sound (usually bass or treble) on domestic equipment.

Top. High frequencies in the audio range, particularly in the range 8–16 kHz.

Top response. Ability of a component to handle frequencies at the higher end of the audio range.

Track. (a) To move a camera or boom toward or away from the performance area. (b) An individual recording among several on a record. (c) One of several (typically, up to eight, 16 or 24) recordings made side by side on multitrack tape.

Trailer. (a) An item in the continuity (link) between programmes that advertises future presentations. (b) Uncoated tape, usually red or alternately red and white, that is cut into a tape to indicate the end of wanted recorded material.

Transducer. A device for converting a signal from one form to another. The system in which the power is generated or transmitted may be acoustic, electrical, mechanical (disc), magnetic (tape), etc. Thus, microphones, loudspeakers, pick-ups, tape heads, etc., are all transducers.

Transient. The initial part of any sound, before any regular waveform is established. The transient is an important part of any sound and in musical instruments helps establish an identifiable character. *Transient response*: Ability of a component to handle and faithfully reproduce sudden, irregular waveforms. See **Diaphragm**.

Transistor. A semiconductor device that performs most of the functions of a valve (vacuum tube), but differs from it principally in that (a) no heater is required, so that the transistor is always ready for immediate use and no significant unproductive power is consumed, (b) it is much smaller in size, (c) input and output circuitry are not so isolated as in a valve, and (d) there is normally no phase reversal in a transistor, as the control voltage is used to promote flow of current, not to reduce it. A range of designs and functions of transistors (as with valves) is available. The circuit associated with a transistor is somewhat different from that for a valve. Transistors are more sensitive to changes of temperature than valves. Power transistors, which generate heat, need to be well ventilated. See also **Tube**.

Transport. Tape transport (on a tape recorder and playout machine) is controlled by a series of buttons, usually marked with conventional symbols, to record, play, pause, stop, fast forward or rewind a tape. There is often a counter showing elapsed time. It might be possible to run tape past the playhead in contact with it, in order to find a place more quickly or more precisely. On digital editors these and related functions are controlled from a 'transport' window or the transport buttons on a control surface, on which the same conventional symbols are used. The counter becomes time code. With non-linear digital editing, there are extra controls, to jump to the beginning or end or to any specified point or time in between. 'Transport' is used to load (record) the individual tracks. There is often also provision for naming them.

Tremolo. A regular variation in the amplitude of a sound, e.g. of electric guitar, generally at a frequency between 3 and 30 Hz. Sometimes confused with vibrato (q.v.).

Tube (*US*). Vacuum tube, or valve (q.v.). When used as a component of a head amplifier of archaic design in an early condenser microphone or replica, the term has been applied to the microphone itself.

Tunnel effect. Monophonic reverberation associated with an individual source in stereo.

Turntable. The rotating plate of a record player. Besides supporting and gripping the record, it acts as a flywheel, reducing any tendency to wow and flutter. It should therefore be well balanced, with a high moment of inertia (this is highest for a given weight if most of the mass is concentrated at the outer edge).

Tweeter. High-frequency loudspeaker used in combination with a low-frequency unit or 'woofer' and possibly also a mid-range unit (sometimes called a 'squawker'). The problems of loudspeaker design are different at the two ends of the audio spectrum and in some are easier to solve if handled separately. A single unit becomes progressively more directional at higher frequencies: a small separate unit for HF can maintain a broad radiating pattern.

Undermodulation. Allowing a recorded or broadcast signal to take too low a volume, so it competes to an unnecessary degree with the noise of the medium (and has to undergo greater amplification on reproduction, risking greater noise at that stage also).

Unidirectional microphone. This may refer either to a cardioid (q.v.) or near-cardioid type of response that is live on one face and substantially dead on the other, or a microphone with a strongly directional forward lobe. See also **Parabolic reflector**; **Gun microphone**.

User bits. Parts of a digital code that are reserved for allocation by the user. In time code they can be used to indicate a reel or source number.

Valve (*US*: **vacuum tube**). Used in electronic equipment before the advent of the transistor, and still found in some older or repro microphones, this is an almost completely evacuated glass envelope within which electrons released by a heated electrode (the cathode) are collected by a second positively charged electrode (the anode). In this, its simplest form, the valve is a *diode* and it conducts electricity whenever there is a flow of electrons into the cathode (on every other half-cycle of an alternating signal). In the *triode* a grid is placed in the path of the electrons and variation of voltage on this controls corresponding fluctuations in the flow of electrons, producing an amplified version of the signal on the control grid. Additional grids may be added (*tetrode, pentode*, etc.) and these further modify the characteristics of the electron flow.

Velocity microphone (*US*). Ribbon microphone operating in its more common pressure gradient mode.

Velocity (or speed) of sound. In air at room temperature this is about 344 m/s (1130 ft/s) at 20°C (68°F), increasing with temperature – or, more precisely, 331.4 m/s plus 0.607 for each degree Celsius. For Imperial measures it can be calculated roughly as $1087 + 1.1T$ ft/s, where T is degrees Fahrenheit. Humidity also makes a slight difference: in fully saturated damp air it is about 1 m/s (3 ft/s) faster than in very dry air. In liquids and solids it is much faster than in air.

VHF. Very High Frequency. See **Radio transmission**.

Vibrato. Rapid cyclic variation in pitch at a rate of about 5–8 Hz, used by musicians to enrich the quality of sustained notes.

Video clock. Electronic clock on a video screen showing programme details before the start of a television recording. It is shown in vision between minus 30 seconds (or more) and minus 3 seconds, after which the picture cuts to black. Reference tone is recorded between minus 20 and minus 10 seconds, followed by tone pips on each second from nine to three.

Virtual studio. Fully practical computer simulation of the operational components of a studio chain, often with reattached physical components such as a MIDI keyboard or a hardware controller.

Vocal stop. A short break in vocalized sound that precedes certain consonants. These help in the exact location of editing points on tape, and sometimes provide a useful place to cut. It is sometimes possible to lose the sound after a stop completely, e.g. when cutting 'bu/' (for 'but') on to the beginning of a sentence.

Volume control. See **Control**; **Compression**; **Fade**.

Volume meter, VU meter. See **Meter**.

Volume of sound. See **Decibel**; **Level**.

Wave (sound wave). A succession of compressions and rarefactions transmitted through a medium at a constant velocity. In representing this graphically, displacement is plotted against time, and the resulting display has the appearance of a transverse wave (like the ripples on a pool). This diagram shows the *waveform*. The distance between corresponding points on successive cycles (or ripples) is the *wavelength* (often symbolized by the Greek symbol lambda, λ) and this is related to the frequency (f; the rate at which ripples pass a particular point) and sound velocity (c) through the relation $f\lambda = c$. As sound radiates out from a source it forms a *spherical wave* in which intensity is inversely proportional to the square of the distance. When the distance from the source is very large compared with the physical dimensions of any objects encountered, this effect becomes less important: the loss in intensity is small and the wave is said to act as a *plane wave*. In a *standing wave* the nodes occupy fixed positions in space. In a *progressive wave* the whole waveform moves onward at a steady rate. See also **Velocity of sound**; **Bass tip-up**.

Wave, WAV. An uncompressed PC audio file format suitable for broadcast-standard recording on hard disk (usually, 16-bit, 44.1 kHz or more). The file suffix is '.wav'. Broadcast Wave (as defined by EBU) has added data on file content, originator, date and time of recording, etc. Another professionally-used format is the Audio Interchange File Format ('.aif') for Mac.

Weighted and **unweighted**. Different ways of indicating levels of noise or hum relative to signal, with particular reference to their bass content. Programme meters do not normally give any frequency information, and because the ear discriminates against bass low frequencies these do not sound as loud as a VU meter or PPM indicates. For this reason, quoted noise levels are sometimes 'weighted' against bass according to standard loudness contours. Weighted and unweighted measurements may differ by 20 dB or more at low frequencies.

White noise. See **Noise**.

Wiggle or **flicker**. The momentary displacement of a stereo image that occurs when modulated audiotape is cut at an angle or when one of the faders in a stereo pair is not making perfect contact. In equipment with studded faders it occurs when the sliders do not make simultaneous contact with successive studs.

Wildtrack. Film sound recorded without picture.

Windshield (*US*: **windscreen**). Shield that fits over the microphone and protects the diaphragm from 'rattling' by wind, and also contours the microphone for smoother airflow round it.

Wireless microphone. A term that is more often used in America for radio microphone. (In Britain, 'wireless' was the early word for radio, but is now felt to be archaic.)

Woofer. Low-frequency unit in a loudspeaker. See also **Tweeter**.

Woolly. Sound that lacks clarity at high frequencies and tends to be relatively boomy at low.

Wow. Cyclic fluctuation in pitch due to mechanical faults in recording or reproducing equipment (or physical fault in a disc). The frequency of the variation is below, say, 5 Hz. See also **Flutter**; **Vibrato**; **Tremolo**.

'X' and 'Y' signals. Sometimes used instead of A and B. (*US*: **X/Y technique** means coincident microphone pair). See also **'A' signal**; **'B' signal**.

XLR connector. A rugged type of connector that is widely used in studios.

Zero level. See **Level**.

Zip. A data file format that can be used for instant replay of radio station ID, signature tunes and jingles, commercials, trailers, standard announcements, etc.; 100, 250 (or possibly 750) MB disks are used.

Bibliography

Bartlett, Bruce and Jenny. *Practical Recording Techniques* (Third Edition), Focal Press, Oxford (2001). A comprehensive introduction to the equipment required and problems encountered when setting up a studio.

Borwick, John (Ed.). *Loudspeaker and Headphone Handbook* (Third Edition), Focal Press, Oxford (1990). Includes studio monitoring and equipment for electric instruments.

Collins, Mike. *Pro Tools for Music Production*, Focal Press, Oxford (2001). Rather more than just a guide to a widely-used MIDI plus audio workstation, this gives the popular music background to the virtual studio. It also gives advice on the infuriating problems encountered when working with computers.

Eargle, John. *The Microphone Book*, Focal Press, Oxford (2001). A detailed account of the principles underlying a wide variety of microphones.

Holman, Tomlinson. *5.1 Surround Sound, Up and Running*, Focal Press, Oxford (1999). From the film academic who numbered its decimal component, a broad introduction to surround.

Holman, Tomlinson. *Sound for Film and Television* (Second Edition), Focal Press, Oxford (2001). A Hollywood angle on audio. Includes CD illustration of many fundamental audio concepts.

Huber, David Miles. *The MIDI Manual* (Second Edition), Focal Press, Oxford (1998). A comprehensive guide to one of the precursors (and now an essential component) of the virtual studio.

Huber, David Miles and Runstein, Robert E. *Modern Recording Techniques* (Fifth Edition), Focal Press, Oxford (2001). A guide to techniques used in the music recording industry. American terminology.

Millerson, Gerald. *Television Production* (Thirteenth Edition), Focal Press, Oxford (1999). A highly analytical study of the medium, including the contribution of sound, recording media, electronic effects and on-location news collecting.

Potter, Kopp and Green-Kopp. *Visible Speech*, Dover Publications, New York, and Constable, London (1966). Useful when considering how to record, treat and edit speech and song.

Robertson, A. E. *Microphones* (Second Edition), Iliffe Books, London (1963). Uses idiosyncratic (but helpful) mathematics to describe the engineering principles involved in the use of microphones.

Rumsey, Francis. *Spatial Audio*, Focal Press, Oxford (2001). A (literally) in-depth, if somewhat academic study of two-dimensional sound, with a short excursion into the third.

Stockhausen, Karlheinz. *Study II* (Score) Universal Edition. See this and the scores of other works of a similar nature for illustration of the problems (and purpose?) of the notation of electronic music.

Wood, A. B. *The Physics of Music* (Seventh Edition), Methuen (1975). A useful primer.

Index

Note: Principal entries are shown in bold; entries relating to illustrations are shown in italic.